SEMICONDUCTORS

SEMICONDUCTORS

Helmut F. Wolf
Signetics Corporation
Sunnyvale, California

WILEY-INTERSCIENCE
a division of John Wiley & Sons, Inc.
New York   London   Sydney   Toronto.

Copyright © 1971, by John Wiley & Sons, Inc.

All rights reserved. Published simultaneously in Canada.

No part of this book may be reproduced by any means, nor transmitted, nor translated into a machine language without the written permission of the publisher.

Library of Congress Catalog Card Number: 77-159286

ISBN 0-471-95949-9

Printed in the United States of America.

10 9 8 7 6 5 4 3 2 1

Preface

For the last few years the market for integrated circuits has grown at a more rapid rate than the gross national product. Compared to 1966, the GNP in 1970 has increased by 32%, whereas the semiconductor IC market has increased by 254%. The worldwide market for semiconductor products will increase from 2.5 billion dollars in 1970 to 4.0 billion dollars annually in 1980. These figures indicate the importance and the impact of the semiconductor industry, although the integrated circuit sector represents only 0.02% of the GNP.

The basis for the generation and fabrication of integrated circuits and other semiconductor products lies in an advanced semiconductor technology. Since the advent of the transistor in 1948, semiconductor technology has progressed at a very rapid rate. In part this has been made possible by solving the problem of obtaining hyperpure semiconductor materials. Silicon continues to be the dominant material used for discrete devices and integrated circuits. Other semiconductors, notably germanium and gallium arsenide, have established themselves as significant materials, although their percental usage is relatively small.

The purpose of this book is to give in concise form the most significant information on most semiconductors of practical interest, mainly in semiconductor technology and semiconductor device fabrication. In its organization it falls almost exactly in the middle between a handbook and a textbook, although the distinction usually is rather fundamental. It is certainly intended for reference use and should serve as a useful handbook for people involved in semiconductor research and development. On the other hand, it can serve as a brief text since it has more or less complete continuity and can be read completely. It does not try to be all exhaustive or all-inclusive but attempts to be a practical book which gives answers to the every-day questions of semiconductor workers.

The purpose of the book is twofold. First, it summarizes the theoretical aspects of semiconductor behavior, generally without resorting to derivations of equations; for deeper background information the reader is referred to the various available textbooks. Second, it gives a selection of experimental and other data necessary for the design and understanding of the operation of semiconductor devices. However, a detailed discussion of semiconductor devices is not given here, except for simple *p-n* junctions, mainly because these are special cases of *p-n* junctions, surface or bulk structures, whereas the topics selected for this book have been of more fundamental and general nature. It is intended to complement the author's book "Silicon Semiconductor Data"* where the emphasis has been on a graphical presentation of mostly experimental data on silicon. In this book the emphasis is on an understanding of the nature of semiconductor phenomena and related subjects.

Structurally the book starts with a discussion of general semiconductor properties including a review of crystals as far as necessary for an understanding of semiconductor behavior, the energy band structure, and characteristics of degenerate and nondegenerate semiconductors. Then the properties of various semiconductors are compared. Thermal, optical, and etching characteristics of semiconductors are then discussed. Doping techniques are described in the second chapter which treats ion implantion as an alternative to solid-state diffusion. It includes also the growth of epitaxial films. In the third chapter the electrical behavior of semiconductors is reviewed. Since the properties of semiconductors change with the addition of foreign atoms, their effect on carrier density, mobility, resistivity, and others is discussed. The properties of silicon dioxide, still the most widely used insulator in semiconductor technology, are given in the fourth chapter. This includes the oxidation of silicon surfaces, impurity redistribution at the semiconductor surface during thermal oxidation, diffusion of impurities in silicon dioxide, and its masking capabilities. Chapter five gives the properties of semiconductor surfaces and of metal-semiconductor interfaces and discusses the formation of barriers at the interfaces. The sixth chapter describes the properties of *p-n* junctions, i.e., depletion layer, avalanche breakdown, capacitance, electric field, and others. In the seventh chapter a few common semiconductor measurement techniques are discussed.

The numerical values of properties given in this book do not necessarily represent the most accurate or the most complete values obtained, although every effort has been made to include the latest available figures. For more extensive information the reader is referred to the excellent compilation of data on various semiconductors by Mrs. Meta Neuberger at the Electronic Properties Information Center (EPIC), Hughes Aircraft Company.

The critical review of the manuscript by Dr. A. Ballonoff and the constructive criticism of the chapter on Metal-Semiconductor Contacts by Dr. A. N. Saxena are greatly appreciated.

Sunnyvale, California   HELMUT F. WOLF
August 1971

---

* Published by Pergamon Press, Oxford, 1969.

Contents

1
GENERAL PROPERTIES OF SEMICONDUCTORS — 1

- 1–1 Crystal Structure and Chemical Bond — 15
  - 1. CRYSTAL STRUCTURE OF SEMICONDUCTORS — 15
  - 2. CHEMICAL BONDS IN A SOLID — 16
  - 3. HETEROPOLAR AND HOMOPOLAR BONDS — 18
  - 4. AMORPHOUS AND POLYCRYSTALLINE SEMICONDUCTORS — 21
  - 5. REFERENCES — 24
- 1–2 Properties of Compound Semiconductors Compared to Elemental Semiconductors — 25
  - 1. ELEMENTAL SEMICONDUCTORS — 25
  - 2. COMPOUND SEMICONDUCTORS — 26
  - 3. CLASSIFICATION OF SEMICONDUCTORS — 28
  - 4. BAND STRUCTURE — 30
  - 5. CARRIER MOBILITY AND EFFECTIVE MASS — 31
  - 6. IMPURITIES AND DEFECTS — 31
  - 7. SUMMARY OF THE MOST IMPORTANT PROPERTIES — 32
  - 8. PROPERTIES OF GERMANIUM — 38
  - 9. PROPERTIES OF GALLIUM ARSENIDE — 39
  - 10. REFERENCES — 40
- 1–3 Summary of Properties of Silicon — 41
  - 1. GENERAL PROPERTIES — 41
  - 2. REFERENCES — 47
- 1–4 Energy Band Structure — 48
  - 1. SEMICONDUCTOR ENERGY BAND STRUCTURE — 48
  - 2. ELECTRON ENERGY BANDS — 49
  - 3. ENERGY GAP — 52

		4. DEPENDENCE OF ENERGY GAP ON TEMPERATURE	53
		5. DEPENDENCE OF ENERGY GAP ON PRESSURE	55
		6. MAXIMUM TEMPERATURE OF OPERATION	56
		7. REFERENCES	57
	1–5	Degenerate and Nondegenerate Semiconductors	59
		1. DISTINCTION BETWEEN LIGHTLY DOPED AND HEAVILY DOPED SEMICONDUCTORS	59
		2. NONDEGENERATE SEMICONDUCTOR	61
		3. DEGENERATE SEMICONDUCTOR	61
		4. DEGENERACY CRITERION	64
		5. REFERENCES	69
	1–6	Fermi Level and Impurity Energy Levels	70
		1. FERMI LEVEL (FERMI ENERGY)	70
		2. INTRINSIC SEMICONDUCTOR	74
		3. EXTRINSIC SEMICONDUCTOR	76
		4. FERMI POTENTIAL	79
		5. IMPURITY ENERGY LEVELS	80
		6. EQUILIBRIUM AND NONEQUILIBRIUM CONDITIONS	81
		7. REFERENCES	82
	1–7	Thermal Characteristics	83
		1. THERMAL CONDUCTION	83
		2. CONTRIBUTIONS TO THERMAL CONDUCTIVITY	87
		3. PHONON THERMAL CONDUCTIVITY	88
		4. CARRIER THERMAL CONDUCTIVITY	91
		5. THERMOELECTRIC POWER GENERATION	93
		6. THERMAL RESISTANCE	96
		7. TEMPERATURE INCREASE OWING TO THERMAL RADIATION	97
		8. PHONON (SOUND) VELOCITY	97
		9. DEBYE TEMPERATURE	98
		10. THERMAL EXPANSION	99
		11. VAPOR PRESSURE	101
		12. REFERENCES	103
	1–8	Optical and Dielectric Characteristics	104
		1. REFRACTIVE INDEX AND ABSORPTION COEFFICIENT	104
		2. FREE CARRIER ABSORPTION	109

       3. REFRACTIVE INDEX AND DIELECTRIC
          CONSTANT   111
       4. OPTICAL PENETRATION DEPTH   111
       5. TRANSMISSION AND REFLECTION   113
       6. DEBYE LENGTH   114
       7. DIELECTRIC CONSTANT AND
          RELAXATION TIME   115
       8. PHOTOCONDUCTIVITY   116
       9. PHOTOVOLTAIC EFFECT   118
     10. REFERENCES   119
1–9   Stress in a Semiconductor   120
       1. EFFECT OF STRESS ON SEMICONDUCTOR
          PROPERTIES   120
       2. STRESS OWING TO THE PRESENCE OF
          IMPURITIES   123
       3. VELOCITY OF A DISLOCATION LINE
          UNDER STRESS   125
       4. ELASTICITY OF A SEMICONDUCTOR   127
       5. ACOUSTOELECTRIC EFFECT   129
       6. REFERENCES   129
1–10 Etching   130
       1. ETCHING MECHANISMS   130
       2. ETCH PROCESSES   131
       3. ETCH RATE   131
       4. REFERENCES   135
1–11 Examples   136

2
IMPURITIES IN SEMICONDUCTORS AND EPITAXIAL GROWTH   139

2–1   Diffusion   144
       1. SOLID-STATE DIFFUSION   144
       2. IMPURITIES IN SEMICONDUCTORS   154
       3. MAXIMUM SOLID SOLUBILITY   156
       4. LIQUID SEMICONDUCTORS   158
       5. REFERENCES   162
2–2   Impurity Concentration Profile   163
       1. MATHEMATICAL DESCRIPTION OF
          IMPURITY PROFILES   163

2. CONCENTRATION DEPENDENCE OF DIFFUSION COEFFICIENT 169
3. TWO-STEP DIFFUSION 170
4. THE COMPLEMENTARY ERROR FUNCTION 172
5. IMPURITY DOSAGE AND IMPURITY GRADIENT 173
6. ADJUSTMENT OF SURFACE CONCENTRATION (EXTERNAL RATE LIMITATION) 177
7. DIFFUSION FROM TWO SIDES OF A SEMICONDUCTOR 179
8. IN-DIFFUSING AND OUT-DIFFUSING IMPURITIES 179
9. EFFECT OF BUILT-IN ELECTRIC FIELD ON IMPURITY DISTRIBUTION 180
10. DIFFUSION IN THE PROXIMITY OF A MASK WINDOW 184
11. JUNCTION DEPTH 187
12. LAYER RESISTIVITY (ρ) 189
13. SURFACE RESISTANCE (V/I) 193
14. REFERENCES 193

2–3 Properties of Impurities Used in Silicon Technology 195
1. n-TYPE AND p-TYPE DIFFUSION SOURCES AND CHEMICAL REACTIONS 195
2. BORON AND PHOSPHORUS DIFFUSION SYSTEMS 196
3. REFERENCES 200

2–4 Gold in Silicon 201
1. GOLD DIFFUSION MECHANISMS 201
2. SOLUBILITY ENHANCEMENT 201
3. GOLD DISTRIBUTION IN SILICON 204
4. RESISTIVITY INCREASE OWING TO DEEP-LYING IMPURITIES 206
5. REFERENCES 208

2–5 Ion Implantation 209
1. GENERAL CHARACTERISTICS 209
2. ENERGY LOSS OF PROJECTILE 211
3. ENERGY TRANSFER DURING COLLISION 212
4. RANGE OF PROJECTILE 212
5. IMPURITY DISTRIBUTION 217

	6. PEAK IMPURITY CONCENTRATION	217
	7. IMPURITY CONCENTRATION GRADIENT	219
	8. REFERENCES	222
2–6	Energy Levels of Impurities	223
	1. ACTIVATION ENERGY OF SHALLOW IMPURITIES	223
	2. ACTIVATION ENERGY OF DEEP IMPURITIES	227
	3. CARRIER DENSITY AND ACTIVATION ENERGY	227
	4. DECREASE OF ACTIVATION ENERGY WITH IMPURITY CONCENTRATION	228
	5. REFERENCES	228
2–7	Growth of Epitaxial Film	229
	1. FILM GROWTH CONDITIONS	229
	2. GROWTH OF INTRINSIC AND EXTRINSIC FILMS	232
	3. GAS-PHASE MASS-TRANSFER COEFFICIENT	234
	4. REDISTRIBUTION OF IMPURITIES NEAR FILM-SUBSTRATE INTERFACE DURING EPITAXIAL GROWTH	236
	5. REFERENCES	239
2–8	Phase Diagrams	240
	1. PHASE DIAGRAMS	240
	2. REFERENCES	242
2–9	Examples	243

3
ELECTRICAL BEHAVIOR OF SEMICONDUCTORS 249

3–1	Carrier Density	254
	1. CARRIER CONCENTRATION IN AN INTRINSIC SEMICONDUCTOR	254
	2. CARRIER CONCENTRATION IN AN EXTRINSIC SEMICONDUCTOR	259
	3. NEUTRALITY CONDITION	263
	4. REFERENCES	264
3–2	Minority Carrier Drift Velocity and Ionization Rate	265
	1. CARRIERS IN AN ELECTRIC FIELD	265
	2. CARRIER MOBILITY AND DRIFT VELOCITY	265

xii CONTENTS

	3. IONIZATION	268
	4. REFERENCES	272
3–3	Carrier Mobility and Carrier Diffusion Coefficient	274
	1. CARRIER MOBILITY AND MEAN FREE PATH	274
	2. SCATTERING MECHANISMS	278
	3. TOTAL CARRIER MOBILITY	281
	4. TEMPERATURE DEPENDENCE OF MOBILITY	281
	5. IMPURITY CONCENTRATION DEPENDENCE OF MOBILITY	283
	6. MOBILITY IN A DEGENERATE SEMICONDUCTOR	284
	7. MOBILITY IN A THIN FILM	284
	8. CARRIER DIFFUSION COEFFICIENT	289
	9. CARRIER DIFFUSION CURRENT	290
	10. DEPENDENCE OF DRIFT VELOCITY AND MOBILITY ON ELECTRIC FIELD	292
	11. REFERENCES	296
3–4	Hall Coefficient and Hall Mobility	297
	1. HALL EFFECT	297
	2. HALL COEFFICIENT	299
	3. VARIATION OF HALL COEFFICIENT WITH IMPURITY CONCENTRATION	300
	4. VARIATION OF HALL COEFFICIENT WITH MAGNETIC FIELD	301
	5. HALL MOBILITY	301
	6. HALL MOBILITY IN THIN FILMS	302
	7. TEMPERATURE DEPENDENCE OF HALL MOBILITY	302
	8. SENSITIVITY OF HALL GENERATOR	303
	9. REFERENCES	306
3–5	Semiconductor Resistivity	307
	1. RESISTIVITY	307
	2. IMPURITY CONCENTRATION DEPENDENCE OF RESISTIVITY	308
	3. TEMPERATURE DEPENDENCE OF RESISTIVITY	310
	4. AVERAGE RESISTIVITY AND SHEET RESISTIVITY OF SURFACE AND SUBSURFACE LAYERS	312
	5. REFERENCES	316

3–6	Minority Carrier Recombination and Lifetime	318
	1. CARRIER RECOMBINATION	318
	2. BULK RECOMBINATION	321
	3. SURFACE RECOMBINATION	329
	4. RECOMBINATION RADIATION	330
	5. DEFINITIONS	331
	6. REFERENCES	332
3–7	Examples	333

4
PROPERTIES OF SILICON DIOXIDE   335

4–1	Atomic Structure and Energy Diagram	339
	1. NONCRYSTALLINE SOLIDS	339
	2. STRUCTURE OF SiO_2	339
	3. ENERGY BANDS IN SiO_2	341
	4. REFERENCES	341
4–2	Thermal Oxidation	342
	1. METHODS OF OXIDE FORMATION	342
	2. THERMAL OXIDATION	343
	3. CHEMISTRY AND KINETICS OF THERMAL OXIDATION	345
	4. GENERAL RELATIONSHIP FOR THERMAL OXIDATION	346
	5. RATE OF OXIDE GROWTH	348
	6. OXIDANT CONCENTRATION WITHIN OXIDE	349
	7. OXIDATION RATE CONSTANTS AND TIME CONSTANTS	350
	8. DEPENDENCE OF RATE CONSTANTS ON IMPURITY CONCENTRATION, CRYSTAL ORIENTATION, AND PARTIAL PRESSURE	351
	9. OXIDATION COEFFICIENT	354
	10. REDISTRIBUTION OF IMPURITIES DURING SEMICONDUCTOR OXIDATION	356
	11. SEGREGATION COEFFICIENT	359
	12. SURFACE EFFECTS	361
	13. HENRY'S LAW	361
	14. REFERENCES	361
4–3	Diffusion in SiO_2	363
	1. DIFFUSION MECHANISMS	363

xiv   CONTENTS

        2. COMPARISON OF DIFFUSION IN Si AND SiO_2 365
        3. MASKING CAPABILITIES OF PASSIVATING LAYER   366
        4. MASKING CAPABILITY OF SiO_2 369
        5. REFERENCES   370
4–4   Other Characteristics   371
        1. ETCHING OF SiO_2 371
        2. COMPRESSIBILITY   371
        3. VISCOSITY   373
        4. OTHER PROPERTIES OF SiO_2 374
        5. REFERENCES   376
4–5   Examples   377

5
SEMICONDUCTOR SURFACES   379

5–1   The Ideal MIS System   384
        1. OPERATING MODES OF AN MIS SYSTEM   384
        2. REFERENCES   387
5–2   Semiconductor Surface Depletion and Inversion   388
        1. SEMICONDUCTOR SURFACES   388
        2. SURFACE OPERATING CONDITIONS   389
        3. SURFACE STATES   390
        4. OXIDE OR INSULATOR CHARGES   392
        5. SEMICONDUCTOR CHARGES   392
        6. DEPLETION LAYER WIDTH   397
        7. EFFECTIVE IMPURITY CONCENTRATION IN DEPLETION LAYER   398
        8. TEMPERATURE DEPENDENCE OF TURN-ON VOLTAGE   398
        9. SEMICONDUCTOR SURFACE STATE DENSITY AS A FUNCTION OF CRYSTAL ORIENTATION   400
      10. SURFACE MOBILITY   400
      11. ELECTRIC FIELD AS A FUNCTION OF CARRIER CONCENTRATION IN THE SPACE CHARGE REGION   404
      12. REFERENCES   405

CONTENTS   xv

5–3	MIS Capacitance	407
	1. TOTAL CAPACITANCE OF AN MIS STRUCTURE	407
	2. MAXIMUM AND MINIMUM CAPACITANCE	413
	3. OXIDE CAPACITANCE	414
	4. SURFACE CAPACITANCE	415
	5. FREQUENCY DEPENDENCE OF SURFACE CAPACITANCE	416
	6. TEMPERATURE DEPENDENCE OF MIS CAPACITANCE	417
	7. REFERENCES	417
5–4	Metal-Semiconductor Contacts	418
	1. FREE ELECTRONS	418
	2. THE METAL-VACUUM SYSTEM	419
	3. THE METAL-SEMICONDUCTOR SYSTEM	420
	4. METAL AND SEMICONDUCTOR WORK FUNCTIONS	423
	5. PROPERTIES OF THE METAL-SEMICONDUCTOR BARRIER	427
	6. CURRENT DENSITY THROUGH METAL-SEMICONDUCTOR CONTACT	433
	7. METAL-SEMICONDUCTOR INTERFACE FERMI LEVEL	436
	8. ELECTRON AFFINITY AND ELECTRO-NEGATIVITY	437
	9. PHOTOEMISSION	438
	10. REFERENCES	440
5–5	Examples	441

6
p-n JUNCTIONS   443

6–1	Diffusion (Built-in) Voltage	448
	1. ORIGIN OF BUILT-IN VOLTAGE	448
	2. NORMALIZED BUILT-IN VOLTAGE	452
	3. REFERENCES	452
6–2	Depletion Layer Characteristics	453
	1. *p-n* JUNCTIONS	453
	2. TOTAL JUNCTION CAPACITANCE	457

xvi CONTENTS

3. WIDTH AND CAPACITANCE OF DEPLETION LAYER	459
4. FRACTIONAL WIDTH OF DEPLETION LAYER	461
5. CROSS-OVER FROM LINEARLY GRADED TO STEP JUNCTION BEHAVIOR	464
6. DEPLETION LAYER WIDTH AND CAPACITANCE AS A FUNCTION OF JUNCTION GEOMETRY	465
7. ELECTRIC FIELD NEAR $p\text{-}n$ JUNCTION	468
8. REFERENCES	471
6–3 Junction Characteristics at Breakdown	472
1. JUNCTION BREAKDOWN MECHANISMS	472
2. CARRIER MULTIPLICATION AND AVALANCHE BREAKDOWN	473
3. AVALANCHE BREAKDOWN VOLTAGE	477
4. REDUCTION OF BREAKDOWN VOLTAGE	481
5. TEMPERATURE DEPENDENCE OF BREAKDOWN VOLTAGE	485
6. DEPLETION LAYER WIDTH AT BREAKDOWN	487
7. CRITICAL ELECTRIC FIELD	489
8. DIFFUSED JUNCTION	491
9. JUNCTION CURRENT AT FORWARD BIAS	492
10. REFERENCES	493
6–4 Examples	494

7
MEASUREMENT TECHNIQUES 497

7–1 Determination of Oxide Thickness and Junction Depth	498
1. MEASUREMENTS BASED ON LIGHT INTERFERENCE	498
2. OXIDE THICKNESS (x_0)	498
3. JUNCTION DEPTH	501
4. JUNCTION DELINEATION	501
5. EXAMPLES	501
6. REFERENCES	502
7–2 In-Line Four-Point Probe Measurements	503
1. RESISTIVITY OF SEMICONDUCTOR SAMPLE	503
2. REQUIREMENTS FOR CORRECT V/I MEASUREMENT	505

	3. DEVIATIONS FROM IDEAL CONDITIONS	506
	4. EXAMPLE	510
	5. REFERENCES	510
7–3	Other Measurement Techniques	511
	1. GENERAL COMMENTS	511
	2. CRYSTALLOGRAPHIC ORIENTATION	511
	3. CONDUCTIVITY TYPE	511
	4. ENERGY GAP	511
	5. THERMAL CONDUCTIVITY	512
	6. OPTICAL PROPERTIES	512
	7. THIN FILM THICKNESS	512
	8. CARRIER DENSITY	513
	9. CARRIER MOBILITY	513
	10. CARRIER LIFETIME	513
	11. RECOMBINATION RADIATION	514
	12. RESISTIVITY	516
	13. SURFACE STATE DENSITY	517
	14. REFERENCES	517

APPENDIX 519

A–1	Properties of Metals and Other Data	520
	1. SKIN EFFECT	520
	2. ELECTRONIC PROPERTIES	522
	3. REFERENCES	522
A–2	List of Symbols	526
	1. FREQUENTLY AND GENERALLY USED SYMBOLS	526
	2. NOTATIONS	526

AUTHOR INDEX 543

SUBJECT INDEX 547

Recommended References and Textbooks on Solid State Physics and Semiconductor Technology*

(1) W. Shockley, *Electrons and Holes in Semiconductors*, D. Van Nostrand Company, Inc., Princeton, N. J., 1950.
(2) H. K. Henisch, *Rectifying Semi-Conductor Contacts*, Clarendon Press, Oxford, 1957.
(3) W. W. Gärtner, *Transistors—Principles, Design, and Applications*, D. Van Nostrand Company, Inc., Princeton, N. J., 1960.
(4) A. B. Phillips, *Transistor Engineering*, McGraw-Hill Book Co., New York, 1962.
(5) J. S. Blakemore, *Semiconductor Statistics*, International Series of Monographs on Semiconductors, Vol. 3, Pergamon Press, Oxford, 1962.
(6) B. I. Boltaks, *Diffusion in Semiconductors*, Academic Press, New York, 1963.
(7) G. E. Moore, *Microelectronics*, edited by E. Keonjian, McGraw-Hill Book Co., New York, 1963, Chapter 5.
(8) H. Salow, H. Krömer, H. Beneking, and W. von Münch, *Der Transistor*, Springer-Verlag, Berlin, 1963.
(9) R. B. Adler, A. C. Smith, and R. L. Longini, *Introduction to Semiconductor Physics*, John Wiley and Sons, New York, 1964.
(10) J. L. Moll, *Physics of Semiconductors*, McGraw-Hill Book Co., New York, 1964.
(11) R. G. Rhodes, *Imperfections and Active Centers in Semiconductors*, International Series of Monographs on Semiconductors, Vol. 6, Pergamon Press, Oxford, 1964.

* References are listed in chronological order.

REFERENCES

(12) W. R. Runyan, *Silicon Semiconductor Technology*, McGraw-Hill Book Co., New York, 1965.
(13) L. P. Hunter, *Semiconductor Phenomena and Devices*, Addison-Wesley, Reading, Mass., 1966.
(14) J. T. Wallmark and H. Johnson, *Field Effect Transistors—Physics, Technology and Applications*, Prentice-Hall, Englewood Cliffs, N. J., 1966.
(15) A. S. Grove, *Physics and Technology of Semiconductor Devices*, John Wiley and Sons, New York, 1967.
(16) H. C. Lin, *Integrated Electronics*, Holden-Day, San Francisco, 1967.
(17) D. R. Frankl, *Electrical Properties of Semiconductor Surfaces*, International Series of Monographs on Semiconductors, Vol. 7, Pergamon Press, Oxford, 1967.
(18) N. B. Hannay, *Solid-State Chemistry*, Prentice-Hall, Inc., Englewood Cliffs, N. J., 1967.
(19) R. K. Willardson and A. C. Beer, Editors, *Semiconductors and Semimetals*, Volumes I–IV, Academic Press, New York, 1966–1968.
(20) R.M. Burger and R. P. Donovan, *Fundamentals of Silicon Integrated Device Technology*, Volumes 1, 2, and 3, Prentice-Hall, Inc., Englewood Cliffs, N. J., 1968.
(21) S. K. Ghandhi, *The Theory and Practice of Microelectronics*, John Wiley and Sons, New York, 1968.
(22) C. Kittel, *Introduction to Solid State Physics*, John Wiley and Sons, New York, 1968.
(23) J. S. Blakemore, *Solid State Physics*, W. B. Saunders Company, Philadelphia, 1969.
(24) S. M. Sze, *Physics of Semiconductor Devices*, John Wiley and Sons, New York, 1969.
(25) H. F. Wolf, *Silicon Semiconductor Data*, International Series of Monographs on Semiconductors, Vol. 9, Pergamon Press, Oxford, 1969.
(26) J. F. Gibbons, *Semiconductor Electronics*, McGraw-Hill Book Co., New York, 1969.
(27) L. P. Hunter, Editor, *Handbook of Semiconductor Electronics*, McGraw-Hill Book Co., New York, 1970.
(28) J. W. Mayer, L. Eriksson, and J. A. Davies, *Ion Implantation in Semiconductors—Silicon and Germanium*, Academic Press, New York, 1970.
(29) H. H. Wieder, *Intermetallic Semiconducting Films*, International Series of Monographs on Semiconductors, Vol. 10, Pergamon Press, Oxford, 1970.

SEMICONDUCTORS

1
GENERAL PROPERTIES OF SEMICONDUCTORS

1–1 Crystal Structure and Chemical Bond
1–2 Properties of Compound Semiconductors Compared to Elemental Semiconductors
1–3 Summary of Properties of Silicon
1–4 Energy Band Structure
1–5 Degenerate and Nondegenerate Semiconductors
1–6 Fermi Level and Impurity Energy Levels
1–7 Thermal Characteristics
1–8 Optical and Dielectric Characteristics
1–9 Stress in a Semiconductor
1–10 Etching
1–11 Examples

Semiconductors of greatest present technological interest are the elemental semiconductors which are the elements of Group IV of the Periodic System (A^{IV}) and the compound semiconductors which are compounds of Groups III and V ($A^{III}B^{V}$) and of Groups II and VI ($A^{II}B^{VI}$) and some compounds of the composition $A^{III}B^{V}_{1-x}C^{V}_{x}$.

Except for semiconducting glasses (where there is, however, a short-range order or microcrystallinity), all semiconductors of scientific or technological interest have a crystalline structure. The nature of the crystalline structure is, in turn, of significant influence on the semiconductor properties. Amorphous semiconductors which have short-range order can be considered to be quasi-periodic where the macroscopic disorder can be taken care of by a coordinate transformation; if carrier transport is involved, however, such semiconductors display poor characteristics of mobility, lifetime, diffusion length, etc.

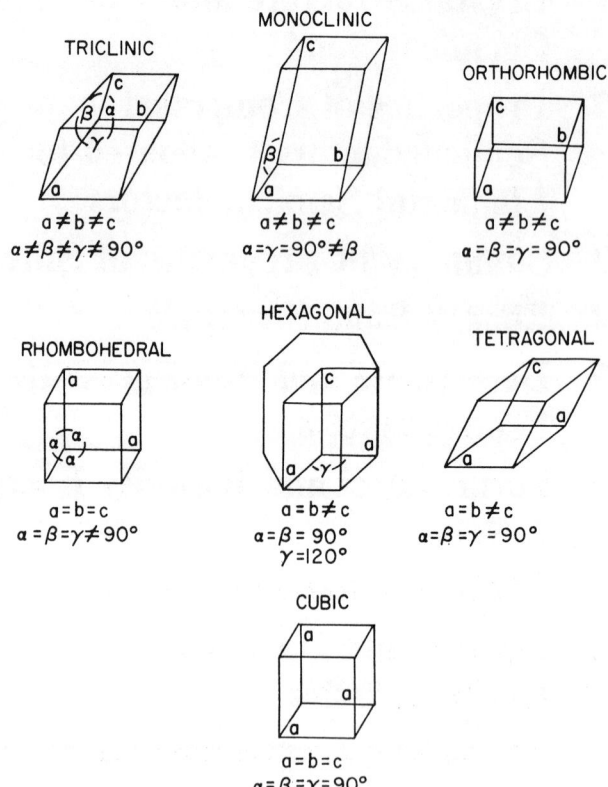

FIGURE 1.1

The seven crystal systems arranged according to the lengths of the axes and the angles between them.

The following review of crystallographic properties is intended to give semiconductors their proper place within the family of crystals.

In an ideal crystal every lattice point repeats itself periodically in identical surroundings. Along any straight line drawn through the crosspoints of such a lattice, points form a periodic sequence that is identical for any parallel line drawn at a repeat distance.

Mathematically there are 32 classes of crystals which can be grouped into 7 systems according to the relative lengths of the three axes and the angles between the axes. These crystal systems can be described as follows (see Figures 1.1 and 1.2).

(a) Triclinic:
Three axes of unequal length not at right angles to each other.
(b) Monoclinic:
Three axes of unequal length, two of which are perpendicular.
(c) Orthorhombic:
Three axes of unequal length, all of which are at right angles to each other.
(d) Rhombohedral:
Three axes of equal length, all of which are at equal but not at right angles to each other.
(e) Hexagonal:
Three coplanar axes of equal length at 120° to each other and a fourth axis of different length at right angle to the others; or two axes of equal

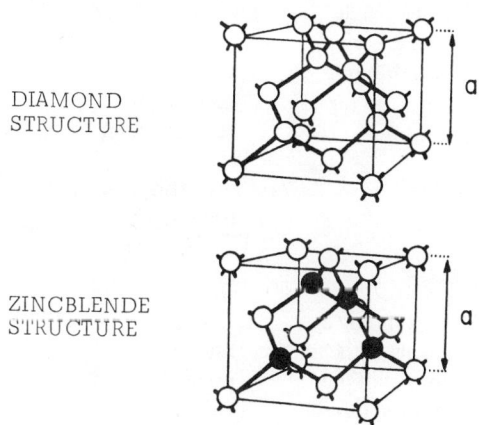

FIGURE 1.2

Lattice structure of diamond-type and zincblende-type crystals. The diamond structure is found in the elemental semiconductors C, Ge, Si, Sn; the zincblende structure in compound semiconductors as GaAs, GaP, InSb, and others. The lattice constant a is indicated.

4   GENERAL PROPERTIES OF SEMICONDUCTORS

length at 120° angle to each other and perpendicular to a third axis of different length.
(f) Tetragonal:
Three mutually perpendicular axes with two of them of equal length.
(g) Cubic:
Three axes of equal length at right angles to each other.

This classification, however, gives only limited indication of the physical properties of the crystals. In order to describe the electrical and some other properties of semiconductors, a classification according to the cohesive energies between the crystal elements is more appropriate (see Figure 1.3 and Table 1.1).

The cohesive or lattice energy is defined as the energy required to separate the atoms of 1 gram atom or 1 mole of a crystal. In this way, crystals can be divided into five groups. This classification is not quite unambiguous, e.g., silicon and germanium display valence cohesion but their conduction mecha-

TABLE 1.1

Classification of Chemical Bonds

Type of cohesion (bond)	Cohesive forces	Characteristics of crystal	Examples	Cohesive energy [eV/molecule]
1. Van der Waals cohesion (molecular crystals)	Van der Waals forces; very weak binding	Low melting and boiling point; small hardness; easily compressible; electrically insulating; transparent down to far UV	A Organic compounds H	0.08 0.01
2. Dipole cohesion (hydrogen bond)	Weak binding by means of the fixed dipoles of the molecular building blocks or hydrogen bridges; nonlocalized bonds	Tendency to build large molecules; electrically insulating; dielectric activity; optically transparent	HF Ice	0.30 0.52

TABLE 1.1 (*Continued*)

Type of cohesion (bond)	Cohesive forces	Characteristics of crystal	Examples	Cohesive energy [eV/molecule]
3. Metallic cohesion	Binding through free electrons which are separated from atoms during crystal formation; non-localized bonds	High electric conductivity by free electrons; opaque and highly reflecting in infrared and visible light, transparent in UV	Na Ag Fe	1.13 3.00 4.08
4. Covalent cohesion	The same forces which bind identical atoms in a molecule; each of the four valence electrons of every atom forms a strongly localized bond	At low temperature negligible conductivity of pure crystal; transparent in IR range; high hardness; often high melting points; electrical insulator or semiconductor with one valence electron of one of the four neighboring atoms; chemical inertness	Semiconductors Si InSb Diamond	 3.70 3.40 7.36
5. Ionic cohesion	Strong electrostatic binding; crystal is built of positive and negative ions	High IR absorption; electrolytic conduction, increasing with temperature	NaCl LiF Alkalis	7.80 10.41

6   GENERAL PROPERTIES OF SEMICONDUCTORS

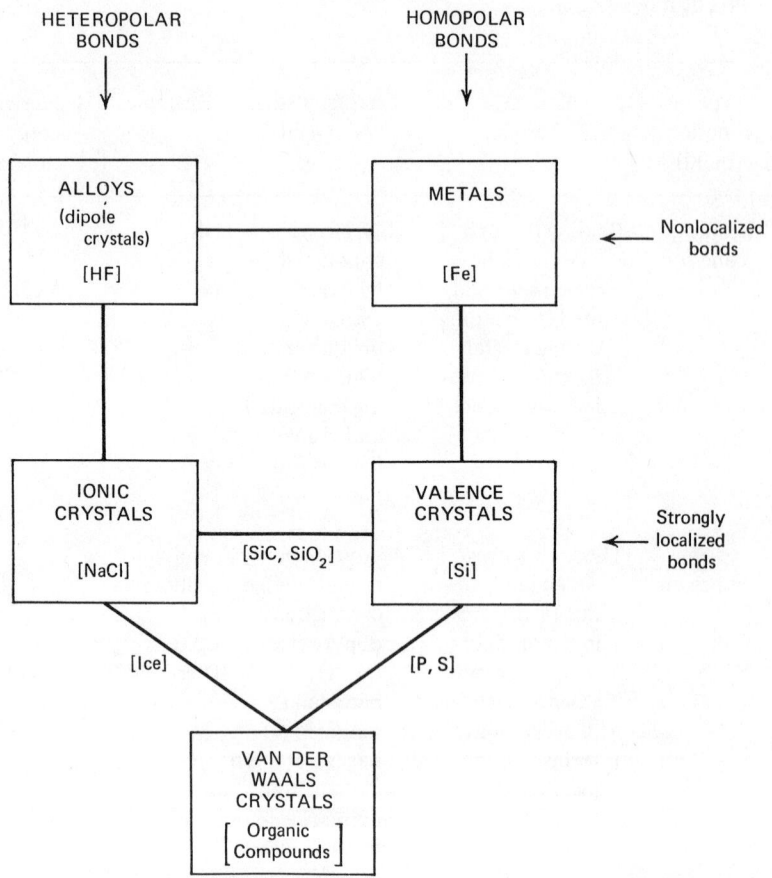

FIGURE 1.3

The five crystal groups arranged according to the cohesive energies between lattice elements.

nism is similar to that of metals, i.e., by free carriers. These groups represent limiting cases with many transitional types in between.

The planes that pass through a crystal lattice are described by the Miller indices which are defined as the reciprocals of the intercepts of the plane considered with the three crystallographic axes. These reciprocals are expressed as the smallest possible integers having the same ratio.

An ideal crystal is defined by a regular arrangement of atoms or ions where the cohesive forces are a function of the properties of the lattice elements and of the nature of the bond. The properties of a real crystal deviate considerably from those of an ideal crystal. In addition to the random thermal motion of atoms about their resting point, there are essentially three types of lattice defects. The presence of these defects affects the mechanical, electrical, and other characteristics of a crystal to a high degree. The most significant effects are related to resistivity, mobility, and carrier lifetime. The presence of crystal defects changes both the internal energy of the crystal and its entropy. Consequently, their equilibrium concentration depends upon the energy of their formation and upon the equilibrium temperature.

The main types of lattice defects are the following.

(a) *Thermal defects*:

These are either vacancies or interstitial atoms or ions and are intrinsic defects because the crystal can form these by thermal activation alone. The diffusion of lattice constituents and the low-temperature ionic conductivity are due to thermal lattice defects and are therefore structure-sensitive crystal properties. Each real crystal has a number of macroscopic (e.g., grain boundaries, cracks, dislocations) and microscopic (e.g., empty lattice sites, i.e., lattice vacancies, ions in interstitial lattice sites) lattice defects. The defect density increases with temperature. Defects enable lattice constituents to have a much greater mobility than in an ideal crystal since the diffusion can progress stepwise from vacancy to vacancy. There are three main types of thermal defects.

   (i) *Frenkel defect:*

It occurs if a lattice ion obtains enough kinetic energy so that it can leave its original lattice position. As a result, a normal lattice ion sits in an interstitial place (i.e., has not found its correct potential well but, by deformation of the crystal, has made room for itself in the neighborhood of its correct position), thus producing a lattice vacancy and an interstitial ion.

   (ii) *Schottky defect:*

It occurs if ions of both signs migrate to the crystal surface and thus produce vacancies only. In this case crystal vacancies are moved from their original position by thermal energy.

   (iii) *Anti-Schottky defect:*

It occurs if ions of both signs move to the crystal surface so that interstitial ions are moved from their original position by thermal energy.

Which of these defect types will be dominant depends upon the activation energies which are approximately 0.5 to 2.0 eV per atom. If E_a^* is the energy required to form a defect, C_L is the density of lattice atoms, and z_i is the density of available interstitial sites, then the equilibrium density of Schottky defects is

$$z_S = C_L \exp(-E_a^*/kT) \qquad (1.1)$$

8   GENERAL PROPERTIES OF SEMICONDUCTORS

and that of Frenkel defects

$$z_F = (C_L z_i)^{1/2} \exp(-E_a^*/2kT). \tag{1.2}$$

The factor 2 in the exponent of Frenkel defects is due to the fact that migration of one ion into an interstitial site produces two lattice defects. Frenkel defects may disappear by recombination of the interstitial lattice ions with lattice vacancies if the crystal is annealed. Schottky defects which result from the migration of ions to the crystal surface cannot be annealed. The generation of Schottky defects is associated with an increase of the crystal volume; the generation of Frenkel defects is not.

(b) Chemical defects:

These are impurity atoms at substitutional or interstitial crystal sites and are extrinsic defects because they are introduced externally (e.g., during crystal growth, diffusion, ion implantation, plastic deformation, etc.). In the case of a substitutional defect, a foreign atom occupies a site which would ideally be occupied by a crystal atom; in the case of an interstitial defect, a foreign atom occupies the space between two lattice atoms, particularly if the diameter of the foreign atoms is much smaller than that of the crystal atoms.

(c) Dislocation or line defects:

These defects are also extrinsic and are introduced externally. Dislocations are one-dimensional line defects in an otherwise perfect crystal and result in geometric faults in the lattice. Dislocation defects occur when the crystal is subjected to stresses in excess of the elastic limit, e.g., during its growth from the melt. There are two types of dislocation defects: screw dislocations and edge dislocations. Dislocation defects cannot be formed by thermal activation alone since the energy available below the crystal melting point is insufficient. Dislocations may move completely through a crystal (under proper conditions they will even move out of the crystal); the mechanism for such a movement is called a slip. It results in movement along planes of high atomic density where opposing forces are at a minimum. The energy of movement of a dislocation is approximately 0.15 eV per atom for silicon, whereas the energy of formation of a dislocation is about 10–20 eV; i.e., although it is very difficult to form a dislocation, it is very easy to induce a dislocation to move in a crystal.

The variation of potential (electrostatic) energy of an electron as a function of distance from the nucleus (assumed to be of infinitely small diameter) of an atom follows a hyperbola-shaped curve, except for the influence of the outer electrons. The nucleus is surrounded by a potential well whose bottom is the lowest energy level. If the atoms are equally spaced, the potential has a maximum halfway between the nuclei and it decreases as the nuclei are approached. Electrons close to the nucleus (inner electrons) are not affected by neighboring atoms; electrons of higher potential energy, on the other hand, are affected.

Coupling between two neighboring atoms (in analogy to the coupling of two pendulums) is through two (approximately equal) resonant frequencies (ω_1 and ω_2) instead of one resonant frequency (ω_o) for an independent system (free atom). In a lattice consisting of a number of atoms, corresponding to an equal number of oscillating systems, there are as many resonant frequencies as there are atoms that deviate less from ω_o the further away they are from the reference atom. Since the neighboring atoms correspond to frequencies ω_1 and ω_2, all other resonant frequencies lie between them (see also the inset of Figure 3.3). Therefore, the electron energy in a crystal corresponds to an energy band whose width is determined by the difference between ω_1 and ω_2; i.e., while in a free atom the electrons are allowed to have only certain (discrete) energy values (corresponding to frequency ω_o), in a crystal they are allowed to have certain energy ranges (energy bands). Consequently, there are—for both free atoms and atoms in a crystal—certain energy levels which electrons are not allowed to occupy (forbidden regions, energy gaps).

Metals, semiconductors, and insulators can be theoretically treated from the point of view of an energy band model. In a metal, the energy band containing the valence electrons is empty, i.e., there exist more energy states in the band than there are electrons to occupy them. Consequently, a metal is a good electric conductor since the electrons are free to increase their kinetic energies under the influence of external fields and may acquire net momentum which manifests itself as electric current.

In an insulator, the valence band is completely filled and the number of valence electrons is equal to the number of available energy states. In order for an insulator to conduct, the electrons must acquire energy, and this can be done only by moving them to higher energy levels. Since the nearest allowed higher energy level is at least the width of the forbidden energy gap away (on the order of several eV), insulators are poor electric conductors.

Semiconductors have an energy band structure similar to that of insulators, however, the energy gap is much smaller (generally of the order of 1 eV). Thermal energies imparted to electrons in the semiconductor at room temperature ($kT = 2.6 \cdot 10^{-2}$ eV at 25°C) are sufficient so that a significant number are able to cross the energy gap and are available for conduction. The conductivity of a semiconductor or an insulator is proportional to the numbers of electrons and holes in the conduction and valence bands, respectively, and to the ease with which carriers are moved by an electric field, i.e., to the carrier mobility. The number of free electrons and holes available for conduction is a function of their energy distribution, which follows the laws of statistical mechanics. A density-of-states function $g(E)$ may be defined so that $g(E)\,dE$

gives the number of available energy states occurring within an energy interval between E and $E + dE$, and it is further defined to be continuous within the allowed energy bands of the semiconductor and zero in the forbidden energy gap.

The probability of an energy level E being occupied by an electron is given by the Fermi-Dirac distribution function

$$f(E) = \{1 + \exp[(E - E_F)/kT]\}^{-1}.$$

The probability of a level being unoccupied by an electron (which is equal to the probability of a level being occupied by a hole) is

$$1 - f(E) = \{1 + \exp[(E_F - E)/kT]\}^{-1}.$$

Considering only energies higher than $2\,kT$ above the Fermi level, such that $(E - E_F)/kT > 2$, $f(E)$ may be approximated by

$$f(E) \approx \exp[-(E - E_F)/kT].$$

This function is exactly the Maxwell-Boltzmann distribution which applies to a gas of classical particles. If, then, the edge of the conduction band is greater than $2kT$ above the Fermi level, the electrons obey Boltzmann statistics and the system is referred to as being nondegenerate. In case the Fermi level is less than $2kT$ away from the conduction band edge, the above approximate equation is not valid (the electron gas is degenerate) and the complete Fermi-Dirac distribution function must be used. The density of electrons in the conduction band at thermal equilibrium is given by forming the product $g(E)f(E)$ and integrating over the band. An analogous situation holds for the holes in the valence band, i.e., Maxwell-Boltzmann statistics are applicable if the Fermi level is more than $2kT$ above the valence band edge.

In noncontinuous bands, i.e., in the lower energy bands, the electron motion is mainly restricted to within the potential well; however, a number of these electrons will be able to tunnel through the potential hills. The noncontinuous bands are occupied by a certain number of electrons (n_z). The continuous bands have also a certain number of electrons, but there they belong to the entire crystal and are supplied by all atoms; if the entire crystal contains N^* atoms, the number of electrons in a continuous band is N^*n_z.

The number of allowed energy bands is not restricted; the number of electrons per atom, however, is limited and equals the atomic number so that the higher bands may not be completely filled. Since an electron of a fully occupied band can move through the crystal only if another electron moves in the opposite direction (resulting in zero net current) electrons of fully occupied bands do not contribute to electric conductivity. On the other hand, conductors whose conductivity is due to free electrons must have at least one band (namely the highest one which contains electrons) which is only partially filled with electrons.

GENERAL PROPERTIES OF SEMICONDUCTORS

The point of zero electron potential energy is defined as follows. If an electron far outside the crystal lattice has energy zero and is brought into the crystal (of temperature $T = 0$) it will gain an energy equal to the crystal work function; this energy is taken as the zero point of electron potential energy. An electron volt [eV] is defined as the kinetic energy which an electron receives when falling through the potential difference of one volt.

Among all the characteristics that crystals display, the most significant variation is in electric conductivity; it varies over about 30 orders of magnitude. In order of increasing conductivity, crystals can be grouped as follows.

Insulators	$\sigma = 10^{-22}$ to 10^{-13}	$(\Omega\,\text{cm})^{-1}$
Electrolytes	$\sigma = 10^{-13}$ to 10^{-10}	$(\Omega\,\text{cm})^{-1}$
Semiconductors	$\sigma = 10^{-10}$ to 10^3	$(\Omega\,\text{cm})^{-1}$
Metals	$\sigma = 1$ to 10^3	$(\Omega\,\text{cm})^{-1}$
Superconductors	$\sigma > 10^3$	$(\Omega\,\text{cm})^{-1}$

Electric conductivity is proportional to the product of free carrier density and mobility. High mobility means that carriers will, under the influence of an electric field, reach a high velocity before colliding with the lattice. Since a high carrier velocity will result in a high Lorentz force, semiconductors with high carrier mobility will be effective Hall generators.

The group of semiconductors comprises crystals which do not have metallic cohesion and which, therefore, have a lower conductivity than metals. As in metals, electric conduction in semiconductors is by free electrons; the difference between metals and semiconductors lies in the fact that metals make free electrons available during crystal formation; in semiconductors, external influences (temperature increase, illumination, etc.) are required to generate them. Contrary to what occurs in metals (where the number of free carriers is essentially independent of temperature), the generation of free carriers in semiconductors increases exponentially with temperature.

The valence electrons of a semiconductor are located at so-called electron bridges between atoms; each bridge is occupied by two electrons. Because of the thermal motion, in thermal equilibrium some of these bridges will be open so that a number of electrons will be free for conduction. The bridges lacking electrons will then act as holes; neighboring electrons will move into these holes thus creating new holes of their own. This mechanism is very sensitive to lattice disturbances, i.e., to lattice defects and foreign atoms, so-called impurity atoms. If such a lattice has more electrons than atoms, not all electrons will be needed for the electron bridges, there will be an excess of electrons which are only weakly bound to the atoms and which can easily be separated from these atoms; the crystal will appear to be n-type because of an excess of electrons. If the lattice has less electrons than are necessary to form the electron bridges, some bridges will lack electrons so that neighboring electrons may move into these positions, thereby creating a hole movement; the crystal will appear to be *p*-type.

In a pure (intrinsic) semiconductor electrons and holes are found in equal number. This equality is not affected by temperature. If, however, impurity atoms are added there will be an excess of electrons or holes depending upon the nature of the impurities. Elements of Group III in a Group IV semiconductor act as a sink for electrons (acceptors) so that the semiconductor will become *p*-type and elements of Group V act as a source for electrons (donors) so that the semiconductor will become *n*-type. Compounds of equal amounts of elements of Groups III and V (III–V compounds) are again intrinsic semiconductors and can be made *n*-type or *p*-type by addition of other impurity atoms.

Impurity atoms in a semiconductor do not form continuous bands (as the atoms of the semiconductor do) since they are tied to fixed positions within the crystal. For this reason, they usually form discrete, spatially separated energy levels. Thermal energy is required to activate these donor or acceptor levels but this energy is much less than that needed to bridge the energy gap. Since the density of the impurity atoms is much less than the density of the semiconductor atoms, Boltzmann statistics can usually be used for impurities rather than Fermi statistics.

Dielectrics and pure semiconductors are insulators at sufficiently low temperature; however, dielectrics retain their high resistivities to temperatures at which the crystalline or semicrystalline phase ceases to exist, whereas semiconductors are reasonably good conductors at those temperatures. This is due to the larger band gap in dielectrics than in semiconductors. In addition to affecting the electrical properties, the difference in band structures between dielectrics and semiconductors has significant influence on their optical properties.

The reason for the difference in band structure of semiconductors and dielectrics is due to the covalent character of semiconductors and the strongly ionic character of dielectrics. In semiconductors the valence electrons have a large amplitude over extended regions of the crystal and are more tightly bound to a single lattice atom; in dielectrics the electrons find it energetically more advantageous to complete atomic shells by transferring from one atom to another.

In metals (whose only free carriers are electrons but not holes) the total number of electrons cannot be increased without charging the entire crystal negatively; in a semiconductor, however, the number of available carriers can easily be increased by breakup of the electron bridges.

All atoms contain electrons. Since these may be separated from the atom nuclei by appropriate external forces, each crystal must become conductive under certain circumstances so that the nonconductive state can be defined as the ideal state of a solid. Insulators approach this ideal state closely. They can be polarized in a static electric field, resulting in a noncoincidence of the centers of gravity of electrons and nucleus (i.e., their dielectric constant is higher than unity).

In metals and in most semiconductors the electrons within a partially filled band do not belong to single atoms but to the entire crystal; they can freely move through the crystal, but interact with the lattice during this travel. These conduction or free electrons form an electron gas, exchange momentum and energy with the lattice, and acquire its temperature so that their average energy becomes, in analogy to a gas,

$$m_n v_n^2/2 = (3/2)kT. \tag{1.3}$$

In a small electric field (E) the free electrons will be accelerated by $dv_n/dt = qE/m_n$ which is superimposed on the random thermal motion of the lattice so that the electrons will attain a velocity just prior to a collision

$$v_n = q\tau_c E/m_n \tag{1.4}$$

where τ_c is the time between successive collisions of electrons with the lattice atoms. During the collisions the electrons will transfer their kinetic energy to the lattice whose temperature will consequently rise; after each collision the electrons will have an initial velocity $v_n = 0$ so that the average electron velocity

$$\overline{v_n} = q\tau_c E/2m_n. \tag{1.5}$$

Thus the electric conductivity

$$\sigma = qn\overline{v_n}/E = q^2 \tau_c n/2m_n \tag{1.6}$$

and the thermal conductivity

$$\kappa = n\overline{v_n} k l_c/2 = qkn\tau_c l_c E/4m_n. \tag{1.7}$$

The ratio of thermal to electric conductivity depends upon temperature only. The mean free path (i.e., the average distance an electron travels between two collisions) is

$$l_c = \overline{v_n}\tau_c = qE/2m_n. \tag{1.8}$$

In crystals which have at least some ionic character two types of vibrational modes may occur—optical and acoustical vibrations. The acoustical vibrations produce electric dipole moments which vary with the frequency of vibration; the optical vibrations correspond to frequencies at which radiation is emitted or absorbed. For the same wavelength the frequency of the optical vibrations is larger than that of the acoustical ones.

Lattice vibrations may also be excited locally, for example, when a valence electron of a lattice atom is excited by light absorption or is separated from its ion. This results in a sudden change of the neighboring bonds and the ions begin to vibrate. The locally produced vibrations then propagate through the crystal as waves. Thus, lattice vibrational quanta are generated which, in analogy to light quanta (photons), are called phonons. An excited electron can lose its energy either by emission of a photon or by emission of one or several phonons, i.e., by excitation of lattice vibrations.

The energy of a phonon is

$$E_{ph} = hv_{ph}. \tag{1.9}$$

While photons move even in vacuum and always with the velocity of light (c), phonons are restricted to matter and consequently their velocity is that of sound (c_{ac}); the ratio c/c_{ac} is in the order of 10^4 to 10^5. The momentum of a phonon is

$$p_{ph} = h/\lambda = hv_{ph}/c_{ac}$$
$$\leq hv_o/c_{ac} = h/4a \approx 10^{-19} \text{ g cm/sec}; \tag{1.10}$$

i.e., the momentum of a phonon is 10^4 to 10^5 times higher than the momentum of a photon of the same frequency (i.e., of the same energy). The term a in equation (1.10) is the lattice constant of the crystal.

Two other quantities of general importance in semiconductor analysis are important and are given here. Radius (Bohr's radius, a_o) and ionization energy (E_H) of a hydrogen atom are fundamental atomic quantities used as reference for other atoms:

$$a_o = h^2/4\pi^2 m_o \varepsilon_o^2 = 0.529 \text{ Å}, \tag{1.11}$$

$$E_H = m_o q^4/8h^2\varepsilon_o^2 = q^4/32\pi^2\varepsilon_o^4 a_o = 13.59 \text{ eV}. \tag{1.12}$$

For comparison, the radius of the nth electron orbit in an atom of number Z is

$$r_n = a_o n^2/Z \tag{1.13}$$

(i.e., for hydrogen where $n = 1$, $Z = 1 : r_n = a_o$) and the ionization energy

$$E_Z = q^4/32\pi^2\varepsilon_o^4 r_n = Zq^4/32\pi^2\varepsilon_o^4 a_o n^2. \tag{1.14}$$

GENERAL REFERENCES

(1) A. R. von Hippel, *Molecular Science and Molecular Engineering*, The Technology Press of MIT, 1959.
(2) W. Finkelnburg, *Structure of Matter*, Academic Press, Inc., New York, 1964.
(3) F. A. Cotton and G. Wilkinson, *Advanced Inorganic Chemistry*, Second Edition, Interscience Publishers, New York, 1967.
(4) H. Ehrenreich, *Symposium on the Optical Properties of Dielectric Films*, The Electrochemical Society, Inc., Boston, May 6 and 7, 1968.

1-1 Crystal Structure and Chemical Bond

1. CRYSTAL STRUCTURE OF SEMICONDUCTORS

Elemental and binary compound semiconductors which average four valence electrons per atom preferentially form a tetrahedral phase. Each atom is surrounded by four equally distant nearest neighbors which lie at the corners of a tetrahedron. The bond between two nearest neighbors is formed by two electrons of opposite spin. The most important tetrahedral crystal lattices are the diamond, the zincblende, and the wurtzite.

In the diamond lattice each atom lies in the center of a tetrahedron formed by the four nearest neighbors. Two neighboring tetrahedrons are oriented in such a way that the base triangles are rotated by 60° from one another. The diamond lattice can be described as two intertwined, face-centered cubic lattices.

The zincblende lattice is formed in the same way except that the two nearest-neighbor points are occupied by different elements. The diamond structure is therefore restricted to elements, the zincblende structure to binary compounds. The zincblende structure can also be described as two intertwined, face-centered cubic lattices.

In the wurtzite lattice the four nearest-neighbor tetrahedrons are oriented in such a way that the two bases lie exactly over each other. It is not cubic but can be considered as two interpenetrating, closely packed hexagonal lattices. The wurtzite lattice is restricted to binary compounds.

Some compound semiconductors form modifications of these three basic lattice types, whereas the crystal structure of a few others has no relation to the tetrahedral phase. The BN lattice, for example, is formed by distorting the wurtzite lattice in such a way that three of the four nearest neighbors lie in the same plane with the atom, whereas the fourth neighbor is oriented at right angles to this plane. The NaCl lattice can be considered as interpenetrating cubic lattices with coordination number six.

Elemental semiconductors normally have the diamond structure. Si and Ge occur only in the diamond lattice, whereas Sn has two modifications—α-Sn occurs in the diamond structure and β-Sn in a modification of the zincblende structure. The compound SiC has the zincblende structure. Of the III–V compounds, the phosphides, arsenides, and antimonides of boron, aluminum,

TABLE 1.2

Dimensional Characteristics of the Cubic Lattices

	Simple (primitive) lattice	Body-centered lattice	Face-centered lattice	Diamond and zincblende lattice*
Volume of conventional unit cell	a^3	a^3	a^3	a^3
Volume of primitive cell	a^3	$a^3/2$	$a^3/4$	$a^3/4$
Lattice points per cell	1	2	4	4
Lattice points per unit volume	$1/a^3$	$2/a^3$	$4/a^3$	$4/a^3$
Number of nearest lattice points	6	8	12	4
Nearest lattice point distance	a	$(\sqrt{3}/2)a$	$(1/\sqrt{2})a$	$(\sqrt{3}/4)a$
Number of second-nearest lattice points	12	6	6	6
Second-nearest lattice point distance	$\sqrt{2}a$	a	a	$(1/\sqrt{2})a$
Examples	Po	W, Mo	Ni, Cu, Au, Pt	Si, Ge, GaAs, etc.

* Two interpenetrating, face-centered lattices.

gallium, and indium have the zincblende structure. Most of the other III–V compounds have the NaCl structure. Among the II–VI compounds the oxides of beryllium and zinc, and the sulphides, selenides, and tellurides of beryllium, zinc, cadmium, and mercury have the zincblende or wurtzite structures. Among the I–VII compounds only the four copper compounds and AgI have these structures.

2. CHEMICAL BONDS IN A SOLID

Chemical bonds within a crystal lattice are due to forces between atoms or ions. Between two atoms at infinite distance there are no forces, i.e., the potential energy of interaction is zero. If the two atoms are brought together, there are both attractive (binding) and repulsive forces, resulting in positive or negative potential energy of interaction, depending upon the distance between them. At large distance the attractive force is dominant, at small distance the repulsive force begins to dominate. The total potential energy is

$$E_T = -\alpha_a/r^{m_a} + \alpha_r/r^{m_r} \qquad (1.15)$$

where the first term represents attraction and the second one repulsion; r is the atomic distance, α_a and α_r are constants, and m_a and m_r are small integers characteristic of the system. The atomic distance corresponding to the minimum potential energy is the equilibrium separation at $T = 0°\text{K}$, since the net force at this point is zero. The potential energy as function of interatomic spacing is shown in Figure 1.4.

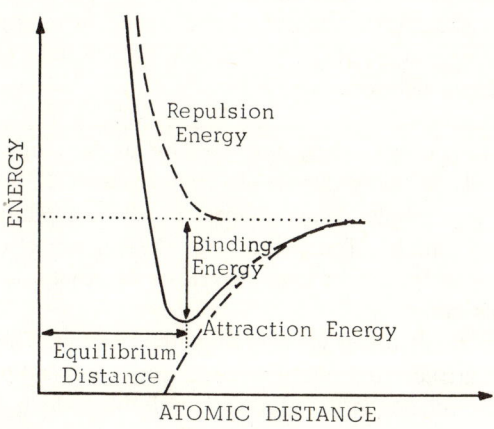

FIGURE 1.4

Repulsive or attractive energy between two atoms as function of atomic distance.

The important types of chemical bonds in a crystal are as follows.

(a) Ionic bond:

If one atom of an interacting pair is electropositive and the other electronegative, so that the first one loses a valence electron to the second, then the attractive force is due to the electrostatic (Coulomb) attraction of two oppositely charged ions and an ionic bond is formed. As a result of this transfer of valence electrons, each ion in an ionic crystal tends to be surrounded by oppositely charged ions while equally charged ions occupy more distant positions. Ionic crystals show high bonding strength due to the strong electrostatic forces between the ions; they normally exhibit high strength and hardness and high melting points. They are brittle because of the directional nature of the bonding forces. Their electrical conductivity is low.

18   GENERAL PROPERTIES OF SEMICONDUCTORS

(b) Covalent bond:
Covalent bonding between two atoms is due to the sharing of electrons, supplied by one or both of the atoms. The sharing results from the overlap of the bonding orbitals, which lowers the energy of the system. The shared electrons usually serve to fill the outer valence shell of each atom. The covalent bond is very strong, resulting in high hardness and high melting point. In organic molecules (which are covalent in character) intramolecular bonds are strong whereas intermolecular bonds (which are noncovalent) are responsible for their low mechanical strength and low melting point.

Bonds which fall between the ionic and covalent types have corresponding characteristics; the degree to which a bond is ionic or covalent is described by the electronegativities of the two atoms, i.e., by their relative abilities to attach an extra electron.

(c) Metallic bond:
The bonding in a metal is best described by all the atoms of the crystal taken collectively, with the valence electrons from all the atoms belonging to the crystal as a whole. Many properties can be explained in terms of the free-electron model. The freedom of the valence electrons to move through the metal leads to a high electric conductivity.

(d) Van der Waals bond:
Weak van der Waals forces can bind together a solid. These forces arise from the weak attraction between instantaneous electric dipoles due to the motion of the electrons in the atoms or molecules which do not have to have a permanent dipole movement because the fluctuating dipoles average to zero. The binding in such solids is quite weak since these forces are small. Consequently these crystals have low melting points and low cohesive strengths.

(e) Dipole bond:
Under certain conditions a hydrogen atom can form a second bond which may be equivalent to the first bond because of resonance between situations where the electron goes to one or the other of the two neighboring atoms. Dipole bonds are usually quite weak.

3. HETEROPOLAR AND HOMOPOLAR BONDS

Depending upon the constituents of a crystal lattice and the forces between them, there are two main types of binding:

(a) Heteropolar (ionic) bond:
It exists between ions of opposite charge (e.g., between Na and Cl in the NaCl lattice) and is caused by Coulomb forces between them. It is

1–1 CRYSTAL STRUCTURE AND CHEMICAL BOND   19

usually found in crystals where one of the two bound atoms has a large and the other one a small electronegativity; the more electronegative atom "attracts" another electron to itself, becoming negative in the process, so that the remaining positive ion is attracted to the negative ion. True heteropolar bonding is usually not found in crystals, but heteropolar (ionic) bond and homopolar (covalent) bond coexist and the valence is of polar nature and has a permanent dipole moment. In this case the polar and covalent portions of the bond are described by the electronegativity. The bond parameter ϕ_B gives the percentage of the two bonds.

$$\phi_B = \alpha^* \varphi_{\text{heteropolar}} + \beta^* \varphi_{\text{homopolar}} \qquad (1.16)$$

so that the heteropolar portion of a bond is $\alpha^{*2} \cdot 100\%$ and the homopolar portion $\beta^{*2} \cdot 100\%$, where $\alpha^{*2} + \beta^{*2} = 1$.

The degree of heteropolar (ionic) binding of the bond A-B is empirically

$$\begin{aligned} I_{A\text{-}B} &= 1 - \exp\left[-(X_{P(A)} - X_{P(B)})^2/4\right] \\ &= 1 - \exp\left(-E_{B(A\text{-}B)}/4\right) \end{aligned} \qquad (1.17)$$

where X_P is the electronegativity of atom A or atom B. The ionic character (in percent) of a bond for the differences of electronegativities of atoms A and B is given in Table 1.3.

TABLE 1.3

Degree of Ionic Binding as Function of Difference of Electronegativities

$X_{P(A)} - X_{P(B)}$	Ionic binding [%]
0	0
0.2	1
0.4	4
0.6	8
1.0	22
1.4	39
1.8	55
2.2	70

The ionic contribution to the binding causes a resonance strengthening of the lattice and consequently an increase of the melting point and a widening of the energy gap. Elements with the diamond lattice are purely homopolar; the lattice constant of an element with the diamond structure is practically identical with that of the isoelectronic III–V compound and the heteropolar contribution to the compound is relatively small. A covalent radius can be assigned to each atom such that the sum of the covalent radii of the two nearest neighbors in IV–IV, III–V, II–VI, and I–VII compounds agrees well with the distance between the two neighbors.

Table 1.4 gives covalent radii of elements of Groups III, IV, and V. The

TABLE 1.4

Covalent Radius of Elements of Groups III, IV, and V (in Å)

B	0.88	C	0.77	N	0.70
Al	1.26	Si	1.17	P	1.10
Ga	1.26	Ge	1.22	As	1.18
In	1.44	Sn	1.40	Sb	1.36
Tl	1.47	Pb	1.46	Bi	1.46

decreasing trend of the covalent radius from left to right and the increasing trend from top to bottom are apparent from this table. The binding energy (in eV) of a heteropolar bond for a binary semiconductor is empirically

$$E_{B(A\text{-}B)} = (X_{P(A)} - X_{P(B)})^2 \tag{1.18}$$

Electronegativities of selected elements are given in Table 1.5.

TABLE 1.5

Electronegativities of Selected Elements

Group III		Group IV		Group V	
B	2.0	C	2.5	N	3.0
Al	1.5	Si	1.8	P	2.1
Ga	1.6	Ge	1.8	As	2.0
In	1.7	Sn	1.8	Sb	1.9

The Coulomb forces and the stability of the lattice are the larger, the larger the ionic charge and the smaller the distance between the atoms.

(b) Homopolar (covalent) bond:

It exists between atoms or ions of the same character which have common valence electrons. A covalent bond contains no permanent electric moment. It occurs mainly in molecules consisting of two identical atoms. The covalent binding is a result of the exchange of electrons between two

1-1 CRYSTAL STRUCTURE AND CHEMICAL BOND

equal atom nuclei. This exchange of electrons between atoms is called quantum-mechanical resonance which leads to an energy coupling and hence a chemical bond.

A chemical bond possesses a certain strength, i.e., the binding energy, which is characteristic of the lattice constituents and which is the force required to break a bond. Examples of binding energies are given in Table 1.6.

TABLE 1.6
Binding Energies of Selected Bonds

Bond	Binding energy $E_{B(A-B)}$ [eV]
Si-Si	1.84
Ge-Ge	1.84
Ga-As	0.16
Ga-P	0.25
In-P	0.25
In-As	0.09
Si-O	2.90

4. AMORPHOUS AND POLYCRYSTALLINE SEMICONDUCTORS

Crystalline semiconductors (in which the position of atoms is periodic) represent one end of the spectrum of semiconducting materials, whereas amorphous semiconductors (in which the average position of atoms is not fixed into a definite structure) represent the opposite end. Between these extremes are found the polycrystalline semiconductors which, depending upon the nature of their structural disorder, display characteristics that fall between these extremes. Crystalline dielectrics and amorphous semiconductors have certain characteristics in common, e.g., their electric conductivity is substantially less than that of crystalline semiconductors. However, there are distinct differences between amorphous and polycrystalline semiconductors and crystalline dielectrics. In a crystal lattice the arrangement of lattice atoms has a three-dimensional order, the spacing between atomic centers remains the same throughout the crystal, and the electron diffraction pattern shows periodicity. By contrast, amorphous semiconductors have a random atomic structure with short-range order, indicated by a diffuse diffraction pattern.

22 GENERAL PROPERTIES OF SEMICONDUCTORS

A brief comparison of the electric behavior of crystalline dielectrics and amorphous and polycrystalline semiconductors is as follows.

(a) Crystalline dielectrics:
Crystalline dielectrics have relatively large energy gaps (of the order of several eV) and relatively high carrier mobilities (of the order of 10 to 10^3 cm^2/V sec), but very small thermal carrier densities. They normally have high densities of localized trapping levels, usually widely distributed over the energy gap. In the presence of an injecting contact there is space-charge-limited carrier flow, and in the presence of a blocking contact there is Schottky thermionic emission.

(b) Amorphous semiconductors:
The electric behavior of amorphous semiconductors can be described only if the concepts of energy bands, energy gap, conduction mechanisms, etc. are modified. In these semiconductors the energy band structure depends mainly on nearest-neighbor arrangements. The band edges are diffuse, resulting in the gradual transition from conducting states in which carriers are free to move to deep trapping levels in which carriers are strictly localized. Electronic conduction does not occur by free carriers but by more or less localized carriers "hopping," i.e., tunneling, between sites, usually with thermal activation. The activation energy is not clearly defined since the hopping sites are randomly distributed both in space and in energy. The activation energy increases with temperature as more and more carriers become excited into the conduction process. The resistivity is mainly a function of the type of disorder and is relatively insensitive to the nature of the semiconductor.

(c) Polycrystalline semiconductors:
The electric behavior of polycrystalline semiconductors shows similarities to that of single-crystalline materials since the microscopic regions of crystallinity (domains) have a short-range order or a microscopic periodicity. Polycrystalline semiconductors can, therefore, be considered to be quasi-periodic. Although they have basically some properties similar to crystalline materials, the state densities are undefined and washed out. The electric conductivity is significantly lower since, in addition to carrier scattering processes present in single-crystalline materials, scattering at the domain boundaries affects the carrier mobility.

In crystalline semiconductors (as well as in crystalline, polycrystalline, and amorphous metals) carrier mobility is not thermally activated and a mean free path can be defined. In amorphous semiconductors, on the other hand, carrier mobility results from thermally activated hopping in which an electron jumps from one localized state to another, accompanied by phonon emission or absorption. Since the electron states are localized, the conductivity tends to zero with temperature because thermal activation is required, although the density of states is everywhere finite. In this case a mean free path cannot be defined.

Amorphous semiconductors can be classified in three categories depending upon the nature of the structural disorder.

(a) Elemental semiconductors:
This group includes Si, Ge, and Sn. The dominant types of disorder are of translational and configurational nature. The distribution of localized states peaks around one or a few energies within the energy gap. The short-range order extends over long distances.

(b) Covalent compound semiconductors:
This group includes the simplest amorphous compound semiconductors, i.e., the multicomponent alloy glasses mainly formed of Group IV, V, and VI elements (chalcogenide glasses). The dominant types of disorder are of translational, configurational, and compositional nature. The compositional disorder makes it possible that practically all atoms locally satisfy their valence bond requirements. The distribution of localized states is continuous throughout the energy gap.

(c) Ionic or tightly bound materials:
This group includes silicate glasses, SiO, Al_2O_3, Ta_2O_5, etc. The energy gap of these materials is usually greater than 2 eV. The dominant type of disorder is of translational nature, although impurity atoms or deviations from stoichiometry can cause significant disorder. The localized states have donor- or acceptor-like characteristics and form a narrow band within the energy gap.

The tunneling (hopping) mechanism in amorphous semiconductors is similar to the tunneling between atoms in a heavily doped crystalline semiconductor (impurity conduction). The electron hopping mobility increases rapidly near the valence and conduction band edges. The hopping mobility is

$$\mu = qa^2 v_{el}/6kT = (qh/12\pi)/m_n kT$$
$$\approx 10 \text{ cm}^2/\text{V sec at } 300°\text{K}$$

and the electric conductivity

$$\sigma = \sigma_0 \exp(-E_a/2kT),$$

where a is the average interatomic separation, E_a is an activation energy of the order of 1 eV, and v_{el} is an electronic frequency given by

$$v_{el} = h/2\pi m_n a^2 \approx 10^{15} \text{ sec}^{-1}.$$

The asymptotic high-temperature conductivity σ_0 is of the order of 10^4 Ω^{-1} cm^{-1} and differs little from the value for crystalline semiconductors.

In a crystalline semiconductor the energy gap is the energy difference between the densities of states, the carrier mean free path is rather long, and the carrier mobility is given by $\mu = q\tau_c/m_n$. In contrast, in a noncrystalline semiconductor the energy gap is the energy difference between mobility edges, with a mean free path on the order of the interatomic distances. Sufficiently deep into the bands the mobility is similar to that in crystalline material, i.e., the mean free path is relatively long.

The optical properties (reflection and absorption) of amorphous semiconductors are similar to those of crystalline materials since in both cases they depend upon nearest-neighbor interactions which are similar in both.

The solid solubilities of impurities in amorphous semiconductors may reach several atomic percent and are higher (by one or two orders of magnitude) than those in crystalline semiconductors where they reach only about 0.1 atomic percent. At impurity concentrations higher than the solid solubility the impurity atoms in a crystalline semiconductor are not electrically active and form a separate phase that precipitates in the lattice, whereas in an amorphous semiconductor this is not the case. The electrical effectiveness of defects in an amorphous material is significantly reduced since these defects are usually trapped in the deep levels that are present within the energy gap as a result of the disorder.

5. REFERENCES

(1) O. Madelung, *Physics of III–V Compounds*, John Wiley and Sons, New York, 1964.
(2) N. B. Hannay, *Solid State Chemistry*, Prentice-Hall, Inc., Englewood Cliffs, N.J., 1967.
(3) C. Kittel, *Introduction to Solid State Physics*, John Wiley and Sons, New York, 1968.
(4) A. K. Jonscher, *J. El. Chem. Soc.*, **116**, 217C (1969).
(5) H. Fritzsche and S. R. Ovshinsky, *J. Non-Cryst. Solids*, **2**, 393 (1970).
(6) E. A. Davis and R. F. Shaw, *J. Non-Cryst. Solids*, **2**, 406 (1970).
(7) M. H. Cohen, *J. Non-Cryst. Solids*, **2**, 432 (1970).

1-2 Properties of Compound Semiconductors Compared to Elemental Semiconductors

1. ELEMENTAL SEMICONDUCTORS

The semiconductors of Group IV, i.e., the elemental semiconductors, have the diamond (cubic) crystal structure which is characterized by an extremely high stability since any forces attacking a single lattice atom are distributed in four different directions and are thus less effective. The crystal atoms are bound by homopolar cohesive forces which act as bridges between neighboring atoms. The cohesive forces are due to two electrons of the outer shell per atom which do not participate in electric conduction. The nature of the homopolar bond as well as the small effective carrier mass due to the homopolar cohesion are responsible for the high carrier mobilities in the diamond lattice. Two other important properties of semiconductors are the energy band gap between conduction and valence bands and the melting point. These properties are listed in Table 1.7. The trend from electronegative to electropositive character with increasing atomic number is evident in Group IV elements; C is strongly nonmetallic, Si is chemically essentially nonmetallic, Ge is metalloid, α-Sn is metallic.

TABLE 1.7

Characteristics of Group IV Semiconductors

Element	Electron mobility [cm²/V sec]	Band gap [eV]	Melting point [°C]	First ionization potential [eV]	Electronegativity	Covalent radius [Å]
C	1800	5.30	3800	11.26	2.5	0.77
Si	1350	1.12	1417	8.15	1.8	1.17
Ge	3900	0.67	937	7.88	1.8	1.22
α-Sn	1600	0.08	232	7.33	1.8	1.40

GENERAL PROPERTIES OF SEMICONDUCTORS

The strength of single covalent bonds between atoms of Group IV and those of other Groups generally decreases with increasing atomic number; at high atomic number, however, a slight decrease is observed. These energies do not reflect the ease of heterolytic breaking of bonds which is the usual way in chemical reactions. Table 1.8 lists some average bond energies.

TABLE 1.8
Bond Energies of Group IV Semiconductors

Element	Energy of bond [eV] with				
	H	C	Cl	Br	O
C	4.29	3.60	3.43	2.86	3.56
Si	3.04	2.99	3.73	2.99	3.82
Ge	3.21	3.08	3.69	2.95	
α-Sn	3.08	2.95	3.56	2.82	

Another important characteristic of Group IV elements is their ability to be doped with other elements, notably with elements of Groups III and V. Because these impurities contribute to an excess or a deficiency of one electron per atom their behavior can be described by a quasi-hydrogen model; using the hydrogen model, these impurities have theoretical energy levels which agree well with those experimentally found (≈ 0.05 eV for Si, ≈ 0.01 eV for Ge). Because of the large band gap of diamond and the small band gap and low melting point of α-Sn these elements have not found use as practical semiconductor materials.

2. COMPOUND SEMICONDUCTORS

Compound semiconductors are mainly those formed of elements Al, Ga, In (Group III) and of elements P, As, Sb (Group V). The nine compounds formed from these (AlP, AlAs, AlSb, GaP, GaAs, GaSb, InP, InAs, InSb) have the zincblende structure. In a first-order approximation the cohesion of III–V compounds is homopolar, similar to the Group IV semiconductors; however, elements of Group III are slightly more electropositive and elements of Group V are slightly more electronegative than elements of Group IV so that III–V compounds have a slightly more heteropolar bond than Group IV elements. As a result, the cohesive forces of these compounds are slightly higher by an additional ionic term. For example, InSb and Sn have the same lattice spacing and, in the average, the same atomic weight and the same density; because of the quantum-mechanical resonance effect the cohesive

TABLE 1.9

Characteristics of Semiconductors of the Same Density and Atomic Spacing

	Group	Electron mobility [cm²/V sec]	Band gap [eV]	Melting point [°C]
α-Sn	IV	1,600	0.08	232
InSb	III–V	80,000	0.17	525
CdTe	II–VI	300	1.5	1098
AgI	I–VII	30	2.8	555

forces in the compound are higher than those in the element; therefore, thermal lattice vibrations are less in the compound than in the element. Hence the electron mobility in InSb is considerably higher than in α-Sn (for InSb μ_n = 80,000 cm²/V sec, for α-Sn μ_n = 1600 cm²/V sec).

Compound semiconductors differ from elemental semiconductors mainly because a certain fraction of the lattice bonds is of ionic nature. Therefore,

TABLE 1.10

Comparison of Lattice Spacings (a), Average Atomic Weights (w_{at}), and Energy Gaps (E_G) of Semiconductors of the Same Groups and of the Same Horizontal Lines of the Periodic System*

	IV		III–V		II–VI		I–VII	
a		3.65		3.62		2.70		
w_{at}	C	12.01	BN	12.40	BeO	12.50	LiF	12.97
E_G		5.3		4.6				
a		5.43		5.46		5.19		5.63
w_{at}	Si	28.09	AlP	28.98	MgS	28.19	NaCl	29.72
E_G		1.1		3.0				
a		5.66		5.65		5.67		5.69
w_{at}	Ge	72.59	GaAs	72.32	ZnSe	72.17	CuBr	71.73
E_G		0.7		1.4		2.7		
a		6.49		6.48		6.48		6.47
w_{at}	Sn	118.69	InSb	118.29	CdTe	120.00	AgI	117.39
E_G		0.1		0.2		1.5		2.8

* The lattice spacing is given in angströms, the energy gap in electron volts.

the scattering of electrons by optical phonons may play an important role in transport phenomena in these compounds. They also differ from elemental semiconductors because their conduction bands are nonparabolic.

In making the transition from Group IV elements to III–V, II–VI, and I–VII compounds, heteropolarity increases resulting first in a mobility increase and then in a drastic mobility decrease. Table 1.9 illustrates this for semiconductors whose density and atomic spacing are nearly identical.

Other II–VI and I–VII semiconductors have still lower mobilities (e.g., for CdS $\mu_n \approx 10^2$ cm^2/V sec, for NaCl $\mu_n \approx 10^{-3}$ cm^2/V sec), so that only III–V compounds have semiconducting properties comparable to those of Group IV elements.

In Table 1.10 some properties of semiconductors having approximately the same average atomic weight are compared; these are the semiconductors of Groups IV, III–V, II–VI, and I–VII which correspond to the same horizontal line of the Periodic System.

3. CLASSIFICATION OF SEMICONDUCTORS

The III–V compound semiconductors form a link between the semiconductors of Group IV and the II–VI and I–VII compound semiconductors. Consequently, it is possible to deduce the characteristics of the III–V compounds from the characteristics of the neighboring materials. This is justified also because the corresponding compounds and elements have almost identical lattice constants. Even the corresponding II–VI and I–VII compounds have only slightly larger lattice constants.

The diagram of Table 1.11 can be obtained if the isoelectronic III–V compounds are assigned to their Group IV semiconductors; it also shows the tendencies of several properties.

Table 1.11 indicates two regularities ("Welker's Rule").

(a) Vertical direction:
Binding energy and energy gap decrease downwards. They are largest for C and BN and smallest for α-Sn and InSb. Electron mobility shows the opposite tendency. The increase in binding energy in the downward direction is due to the increasing atomic number which results in a more and more loose binding of the valence electrons which form the bonds. With decreasing binding the energy bands become wider and the forbidden energy gap therefore narrower. Wide energy bands are associated with small effective carrier masses at the band edge, resulting in high carrier mobility.

(b) Horizontal direction:
Binding energy, energy gap, and electron mobility increase in the horizontal transition from elemental to compound semiconductor. The in-

TABLE 1.11
Classification of Semiconductors

Elemental semiconductors	Compound semiconductors
C	BN
SiC	BP AlN
Si	BAs AlP GaN
$Si_{0.5}Ge_{0.5}$	BSb AlAs GaP InN
Ge	AlSb GaAs InP
	GaSb InAs
Sn	InSb

Binding Energy* ↑⌐→	Melting point	↑
Energy gap ↑⌐→	Refractive index	↓
Electron mobility ⌐→	Transparent	↑
Hole mobility ←⌐	Opaque	↓

* The direction of the arrows indicates an increasing value of a property in that direction.

crease in binding energy from left to right is due to the valence electrons which form the bonds being less loosely bound. With increasing binding the individual energy bands become narrower and the energy gap therefore wider. A wide energy band, in turn, is associated with a small effective mass at the band edge which results in high carrier mobility. The high electron mobility is also due to the fact that in the most important III–V compounds the (000) minimum forms the bottom of the conduction band. The decreasing hole mobility from left to right is due, on one hand, to the increasing binding energy and, on the other hand, to the fact that with increasing polarization of the electron bonds the valence electrons and therefore the holes concentrate more and more around the anions, which makes it more difficult for a valence electron to transfer from one anion to another and thereby decreases the hole mobility.

These tendencies in the vertical and particularly in the horizontal direction are only approximate. In the horizontal direction they are mainly valid in the transition from elemental to compound semiconductors and less accurate within the compound semiconductor group.

4. BAND STRUCTURE

The most important property of a semiconductor is its energy gap. For Group IV semiconductors the energy gap decreases with increasing atomic weight, i.e., they range from an almost pure insulator to a true metal.

Some important band structure parameters of selected compound semiconductors are listed in Table 1.12 which combines theoretical and experimental data.

TABLE 1.12
Band Structure Parameters of Selected Semiconductors

Parameter	Semiconductor						
	InSb	InAs	InP	GaSb	GaAs	GaP	AlSb
Lowest conduction band minimum at	(000)	(000)	(000)	(000)	(000)	(100)	(100)
Higher subband minima at		(100)	(111) (100)	(100)	(000)	(000)	
Spin orbit splitting (Δ) [eV]	0.98	0.43	0.18	0.81	0.33	0.13	0.75
Band gap [eV]							
$E_{G,\,opt}$ (300°K)	0.18	0.36	1.26	0.70	1.43	2.24	1.60
$E_{G,\,opt}$ (4°K)	0.24	0.43	1.42	0.81	1.52	2.33	1.70
$E_{G,}$ (0°K)	0.27	0.47	1.34	0.80	1.40		1.65
Effective mass (300°K)							
$m_{n,\,opt}/m_o$	0.012	0.025	0.077	0.047	0.057	0.340	0.390
$m_{n,\,th}/m_o$	0.015	0.020	0.050	0.900	1.200		
m_{el}/m_o	0.015	0.026	0.072	0.046	0.084	0.130	0.110
$m_{p1,\,opt}/m_o$	0.450	0.400		0.230	0.680		
$m_{p1,\,el}/m_o$	0.500	0.300	0.200	0.390	0.500	0.500	0.650
$m_{p1,\,th}/m_o$	0.530	0.820	1.000	0.710	1.000	0.560	1.100
$m_{p2,\,th}/m_o$	0.016	0.031	0.086	0.053	0.100	0.130	0.120
$m_{p3,\,th}/m_o$	0.120	0.110	0.180	0.160	0.210	0.220	0.270
Effective charge (q^*/q)	0.42	0.56	0.68	0.33	0.48	0.58	0.48

The effective charge, i.e., the charge difference between the Group III atom with its surrounding electron cloud and the Group V atom with its surrounding electron cloud, determines the transport properties of a III–V semiconductor.

5. CARRIER MOBILITY AND EFFECTIVE MASS

Carrier mobility is affected by two factors: the effective mass of the charge carriers and their interaction with the lattice. The lattice of the III–V compounds is more tightly bound because of a small amount of ionic character in the binding which could result in a weaker interaction between carriers and the lattice and consequently in a larger mobility in III–V compounds compared to elemental semiconductors.

The effective electron and hole masses can be estimated from

$$m_n/m_o \approx 1/\{1 + (20/3)[(2/E_G) + 1/(E_G + \Delta)]\}$$
$$\approx E_G/20 \quad \text{for} \quad \Delta \ll E_G \ll 20 \text{ eV}, \tag{1.19}$$

$$m_{p2}/m_o \approx 1/[(40/3\ E_G) - 1]$$
$$\approx (3/40)E_G \approx E_G/13, \tag{1.20}$$

$$m_{p3}/m_o \approx 1/\{20/[3(E_G + \Delta)] - 1\}$$
$$\approx (3/20)(E_G + \Delta); \tag{1.21}$$

where E_G is the energy gap and Δ is the spin orbit splitting.

6. IMPURITIES AND DEFECTS

The most important acceptors in III–V semiconductors are atoms from Group II which are located substitutionally in the sublattice of the trivalent lattice atoms. The most important donors are atoms from Group VI in the sublattice of the pentavalent lattice atoms. Atoms of Group IV may be donors if they are built into the III sublattice or acceptors if they are built into the V sublattice. This is shown in Table 1.13, where A means acceptor and D donor.

Although the incorporation of foreign atoms of Group III or V into a III–V semiconductor will not affect the carrier concentration, there will be an

TABLE 1.13

Impurity Types in Compound Semiconductors

Group IV impurity	Semiconductor					
	InSb	InAs	InP	GaSb	GaAs	AlSb
Si	A	D	D	A	D	A
Ge	A	D	D	A	D	A
Sn	D	D	D	A	D	A+D

32 GENERAL PROPERTIES OF SEMICONDUCTORS

effect on the physical properties of the host lattice. An impurity atom built into the wrong sublattice, for example the replacement of a Ga atom in GaAs by an As atom, will cause a slight deviation from stoichiometry, although such deviations are more pronounced in II–VI semiconductors. Furthermore, compound semiconductors may possess lattice defects, such as vacancies.

The distribution coefficient, i.e., the ratio of impurity concentration in the solid (at the interface) to the concentration in the melt, for some elements in III–V compounds is given in Table 1.14.

TABLE 1.14
Distribution Coefficient of Impurities in Compound Semiconductors

Impurity	Semiconductor					
	InSb	InAs	InP	GaSb	GaAs	AlSb
Cd	0.26	0.13		0.02	0.02	0.002
Ge	0.05	0.07	0.05	0.32	0.03	0.03
Se	0.50	0.93	0.60	0.18	0.50	0.003
Si		0.40		1.00	0.10	0.05
Sn	0.06	0.09	0.03	0.01	0.03	0.0005
Te	3.50	0.44		0.40	0.30	
Zn	10.0	0.77		0.30	0.10	0.02

The nature of dominant lattice defects differs between Group IV and Group II–VI semiconductors. Departures from stoichiometry are found in the compound semiconductors; an excess of either of the two components forms lattice defects which affect the carrier mobility. Such nonstoichiometric behavior does not exist in the case of Group IV elements. Group III–V semiconductors show very few deviations from stoichiometry; the most important impurities are built into the lattice substitutionally. Consequently, the III–V compounds combine the advantages of elemental semiconductors with a wide variety of other interesting properties, which explains the physical and technological interest in these semiconductors.

7. SUMMARY OF THE MOST IMPORTANT PROPERTIES

Table 1.15 lists the important properties of various lightly doped semiconductors at 300°K.

TABLE 1.15

Summary of Properties of Lightly Doped Semiconductors*

Group	Semi-conductor	E_G [eV]	μ_n [cm²/Vsec]	μ_p	m_n/m_o	m_p/m_o	a [Å]	T_m [°C]	ε_s	d_s [g/cm³]
IV	C	5.3	1800	1600			3.56	3800	5.8	3.51
	Si	1.1	1350	475	0.23	0.12	5.43	1417	11.7	2.33
	Ge	0.7	3900	1900	0.03	0.08	5.66	937	16.0	5.33
	SiC	2.8	400	50	0.60	1.20	4.36	2830	10.0	3.22
III–V	AlAs	2.2	180				5.66	1600	8.5	3.79
	AlP	3.0	80				5.46	1500	11.6	2.38
	AlSb	1.6	200	420	0.30	0.40	6.14	1050	10.1	4.26
	BN	4.6					3.62	3000	7.1	2.20
	BP	6.0		300			4.54	1250	11.6	2.97
	GaAs	1.4	8500	400	0.07	0.09	5.65	1237	10.4	5.32
	GaP	2.3	110	75	0.12	0.50	5.45	1465	8.5	4.13
	GaSb	0.7	4000	1400	0.20	0.39	6.10	712	14.0	5.60
	InAs	0.4	33000	460	0.03	0.02	6.06	942	11.7	5.66
	InP	1.3	4600	150	0.07	0.69	5.07	1070	10.3	4.78
	InSb	0.2	80000	750	0.01	0.18	6.48	525	15.6	5.77
II–VI	CdS	2.6	340	18	0.21	0.80	5.83	1750	5.4	4.84
	CdSe	1.7	600		0.13	0.45	6.05	1350	10.0	5.74
	CdTe	1.5	300	65	0.14	0.37	6.48	1098	11.0	5.86
	ZnS	3.6	120	5	0.40		5.41	1850	5.2	4.09
	ZnSe	2.7	530	16	0.10	0.60	5.67	1515	8.4	5.26
	ZnTe	2.3	530	900	0.10	0.60	6.09	1238	9.0	5.70
IV–VI	PbS	0.4	600	200	0.25	0.25	5.94	1077	17.0	7.50
	PbSe	0.3	1400	1400	0.33	0.34	6.15	1062	23.6	8.10
	PbTe	0.3	6000	4000	0.22	0.29	6.46	904	30.0	8.16
II–IV	Mg₂Ge	0.7	530	110				1115		
	Mg₂Si	0.8	370	65		0.46	6.34	1102		1.94
	Mg₂Sn	0.4	210	150			6.75	778		3.66
II–V	Cd₃As₂	0.1		15000	0.05			721		6.21
	CdSb	0.5	300	1000	0.16	0.10	6.47	456		6.92
	Zn₃As₂	0.9		10				1015		5.53
	ZnSb	0.5	10	350	0.15			546		6.33

*Explanation of symbols: E_G = energy gap; μ_n, μ_p = electron, hole mobility; m_n/m_o = effective electron mass; m_p/m_o = effective hole mass; a = lattice constant; T_m = melting point; ε_s = dielectric constant; d_s = density.

TABLE 1.15 (*Continued*)

Group	Semi-conductor	E_G [eV]	μ_n [cm²/Vsec]	μ_p	m_n/m_o	m_p/m_o	a [Å]	T_m [°C]	ε_s	d_s [g/cm³]
V–VI	As₂Se₃	1.6	15	45				608		4.75
	As₂Te₃	1.0	170	80	0.36		14.40	360		6.00
	Bi₂Te₃	0.2	10000	400	0.32	0.21	10.45	580		7.70
	Sb₂Se₃	1.2	15	45			11.68	612		5.81
	Sb₂Te₃	0.3		270		0.34		620		6.50
V–VIII	PtSb₂	0.1	200	1400			6.43	1240		
III–VI	In₂Se₃	1.3	30					890		5.67
	In₂Te₃	1.0	340		0.70	1.23	6.15	667		5.78

Table 1.16 compares the properties of the technologically most important semiconductors with those of SiO₂ where applicable. Semiconductors of low impurity level are considered at 300°K, unless otherwise noted.

The properties of silicon are given in more detail in Section 1-3 and the properties of germanium and gallium arsenide are discussed in more detail below.

Table 1.17 lists some applications of presently used semiconductors.

TABLE 1.16

Properties of the Most Important Semiconductors and SiO₂

Property	Si	Ge	GaAs	SiO₂
Crystal structure	Diamond; 8 atoms per unit cell	Diamond; 8 atoms per unit cell	Zincblende; 8 atoms per unit cell	Random network of SiO₄ tetrahedra; 50% covalent, 50% ionic bonding
Atomic number	14	32	31/33	14/8
Atomic or molecular weight	28.09	72.60	144.63	60.08
Lattice constant [Å]	5.43	5.66	5.65	
Atoms or molecules [10²²/cm³]	5.00	4.42	2.21	2.30

TABLE 1.16 (*Continued*)

Property	Si	Ge	GaAs	SiO$_2$
Density [g/cm^3]	2.33	5.32	5.32	2.27
Energy gap [eV]	1.12	0.57	1.43	8
Dielectric constant	11.7	16.3	10.4	3.9
Melting point [°C]	1417	937	1238	1700
Vapor pressure [Torr]	10^{-7} (1050°C)	10^{-9} (750°C)	1 (1050°C)	10^{-3} (1050°C)
Specific heat [10^7cm^2/sec$^{2\circ}$K]	0.70	0.31	0.35	1.00
Thermal conductivity [W/cm°K]	1.5	0.6	0.8	0.01
Thermal diffusivity [cm^2/sec]	0.90	0.36	0.44	0.006
Coefficient of linear thermal expansion [10^{-6}/°K]	2.5	5.8	5.9	0.5
Intrinsic carrier concentration [cm^{-3}]	$1.45 \cdot 10^{10}$	$2.4 \cdot 10^{13}$	$9.0 \cdot 10^6$	
Lattice electron mobility [cm^2/V sec]	1350	3900	8600	
Lattice hole mobility [cm^2/V sec]	480	1900	250	
Effective density of states in conduction band (N_c) [cm^{-3}]	$2.8 \cdot 10^{19}$	$1.0 \cdot 10^{18}$	$4.7 \cdot 10^{18}$	
Effective density of states in valence band (N_v) [cm^{-3}]	$1.0 \cdot 10^{19}$	$6.0 \cdot 10^{18}$	$7.0 \cdot 10^{18}$	
Electric field at breakdown [V/μm]	30	8	35	600

TABLE 1.16 (*Continued*)

Property	Si	Ge	GaAs	SiO$_2$
Raman phonon energy [eV]	0.063	0.037	0.035	
Work function [V]	4.8	4.4	4.7	
Effective mass electrons (m_n/m_o)	$m_l = 0.97$ $m_t = 0.19$	$m_l = 1.6$ $m_t = 0.08$	0.068	
holes (m_p/m_o)	$m_{lh} = 0.16$ $m_{hh} = 0.50$	$m_{lh} = 0.04$ $m_{hh} = 0.30$	0.12, 0.50	
Electron affinity [eV]	4.05	4.00	4.07	1.00
Average energy loss per phonon scattering [eV]	0.063	0.037	0.035	
Optical phonon mean free path (l_{ph}) [Å]				
electrons	62	65	35	
holes	45	65	35	

TABLE 1.17
Application of Semiconductors

	Effect	Cause	Application	Semiconductors
1.	Transistor effect	Current multiplication	Amplifier, etc.	Si, Ge
2.	Tunnel effect	p-n junction in degenerate semiconductor	High frequency switch and storage element, oscillator, amplifier	Si, Ge, GaAs
3.	Avalanche effect	Carrier generation, hot electrons	Cryogenic switches, high frequency generation, high frequency amplification	Si, Ge, GaAs

TABLE 1.17 (*Continued*)

	Effect	Cause	Application	Semiconductors
4.	Gunn effect	Hot electrons in semiconductors with two different band minima	High frequency generation, high frequency amplification	GaAs, InP CdTe
5.	Piezo effect	Polar cohesion in semiconductor	Electroacoustic amplifier	GaAs, CdS, CdSe
6.	Piezoresistance	Disformation of band structure by pressure	Pressure indicator	Si, Ge
7.	Varactor effect	Voltage-dependent space charge and capacity at $p-n$ junction	Parametric amplification, frequency multiplication, tuning diode	Si, Ge, GaAs
8.	Pair generation	Carrier generation by light or irradiation at a $p-n$ junction	Photocell, solar cell, particle counter	Si, Ge, GaAs
9.	Electroluminescence	Radiative carrier recombination at $p-n$ junction	Light displays, generation of incoherent light by injection	GaAs, GaP, InAs, InSb, SiC
10.	Laser effect	Radiative carrier recombination by injection to degeneracy	Laser diode, generation of coherent light by injection	GaAs, InAs, InSb
11.	Galvanomagnetic effect	Influence of magnetic field on carrier motion	Hall generator, field plate	Si, InAs, InSb
12.	Plasma waves	Interaction between charge carriers and electromagnetic waves in a magnetic field	Gyrator	InSb

8. PROPERTIES OF GERMANIUM

Germanium is of the diamond crystal structure; the lattice constant, i.e., the side of the unit cell, is 5.6575 Å at 300°K; the number of atoms per cm^3 is $4.42 \cdot 10^{22}$. The density is 5.3267 g/cm^3 at 300°K. The thermal expansion at 300°K is $+5.75 \cdot 10^{-6}$ °K^{-1}; it is not quite isotropic and differences up to 4% are observed; the expansion is greatest in the (100) and least in the (111) direction. The Debye temperature is 406°K; the specific heat, 5.47 cal/mole °K at 300°K, varies with temperature as $4.62 + 2.27 \cdot 10^{-3} T$. The latent heat of fusion is 8100 cal/mole and of sublimation 89000 cal/mole at 1150°K. Elastic constants for single-crystalline material are

$$c_{11} = 12.9 \cdot 10^{11} \text{ dyne/cm}^2,$$
$$c_{12} = 4.8 \cdot 10^{11} \text{ dyne/cm}^2,$$
$$c_{44} = 6.7 \cdot 10^{11} \text{ dyne/cm}^2.$$

The piezoelectric coefficients are given in Table 1.18. The velocity of sound is $5.4 \cdot 10^5$ cm/sec for 2 MHz longitudinal waves propagated in the $\langle 111 \rangle$ direction.

TABLE 1.18
Piezoelectric Coefficients of Germanium

	n-Type [cm^2/dyne]	p-Type [cm^2/dyne]
π_{44}	$+138 \cdot 10^{-12}$	$+98 \cdot 10^{-12}$
$\pi_{11} - \pi_{12}$	$+1 \cdot 10^{-12}$	$-11 \cdot 10^{-12}$
$\pi_{11} + 2\pi_{12}$	$-17 \cdot 10^{-12}$	$+3 \cdot 10^{-12}$

The band structure has a valence band maximum centered on $k_w = 0$, i.e., at the center of the first Brillouin zone, whereas the conduction band minima are not located there. The conduction band has four equivalent minima at the zone boundary in the $\langle 111 \rangle$ directions. The constant energy surfaces corresponding to values of k_w near the minima are ellipsoids of revolution with their symmetry axes along the $\langle 111 \rangle$ directions. The values of the longitudinal and transverse components of the effective mass are $m_l/m_o = 1.64$ and $m_t/m_o = 0.82$; the anisotropy coefficient is $m_l/m_t = 2$. The conductivity effective mass is $m_c/m_o = 3(m_l/m_o)(2m_l/m_t + 1) = 0.12$ and the density of states effective mass $m_d/m_o = (m_l/m_t)^{1/3}/m_o = 0.22$. The energy bands are degenerate; if the constant energy surfaces are represented as two sets of spherical surfaces characterized by a simple scalar effective mass, then there is a system of light and heavy holes having effective masses 0.044 m_o and 0.28 m_o, respectively. The constant energy surfaces are not perfectly spherical, the heavy hole band being considerably warped. The value of spin-orbit splitting is 0.28 eV and the hole effective mass in the split-off band is 0.077 m_o. The energy gap at 0°K is 0.780 eV and at 300°K is 0.670 eV.

At 300°K the intrinsic electrical resistivity is 50 Ω cm, corresponding to a charge carrier concentration (electrons plus holes) of $2.4 \cdot 10^{13}$ cm^{-3}. At 300°K, electron mobility is 3900 cm^2/V sec and hole mobility 1900 cm^2/V sec. Thermal conductivity at 300°K is 0.63 W/cm°K. The refractive index is 4.0 at long wavelengths, corresponding to a dielectric constant of 16.0. The reflectivity is 0.36 over a wide range of wavelengths in the infrared.

9. PROPERTIES OF GALLIUM ARSENIDE

GaAs has the zincblende crystal structure with a lattice constant 5.6534 Å at 300°K; the number of atoms per cm^3 is $2.21 \cdot 10^{22}$, the density is 5.32 g/cm^3, and the frequency of lattice vibrations $7.2 \cdot 10^{12}$ Hz. The thermal expansion at 300°K is $+5.7 \cdot 10^{-6}$ °K^{-1}. The Debye temperature is 344°K, the specific heat is $0.31 \cdot 10^7$ cm^2/sec^2 °K. Elastic constants for single-crystalline material at 300°K are

$$c_{11} = 11.81 \cdot 10^{11} \text{ dyne/cm}^2,$$
$$c_{12} = 5.32 \cdot 10^{11} \text{ dyne/cm}^2,$$
$$c_{44} = 5.94 \cdot 10^{11} \text{ dyne/cm}^2.$$

The Debye length is 28.80 Å. The heat of vaporization (between 1000 and 1200°C) is 90 kcal/mole and the heat of formation at 300°K is 18.0 kcal/mole.

The valence band consists of eight subbands if the spin is taken into account or of four subbands if the spin is neglected. Three of the four bands are degenerate at $k_w = 0$ and form the upper edge of the band, the fourth one forms the bottom; spin-orbit interaction results in a splitting of the band at $k_w = 0$. The two upper valence bands are approximately parabolic and have different curvatures, namely the wider heavy-hole band and the narrower light-hole band. The conduction band also consists of several subbands, with the bottom of the bands at $k_w = 0$. The constant energy surface is a sphere at the zone center. The effective masses are $m_n/m_o = 0.068$ and $m_p/m_o = 0.09$. The energy gap is 1.52 eV at 0°K and 1.43 eV at 300°K; its variation with temperature is $-4.3 \cdot 10^{-4}$ eV/°K and its variation with pressure is $+10^{-8}$ eV/g cm^2.

At 300°K the intrinsic electrical resistivity is $7.0 \cdot 10^7$ Ωcm, corresponding to a charge carrier concentration (electrons plus holes) of $1.8 \cdot 10^8$ cm^{-3}. At 300°K electron mobility is 8500 cm^2/V sec and hole mobility 400 cm^2/V sec. The thermal conductivity at 300°K is 0.8 W/cm °K. The refractive index is 3.30, corresponding to a dielectric constant of 10.4. The reflectivity is 0.29. The optical absorption edge is 0.86 μm.

10. REFERENCES

(1) H. Clauser et al., *Encyclopedia of Engineering Materials and Processes*, Reinhold Publishing Company, New York, 1963.
(2) D. M. Hamilton, *SCP and SST*, **7**, 15 (June 1964).
(3) O. Madelung, *Physics of III–V Compounds*, John Wiley and Sons, New York, 1964.
(4) C. A. Hogarth, Editor, *Materials Used in Semiconductor Devices*, John Wiley and Sons, New York, 1965.
(5) E. L. Kern and E. Earleywine, *SCP and SST*, **8**, 29 (Oct. 1965).
(6) R. K. Willardson and A. C. Beer, Editors, *Semiconductors and Metals*, Vol. I, Academic Press, New York, 1967.
(7) A. S. Grove, *Physics and Technology of Semiconductors and Metals*, John Wiley and Sons, New York, 1967.
(8) W. Kleen and W. Heywang, *Siemens-Zeitschrift*, **42**, 79 (1968).
(9) M. Neuberger, private communication.

1-3 Summary of Properties of Silicon

1. GENERAL PROPERTIES

About 25% of the earth's crust consist of silicon, exceeded in abundance only by oxygen. Silicon is found mainly in its oxides and silicates. Most silicon oxides appear as sand, quartz, rock crystal, amethyst, agate, flint, and opal; most silicates as granite, hornblende, asbestos, feldspar, clay, and mica.

Silicon and its derivatives play a significant role in plant and animal life. Diatoms extract SiO_2 from water for construction of their cell walls and SiO_2 is found in bones and the ashes of plants.

In technology Si and SiO_2 have found extensive use. SiO_2 is the main constituent of glasses and, in the form of sand and clay, it is used in bricks and concrete; silicates are used in making enamels and pottery; silicon carbide forms abrasives; silicon is an important constituent of steel. One of the more recent uses of silicon is in semiconductor technology and microelectronics where silicon has become by far the most dominant semiconductor material. Its widespread use in microelectronics is mainly due to the advent of hyper-pure semiconductors and of a technology which allowed the controlled addition of foreign atoms in well-specified quantities.

The electrical properties of silicon can be affected significantly by the addition of foreign atoms (impurities); for example, the electrical conductivity of silicon can be increased by several orders of magnitude compared to the intrinsic material.

Amorphous silicon (which, in reality, is microcrystalline) is a brown, inflammable powder which can easily be melted or vaporized. Its specific gravity is 2.35. It can be easily reduced. It is soluble in HF to SiF_4 (silicon fluoride).

Crystalline silicon is found, similar to carbon, in two modifications, in very hard black octahedrons and in gray tetrahedrons (similar to graphite). The crystallinic form has a specific gravity of 2.50; it does not change its constitution even in oxygen. It is soluble in a mixture of HF and HNO_3 to SiF_4.

Both modifications have a melting point of 1417°C and vaporize at high temperature.

Hyper-pure silicon can be obtained by the reaction of $SiHCl_3$ (trichlorosilane) with hydrogen; the reduction of $SiCl_4$ (silicon tetrachloride) with Zn,

Cd, or H_2; the pyrolytic decomposition of SiH_4 (silane); the reduction of SiI_4 (silicon tetraiodide) and $SiBr_4$ (silicon tetrabromide) with H_2.

A common three-step purification method is as follows. First, ordinary, chemically pure Si is converted to a silicon halide or to $SiHCl_3$ which is then purified by fractional distillation. It is then reconverted to elemental silicon by high-temperature reduction with hydrogen or by direct high-temperature thermal decomposition. This pure silicon is then made hyper-pure (less than 10^{-9} impurity atom percent) by zone refining; in this process a localized cross-sectional wafer is molten, the molten zone is then forced to move along the silicon crystal by motion of the heat source thus carrying the impurities to the crystal end since impurities are more soluble in the melt than in the solid so that they concentrate in the melt.

Silicon is relatively inert to most elements. It is attacked by halogens and alkali, but not by most acids except hydrofluoric acid. Because of the strong affinity of silicon to oxygen almost all silicon compounds are converted into derivatives of SiO_2 upon contact with H_2O or O_2.

The atoms of single-crystalline silicon usually have the diamond structure of the cubic crystallographic system. Silicon belongs to the hexoctahedral class, i.e., the preferred growth habit is the octahedron. Silicon crystals deposited or grown without constraints grow in the $\langle 111 \rangle$ direction and usually develop large facets due to the slower growth rate of the (111) planes as compared with other crystallographic orientations. The (111) plane provides the highest atomic packing density and is in part responsible for this slow growth and it is the slowest dissolving or etching plane. This is significant in vapor-phase etching; e.g., vapor etch rates of different silicon orientations for 5% HCl in H_2 at 1200°C are

(100): 3.4 μm/min,
(110): 3.0 μm/min,
(111): 1.5 μm/min.

Similarly, the wet thermal oxidation of silicon is orientation-dependent, e.g., in the linear portion of the growth rate at 1100°C as follows

(100): $3.02 \cdot 10^{-3}$ μm/min,
(110): $3.46 \cdot 10^{-3}$ μm/min,
(111): $3.48 \cdot 10^{-3}$ μm/min.

Externally imposed constraints usually prevent well-defined facets from forming. Vapor-grown crystals have the fewest of these constraints and often develop large facets. During selective epitaxial deposition small octahedrons are sometimes formed from nucleation on oxide surfaces.

Silicon processing is partially influenced by the crystallographic orientation due to the orientation dependence of physical and chemical properties, e.g., etch rates, Young's modulus, impurity diffusion coefficients, and others. Table 1.19 summarizes the orientation dependence of some properties of silicon.

TABLE 1.19
Orientation Dependence of Selected Properties of Silicon

Property	Orientation dependence	Remarks
Density	No	
Hardness	Yes	
Breaking strength (cleavage)	Yes	
Bulk and Young's modulus	Yes	
Elastic constants	Yes	
Epitaxial deposition rate	Yes	
Heat capacity	No	
Thermal conductivity	No	
Thermal expansion	No	
Optical absorption coefficient, refractive index, dielectric constant	No	High dislocation density can increase scattering in some directions
Etch rate	Slight	
Electric mobility and resistivity	No	High dislocation density can reduce carrier mobility in some directions
Impurity diffusion coefficient	Yes	Observed differences in junction depth with orientation are probably related to surface effects or oxidation
Segregation coefficient	Yes	
Thermal oxidation rate	Yes	Under some conditions the apparent orientation dependence is related to surface effects

Other general properties of silicon are as follows. The lattice parameter (atomic spacing) is 5.4307 Å at 300°K. This determines the number of atoms per cm^3 as $5 \cdot 10^{22}$. The density at 300°K is 2.329 gcm^{-3} and the coefficient of linear thermal expansion at 300°K is $+2.33 \cdot 10^{-6}$°C^{-1}. Around 300°K the specific heat is $5.74 + 0.617 \cdot 10^{-3} T - 1.01 \cdot 10^9 T^{-2}$ cal/mole °K. At 1550°K the latent heat of fusion is $1.195 \cdot 10^4$ cal/mole and the latent heat of sublimation is $1.050 \cdot 10^5$ cal/mole. The velocity of sound is $9.15 \cdot 10^5$ cm/sec for longitudinal waves propagating in the $\langle 110 \rangle$ direction. At 300°K the

44 GENERAL PROPERTIES OF SEMICONDUCTORS

TABLE 1.20

Summary of Properties of Silicon (Lightly doped; 300°K, unless otherwise noted)

Property	Symbol	
Atomic properties		
Atomic number	Z	14
Atomic weight	w_{at}	28.09
Atomic density [cm^{-3}]	C_L	$5.00 \cdot 10^{22}$
Crystal structure		cubic
Lattice constant (cube edge) [Å]	a	5.43
Tetrahedral radius [Å]	r_a	1.17
Atomic volume [cm^3]	u_{Si}	$2.0 \cdot 10^{-23}$
$u_{Si} = 1/C_L = (4\pi/3)r_a^3$		
Electric properties		
Energy gap (0°K) [eV]	E_G	1.153
(300°K) [eV]		1.119
Temperature coefficient of E_G [eV/°C]	dE_G/dT	$-2.3 \cdot 10^{-4}$
Pressure coefficient of E_G [eV/atm]	dE_G/dp	$-1.5 \cdot 10^{-6}$
Product of electron and hole density [cm^{-6}]	n_i^2	$1.5 \cdot 10^{33} T^3 \exp(-14028/T)$
[cm^{-3}]	n_i	$1.5 \cdot 10^{10}$
Lattice electron drift mobility [cm^2/V sec]	μ_n	1350
Lattice hole drift mobility [cm^2/V sec]	μ_p	475
Temperature coefficient of lattice electron mobility [cm^2/V sec]	$d\mu_n/dT$	-11.6
[%/°C]		-0.86
Temperature coefficient of lattice hole mobility [cm^2/V sec]	$d\mu_p/dT$	-4.3
[%/°C]		-0.90
Electron diffusion coefficient [cm^2/sec]	D_n	34.6
Hole diffusion coefficient [cm^2/sec]	D_p	12.3
Intrinsic resistivity [Ω cm]	ρ_i	$2.3 \cdot 10^5$
Photoemission work function [eV]	ϕ_{ph}	5.05
Dielectric constant	ε_s	11.7

TABLE 1.20 (*Continued*)

Property	Symbol	
Critical electric field—electrons [V/cm]	$E_{crit,\,n}$	2500
Critical electric field—holes [V/cm]	$E_{crit,\,p}$	7500
Magnetic susceptibility	κ_m	$-0.13 \cdot 10^{-6}$
Mechanical and optical properties		
Density [g/cm²]	d_s	2.329
Hardness [Mohs]	H	7.0
Elastic constants [dyne/cm²]	c_{11}	$1.674 \cdot 10^{12}$
	c_{12}	$0.652 \cdot 10^{12}$
	c_{44}	$0.796 \cdot 10^{12}$
Temperature coefficient of elastic constants [dyne/cm²°C]	dc_{11}/dT	$-75.0 \cdot 10^{-6}$
	dc_{12}/dT	$-24.5 \cdot 10^{-6}$
	dc_{44}/dT	$-55.5 \cdot 10^{-6}$
Volume compressibility [cm²/dyne]	du_v/dp	$0.98 \cdot 10^{-12}$
Young's modulus ($\langle 111 \rangle$ direction) [dyne/cm²]	Y_l	$1.9 \cdot 10^{12}$
Bulk modulus [dyne/cm²]	Y_B	$7.7 \cdot 10^{11}$
Raman phonon energy [eV]	E_{ph}	0.063
Surface tension (at freezing point) [dyne/cm²]	σ_o	720
Critical pressure [atm]	p_c	1450
Refractive index (at $\lambda = 6\ \mu m$)	n^*	3.42
Wavelength of dominant emission [μm]	λ_E	1.04
Thermal properties		
Melting point [°C]	T_m	1417
Boiling point [°C]	T_B	2600
Debye temperature [°K]	T_D	658
Latent heat of fusion [eV]	H_F	0.41
Heat of vaporization (at boiling point) [eV]	H_V	3.08
Specific heat [cal/g°C]	c_p	0.166
[cal/cm³ °C]		0.386
Thermal conductivity [W/cm °C]	κ	1.5
[cal/sec cm °C]		0.20

46 GENERAL PROPERTIES OF SEMICONDUCTORS

TABLE 1.20 (*Continued*)

Property	Symbol	
Thermal diffusivity [cm^2/sec]	D_{th}	0.90
Coefficient of linear thermal expansion [°C^{-1}]	α'	$+2.33 \cdot 10^{-6}$
Expansion on freezing [%]	du_v	$+9.0$
Critical temperature [°C]	T_c	4920
Frequency of lattice vibrations [Hz]	ν_o	$1.39 \cdot 10^{13}$

intrinsic electrical resistivity is $2.3 \cdot 10^5$ Ωcm corresponding to a charge carrier concentration of $3 \cdot 10^{10}$ cm^{-3} (electrons plus holes). The refractive index is 3.42, corresponding to a dielectric constant of 11.7 assuming no dispersion. The optical reflectivity is 0.30 for wavelengths up to 150 μm.

The band structure of silicon characterizes it as a many-valley semiconductor. The conduction band has six equivalent minima occurring at points lying in $\langle 100 \rangle$ directions with a k_w equal to $0.8\, k_{w\,\text{max}}$ where $k_{w\,\text{max}}$ is the value corresponding to the zone boundary in these directions. The conduction band has constant energy surfaces corresponding to values of k_w near the minima which are ellipsoids of revolution with their axes of symmetry along the $\langle 100 \rangle$ directions. Values of the longitudinal (m_l/m_o) and transverse (m_t/m_o) components of the effective electron mass are 0.97 and 0.19. The conductivity effective mass is $0.25\, m_o$ and the density-of-states effective mass is $0.33\, m_o$.

The values of the effective masses of electrons and holes $(m_{n,\,p}/m_o)$ in silicon can be summarized as follows:

Longitudinal electron effective mass	$m_{nl}/m_o = 0.97$
Transverse electron effective mass	$m_{nt}/m_o = 0.19$
Light hole effective mass	$m_{pl}/m_o = 0.16$
Heavy hole effective mass	$m_{ph}/m_o = 0.50$

The valence band has an energy maximum at the center of the first Brillouin zone, i.e., at $k_w = 0$. Longitudinal and transverse components of the effective hole mass are $0.16\, m_o$ and $0.49\, m_o$. The constant-energy surfaces are not perfectly spherical. Spin-orbit splitting of 0.035 eV prevents triple degeneracy of the valence bands at $k_w = 0$. The hole effective mass in the split-off hole band is $0.25\, m_o$.

Table 1.20 lists the most important properties of silicon.

2. REFERENCES

(1) E. M. Conwell, *Proc. IRE* **46**, 1281 (1958)
(2) C. G. Currin and E. Earleywine, *SCP and SST*, **7**, 21 (June 1964)
(3) E. L. Kern and E. Earleywine, *SCP and SST*, **8**, 29 (October 1965)
(4) E. L. Kern and L. A. Teichthesen, *SCP and SST*, **8**, 43 (October 1965)
(5) W. R. Runyan, *Silicon Semiconductor Technology*, McGraw-Hill Book Co., New York, 1965.
(6) *Materials Used in Semiconductor Devices*, C. A. Hogarth, Editor, John Wiley and Sons, New York, 1965.
(7) E. Spenke, *History and Future Needs in Silicon Technology*, First International Symposium on Silicon Materials, Science and Technology, New York, May 5–9, 1969.
(8) W. A. Adcock, *Silicon Devices and the Future*, First International Symposium on Silicon Materials, Science and Technology, New York, May 5–9, 1969.
(9) K. E. Bean and P. S. Gleim, *Proc. IEEE*, **57**, 1469 (1969).

1-4 Energy Band Structure

1. SEMICONDUCTOR ENERGY BAND STRUCTURE

The band structures of semiconductors can be classified into two types if only band-to-band or exciton transitions in near-intrinsic semiconductors are considered; if the impurity concentration is high, the momentum is not necessarily conserved and variations in the transition process take place.

(a) Direct band gap semiconductors:
   The lowest conduction band minimum and the highest valence band maximum are at the same wave vector in the Brillouin zone. The momentum is automatically conserved (first-order process) and available optical gain and radiative transitions are strong.
   Examples: GaAs, GaSb, InAs, InP, InSb, CdS.

TABLE 1.21
Band Structures of Semiconductors

Semiconductor	Band structure	
	Energy gap	Energy bands
Si	Wide, indirect	Complex band edge structures (many-valley, warped surfaces)
Ge	Moderately wide, indirect	Complex band edge structures
GaAs, InP, CdTe	Wide, direct	Subsidiary conduction band minima; complex valence band edge structure (warped surfaces)
GaP	Very wide, indirect	Complex band edge structures (like those of Si)
GaSb	Moderately wide, direct	Low-lying subsidiary conduction band minima; complex valence band edge structure (warped surfaces)
InAs, InSb	Narrow, direct	Complex valence band edge structure (warped surfaces)
PbTe	Narrow, direct	Complex band edge structures (many-valley)

(b) Indirect band gap semiconductors:
The lowest conduction band minimum and the highest valence band maximum are at different wave vectors. The momentum is not conserved (second-order process) and the optical transition involves phonons or other scattering centers in order to conserve momentum and energy. Available optical gain and radiative transitions are weak.

Examples: Si, Ge, AlAs, AlP, AlSb, GaP, SiC.

The important characteristics of the band structures of common semiconductors are given in Table 1.21, where features that make these semiconductors unique in relation to others are compared.

2. ELECTRON ENERGY BANDS

Figure 1.5 shows the electron energy as function of the wave vector k_w for a free electron and an electron in a lattice of lattice constant a. The energy gap is associated with Bragg reflections at $\pm n_B \pi/a$, where n_B is an integer ($n_B = 1, 2, 3, \ldots$). The region in k_w-space between $-\pi/a$ and $+\pi/a$ is the first Brillouin zone of the lattice, i.e., for silicon between ± 0.5775 Å^{-1}. Forbidden

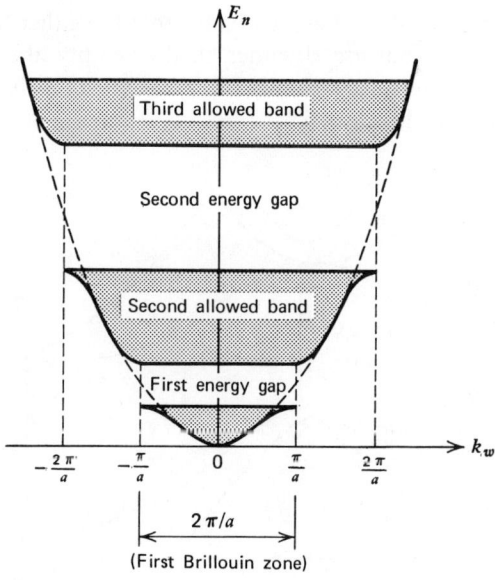

FIGURE 1.5

Electron energy (E_n) as function of wave vector (k_w), showing the energy of a free electron (dashed lines) and of an electron in a monatomic linear lattice of lattice constant a (solid lines). The first Brillouin zone has a width $2\pi/a$, the second Brillouin zone a width $4\pi/a$, etc.

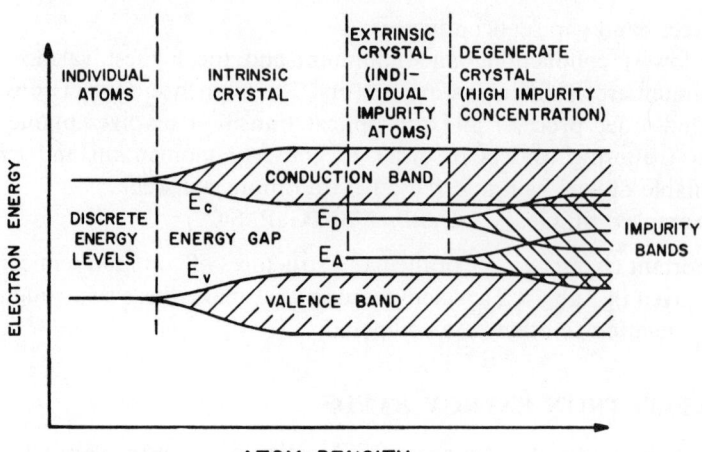

FIGURE 1.6

Electron energy as function of separation of atoms in a lattice containing impurity atoms.

regions, i.e., energy levels for which no electron energy states are allowed (energy gaps), separate allowed energy bands that are either filled, empty, or partially filled. If the bands are all either filled or empty, the crystal behaves like an insulator. If bands are partially filled, the crystal behaves like a metal or a semiconductor (semimetal).

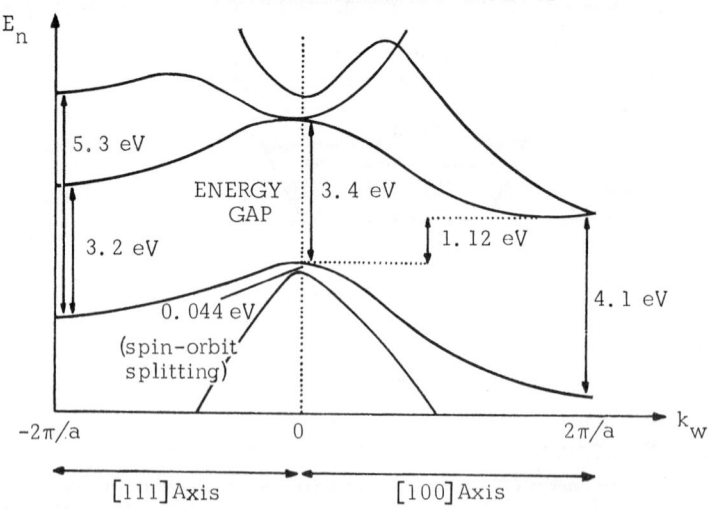

FIGURE 1.7

Electron energy (E_n) as function of wave vector (k_w) for silicon at 300°K.

The degenerate valence band level (several levels coincide) located at $k_w = (000)$ is the maximum valence band level. In the unperturbed crystal all conduction band minima have the same level.

The forbidden band (energy gap) is the difference between the lowest conduction band minimum and the valence band maximum (shaded area).

Figure 1.6 shows in a simplified energy diagram the influence of atomic density on the band structure of a semiconductor. At low atomic density, interaction between atoms is negligible so that electrons occupy discrete

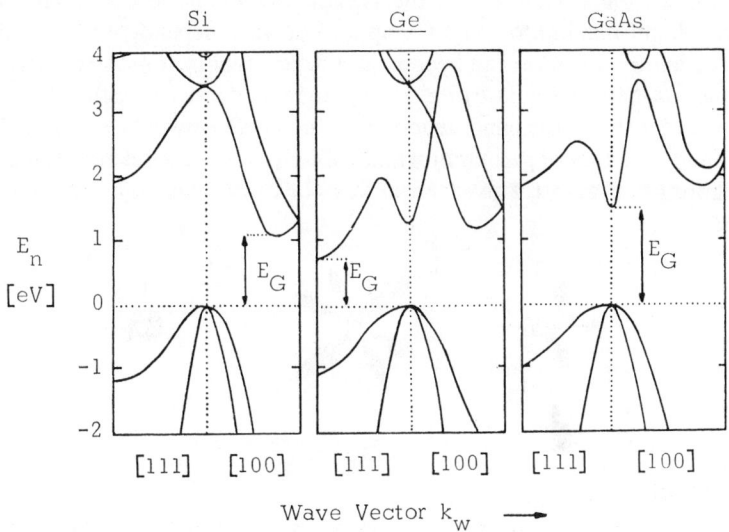

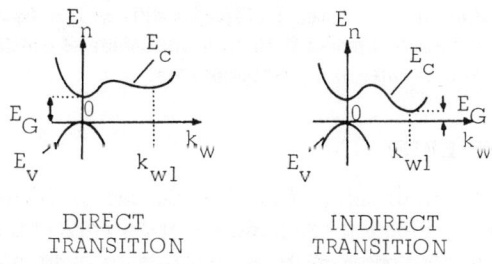

FIGURE 1.8

Energy bands of Si, Ge, and GaAs. The upper diagram shows the electron energy (E_n) as function of the wave vector (momentum; k_w). The lower part of the illustration gives a comparison between direct and indirect energy gap semiconductors. For direct transition $E_c(0) < E_c(k_{w1})$, for indirect transition $E_c(0) > E_c(k_{w1})$. Si and Ge are indirect semiconductors, GaAs is a direct semiconductor.

52 GENERAL PROPERTIES OF SEMICONDUCTORS

energy levels. As the spacing between atoms decreases, interaction will become significant, resulting in a broadening of the discrete energy levels to bands. If the semiconductor crystal density is further increased by the addition of impurity atoms, additional energy levels corresponding to these impurities will be found. As the impurity density increases these impurity energy levels will also broaden and form noncontinuous energy bands. Only if these levels are within a forbidden region will they be of significant influence on the semiconductor properties.

Stress on the crystal affects the energy levels. The levels making up the valence band maximum can shift up and down independently, and thus the valence band maximum can become nondegenerate in some stress states. The conduction band minima can also shift independently of each other.

Figure 1.7 shows the band structure of silicon, a many-valley semiconductor. The band structure of germanium and gallium arsenide is schematically shown in Figure 1.8. Figure 1.9 gives the surfaces of constant energies for Si, Ge, and GaAs.

Si   Ge   GaAs

FIGURE 1.9

Shapes of constant energy surfaces of Si, Ge, and GaAs. For silicon there are six ellipsoids of constant energy surface located from the center of the Brillouin zone at about three-quarters of the distance to the zone boundary. For Ge there are eight ellipsoids with the zone boundaries at the center of the ellipsoids. For GaAs there is one sphere of constant energy surface located at the center of the Brillouin zone.

3. ENERGY GAP

The energy (band) gap is defined as the energy difference between the lower edge of the conduction band and the upper edge of the valence band, i.e., the energy difference between the $k_w = 0$ valence band maximum and the various conduction band minima. The value E_G/kT is indicative of thermal agitation of electrons across the energy gap since the intrinsic carrier concentration depends exponentially upon E_G/kT.

The energy gap of a ternary semiconductor lies between the energy gaps of its binary components and is determined by its composition. For example, $GaAs_{1-x}P_x$ has a band gap between those of GaP and GaAs.

4. DEPENDENCE OF ENERGY GAP ON TEMPERATURE

Within a limited temperature range the temperature dependence of the energy gap of most semiconductors can be represented by

$$E_G(T) = E_{Go} - \alpha_G kT$$
$$= E_{Go} - (dE_G/dT)(T - 300°K) \quad (1.22)$$

where E_{Go} is the energy gap at the reference temperature (300°K) and α_G is an empirical constant characteristic of the semiconductor.

$$\alpha_G = (dE_G/dT)(T - 300°K)/kT \quad (1.23)$$

The temperature dependence of the intrinsic carrier concentration is related to the same parameters by

$$n_i(T) = a_i T^3 \exp(-E_{Go}/kT) \quad (1.24)$$

where

$$a_i = 4(2\pi k/h^2)^3 (m_n m_p)^{3/2} \exp(-\alpha_G). \quad (1.25)$$

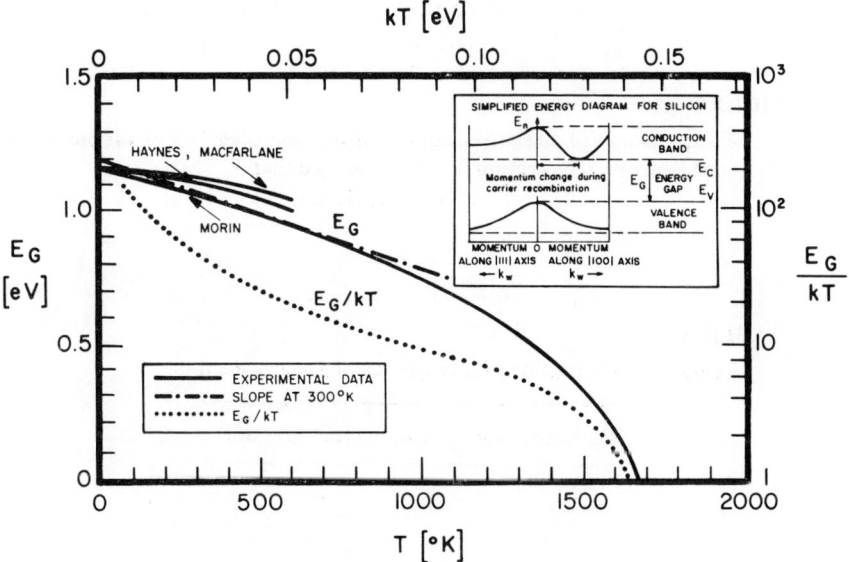

FIGURE 1.10

Energy gap (E_G) of silicon vs. temperature (T). Atmospheric pressure, nondegenerate semiconductor. Curves are based on experimental data. The solid curve given by Morin and Maita is based on indirect measurements; the dashed line corresponds to dE_G/dT at 300°K. The data of MacFarlane and Haynes are based on direct measurements.

54 GENERAL PROPERTIES OF SEMICONDUCTORS

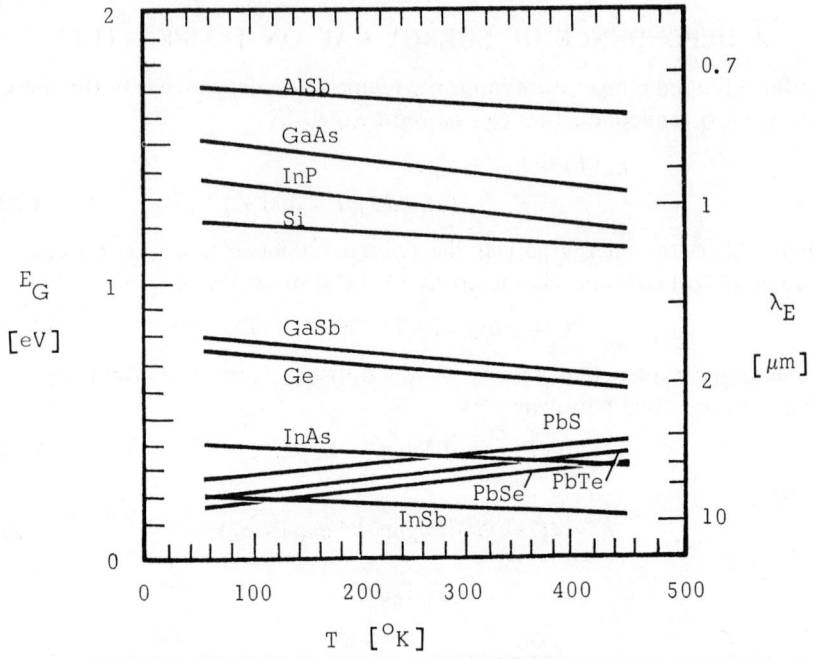

FIGURE 1.11

Energy gap (E_G) and wavelength of absorption edge (λ_E) of various semiconductors vs. temperature (T). It is assumed that
$$\lambda_E[\mu m] = 1.237/E_G[eV].$$

TABLE 1.22

Temperature Variation of Energy Gap of Semiconductors

Semiconductor	$dE_G/dT[10^{-4} eV/°K]$
Si	−2.3
Ge	−3.7
AlSb	−4.1
GaAs	−4.3
GaP	−5.4
GaSb	−4.3
InAs	−3.5
InP	−4.5
InSb	−2.9

In the temperature range between 100 and 400°K, the energy gap varies nearly linearly with temperature. The empirically determined temperature dependence of E_G of selected semiconductors around 300°K is given in Table 1.22. Some lead-containing semiconductors have a positive temperature coefficient of E_G, e.g., PbS, PbSe, PbTe (see Figure 1.11).

5. DEPENDENCE OF ENERGY GAP ON PRESSURE

The energy gap of silicon decreases with increasing hydrostatic pressure by

$$dE_G/dp = -2.4 \cdot 10^{-9} \text{ eV/g cm}^2,$$

whereas the energy gaps of Ge and GaAs increase:

$$\text{for Ge,} \quad dE_G/dp = +5 \cdot 10^{-9} \text{ eV/g cm}^2,$$
$$\text{for GaAs,} \quad dE_G/dp = +9 \cdot 10^{-9} \text{ eV/g cm}^2.$$

Table 1.23 lists the pressure coefficients for other semiconductors. For

TABLE 1.23
Pressure Coefficients of Selected Semiconductors

Semiconductor	dE_G/dp $[10^{-9}\text{eV/g cm}^2]$	Pressure range $[10^7 \text{ g/cm}^2]$	Minimum assigned to that pressure range
Si	−2.4		(111)
Ge	5.0		(111)
	12.0		(000)
	−2.0		(100)
AlSb	−1.6	<5	(100)
GaAs	9–12	<6	(000)
	−8.7	>6	(100)
GaP	−1.7	<5	(100)
GaSb	12.0	<2	(000)
	7.3	>2	(111)
InAs	5–8	<2	(000)
	3.2	>2	(111)
InP	4.6	<4	(000), (111)
	−10.0	>4	(100)
InSb	14–15	<3	(000)

germanium it is given for the three conduction band minima relative to the upper edge of the valence band; for the III–V compounds it is given for the absorption edge.

56 GENERAL PROPERTIES OF SEMICONDUCTORS

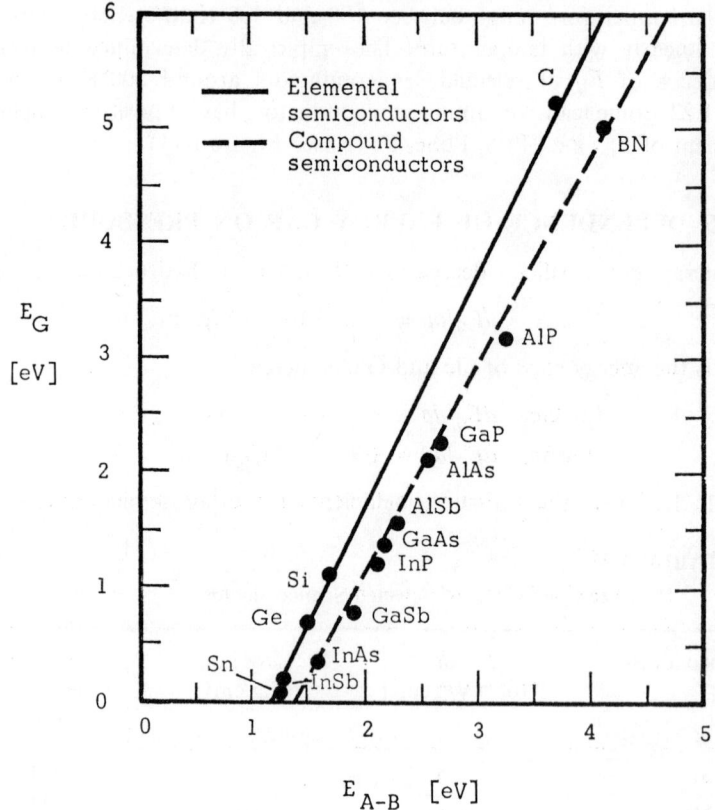

FIGURE 1.12

Energy gap (E_G) of various semiconductors as function of single bond energy (E_{A-B}). 300°K.

The energy gap is a function of the bond energy. This is shown in Figure 1.12 where a distinction between elemental and compound semiconductors is made.

6. MAXIMUM TEMPERATURE OF OPERATION

Figure 1.13 shows the maximum temperature of useful operation (T_{max}) of various semiconductors as function of the energy gap. Above this temperature electrons acquire excessive thermal energy and are raised in great number from valence to conduction band. Two distinct ranges with identical slopes are distinguished, with Ge and GaSb forming a transition region.

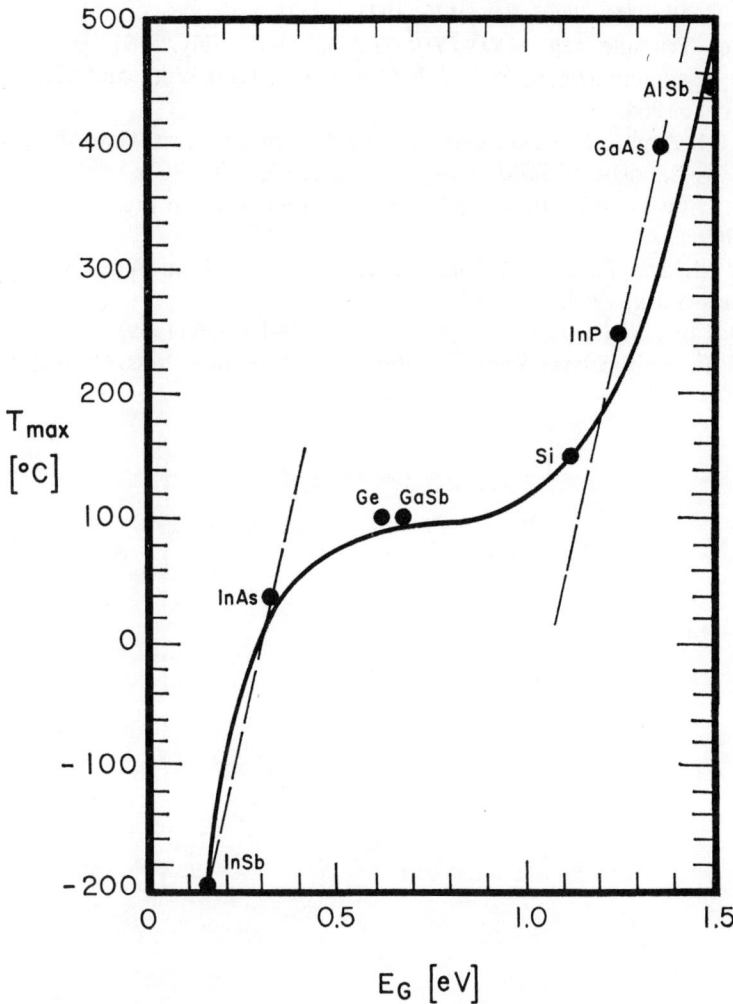

FIGURE 1.13

Maximum temperature of useful operation (T_{max}) of various semiconductors vs. energy gap (E_G).

7. REFERENCES

(1) F. J. Morin and J. P. Maita, *Phys. Rev.*, **96**, 28 (1954).
(2) G. G. MacFarlane et al., *Phys. Rev.*, **111**, 1245 (1958).
(3) J. R. Haynes et al., *J. Phys. Chem. Solids*, **8**, 392 (1959).
(4) M. L. Cohen and T. K. Bergstresser, *Phys. Rev.*, **141**, 789 (1966).

(5) J. C. Phillips, *Phys. Rev.*, **125**, 1931 (1962).
(6) Res. Triangle Rep., *ASD-TDR-63-316*, Vol. V, July, 1964.
(7) O. Madelung, *Physics of III–V Compounds*, John Wiley and Sons, New York, 1964.
(8) J. M. Ziman, *Electrons and Phonons*, Clarendon Press, Oxford, 1960.
(9) P. T. Landsberg, *Solid-State Electronics*, **10**, 513 (1967).
(10) V. I. Fistul', *Heavily Doped Semiconductors*, Plenum Press, New York, 1969.
(11) S. M. Sze, *Physics of Semiconductor Devices*, John Wiley and Sons, New York, 1969.
(12) D. Long, *IEEE Transact. Electr. Dev.*, **ED-16**, 836 (1969)
(13) H. F. Wolf, *Silicon Semiconductor Data*, Pergamon Press, Oxford, 1969.

1-5 Degenerate and Nondegenerate Semiconductors

1. DISTINCTION BETWEEN LIGHTLY DOPED AND HEAVILY DOPED SEMICONDUCTORS

The differences between lightly doped and heavily doped semiconductors are of electronic and of atomic nature.

(a) Electronic differences:
The carrier density in a semiconductor increases with increasing donor and acceptor (impurity) concentration. At low impurity concentration local energy states are formed in the forbidden band. Since the impurity atoms are far apart (their density is several orders of magnitude less than the density of crystal atoms of the semiconductor host), there is no interaction between them, and the carriers obey Boltzmann statistics, which ignores the quantum indistinguishability of particles.
At higher impurity concentration, however, interaction of impurity atoms is not negligible since the distance between impurity atoms is much less; the wave functions of electrons of neighboring impurity centers overlap and the local energy levels broaden symmetrically above and below their original positions and form impurity bands. This results in a decrease in ionization energy of the impurity atoms.
At very high impurity concentration the ionization energy becomes insignificant, i.e., the impurity band merges with conduction or valence band and a single allowed band is formed in the semiconductor. In such a heavily doped semiconductor the high density of carriers requires the application of Fermi-Dirac statistics rather than Boltzmann statistics and the semiconductor may be called degenerate. However, temperature increase increases the intrinsic carrier concentration substantially so that a semiconductor which is degenerate at room temperature may become intrinsic at high temperature although the impurity concentration remains unchanged.
Impurity atoms are fully ionized in the case of high doping level and the

large number of ions affects the behavior of carriers, partially because of scattering.

When the impurity concentration is increased considerably, the density of free carriers increases also and consequently the screening of the impurity centers by electrons becomes stronger. As a result of the screening, the impurity levels may disappear altogether without the formation of an impurity band. In this case an impurity band is regarded as a range of energies in the forbidden band in which the density of states differs from zero.

(b) Atomic differences:

An increase in impurity concentration to very high levels, in addition to the formation of impurity bands due to the increased interaction between neighboring atoms, may result also in the formation of aggregates of impurity atoms or a combination of impurity atoms and host matrix atoms which may be precipitated as second-phase occlusions. If the impurity concentration is insufficient for such a precipitation, the aggregates present in a single-phase solution may have a considerable influence on the crystal properties due to the interaction structure defects or due to the formation of electrically inactive centers.

In a perfect infinite crystal there are no electron energy levels in the forbidden band. However, in a real crystal there are always some departures from a perfect periodic field. Such localized departures may be due to a variety of causes, one of the simplest being foreign atoms with which the crystal is doped. Such impurity atoms, although their density is significantly lower than that of the crystal atoms, even in the degenerate state, have considerable effect on the properties of the semiconductor.

Heavy doping affects the spacing of crystal atoms (lattice constant) in the same way as omnidirectional compression due to the difference of tetrahedral radii of impurity and host. For example, the presence of boron atoms in silicon affects lattice constant (a) and energy gap (E_G) as shown in Table 1.24.

TABLE 1.24

Variation of Lattice Spacing and Energy Gap of Silicon with Boron Concentration

Boron concentration [cm^{-3}]	a [Å]	ΔE_G [eV]
10^{14}	5.4295	0
10^{19}	5.4291	$-0.3 \cdot 10^{-3}$
10^{20}	5.4257	$-3.5 \cdot 10^{-3}$
$2 \cdot 10^{20}$	5.4241	$-5.0 \cdot 10^{-3}$

1-5 DEGENERATE AND NONDEGENERATE SEMICONDUCTORS

The presence of phosphorus atoms in silicon has a much smaller effect on lattice spacing and energy gap because of the better matching of tetrahedral radii.

2. NONDEGENERATE SEMICONDUCTOR

The density of free carriers in a nondegenerate semiconductor is derived from Boltzmann statistics and is for electrons in an n-type semiconductor

$$n = N_c \exp \eta_n \quad (1.26a)$$

and for holes in a p-type semiconductor

$$p = N_v \exp \eta_p. \quad (1.26b)$$

The reduced energies are defined as follows:

$$\eta_n = (E_F - E_c)/kT, \quad \eta_p = (E_v - E_F)/kT \quad (1.27)$$

$$\eta_c = (E - E_c)/kT, \quad \eta_v = (E_v - E)/kT. \quad (1.28)$$

The ratio of carrier concentration to the intrinsic carrier concentration (n_i) is determined by the ratio of interstitial impurity solubilities in an extrinsic semiconductor to that in the intrinsic semiconductor:

$$n/n_i = C_e/C_i \quad (1.29a)$$

or

$$p/n_i = C_e/C_i. \quad (1.29b)$$

The maximum solid solubility (C_{Bmax}) is the sum of interstitial and substitutional solubilities if $C^{(i)} \ll n_i$ and $C^{(s)} \ll n_i$ at the temperature of impurity addition:

$$C_{Bmax} = C^{(i)} + C^{(s)}. \quad (1.30)$$

If $C^{(i)} \gg n_i$ or $C^{(s)} \gg n_i$, the free carrier concentration is significantly disturbed by the addition of these impurities. Subscripts i and e refer to intrinsic and extrinsic conditions, superscripts (i) and (s) to interstitial and substitutional solubility.

3. DEGENERATE SEMICONDUCTOR

Heavily doped semiconductors can be regarded as poorly conducting metals. In such metals and semiconductors the Fermi level lies within the conduction band. Since only those electrons are scattered which have an energy equal to the Fermi energy, the average energy of electrons in heavily doped semiconductors is equal to $(E_c - E_F)$ or $(E_F - E_v)$, not to kT. In a heavily doped (degenerate) semiconductor (i.e., in the emitter of a transistor structure) interaction of impurity atoms with one another cannot be neglected. In the degenerate case Fermi-Dirac statistics rather than Maxwell-Boltzmann

62 GENERAL PROPERTIES OF SEMICONDUCTORS

statistics apply (see Figure 1.14). The density of electrons in a heavily doped n-type semiconductor is

$$n = \frac{4}{\sqrt{\pi}} \left(\frac{2\pi k T m_n}{h^2}\right)^{3/2} \int_0^\infty \frac{\eta_c^{1/2} \, d\eta_c}{1 + \exp(\eta_c - \eta_n)}$$
$$= (2/\sqrt{\pi}) N_c F_{1/2}(\eta_n)$$
$$= N_D \{1 + 2 \exp[(E_F - E_c)/kT]\}^{-1} \quad (1.31a)$$

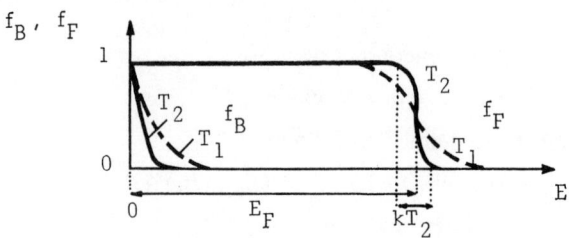

FERMI AND BOLTZMANN ENERGY DISTRIBUTIONS

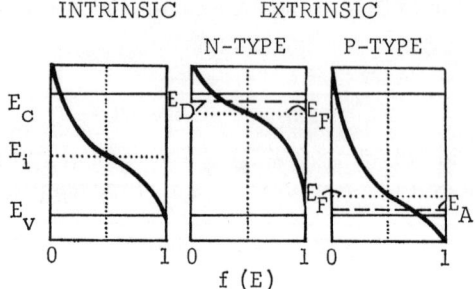

POSITION OF FERMI LEVEL IN INTRINSIC AND EXTRINSIC SEMICONDUCTORS

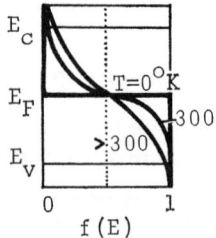

VARIATION OF FERMI LEVEL WITH TEMPERATURE

FIGURE 1.14

Fermi level (E_F) in Fermi and Boltzmann statistics (f_F and f_B) and schematic variation of Fermi level with semiconductor doping and temperature.

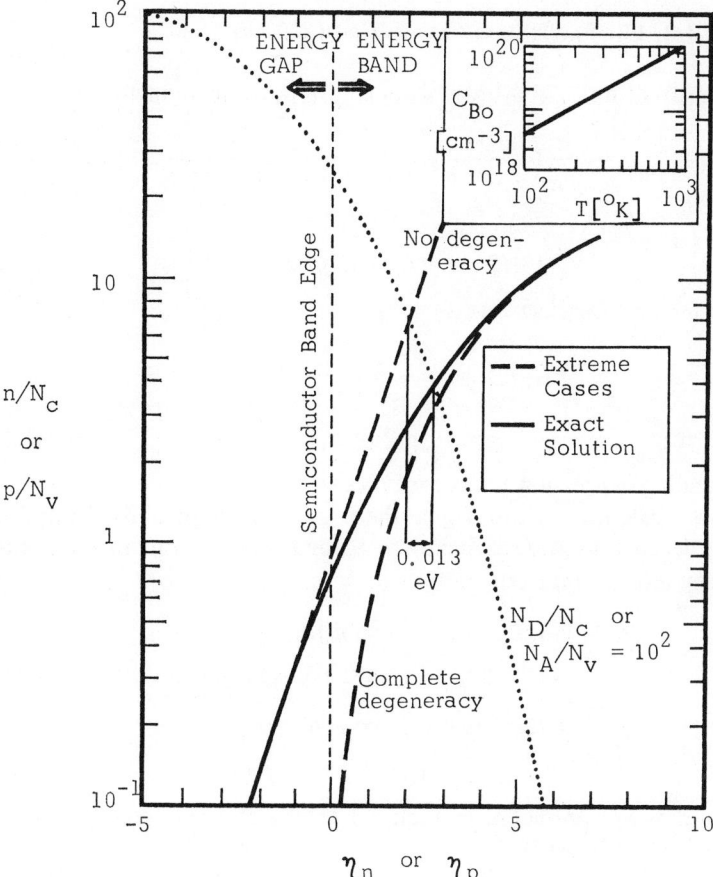

FIGURE 1.15

Relative carrier densities (n/N_c for an n-type semiconductor, p/N_v for a p-type semiconductor) as function of the position of the reduced Fermi level (η_n or η_p) with respect to the band edge (E_c or E_v) and temperature (T). The dashed curves correspond to the extreme cases in which is assumed either
(a) no degeneracy:
$n/N_c = \exp[-(E_c - E_F)/kT]$ and $p/N_v = \exp[-(E_F - E_v)/kT]$ or
(b) complete degeneracy:
$n/N_c = (4/3\sqrt{\pi})[-(E_c - E_F)/kT]^{3/2}$ and
$$p/N_v = (4/3\sqrt{\pi})[-(E_F - E_v)/kT]^{3/2}.$$
These curves represent the Boltzmann approximation. The solid curve corresponds to the exact solution and represents the Fermi integral. It lies between the two extreme cases.

The dotted curve shows the relative carrier concentrations for either $N_D/N_c = 10^2$ or $N_A/N_v = 10^2$.

In the inset the impurity concentration above which degeneracy sets in (C_{Bo}) is given as function of temperature in the case of silicon. If $C_B > C_{Bo}$ at a given temperature or if $T < T_o$ at a given impurity concentration, then the semiconductor is considered to be degenerate.

and the density of holes in a heavily doped p-type semiconductor is

$$p = \frac{4}{\sqrt{\pi}} \left(\frac{2\pi k T m_p}{h^2}\right)^{3/2} \int_0^\infty \frac{\eta_v^{1/2} \, d\eta_v}{1 + \exp(\eta_v - \eta_p)}$$
$$= (2/\sqrt{\pi}) N_v F_{1/2}(\eta_p)$$
$$= N_A \{1 + 2 \exp[(E_v - E_F)/kT]\}^{-1}. \quad (1.31b)$$

The Fermi integral for electrons is

$$F_{1/2}(\eta_n) = \int_0^\infty \frac{\eta_c^{1/2} \, d\eta_c}{1 + \exp(\eta_c - \eta_n)}$$
$$= (\sqrt{\pi}/2)(n/N_c). \quad (1.32)$$

A corresponding expression holds for holes.

The charge balance equations give the total number of ionized impurity atoms in a degenerate semiconductor (n/N_c or p/N_v) in terms of the appropriate energy levels. These equations are:

$$n/N_c = (2/\sqrt{\pi})(1 + 2 \exp \eta_n) F_{1/2}(\eta_n)$$
$$= (1/\xi_n)(2/\sqrt{\pi})(1 + 2 \exp \eta_n) \exp \eta_n, \quad (1.33a)$$
$$p/N_v = (2/\sqrt{\pi})(1 + 2 \exp \eta_p) F_{1/2}(\eta_p)$$
$$= (1/\xi_p)(2/\sqrt{\pi})(1 + 2 \exp \eta_p) \exp \eta_p. \quad (1.33b)$$

These equations are presented in Figure 1.15 for the case $N_D/N_c = 10^2$ or $N_A/N_v = 10^2$ (dotted curve).

4. DEGENERACY CRITERION

An electron gas obeying Fermi-Dirac statistics is called degenerate. The value of the reduced Fermi level η_n or η_p describes the degree of degeneracy. The division between degenerate and nondegenerate states is arbitrary and the transition is continuous.

The occurrence of degeneracy introduces an error in the determination of carrier densities; e.g., the usual degeneracy limit $\eta_n = 0$ or $\eta_p = 0$ results in an error of about 23%. The degeneracy correction factor (ξ) takes this deviation into account.

The relative error introduced by degeneracy in the determination of carrier densities using nondegenerate expressions is

$$dn = 1 - (2/\sqrt{\pi}) F_{1/2}(\eta_n)/\exp(\eta_n) = 1 - (2/\sqrt{\pi}) \xi_n \quad (1.34a)$$

or

$$dp = 1 - (2/\sqrt{\pi}) F_{1/2}(\eta_p)/\exp(\eta_p) = 1 - (2/\sqrt{\pi}) \xi_p. \quad (1.34b)$$

1-5 DEGENERATE AND NONDEGENERATE SEMICONDUCTORS

For example, if degeneracy is assumed to set in at $\eta_n = 0$, then

$$dn = 1 - 0.764 = 0.236.$$

Table 1.25 lists values of the Fermi integrals $F_{1/2}$ and $F_{-1/2}$, of the expo-

TABLE 1.25
Values of Fermi Integrals, Fermi Distribution, and Degeneracy Correction Factor

η_n or η_p	$F_{1/2}(\eta)$	$F_{-1/2}(\eta)$	$\exp(\eta)$	$f(E)$	ξ	$1/\xi$
−4	0.016	0.032	0.013	0.982	1.125	0.890
−3	0.043	0.085	0.050	0.953	1.152	0.867
−2	0.115	0.219	0.135	0.881	1.175	0.851
−1	0.291	0.521	0.368	0.731	1.265	0.790
0	0.678	1.072	1.000	0.500	1.475	0.678
1	1.396	1.820	2.718	0.269	1.945	0.514
2	2.502	2.595	7.389	0.119	2.950	0.339
3	3.977	3.285	20.085	0.047	5.025	0.199
4	5.771	3.874	54.598	0.018	9.460	0.106
5	7.838	4.383	148.413	0.007	18.95	0.053
6	10.144	4.834	403.429	0.0025	39.70	0.025
8	15.381	5.617	2981.958	0.0003	194.0	0.005
10	21.345	6.297	22026.466	0.00004	1034.0	0.001

nential values of the Fermi energy, and of the Fermi-Dirac probability of occupancy $f(E)$ which, for electrons in a degenerate semiconductor, is defined as

$$f(E) = [1 + \exp(-\eta_n)]^{-1}. \tag{1.35}$$

The degree of degeneracy can be conveniently described by the degeneracy correction factor ξ.

$$\begin{aligned}\xi_n &= \exp \eta_n / F_{1/2}(\eta_n) \\ &= (2/\sqrt{\pi})(N_c/n) \exp[(E_F - E_c)/kT] \\ &= (2/\sqrt{\pi})(N_c/n) \exp \eta_n \end{aligned} \tag{1.36a}$$

$$\begin{aligned}\xi_p &= \exp \eta_p / F_{1/2}(\eta_p) \\ &= (2/\sqrt{\pi})(N_v/p) \exp[(E_v - E_F)/kT] \\ &= (2/\sqrt{\pi})(N_v/p) \exp \eta_p. \end{aligned} \tag{1.36b}$$

66 GENERAL PROPERTIES OF SEMICONDUCTORS

This correction factor increases with increasing degeneracy and thus effectively describes the reduction of the number of free carriers in a degenerate semiconductor.

Figure 1.15 gives the degeneracy rate (i.e., the ratio n/N_v) as a function of the location of the Fermi level. The broken lines represent the two extreme cases; i.e., if no degeneracy is assumed (Boltzmann approximation) the Fermi level may be located within an energy band for high impurity concentration, and if complete degeneracy is assumed (Fermi approximation) the Fermi level will always be within one of the energy bands. The true dependence of the carrier concentration upon the location of the Fermi level is given by the solid curve which corresponds to a more exact solution. In actuality degeneracy begins to set in when the Fermi level is approximately 1 to 2 kT away from the band edge within the forbidden gap.

Figure 1.16 shows the Fermi level as a function of impurity activation energy, temperature, and impurity concentration. Figure 1.17 gives the carrier densities in the allowed bands of a semiconductor as a function of temperature and effective carrier masses and of the reduced Fermi energy.

The distinction between the degenerate and the nondegenerate case is made at temperature T_o (see inset of Figure 1.15) at which the kinetic energy (E_n) of a carrier corresponding to the energy at the surface of the Fermi distribution is equal to kT_o; this corresponds approximately to the impurity concentration at which $E_a = 0$. At $T > T_o$ the semiconductor is nondegenerate, at $T < T_o$ it is degenerate, i.e., it behaves like a metal and not all impurities are ionized.

$$E_n = (h^2/8m_o)(3n/\pi)^{2/3} = kT_o, \tag{1.37}$$

$$T_o = (h^2/8km_o)(3/\pi)^{2/3} C_B^{2/3}$$
$$= 4.2 \cdot 10^{-11} C_B^{2/3}, \tag{1.38}$$

where m_o is the rest mass of an electron in free space ($m_o = 9.11 \cdot 10^{-28}$ g).

At 25°C the transition from nondegenerate to degenerate behavior of silicon takes place at

$$C_{Bo} \approx 1.8 \cdot 10^{19} \text{ cm}^{-3}. \tag{1.39}$$

In a degenerate semiconductor the ratio of free carrier density to intrinsic carrier density is

$$n/n_i = (1/\xi_n)(C_e/C_i) \tag{1.40a}$$

or

$$p/n_i = (1/\xi_p)(C_e/C_i). \tag{1.40b}$$

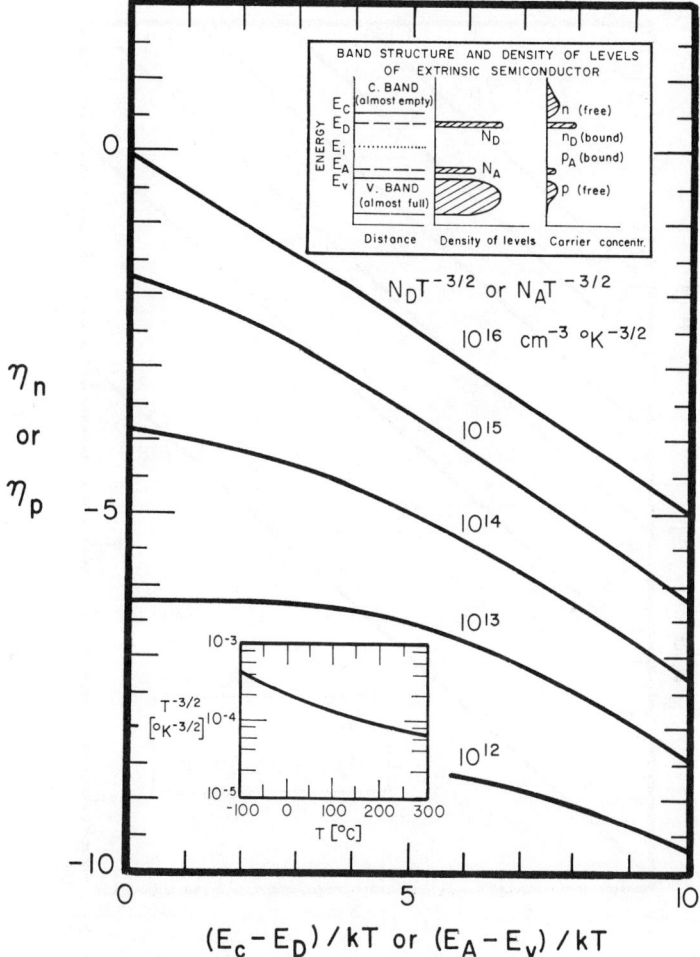

FIGURE 1.16

Position of Fermi level (E_F) with respect to band edge (E_c or E_v) vs. position of impurity level (E_D or E_A) with respect to band edge, impurity concentration (N_D or N_A), and temperature (T). The upper inset shows the band structure and the density of levels of an intrinsic semiconductor. The lower inset shows the temperature conversion T vs. $T^{-3/2}$.

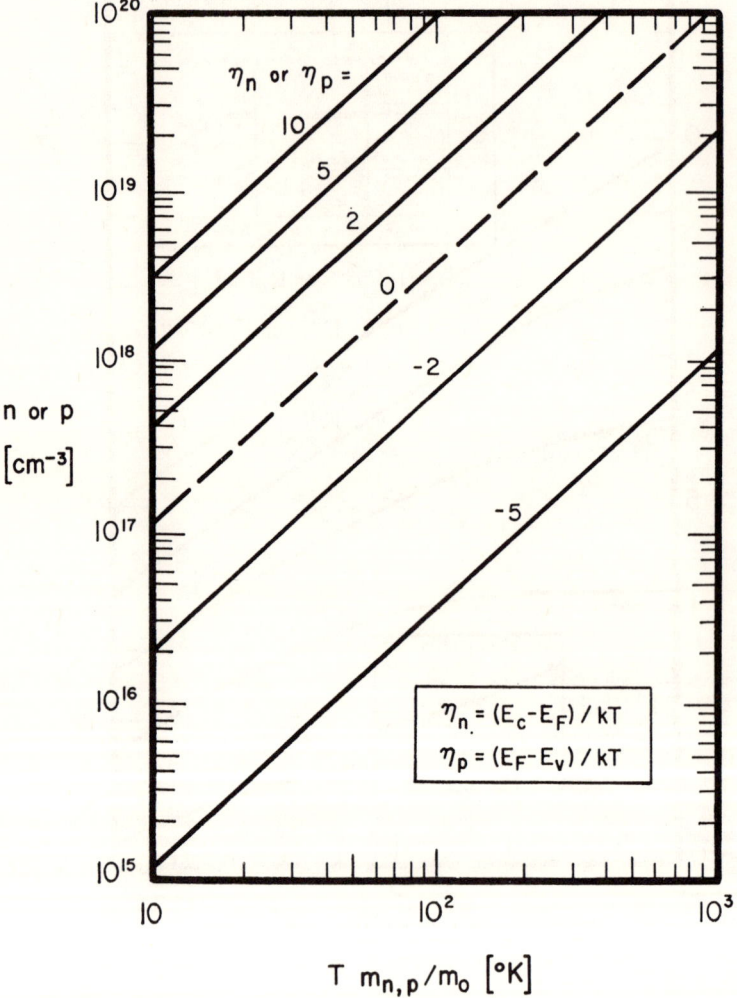

FIGURE 1.17

Carrier density (n or p) within conduction or valence band vs. product of temperature (T), effective carrier mass ($m_{n,p}/m_o$), and reduced Fermi level (η_n or η_p). Arbitrary semiconductor.

1-5 DEGENERATE AND NONDEGENERATE SEMICONDUCTORS

The decrease of energy gap with impurity concentration can be taken into account in an exponential term:

$$n/n_i = (1/\xi_n)(C_e/C_i) \exp(-\Delta E_G/kT) \tag{1.41a}$$

or

$$p/n_i = (1/\xi_p)(C_e/C_i) \exp(-\Delta E_G/kT), \tag{1.41b}$$

where n_i, C_i, and C_e are the values for undisturbed material, and ΔE_G is the decrease in energy gap due to high impurity concentration.

5. REFERENCES

(1) W. Shockley, *Electrons and Holes in Semiconductors*, D. Van Nostrand, New York, 1950.
(2) A. J. Rosenberg, *J. Chem. Phys.*, **33**, 665 (1960).
(3) A. B. Phillips, *Transistor Engineering*, McGraw-Hill Book Co., New York, 1962.
(4) J. S. Blakemore, *Semiconductor Statistics*, Pergamon Press, New York, 1962.
(5) R. N. Hall and J. H. Racette, *J. Appl. Phys.*, **35**, 379 (1964).
(6) V. I. Fistul', *Heavily Doped Semiconductors*, Plenum Press, New York, 1969.
(7) S. J. Brient, Jr. and C. L. Wilson, *IEEE Transact. Electr. Dev.*, **ED-16**, 177 (1969).

1–6 Fermi Level and Impurity Energy Levels

1. FERMI LEVEL (FERMI ENERGY)

The band theory of solids allows general insight into the nature and behavior of semiconductors. This theory, which is based on Drude's, Lorentz's, and Sommerfeld's atomic models, is a result of the broadening of the discrete quantized energy levels of an isolated atom and the considerations of speed and momentum of free electrons. Many properties of a semiconductor depend upon the densities of electrons and holes in the various bands, mainly the conduction and valence bands.

The Fermi level is one of the important energy levels in a semiconductor since its position with respect to the edge of conduction or valence band determines the availability of free carriers (n or p), i.e.,

$$E_c - E_F = kT \ln (N_c/n) \tag{1.42a}$$

or

$$E_F - E_v = kT \ln (N_v/p). \tag{1.42b}$$

The Fermi energy (Fermi level), E_F, is defined as the energy at which exactly half of all possible energy states are occupied by electrons, i.e., as the energy at which the probability of occupancy of an energy state by an electron is 1/2.

$$E_F = h^2 n^{2/3}/8m_n. \tag{1.43}$$

The Fermi energy for a nondegenerate impurity semiconductor at high temperature

$$\begin{aligned} E_F &\approx E_c - kT \ln (N_c/N_D) \quad (n\text{-type}), \\ &\approx E_v + kT \ln (N_v/N_A) \quad (p\text{-type}) \end{aligned} \tag{1.44}$$

and for an intrinsic semiconductor:

$$E_i = (1/2)[(E_c + E_v) + kT \ln (N_v/N_c)] \approx (E_c + E_v)/2. \tag{1.45}$$

1-6 FERMI LEVEL AND IMPURITY ENERGY LEVELS

In an n-type semiconductor:

$$E_F - E_i \approx (E_c - E_v)/2 - kT \ln (N_c/N_D)$$
$$= E_G/2 - kT \ln (N_c/N_D) > 0. \qquad (1.46a)$$

In a p-type semiconductor:

$$E_F - E_i \approx (E_v - E_c)/2 + kT \ln (N_v/N_A)$$
$$= -E_G/2 + kT \ln (N_v/N_A) < 0. \qquad (1.46b)$$

If in an intrinsic semiconductor (of carrier density n_i) the Fermi level is given by E_i, then for an extrinsic semiconductor the deviation of the Fermi level from its intrinsic value

$$E_F - E_i = kT \ln (n/n_i) \qquad (1.47a)$$

or

$$E_F - E_i = kT \ln (p/n_i). \qquad (1.47b)$$

The intrinsic Fermi level (E_i) of the semiconductor bulk is usually considered as a reference energy from which energies are reckoned. The intrinsic Fermi level is close to the center of the energy band gap. At thermal equilibrium (no externally applied voltage) the Fermi level throughout a crystal must be a constant independent of position.

Figure 1.18 shows that with increasing temperature a semiconductor approaches intrinsic characteristics, i.e., the difference $E_F - E_i$ becomes smaller. At a given temperature the difference $E_F - E_i$ becomes larger with increasing impurity concentration. The variation of the energy gap ($E_G = E_c - E_v$) with temperature represents the difference between the two dashed lines in Figure 1.18.

The position of the Fermi level with respect to the band edge is for an n-type semiconductor

$$\eta_n = (E_F - E_c)/kT = \ln (n/N_c) \qquad (1.48a)$$

and for a p-type semiconductor

$$\eta_p = (E_v - E_F)/kT = \ln (p/N_v). \qquad (1.48b)$$

The Fermi level is a function of the impurity concentration of the semiconductor. The deviation from its value in the semiconductor in its intrinsic state is shown in Table 1.26 and in Figure 1.19 for Si, Ge, and GaAs.

At $T = 0°K$ all states with an energy $E < E_F$ are empty, all states for which $E > E_F$ are occupied. At $T > 0°K$ some states may be occupied at $E > E_F$ and some states may be empty at $E < E_F$.

72 GENERAL PROPERTIES OF SEMICONDUCTORS

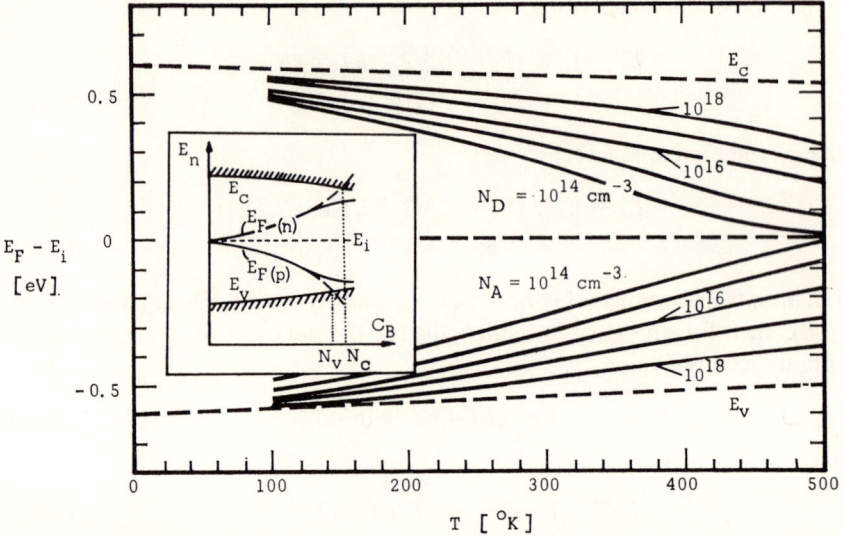

FIGURE 1.18

Variation of extrinsic Fermi level (E_F) from its intrinsic value (E_i) as function of temperature (T) and impurity concentration (C_B; N_D, N_A). Silicon. The inset shows schematically the variation of Fermi energy and band edges (E_c, E_v) with impurity concentration and gives a definition of the densities of state (N_c, N_v).

TABLE 1.26

Position of Fermi Level as Function of Impurity Concentration (300°K)

C_B[cm^{-3}]	$E_F - E_i$[eV]		
	Si	Ge	GaAs
n_i	0	0	0
10^{14}	0.24	0.04	0.36
10^{15}	0.30	0.10	0.42
10^{16}	0.36	0.16	0.48
10^{17}	0.42	0.22	0.54
10^{18}	0.48	0.28	0.60
$E_G/2$[eV]	0.56	0.34	0.72

The Fermi distribution function relates the probability $f(E)$ that a certain state of energy E is occupied by an electron to the Fermi level:

1-6 FERMI LEVEL AND IMPURITY ENERGY LEVELS

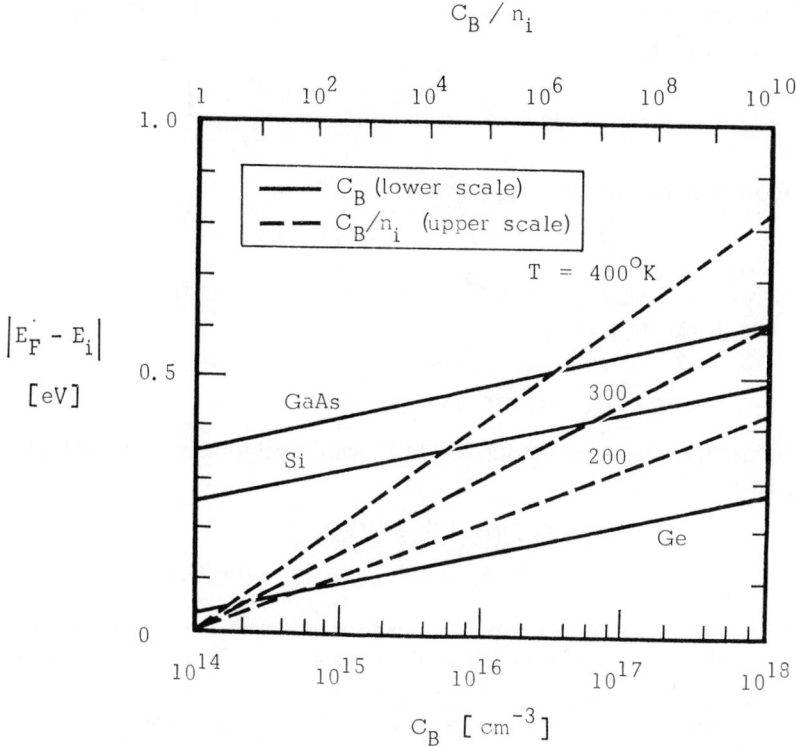

FIGURE 1.19

Variation of Fermi level (E_F) from its intrinsic value (E_i) as function of impurity concentration (C_B) for Si, Ge, and GaAs at 300°K (solid curves, lower scale) and as function of ratio C_B/n_i for arbitrary semiconductor at 200, 300, and 400°K (dashed curves, upper scale). The curves are based on the equation $E_F - E_i = kT \ln(n/n_i)$; for an n-type semiconductor $E_F > E_i$, for a p-type semiconductor $E_F < E_i$. The validity of the curves is impaired at the onset of degeneracy.

$$f(E) = \frac{1}{1 + \exp[(E - E_F)/kT]} \quad (1.49)$$

i.e., at $E = E_F$ and $T > 0$: $f(E) = 0.5$. For $E - E_F \gg kT$

$$f(E) = \exp[-(E - E_F)/kT], \quad (1.50)$$

i.e., Fermi statistics become identical with Boltzmann statistics.

74   GENERAL PROPERTIES OF SEMICONDUCTORS

The probability that a state is empty (=occupied by a hole) is given in Fermi statistics by

$$f_p(E) = 1 - f(E) = \frac{\exp\left[-(E - E_F)/kT\right]}{1 + \exp\left[-(E - E_F)/kT\right]} \quad (1.51)$$

and in Boltzmann statistics $(E - E_F \gg kT)$ by

$$f_p(E) = 1 - f(E) = 1 - \exp\left[-(E - E_F)/kT\right]. \quad (1.52)$$

The variation of $f(E)$ in Boltzmann and in Fermi statistics (f_B and f_F) is shown in Figure 1.14.

2. INTRINSIC SEMICONDUCTOR

For an intrinsic semiconductor the Fermi level (intrinsic Fermi level) is given by

$$\begin{aligned} E_i &= (1/2)(E_c + E_v) - (1/2)kT \ln(N_c/N_v) \\ &= (1/2)(E_c + E_v) - (3/4)kT \ln(m_n/m_p). \end{aligned} \quad (1.53)$$

Since $m_n \approx m_p$ and therefore $N_c \approx N_v$ the intrinsic Fermi level lies in the center of the energy gap so that $E_i \approx (E_c + E_v)/2$. The second term in the above equation is approximately 0.01 eV for silicon at 300°K; i.e., the Fermi level in intrinsic silicon at 300°K is 0.01 eV below the center of the energy gap.

If the density of possible states per unit volume at energy E is $g(E)$, then the number of states in the energy interval between E and $(E + dE)$ is $g(E)\, dE$. Per definition $g(E) = 0$ within the energy gap.

The number of free electrons (n) at an energy $E \geq E_c$ (in the conduction band) is

$$n = \int_{E_c}^{\infty} f(E)g(E)\, dE. \quad (1.54a)$$

and that of free holes (p) at an energy $E \leq E_v$ (in the valence band) is

$$p = \int_{0}^{E_v} f_p(E)g(E)\, dE. \quad (1.54b)$$

Since usually $E_c - E_F \gg kT$ or $E_F - E_v \gg kT$ (i.e., the Fermi energy is far away from the band edges) the following simplifications can be made:

$$n = N_c \exp\left[-(E_c - E_F)/kT\right] \quad (1.55a)$$

$$p = N_v \exp\left[-(E_F - E_v)/kT\right] \quad (1.55b)$$

1-6 FERMI LEVEL AND IMPURITY ENERGY LEVELS

and the effective state densities

$$N_c = \int_{E_c}^{\infty} \exp\left[-(E - E_c)/kT\right]g(E)\,dE. \quad (1.56a)$$

$$N_v = \int_{0}^{E_v} \exp\left[-(E_v - E)/kT\right)]g(E)\,dE. \quad (1.56b)$$

Since n and p are independent of E it can be assumed under the above condition that all electrons of the conduction band have an energy E_c and all holes of the valence band an energy E_v regardless of their actual energies within their corresponding bands. Then the state densities can be described by

$$g_c(E) = (4\pi/h^3)(2m_n)^{3/2}(E - E_c)^{1/2} \quad (1.57a)$$

$$g_v(E) = (4\pi/h^3)(2m_p)^{3/2}(E_v - E)^{1/2} \quad (1.57b)$$

and the effective state densities (Table 1.27) by

$$N_c = 2(2\pi m_n kT/h^2)^{3/2} = 4.82 \cdot 10^{15} T^{3/2} (m_n/m_o)^{3/2} \quad (1.58a)$$

$$N_v = 2(2\pi m_p kT/h^2)^{3/2} = 4.82 \cdot 10^{15} T^{3/2} (m_p/m_o)^{3/2} \quad (1.58b)$$

TABLE 1.27

Effective Carrier Masses and Effective State Densities in Conduction and Valence Bands (300°K)

Semi-conductor	m_n/m_o*	m_p/m_o*	N_c [cm^{-3}]	N_v [cm^{-3}]
Si	0.23	0.12	$2.8 \cdot 10^{19}$	$1.0 \cdot 10^{19}$
Ge	0.03	0.08	$1.0 \cdot 10^{18}$	$6.0 \cdot 10^{18}$
AlSb	0.30	0.40	$4.1 \cdot 10^{19}$	$6.3 \cdot 10^{19}$
GaAs	0.07	0.09	$4.7 \cdot 10^{18}$	$7.0 \cdot 10^{18}$
GaP	0.12	0.50	$1.0 \cdot 10^{19}$	$8.9 \cdot 10^{19}$
GaSb	0.20	0.39	$2.3 \cdot 10^{19}$	$6.1 \cdot 10^{19}$
InAs	0.03	0.02	$1.0 \cdot 10^{18}$	$6.9 \cdot 10^{18}$
InP	0.07	0.69	$4.6 \cdot 10^{18}$	$1.4 \cdot 10^{20}$
InSb	0.01	0.18	$2.5 \cdot 10^{18}$	$1.9 \cdot 10^{19}$

* Here, $m_o = 9.12 \cdot 10^{-28}$ g.

The product of the carrier concentrations,

$$np = N_c N_v \exp\left[-(E_c - E_v)/kT\right] = N_c N_v \exp(-E_G/kT)$$
$$= 4(2\pi kT/h^2)^3 (m_n m_p)^{3/2} \exp(-E_G/kT) = n_i^2, \quad (1.59)$$

76 GENERAL PROPERTIES OF SEMICONDUCTORS

is independent of E_F, E_c, and E_v but dependent upon the energy gap E_G (which is temperature-dependent), the effective masses of the two carriers, and temperature; this relationship is valid for $n = p$ or $n \neq p$.

3. EXTRINSIC SEMICONDUCTOR

An extrinsic semiconductor contains an excess of impurities, i.e., $n \neq p$, so that $n = N_D - N_A + p$ in n-type material and $p = N_A - N_D + n$ in p-type material. Impurity atoms introduce energy levels within the energy gap and may act as donors (energy E_D, concentration N_D) or acceptors (energy E_A, concentration N_A) depending upon their position with respect to the band edges and the Fermi level. Whether carriers can be donated or accepted by these impurities or not depends upon the position of the Fermi level with respect to these impurity levels. If the impurity level lies between the center of the band gap and the Fermi level, no significant contribution to the number of electrically active (free) carriers will be made by the impurity atom. If, however, the impurity level lies between the Fermi level and the corresponding band edge, then the contribution will be significant.

If $N_D - N_A \ll N_c$ or $N_A - N_D \ll N_v$, then the Fermi level lies

(a) in an n-type semiconductor
close to the conduction band with a separation given by

$$E_c - E_F = kT \ln [N_c/(N_D - N_A)].$$

(b) in a p-type semiconductor
close to the valence band with a separation given by

$$E_F - E_v = kT \ln [N_v/(N_A - N_D)].$$

In most cases $N_A \gg N_D$ or $N_D \gg N_A$ so that $N_A = C_B$ or $N_D = C_B$. If $N_A \approx N_D$, i.e., in a compensated semiconductor, the full term $(N_A - N_D)$ must be used.

At high impurity concentrations, i.e., at $N_D - N_A \geq N_c$ or $N_A - N_D \geq N_v$ (degenerate semiconductor), the separation between Fermi level and band edge becomes small compared to kT (i.e., $E_c - E_F \ll kT$ or $E_F - E_v \ll kT$; at 300°K the thermal energy $kT = 0.026$ eV), and deviations from the expressions given for the carrier concentrations become significant since the unit term in the denominator of the Fermi function cannot be neglected. If the impurity concentration approaches the effective state density, i.e., if $N_D - N_A = N_c$ or $N_A - N_D = N_v$, then the band edge coincides with the projected Fermi level—a simplification which is not quite correct since a bending of the Fermi level away from the band edge will occur at high impurity concentration (see inset of Figure 1.18).

1-6 FERMI LEVEL AND IMPURITY ENERGY LEVELS

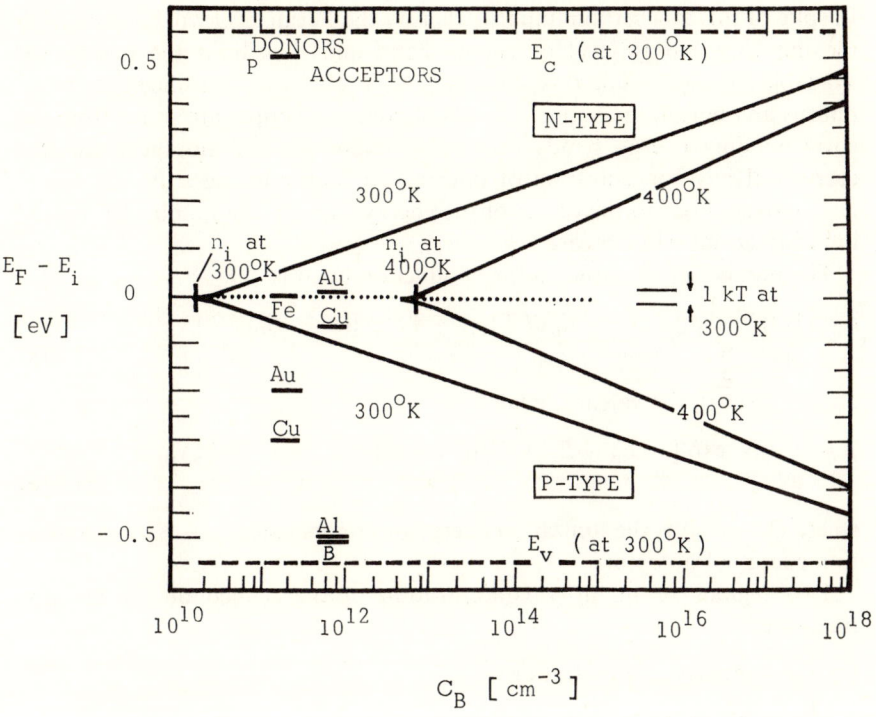

FIGURE 1.20

Variation of Fermi level (E_F) from its intrinsic value (E_i) as function of semiconductor impurity concentration (C_B). Silicon, 300 and 400°K. In the left part of the illustration a few donor and acceptor levels are shown with respect to the band edges.

The highest impurity concentration at which full ionization takes place (i.e., at which the impurity concentration is equal to the carrier density) is

$$N_D \approx (1/4)N_c \exp\left[-(E_c - E_D)/kT\right], \quad (1.60a)$$

$$N_A \approx (1/4)N_v \exp\left[-(E_A - E_v)/kT\right]. \quad (1.60b)$$

Values of the effective state densities N_c and N_v in selected semiconductors are given in Table 1.27. Impurity concentrations in excess of N_c or N_v do not increase the number of free carriers significantly although they affect carrier mobility.

At high impurity concentrations a widening of the discrete impurity levels to energy bands will occur; this means that the ionization energy will decrease correspondingly (see Figure 2.36) which will aid the generation of carriers and partially offset the reduction of the number of carriers discussed above.

78   GENERAL PROPERTIES OF SEMICONDUCTORS

In spite of this at a certain impurity density the Fermi level will coincide with the impurity energy level or energy band and thus limit the number of available carriers. While this value is at $C_B \approx 10^{18}$ cm^{-3} in silicon for phosphorus and boron, the impurity levels of gold and copper are so far from the corresponding energy bands that these impurities will not generate free carriers at any concentration of interest. At higher temperature the Fermi level moves closer to the center of the energy gap and the number of ionized impurity atoms will increase.

The number of electrons in the conduction band is

$$n = (1/2)N_c \exp[-(E_c - E_D)/kT]\{(1 + 4\exp[(E_c - E_D)/kT]N_D/N_c)^{1/2} - 1\} \quad (1.61a)$$

and of holes in the valence band

$$p = (1/2)N_v \exp[-(E_A - E_v)/kT]\{(1 + 4\exp[(E_A - E_v)/kT]N_A/N_v)^{1/2} - 1\} \quad (1.61b)$$

where $(E_c - E_D)$ is the ionization energy of donors and $(E_A - E_v)$ the ionization energy of acceptors.

Two regions of impurity concentration of the semiconductor are distinguished:

(a) Low impurity concentration:
If
$$N_D \ll (1/4)N_c \exp[-(E_c - E_D)/kT],$$
then
$$n = N_D; \quad (1.62a)$$
and if
$$N_A \ll (1/4)N_v \exp[-(E_A - E_v)/kT]$$
then
$$p = N_A; \quad (1.62b)$$

i.e., at small impurity concentration complete ionization takes place and the number of free carriers is equal to the number of impurity atoms.

(b) High impurity concentration:
If
$$N_D \gg (1/4)N_c \exp[-(E_c - E_D)/kT]$$
then
$$n = (N_c N_D)^{1/2} \exp[-(E_c - E_D)/2kT]; \quad (1.63a)$$

and if

$$N_A \gg (1/4)N_v \exp[-(E_A - E_v)/kT]$$

then

$$p = (N_v N_A)^{1/2} \exp[-(E_A - E_v)/2kT]; \quad (1.63b)$$

i.e., at high impurity concentration only partial ionization takes place and the number of free carriers is proportional to $C_B^{1/2}$.

4. FERMI POTENTIAL

General expressions for the Fermi potential in terms of the Fermi level are (see Figure 1.21):

$$u_F = |E_F - E_i|/kT \quad (1.64)$$

and

$$\phi_F = -(E_F - E_i)/q. \quad (1.65)$$

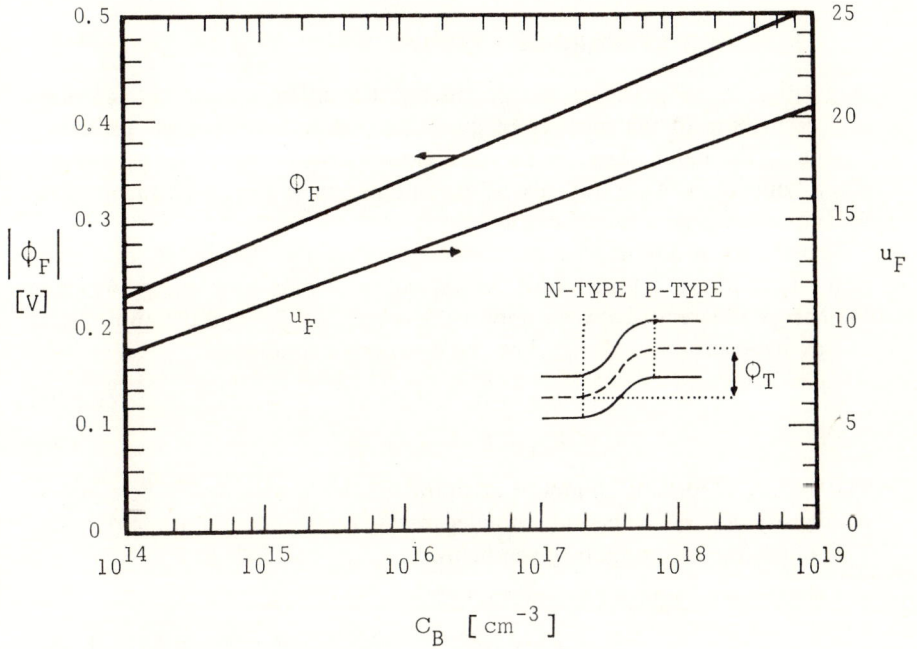

FIGURE 1.21

Fermi potential (u_F and ϕ_F) vs. semiconductor impurity concentration (C_B). Silicon, 300°K. The total Fermi potential (ϕ_T) is the sum of the Fermi potentials on the n- and p-sides of a p-n junction.

80 GENERAL PROPERTIES OF SEMICONDUCTORS

Specifically for the *n*-type region:

$$\phi_{Fn} = -(kT/q) \ln (N_D/n_i) < 0, \qquad (1.66a)$$

and for the *p*-type region:

$$\phi_{Fp} = (kT/q) \ln (N_A/n_i) > 0. \qquad (1.66b)$$

The total potential variation across a *p-n* junction in equilibrium is

$$\phi_T = |\phi_{Fn}| + \phi_{Fp}; \qquad (1.67)$$

ϕ_T is equal to the built-in voltage (V_D) of the *p-n* junction. If a reverse bias (V_a) is applied across the junction, then

$$\phi_T + qV_a = \phi_b; \qquad (1.68a)$$

if $V_a = 0$, then

$$\phi_T = \phi_b; \qquad (1.68b)$$

where ϕ_b is the barrier height of the junction.

5. IMPURITY ENERGY LEVELS

Impurities in an extrinsic semiconductor are either donors or acceptors and contribute to the number of available carriers due to their proximity to one of the band edges, or they are close to the center of the energy gap (deep-lying impurities) and act as recombination centers but do not contribute to the generation of carriers.

The density of donors (N_D) or of acceptors (N_A), i.e., the number of ionized impurity atoms, is a function of the activation or ionization energy which is the energy difference between impurity level (E_D or E_A) and the appropriate energy band edge (E_c or E_v). For nondegenerate conditions

$$E_{aD} = E_c - E_D > kT, \qquad (1.69a)$$

$$E_{aA} = E_A - E_v > kT. \qquad (1.69b)$$

The density of ionized donors or acceptors (which is assumed to be equal to the density of free carriers) in a nondegenerate semiconductor at temperature T is, under the assumptions given below,

$$N_D T^{-3/2} = 2(2\pi m_n kT/h^2)^{3/2} 2\pi^{1/2} \cdot$$
$$\cdot \{1 + 2 \exp [(E_c - E_D)/kT]\} \cdot \exp \eta_n F_{1/2}(\eta_n) \qquad (1.70a)$$

and

$$N_A T^{-3/2} = 2(2\pi m_p kT/h^2)^{3/2} 2\pi^{1/2} \cdot$$
$$\cdot \{1 + 2 \exp [(E_A - E_v)/kT]\} \cdot \exp \eta_p F_{1/2}(\eta_p), \qquad (1.70b)$$

where

$$F_{1/2}(\eta_n) = \int_0^\infty \frac{\eta_c^{1/2}\, d\eta_c}{1 + \exp(\eta_c - \eta_n)} \quad (1.71a)$$

$$F_{1/2}(\eta_p) = \int_0^\infty \frac{\eta_v^{1/2}\, d\eta_v}{1 + \exp(\eta_v - \eta_p)} \quad (1.71b)$$

$$\eta_c = (E - E_c)/kT \quad (1.72a)$$

$$\eta_v = (E_v - E)/kT \quad (1.72b)$$

Assumptions:

(a) All impurity centers are ionized.
(b) The impurity level lies between band edge and Fermi level.
(c) The intrinsic carrier concentration is negligible ($N_A \gg n_i$ or $N_D \gg n_i$).
(d) There is only one type of impurity present (noncompensated semiconductor).

6. EQUILIBRIUM AND NONEQUILIBRIUM CONDITIONS

Equilibrium is defined as the state of an isolated system in which there is no tendency for any of its macroscopic properties to change with time. This corresponds to a minimum of free energy, i.e., to a maximum randomness in the distribution of the particles of the system. If two different systems (for example, an *n*-type and a *p*-type semiconductor region) are brought into contact, then the combined system is in equilibrium only if the Fermi level is the same in both systems, because in equilibrium all levels at a given energy must have the same occupation probabilities. If the Fermi level on both sides is initially different, then there will be a flow of electrons from the system whose Fermi level is higher to the other until equilibrium is established.

In equilibrium the electron distribution over all energy levels can be described by the Fermi-Dirac distribution

$$f_F(E) = \{\exp[(E - E_F)/kT] + 1\}^{-1}. \quad (1.73)$$

The state of equilibrium has to be distinguished from the steady state which is usually a state of nonequilibrium. A system is in a state of nonequilibrium when the electron distribution over all energy levels cannot be described by the Fermi-Dirac distribution. Nonequilibrium conditions may be due to exposure of the crystal to some external stimulus, e.g., radiation other than black-body radiation at the same temperature as the crystal. When the stimulus is removed the system tends to return to equilibrium conditions. Usually population changes are restricted to a small region or they occur at a very fast rate so that the problem of nonequilibrium is less complex.

7. REFERENCES

(1) J. S. Blakemore, *Electrical Communication*, June 1952, p. 131.
(2) H. K. Henisch, *Rectifying Semi-Conductor Contacts*, Clarendon Press, Oxford, 1957.
(3) A. S. Grove, *Physics and Technology of Semiconductor Devices*, John Wiley and Sons, New York, 1967.
(4) D. R. Frankl, *Electrical Properties of Semiconductor Surfaces*, International Series of Monographs on Semiconductors, Vol. 7, Pergamon Press, Oxford, 1967.

1-7 Thermal Characteristics

1. THERMAL CONDUCTION

Thermal conduction is an energy transformation process which includes the creation and annihilation of phonons and which can best be described as the scattering of phonons by static imperfections or by electrons. Thermal energy is transmitted through a crystal by any of the following carriers:

(a) phonons (which are scattered by other phonons, defects, and electrons);
(b) photons;
(c) free electrons or free holes (which are scattered by phonons and defects);
(d) electron-hole pairs;
(e) excitons (bound electron-hole pairs).

In a metal thermal conductivity is mainly due to electric carriers, i.e., electrons. In a semiconductor the main contribution to thermal conduction is due to lattice vibrations (phonons) at low temperature and due to photons at high temperature.

If there is a temperature gradient (∇T), thermal conductivity (κ) can be defined in terms of ∇T and the rate of energy flow per unit area normal to the gradient (Q_H).

$$Q_H = -\kappa \nabla T. \tag{1.74}$$

The thermal conductivity is a proportionality factor given by

$$\kappa = (1/3) c_v c_{ac} l_{ph} \tag{1.75}$$

and is the amount of heat which flows per unit time between two opposite sides of a cube of sides of unit length whose temperature difference is 1°K assuming negligible heat escape from the other sides. In equation (1.75)

c_v = lattice specific heat (a measure of phonon density)
c_{ac} = phonon velocity (speed of sound)
l_{ph} = phonon mean free path. Near the crystal melting point $l_{ph} \approx 5$ to $10a$ where a is the lattice constant; at very low temperature $l_{ph} \approx 0.1$ cm.

84 GENERAL PROPERTIES OF SEMICONDUCTORS

The inset of Figure 1.22 shows schematically the variation of l_{ph} and consequently of κ with temperature. Table 1.28 compares lattice thermal conductivity, phonon mean free path, and Debye temperature (T_D; as obtained from low-temperature specific heat) of some solids, all at 300°K.

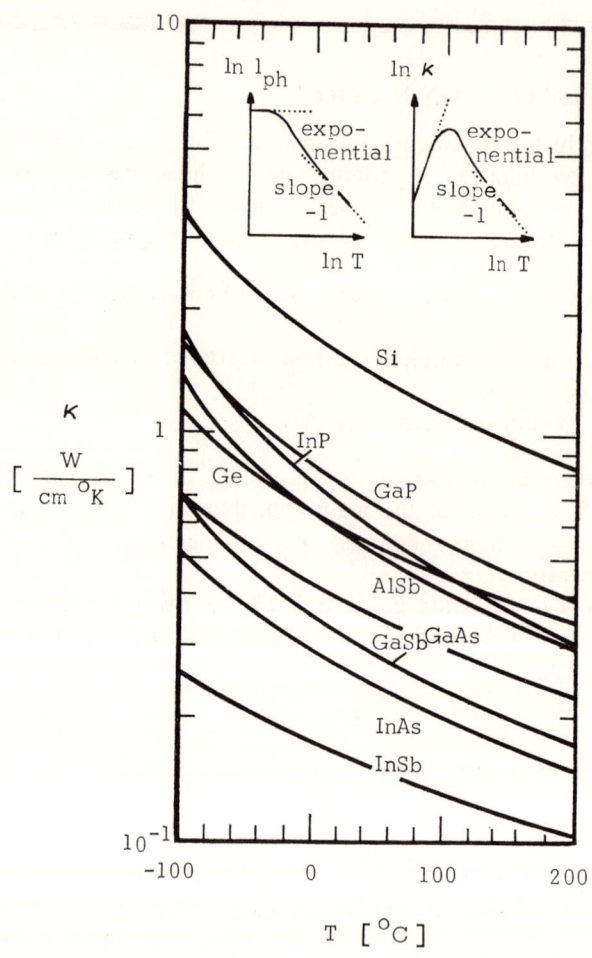

FIGURE 1.22

Thermal conductivity (κ) of various near-intrinsic semiconductors vs. temperature (T). No electric field; empirical data. The inset shows the variation of phonon mean free path (l_{ph}) and thermal conductivity vs. temperature. A decrease of the mean free path sets in when defect and surface scattering become significant.

TABLE 1.28
Thermal Conductivity, Debye Temperature, Density, Bulk Modulus, Phonon Velocity, and Phonon Mean Free Path (300°K)

Material	κ [W/cm °K]	T_D [°K]	d_s [g/cm³]	Y_B [10^{11} dyne/cm²]	c_{ac} [10^5 cm/sec]	l_{ph} [Å]
Si	1.45	658	2.33	9.93	6.53	394
Ge	0.65	378	5.36	7.50	3.74	645
AlAs	0.08	417	3.79			59
AlSb	0.62	292	4.26	5.94	3.72	722
GaAs	0.80	344	5.32	7.49	3.76	864
GaSb	0.33	266	5.60	5.65	3.18	521
InAs	0.27	249	5.66	5.79	3.20	423
InP	0.68	322	4.78			838
InSb	0.19	203	5.77	4.66	2.84	398
Al	3.94	428	2.73	7.61	5.28	4260
BaF₂			4.89	5.64	3.40	
K	1.30	91	0.86	0.33	1.96	4270
LiF	0.10	732	2.60	6.70	5.06	67
NaCl	0.06	321	2.16	2.38	3.34	107
SiO₂	0.14	470	2.65	5.70	4.65	199

The specific heat (defined as the heat a unit volume of a material can absorb or emit if its temperature is raised or reduced by 1°K) is a thermal capacitance and is for solids:

$$c_v = 3R(T_D/T)^2 \exp(T_D/T)/[\exp(T_D/T) - 1]^2$$
$$\approx 3R(T_D/T)^2 \exp(-T_D/T) \quad \text{at} \quad T \ll T_D$$
$$\approx (12\pi^4/5)(kC_L)(T/T_D)^3 \tag{1.76}$$

$$c_p = c_v + \gamma'^2 T \kappa_v^*/d_s \approx c_v. \tag{1.77}$$

In addition to scattering of phonons by each other, phonons may be scattered by:

(a) point defects (impurities);
(b) line defects (dislocations);
(c) vacancies;
(d) grain boundaries in a polycrystalline solid;
(e) outer surface of a single-crystalline solid;
(f) short-range and long-range dissorder in a compound semiconductor;
(g) random distribution of different isotopes of a given chemical species.

All of these scattering centers absorb phonon energy and crystal momentum and thus reduce the phonon mean free path; hence, poor crystals have a very small l_{ph}.

The electron density in a semiconductor is usually too small to produce a very large electronic thermal conductivity. A significant amount of energy can be transported through an intrinsic semiconductor by the diffusion of electron-hole pairs; each pair carries energy slightly smaller than the energy gap (E_G) and diffuses as a neutral complex. The lattice thermal conductivity of a compound semiconductor is considerably smaller than that of solids of either of the terminal compositions, since the lattice periodicity seen by a phonon is impaired, within which atoms of the two species are randomly distributed over equivalent crystallographic sites.

Thermal conductivity is defined as the quantity of heat transmitted per unit time per unit cross-section per unit temperature gradient. It may be based on either or all of three mechanisms:

(a) Thermal agitation of atoms; it results in a mechanical transfer of heat between neighboring atoms.
(b) Thermal radiation between neighboring atoms.
(c) Electron conduction.

In electrical insulators (where there are no free carriers), thermal conduction is mainly by means of lattice vibrations or phonons which are scattered by crystal imperfections or which interact with one another. In the absence of imperfections, thermal conductivity is proportional to $1/T$ at high temperature and to exp $(T_D/2T)$ at low temperature until it is limited by boundary scattering.

Crystals have a higher thermal conductivity than the same materials in their amorphous state; e.g., the thermal conductivity of crystalline quartz is about one order of magnitude higher than that of amorphous quartz. Polycrystalline silicon has a slightly lower thermal conductivity than single-crystalline silicon. The relatively high conductivity of polycrystalline silicon is observed, however, only in the direction parallel to grain growth, but it is considerably lower perpendicular to this direction.

The thermal diffusivity, corresponding to the material diffusion coefficient, is defined as

$$D_h = \kappa/(d_s/c_v). \tag{1.78}$$

In addition to the thermal conductivity, it takes into account the density of the material and the thermal capacitance.

For comparison, thermal and electrical conductivity and linear thermal expansion of selected materials at 300°K are given in Table 1.29. The data for semiconductors are for intrinsic material.

TABLE 1.29
Thermal and Electrical Conductivity and Linear Thermal Expansion Coefficient of Selected Materials (300°K)

Material	κ [W/cm°K]	σ [$(\Omega\,cm)^{-1}$]	α' [$10^{-5}\,°K^{-1}$]
Air	0.0004	$< 10^{-20}$	
Soda-lime glass	0.008	10^{-18}	0.80
He	0.002	$< 10^{-20}$	
H$_2$O	0.01	10^{-7}	4.25
SiO$_2$	0.14	$< 10^{-16}$	0.05
Al$_2$O$_3$	0.16	$< 10^{-16}$	0.92
GaAs	0.80	$3.5 \cdot 10^{-8}$	0.57
Ge	0.58	$4.7 \cdot 10^{-3}$	0.55
Fe	0.78	11.9	0.79
Si	1.35	$8.8 \cdot 10^{-6}$	0.35
Pt	1.70	9.5	0.90
BeO	2.30	$< 10^{-16}$	0.94
W	2.36	18.3	0.46
Al	3.94	35.7	2.32
Au	4.72	45.5	1.39
Cu	6.08	58.9	1.60

2. CONTRIBUTIONS TO THERMAL CONDUCTIVITY

The total thermal conductivity of a semiconductor is

$$\kappa = \kappa_{ph} + \kappa_n + \kappa_{np} \quad (n\text{-type}) \tag{1.79a}$$
$$\kappa = \kappa_{ph} + \kappa_p + \kappa_{np} \quad (p\text{-type}), \tag{1.79b}$$

i.e., it is the sum of:

(a) the lattice or phonon thermal conductivity (κ_{ph}) which is related to the thermal vibrations and which is independent of impurity concentration,
(b) the thermal conductivity of the majority carriers (κ_n or κ_p) which is related to electrical conductivity ($1/\rho$);
(c) a contribution due to mixed conduction (κ_{np}).

More specifically, if a relaxation time can be defined and when the energy bands are spherical or ellipsoidal

$$\kappa = \kappa_{ph} + (5/2 + s^*)k^2T/q^2\rho$$
$$+ (5 + 2s^* + E_G/kT)^2 \frac{np\mu_n\mu_p}{(n\mu_n + p\mu_p)^2} k^2T/q^2\rho. \tag{1.80}$$

88   GENERAL PROPERTIES OF SEMICONDUCTORS

The first term (phonon conductivity) is usually relatively large, while the second term (carrier conductivity) is significant only at high impurity concentration; the third term (mixed conductivity) is very large if $E_G \gg kT$. Due to the various contributions κ will first increase with T at low temperature and then decrease with T at higher temperature. The parameter s^* is a function of the carrier scattering mechanism involved; $s^* = -1/2$ for lattice scattering, $s^* = +3/2$ for impurity scattering.

3. PHONON THERMAL CONDUCTIVITY

If phonon-phonon interactions are the main source of thermal resistance (i.e., at high temperature), the phonon thermal conductivity is given by:

$$\kappa_{ph} = (T_D/\gamma)^2 (T_D/T) w_{at} u_v A_M . \tag{1.81}$$

The phonon thermal conductivity at 300°K as a function of melting temperature (T_m) is shown in Figure 1.23 for various semiconductors having the diamond-cubic crystal structure; it shows that κ_{ph} is proportional to $T_m^{3/2} w_{at}^{-1/2}$. The phonon thermal conductivity is independent of impurity concentration of the semiconductor.

A phonon is the quantum of thermal vibration of the semiconductor lattice, i.e., an electron can gain or lose in a transition an energy of amount hv_o; this phonon energy is equal to the Debye temperature times Boltzmann's constant:

$$E_{ph} = hv_o = kT_D . \tag{1.82}$$

The frequency of lattice vibrations with which the electron interacts is v_o. Interaction between lattice and electron will take place if both wavelengths are comparable, i.e., if

$$m_n v_n \approx m_{ph} v_{ph} ,$$

since

$$\lambda_{\text{electron}} = h/m_n v_n , \tag{1.83}$$

$$\lambda_{\text{phonon}} = h/m_{ph} v_{ph} , \tag{1.84}$$

where m and v are mass and velocity of electron or phonon. The ratio of phonon energy to electron energy is

$$\frac{E_{ph}}{E_n} = \frac{hv_o}{m_n v_n^2/2} \approx \frac{1}{10} \text{ at } 300°K, \tag{1.85}$$

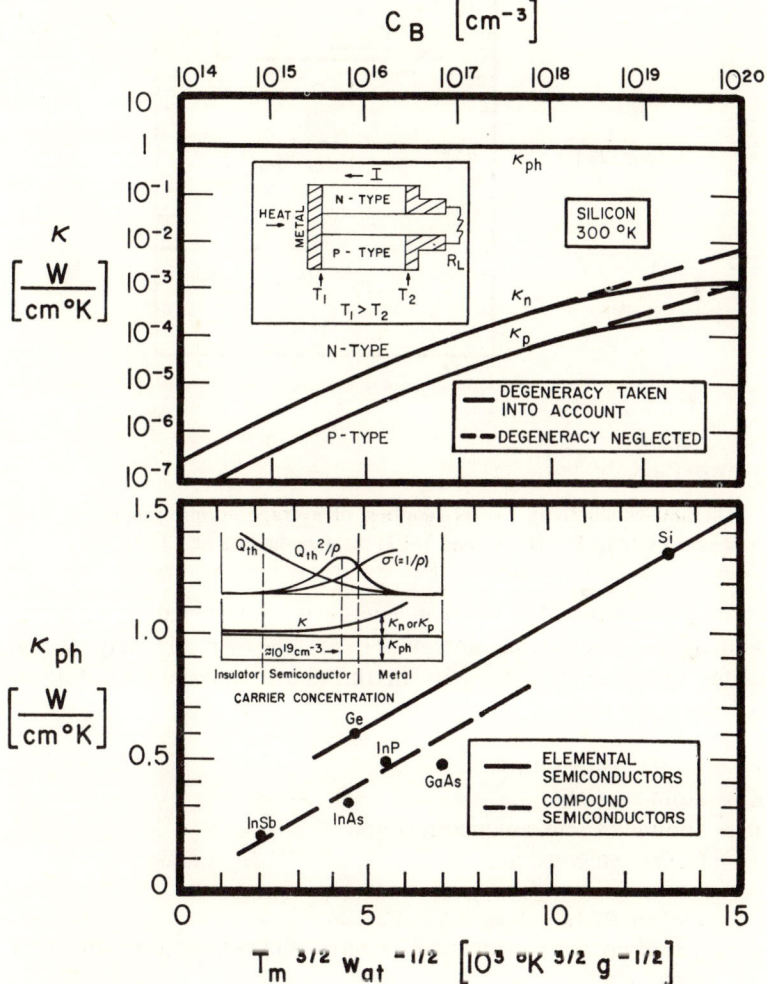

FIGURE 1.23

Thermal conductivity due to carriers (κ_n and κ_p) and due to phonons (κ_{ph}) vs. semiconductor impurity concentration (C_B), melting point of semiconductor (T_m), and atomic weight of semiconductor (w_{at}).

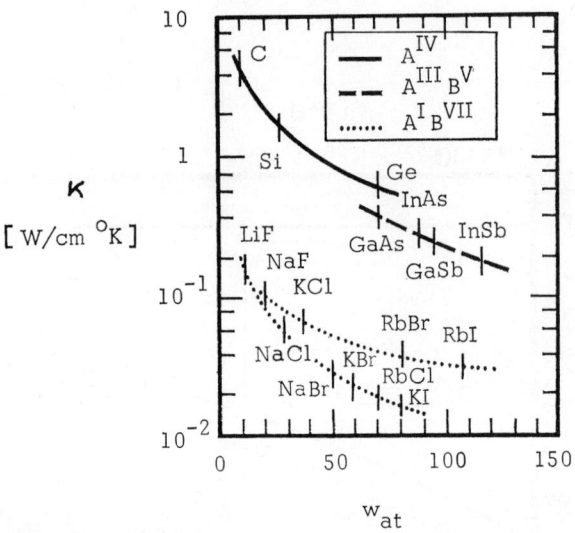

FIGURE 1.24

Thermal conductivity (κ) as function of average atomic weight (w_{at}) for selected Group IV, III–V, and I–VII semiconductors at 300°K.

i.e., at room temperature the electron energy is approximately ten times the phonon energy. The thermal energy (kT) and the thermal voltage (kT/q) are given for reference in Table 1.30. In the above equations the following terms have been used:

T_D = Debye temperature
w_{at} = atomic weight
u_v = volume of semiconductor sample
γ = Grüneisen constant,

$\gamma \equiv -\partial \ln T_D / \partial \ln u_v = \alpha' u_v / \kappa_v^* c_v$;
it is almost independent of material and temperature; for most materials $1 < \gamma < 3$

κ_v^* = thermal compressibility

$\kappa_v^* = -(1/u_v)(\partial u_v / \partial p |_T)$

A_M = material constant
c_p, c_v = specific heat at constant pressure or constant volume
α', γ' = coefficient of linear or volume thermal expansion
R = gas constant

TABLE 1.30

Thermal Energy and Voltage as a Function of Temperature

T [°K]	kT [eV]	kT [erg]	kT/q [V]
0	0	0	0
1	$8.67 \cdot 10^{-5}$	$1.39 \cdot 10^{-16}$	$8.67 \cdot 10^{-5}$
10	$8.67 \cdot 10^{-4}$	$1.39 \cdot 10^{-15}$	$8.67 \cdot 10^{-4}$
100	$8.67 \cdot 10^{-3}$	$1.39 \cdot 10^{-14}$	$8.67 \cdot 10^{-3}$
200	$1.73 \cdot 10^{-2}$	$2.78 \cdot 10^{-14}$	$1.73 \cdot 10^{-2}$
300	$2.60 \cdot 10^{-2}$	$4.17 \cdot 10^{-14}$	$2.60 \cdot 10^{-2}$
400	$3.47 \cdot 10^{-2}$	$5.56 \cdot 10^{-14}$	$3.47 \cdot 10^{-2}$
500	$4.33 \cdot 10^{-2}$	$6.95 \cdot 10^{-14}$	$4.33 \cdot 10^{-2}$
1000	$8.67 \cdot 10^{-2}$	$1.39 \cdot 10^{-13}$	$8.67 \cdot 10^{-2}$
1500	$1.30 \cdot 10^{-1}$	$2.09 \cdot 10^{-13}$	$1.30 \cdot 10^{-1}$

4. CARRIER THERMAL CONDUCTIVITY

The contribution of majority carriers to the thermal conductivity of an n-type or p-type semiconductor is respectively

$$\kappa_n = 1.49 \cdot 10^{-8} T q \mu_n N_c \exp\left[-(E_c - E_F)/kT\right]$$
$$= 0.18(m_n/m_o)^{3/2}(T/300)^{3/2} \mu_n \exp\left[-(E_c - E_F)/kT\right] \quad (1.86a)$$

or

$$\kappa_p = 1.49 \cdot 10^{-8} T q \mu_p N_v \exp\left[-(E_F - E_v)/kT\right]$$
$$= 0.18(m_p/m_o)^{3/2}(T/300)^{3/2} \mu_p \exp\left[-(E_F - E_v)/kT\right] \quad (1.86b)$$

where κ is expressed in [W/cm°K]. The rest mass of the free carrier is m_o and its effective mass is m_n or m_p; μ_n and μ_p are majority carrier mobilities.

On the other hand, the electrical conductivity (inverse of electrical resistivity) of a nondegenerate n-type or p-type semiconductor due to charge carriers is respectively

$$\sigma_n = 1/\rho_n = q\mu_n N_c \exp\left[-(E_c - E_F)/kT\right] \quad (1.87a)$$

or

$$\sigma_p = 1/\rho_p = q\mu_p N_v \exp\left[-(E_F - E_v)/kT\right]. \quad (1.87b)$$

Therefore, the ratio between thermal conductivity due to carriers (κ_n or κ_p) and electrical conductivity ($\sigma = 1/\rho$) in case of thermal scattering in a semiconductor:

$$\kappa_n \rho_n = \kappa_p \rho_p = 2(k^2 T/q^2) = 1.49 \cdot 10^{-8} T [V^2/°K] \quad (1.88)$$

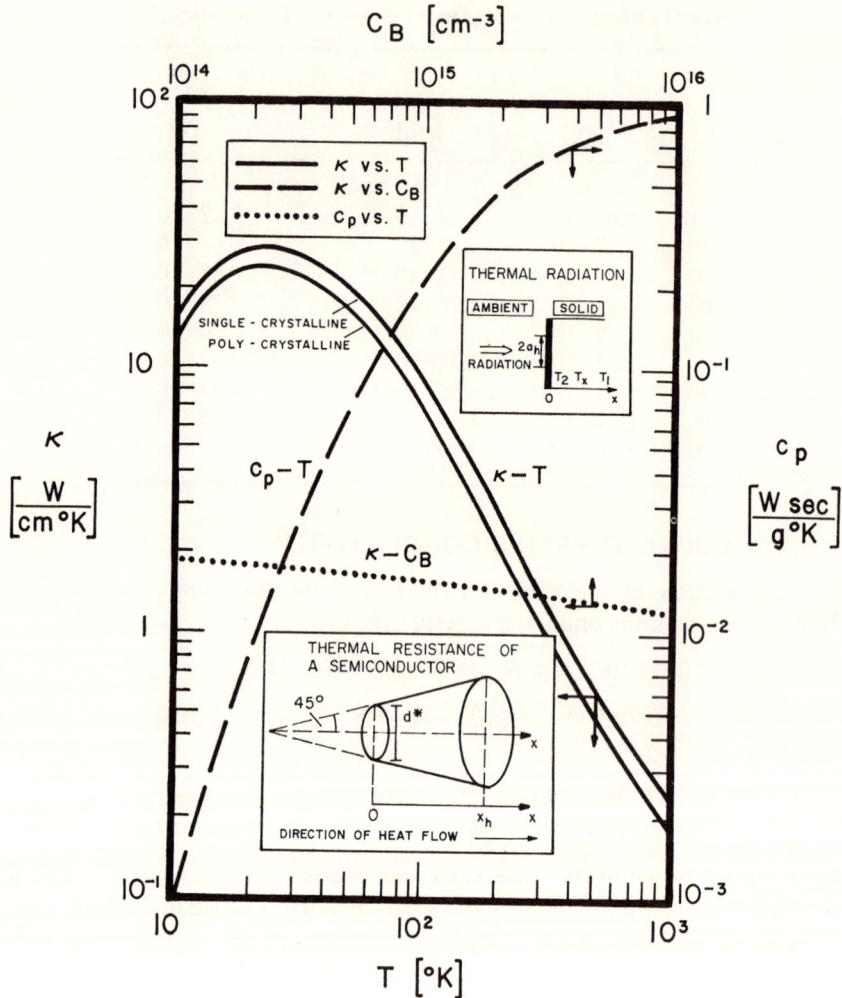

FIGURE 1.25

Thermal conductivity (κ) and specific heat (c_p) of silicon vs. impurity concentration (C_B) or temperature (T). The curve showing the dependence of thermal conductivity upon impurity concentration applies at 300°K. The insets define terms used for describing thermal radiation and thermal resistance of a semiconductor.

and in case of impurity scattering:

$$\kappa_n \rho_n = \kappa_p \rho_p = 4(k^2 T/q^2) = 2.98 \cdot 10^{-8} T \, (\text{V}^2/°\text{K}]. \quad (1.89)$$

For comparison for a metal, regardless of the scattering mechanism and the type of metal (Wiedemann-Franz Law)

$$\kappa_n \rho_n = (\pi^2/3)(k^2 T/q^2) = 2.45 \cdot 10^{-8} T \, [\text{V}^2/°\text{K}]. \quad (1.90)$$

This means that in semiconductors and in metals the ratio of thermal to electrical conductivity due to carriers is proportional to temperature and not dependent upon any other parameter; i.e., in both metals and semiconductors thermal and electrical conductivity depend upon the motion of free carriers.

5. THERMOELECTRIC POWER GENERATION

Generally, the (one-dimensional) electric current density in a semiconductor may be due to two contributions, applied electric field and temperature gradient. It is the sum of these contributions and is

$$I = \sigma[(1/q)(\partial E_F/\partial x) - Q_{th}(\partial T/\partial x)]. \quad (1.91)$$

The term Q_{th} is the Seebeck coefficient or the thermoelectric power and for a nondegenerate semiconductor with a mean free time $\tau \approx E^{s*}$ is given by

$$Q_{th} = -\frac{k}{q} \frac{[(5/2) + s^* + \ln(N_c/n)]n\mu_n - [(5/2) + s^* - \ln(N_v/p)]p\mu_p}{n\mu_n + p\mu_p}. \quad (1.92)$$

In a finite conductor with uniform carrier distribution the carriers diffuse from the hot to the cold junction under the influence of a temperature gradient; they accumulate at the boundary of the conductor and generate an electric field which retards further carriers. This counter potential is the Seebeck coefficient. The temperature gradient results also in a flow of phonons from the hot to the cold junction. They interact with the charge carriers and enhance their drift velocity. (At high temperature, however, this effect is negligible due to phonon-phonon interactions).

Q_{th} is small for intrinsic and for degenerate semiconductors since in both cases the charge is transported approximately equally by electrons above the Fermi level traveling in opposite directions, hence their contributions cancel. The sign of Q_{th} depends on the differences in carrier mobility and density of states on either side of the Fermi level.

In a nondegenerate extrinsic semiconductor where the current is carried by carriers of one type only, the Seebeck coefficient decreases with increasing carrier concentration. Since with increasing impurity concentration the Fermi level moves closer to either of the band edges (see Figure 1.19), a corresponding decrease in resistivity initially compensates for the decrease

94 GENERAL PROPERTIES OF SEMICONDUCTORS

in Q_{th}. Therefore, the product Q_{th}^2/ρ (a figure of merit for thermoelectric power generation) will reach a maximum when the Fermi level approaches the conduction or valence band edge, i.e., when the semiconductor becomes degenerate. The behavior of Q_{th}^2/ρ, Q_{th}, ρ, κ_{ph}, κ_n, and κ_p is shown in the inset of Figure 1.22.

The Seebeck coefficient in an n-type semiconductor

$$Q_{th} = -\frac{k}{q}\left(\frac{5}{2} + s^* + \frac{E_c - E_F}{kT}\right) \tag{1.93a}$$

and in a p-type semiconductor

$$Q_{th} = -\frac{k}{q}\left(\frac{5}{2} + s^* + \frac{E_F - E_v}{kT}\right). \tag{1.93b}$$

In the case of predominantly lattice scattering:

$$Q_{th} = -86[2 + (E_c - E_F)/kT] \, \mu\text{V}/°\text{K} \quad (n\text{-type}) \tag{1.94a}$$

$$Q_{th} = -86[2 + (E_F - E_v)/kT] \, \mu\text{V}/°\text{K} \quad (p\text{-type}) \tag{1.94b}$$

At the onset of degeneracy ($E_F = E_c$ or $E_F = E_v$) for a semiconductor with both types of scattering mechanisms present (at 300°K):

$$Q_{th} \approx -172 \, \mu\text{V}/°\text{K};$$

in a metal (at 300°K) on the other hand:

$$Q_{th} \approx -1 \, \mu\text{V}/°\text{K}.$$

For high thermoelectric power it is necessary to maximize the figure of merit Q_{th}^2/ρ, i.e., to maximize the carrier diffusion and to minimize the phonon diffusion. To this end the semiconductor impurity concentration has to be such that the Fermi level lies within the energy gap at a distance of about 1 kT away from conduction or valence band; on the other hand, a large density of states effective mass ($m_{n,p}/m_o$) and a high carrier mobility (μ) are required. An optimum figure of merit for thermoelectric power generation is therefore obtained in heavily doped, near-degenerate semiconductors, i.e., at an impurity concentration of approximately 10^{19} cm^{-3}.

The value of the Seebeck coefficient which will result in the highest thermoelectric power is

$$Q_{th\,\text{max}} = -\frac{k}{q}\left(\frac{5}{2} + s^* + \frac{E_c - E_F}{kT}\right) = -\frac{2k}{q}\left(1 + \frac{\kappa_n}{\kappa_{ph}}\right) \quad (n\text{-type}), \tag{1.95a}$$

$$Q_{th\,\text{max}} = -\frac{k}{q}\left(\frac{5}{2} + s^* + \frac{E_F - E_v}{kT}\right) = -\frac{2k}{q}\left(1 + \frac{\kappa_p}{\kappa_{ph}}\right) \quad (p\text{-type}). \tag{1.95b}$$

1-7 THERMAL CHARACTERISTICS

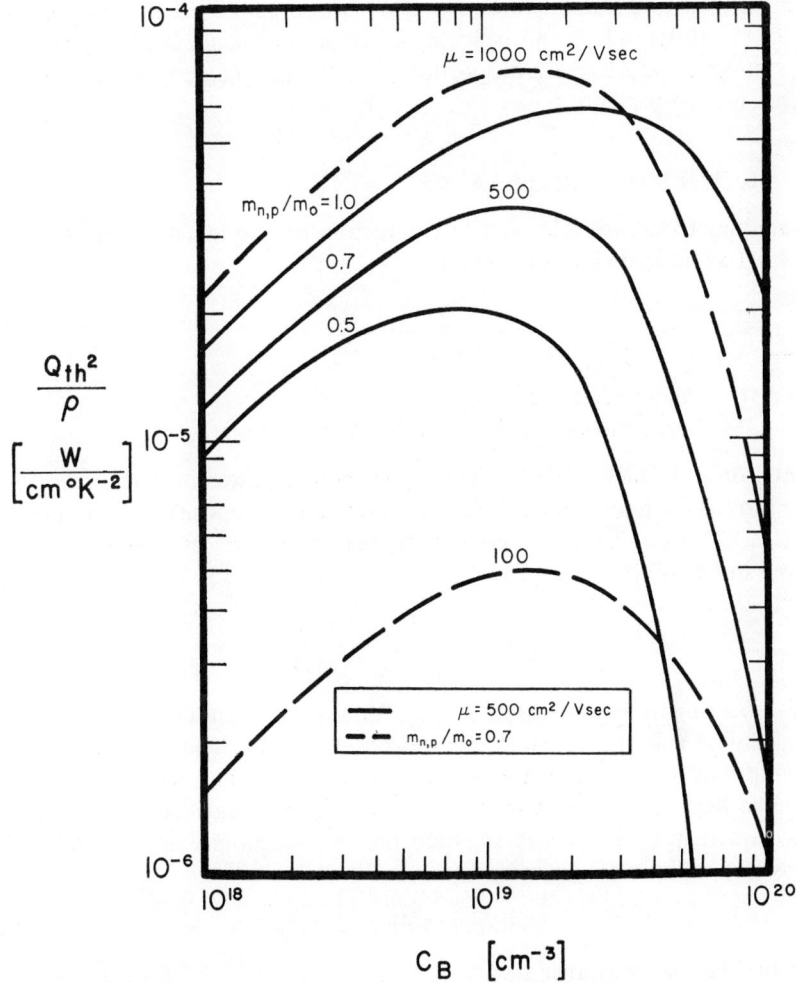

FIGURE 1.26

Thermoelectric figure of merit (Q_{th}^2/ρ) vs. semiconductor impurity concentration (C_B), mass of carrier ($m_{n,p}/m_o$), and carrier mobility (μ) for arbitrary semiconductor at 300°K. It is assumed that lattice scattering is the dominant scattering mechanism.

If $\kappa_{ph} \gg \kappa_n$ or κ_p, the Fermi level coincides with the band edge if thermal scattering dominates. If $\kappa_{ph} \ll \kappa_n$ or κ_p, then the Fermi level lies within the energy gap.

$Q_{th\,max}$ is proportional to $\mu\,(m_{n,p}/m_o)^{3/2}T^{3/2}$. For most semiconductors

$$\mu\,(m_{n,p}/m_o)^{3/2} \leq 10^3 \text{ cm}^2/\text{V sec}$$

because mass and mobility are inversely proportional. Since μ is proportional to $T^{-3/2}$, this product and hence κ_n or κ_p are independent of temperature; κ_{ph}, however, is inversely proportional to T, i.e., $Q_{th\,max}$ will increase with increasing temperature.

6. THERMAL RESISTANCE

In analogy to diffusion laws, heat conduction in a solid can be described by Fick's Law (one-dimensional case):

$$\frac{\partial Q_h}{\partial t} = \kappa A \frac{\partial T}{\partial x} \tag{1.96}$$

where Q_h = amount of heat
A = area through which heat flow is taking place.

If $\partial Q_h/\partial t$ is defined as the electrically generated power in a semiconductor and if ΔT is the temperature difference between two opposite faces of distance x_h (i.e., $\partial T/\partial x = \Delta T/x_h$), then the thermal resistance of a semiconductor (given in °K/W), in analogy to electrical resistance,

$$R_{th} = \frac{\Delta T}{\partial Q_h/\partial x} = \frac{A_h}{\kappa} \frac{x_h}{A} \tag{1.97}$$

This equation represents Ohm's Law for thermal conduction. A_h is a dimensionless material constant; for silicon $A_h = 0.24$. Under the simplifying assumptions that heat is transferred through a cylindrical cross-section, that the heat spreads in a truncated cone with an angle of 45° against the axis, and that x_h is much larger than the cross-sectional diameter (d^*); the cross-sectional area is given by

$$A(x) = (\pi/4)(d^* + x)^2 \tag{1.98}$$

and the thermal resistance by

$$R_{th} = \frac{A_h x_h}{\kappa A(x)} = \frac{\pi}{4} \frac{A_h}{\kappa} \int_0^{x_h} \frac{dx}{(d^* + x)^2}$$

$$= \frac{4}{\pi} \frac{A_h}{\kappa d^{*2}} \arctan \frac{x_h}{d^*}$$

$$\approx \frac{4}{\pi} \frac{A_h x_h}{\kappa d^{*2}} \left[1 - \frac{1}{3}\left(\frac{x_h}{d^*}\right)^2\right]. \tag{1.99}$$

The first term of this equation corresponds to the thermal resistance of a cylindrical disk.

7. TEMPERATURE INCREASE OWING TO THERMAL RADIATION

If the surface of a semi-infinite solid is thermally irradiated, the surface temperature (T_2) will increase; the heat will diffuse from the surface into the solid and the heat flow can be described by the linear differential equation

$$\frac{\partial^2 T}{\partial x^2} = \frac{1}{D_h} \frac{\partial T}{\partial t} \qquad (1.100)$$

with the boundary conditions

$$T_x - T_1 = 0 \quad \text{at} \quad x > 0, \quad t = 0$$
$$T_2 - T_1 = \text{constant} \quad \text{at} \quad x = 0, \quad t > 0$$

and the solutions

(a) Temperature at distance x:

$$T_x - T_1 = T_2 \, \text{erfc}\,(x/2\sqrt{D_h t}). \qquad (1.101\text{a})$$

(b) Temperature at surface ($x = 0$):

$$T_2 - T_1 = 2(F_h/\kappa\pi^{1/2})\sqrt{D_h t}. \qquad (1.101\text{b})$$

If the thermal radiation is confined to a circular spot of diameter a_h

$$T_2 - T_1 = 2(F_h/\kappa\pi^{1/2})\sqrt{D_h t}\,[1/\pi^{1/2} - i\,\text{erfc}\,(a_h/2\sqrt{D_h t})]. \qquad (1.101\text{c})$$

For large spot size or short radiation time the second term becomes negligible and the surface temperature increases to a constant value. The following quantities have been introduced:

D_h = thermal diffusivity, diffusion coefficient
d_s = density of semiconductor
F_h = heat flux at surface upon absorption of thermal radiation
t = time for thermal diffusion to depth x
T_1 = initial uniform temperature of solid
T_2, T_x = temperature of solid at surface, at depth x, respectively.

8. PHONON (SOUND) VELOCITY

The velocity with which a longitudinal wave moves through a liquid is

$$c_{ac} = \lambda v = (Y_B/d_s)^{1/2} \qquad (1.102)$$

where Y_B is the adiabatic elastic bulk modulus (stiffness coefficient). For solids this equation must be modified to take into account the finite rigidity.

98 GENERAL PROPERTIES OF SEMICONDUCTORS

In a crystalline solid the velocity of sound (phonons) depends upon the direction of propagation. Furthermore, solids will sustain the propagation of transverse waves which travel more slowly than longitudinal waves. Despite these modifications, the above equation describes fairly accurately the velocity of sound in crystals; it is in the order of $5 \cdot 10^5$ cm/sec in typical metallic, covalent, and ionic solids (see Table 1.28).

Phonon drag is the transfer of momentum from phonons to free charge carriers. Whenever a temperature gradient exists in a crystal more energetic normal modes of oscillation will be excited at the hot end than at the cold end. Under this condition a net transfer of momentum and energy will take place from hot to cold. This phonon-phonon interaction tends to bring the semiconductor into isothermal equilibrium. Free charge carriers can also interact with the lattice and will be scattered preferentially in the direction of resultant phonon momentum or from hot to cold.

Phonon drag is an important factor in the treatment of electronic effects in which a temperature gradient is present. Some of the important consequences of the existence of the phonon drag are as follows: The magnitude of the Seebeck voltage and Peltier heat is increased because of phonon drag; the effect of phonon drag on the Seebeck voltage or Peltier heat is independent of the charge carrier concentration in a semiconductor at reasonably low concentrations; and the effect of phonon drag is more pronounced at low temperatures.

9. DEBYE TEMPERATURE

Thermal conduction (which is related to the anharmonic coupling of phonons to other phonons) and electric conduction (which is related to the scattering of electrons by phonons) show a different behavior below and above the Debye temperature (T_D). The most probable phonon wavelength is

$$\lambda_{ph} = aT_D/T. \qquad (1.103)$$

In solids with a high Debye temperature (i.e., where interatomic forces are very strong) long wavelength phonons become dominant even at higher temperatures while short wavelength phonons become insignificant.

By definition:

$$T_D = hv_o/k; \qquad T_D/T = hv_o kT. \qquad (1.104)$$

Other expressions for Debye temperature and frequency of lattice vibrations are as follows:

$$T_D = (3/4\pi)^{1/3}(h/k)a(C_L/u_v)^{1/3}(w_{at}\kappa_v^*)^{-1/2}, \qquad (1.105)$$

$$v_o = (3/4\pi)^{1/3}a(C_L/u_v)^{1/3}(w_{at}\kappa_v^*)^{-1/2}. \qquad (1.106)$$

The Debye temperature is a material property and is a function of the elastic constants; it is proportional to

$$T_D \sim (T_m/w_{at})^{1/2}.$$

Debye temperature and the corresponding frequency of lattice vibrations, thermal conductivity, specific heat, lattice spacing, and the most probable phonon wavelength at 300°K are given in Table 1.31.

TABLE 1.31

Debye Temperature, Frequency of Lattice Vibrations, Thermal Conductivity, Specific Heat, Lattice Spacing, and Phonon Wavelength of Selected Near-Intrinsic Semiconductors (300°K)

Semi-conductor	T_D [°K]	v_o [10^{13} Hz]	κ [W/cm°K]	κ_{ph} [W/cm°K]	c_v [10^7 cm²/sec²°K]	a [Å]	λ_{ph} [Å]
Si	658	1.37	1.45	1.3	0.71	5.43	11.9
Ge	378	0.79	0.65	0.6	0.34	5.66	7.1
AlAs	417	0.87	0.08	0.8	0.46	5.66	7.9
AlP	588	1.23	0.90	0.9		5.46	10.7
AlSb	292	0.61	0.62	0.6	0.29	6.14	6.0
GaAs	344	0.72	0.80	0.4	0.31	5.65	6.5
GaP	435	0.91	0.77	0.7		5.45	6.5
GaSb	266	0.54	0.33	0.3	0.25	6.10	5.4
InAs	249	0.52	0.27	0.2	0.25	6.06	5.0
InP	322	0.67	0.68	0.6	0.34	5.87	6.3
InSb	203	0.42	0.19	0.2	0.21	6.48	4.4

10. THERMAL EXPANSION

If the atoms of a crystal experience lattice vibrations (i.e., if $T > 0°K$) and if the bond between two atoms is harmonic, i.e., symmetrical, the forces between these atoms average zero. However, deviations from a harmonic bond cause the existence of repelling forces which result in an expansion of the lattice with temperature.

With few exceptions, all crystals display less than 7% variation in atomic distance and hence in volume between 0°K and their melting points, i.e., the variation in atomic spacing with temperature is always less than the amplitude of thermal vibrations, even near 0°K. If $T \to 0°K$, then the expansion coefficient $\alpha' \to 0$.

The thermal expansion in crystals is related to their melting point. Generally, the linear thermal expansion coefficient of most materials is inversely proportional to their melting point. Table 1.32 gives linear thermal expansion

TABLE 1.32

Thermal Expansion Coefficient and Melting Point

Material	α' [10^{-6}°K^{-1}]	T_m [°K]
Si	3.5	1690
Ge	5.5	1210
α-Sn	5.2	505
AlAs		>1600
AlP		>1500
AlSb		1050
GaAs	5.7	1510
GaP	5.3	1738
GaSb	6.3	985
InAs	4.0	1215
InP	4.5	1343
InSb	5.1	798
SiO$_2$	0.5	1973
Al	24.0	933
Au	14.0	1336
W	4.5	3683

coefficients at 300°K and melting points of selected materials. The linear and cubic thermal expansion coefficients of a few materials are shown in Figure 1.27. Certain crystals expand in one direction and contract in another if the temperature is increased; some crystals (e.g., ice below 100°K) contract in all directions so that the cubic expansion coefficient becomes negative.

(a) One-dimensional (linear) thermal expansion:
 The linear temperature coefficient is given by

$$\alpha' = (1/l_o)(\partial l/\partial T)$$
$$= a'_1 + a'_2 T + a'_3 T^2 + \cdots \quad (1.107)$$

where the constants a'_i are empirically determined quantities and l_o is the original sample length.

 In crystals, α' depends upon the crystal structure and to a lesser degree upon the crystallographic orientation.

(b) Three-dimensional (volume) thermal expansion:

$$\gamma' = (1/u_v)(\partial u_v/\partial T) \quad (1.108)$$

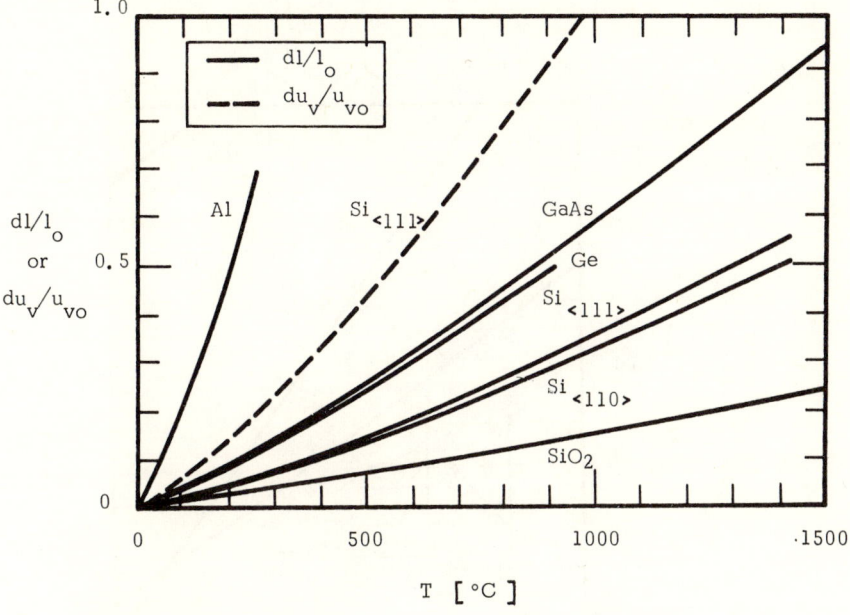

FIGURE 1.27

Linear (dl/l_o) and volume (du_v/u_{vo}) thermal expansion coefficient of several materials as function of temperature (T). The data are referred to 0°C.

where u_v is the original sample volume.
In crystals with cubic structure

$$\gamma' = 3\alpha', \quad (1.109a)$$

in crystals with hexagonal, triagonal, or tetragonal structure

$$\gamma' = 2\alpha'_x + \alpha'_z, \quad (1.109b)$$

in crystals with rhombic structure

$$\gamma' = \alpha'_x + \alpha'_y + \alpha'_z. \quad (1.109c)$$

11. VAPOR PRESSURE

Vapor pressure is the pressure exerted by a solid or a liquid when in equilibrium with its own vapor. It is due to the collision of vapor molecules in a confined space. At the boiling point of solid or liquid vapor pressure and ambient pressure are equal. For most materials vapor pressure is given for a certain temperature range by the Antoine equation

$$p_v = \exp\left[a_1 - a_2/(a_3 + T)\right] \quad (1.110)$$

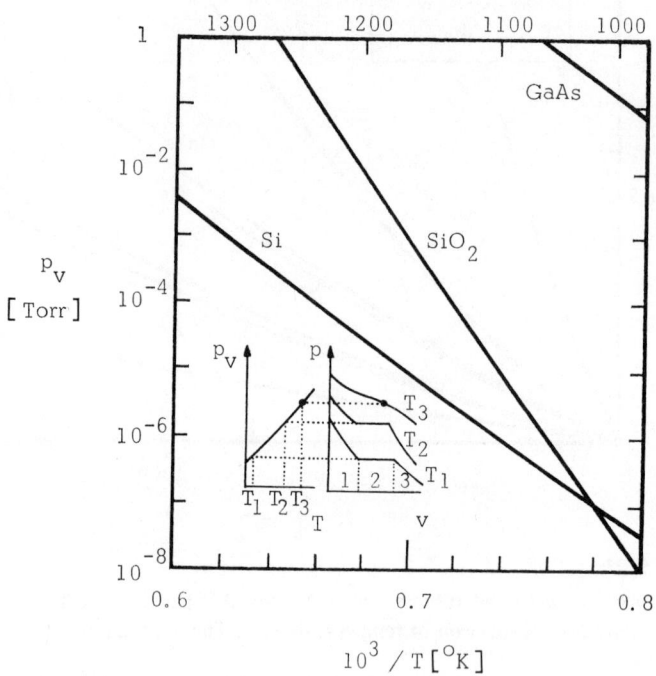

FIGURE 1.28

Vapor pressure (p_v) of Si and SiO_2 vs. temperature (T). The inset shows the temperature dependence of vapor pressure and the corresponding volume dependence of pressure. In region 1 the material is liquid, in region 3 it is vapor, and in region 2 the liquid and the vapor phases coexist. The temperatures $T_1 < T_2 < T_3$.

which for some materials (including Si and SiO_2) where $a_3 = 0$, reduces to

$$p_v = \exp[a_1 - a_2/T]; \tag{1.111}$$

where a_1, a_2, a_3 are empirical constants. Within a limited temperature range at high temperature the relationship

$$p_v = 5.7\, T \exp(A_p - H_E/2.3\, RT) \tag{1.112}$$

has been observed for many materials, where H_E is the heat of evaporation, A_p is a material constant, and R is the gas constant. These equations suggest a strong dependence of p_v upon T, as is evident from Figure 1.28. This temperature dependence is particularly apparent with elements and compounds that have a low melting point.

Vapor pressure is one of the most important parameters in vacuum deposition, especially of metals. In diffusion systems the vapor pressure of the impurity source determines the availability of the impurity at the semiconductor surface and it dictates the optimum temperature of the impurity source.

12. REFERENCES

(1) D. R. Stull, *Ind. Eng. Chem.*, **39**, 517 (1947).
(2) D. F. Gibbons, *J. Phys. Chem. Solids*, **11**, 246 (1959).
(3) H. S. Carslaw and J. C. Jaeger, *Conduction of Heat in Solids*, Oxford University Press, London, 1959.
(4) L. Maissel, *J. Appl. Phys.*, **31**, 211 (1960).
(5) G. A. Slack, *J. Appl. Phys.*, **35**, 3460 (1964).
(6) O. Madelung, *Physics of III–V Compounds*, John Wiley and Sons, New York, 1964.
(7) Research Triangle Report ASD-TDR-63-316, Vol. V, July 1964.
(8) R. H. Fairbanks and C. M. Adams, *Welding J.*, **43**, 97 (1964).
(9) H. Wagini, *Z. Naturfschg.*, **20a**, 494 (1965).
(10) W. R. Runyan, *Silicon Semiconductor Technology*, McGraw-Hill Book Co., New York, 1965.
(11) E. F. Stegmeier and I. Kudman, *Phys. Rev.*, **141**, 767 (1966).
(12) N. A. Lange, *Handbook of Chemistry*, McGraw-Hill Book Co., New York, 1967.
(13) P. D. Maycock, *Solid-State Electronics*, **10**, 161 (1967).
(14) F. D. Rosi, *Solid-State Electronics*, **11**, 831 (1968).
(15) P. S. Nayer et al., *SST*, **12**, 43 (Feb. 1969).
(16) K. E. Bean et al., *First Internat. Symp. on Silicon Mat., Science and Technol.*, Abstr. No. 326, New York, May 5–9, 1969.
(17) J. S. Blakemore, *Solid State Physics*, W. B. Saunders Co., Philadelphia, 1969.
(18) R. E. Honig and D. A. Kramer, *RCA Review*, **30**, 285 (1969).

1-8 Optical and Dielectric Characteristics

1. REFRACTIVE INDEX AND ABSORPTION COEFFICIENT

The attenuation of an electromagnetic wave in a medium is a function of the absorption coefficient (α). The amplitude of the wave at distance x from the surface of a semiconductor is

$$A(x) = A_o \exp(-\alpha x/2)$$
$$= A_o \exp(-2\pi\delta/\lambda) \tag{1.113}$$

where A_o is the amplitude at the surface.

The total absorption coefficient is related to the complex refractive index (N^*) by

$$N^* = n^* - i\alpha\lambda x/4\pi$$
$$= n^* - i\delta \tag{1.114}$$

where n^* is the real part of the refractive index and δ is the extinction coefficient.

The real part of the refractive index is defined as the ratio of velocities of an electromagnetic wave (e.g., light) in vacuum and in the specimen. It is temperature- and wavelength-dependent.

If the wave experiences a transition from an optically thicker medium (n_1^*) to an optically thinner medium (n_2^*), i.e., if $n_1^* > n_2^*$, it will be bent away from the normal to the surface of the medium at an angle α_t; as a consequence, if $\sin \alpha_t > n_2^*/n_1^*$ light cannot enter the optically thinner medium, i.e., it will be totally reflected. Usually $n^* > 1$ (for vacuum $n^* = 1.0$). However, for x-rays $n^* < 1$; also for some metals $n^* < 1$ for visible light (e.g., for silver $n^* = 0.18$).

The imaginary part of the refractive index, the extinction coefficient, is related to the absorption coefficient by

$$\delta = \alpha x\lambda/4\pi. \tag{1.115}$$

It is strongly wavelength-dependent because of the dependence of α upon wavelength and because of the factor λ.

1-8 OPTICAL AND DIELECTRIC CHARACTERISTICS

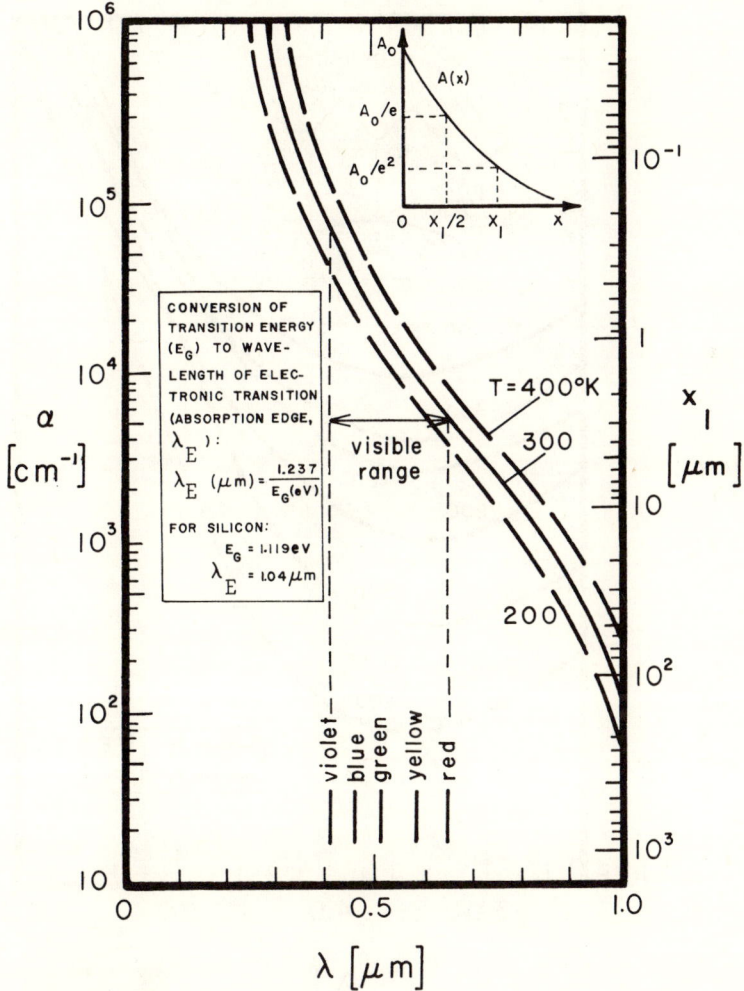

FIGURE 1.29

Absorption coefficient (α) and penetration depth (x_1) vs. wavelength (λ) and temperature (T). Silicon.

In a heavily doped semiconductor, due to its metal-like characteristics, the refractive index cannot be expressed by a real number, but, similar to metals, the complex refractive index must be used. In a semiconductor of low impurity concentration, on the other hand, at $\lambda > \lambda_E$ the absorption coefficient $\alpha \approx 0$ so that $N^* = n^*$. The total absorption coefficient is

$$\alpha = \alpha_{fc} + \alpha' \qquad (1.116)$$

106 GENERAL PROPERTIES OF SEMICONDUCTORS

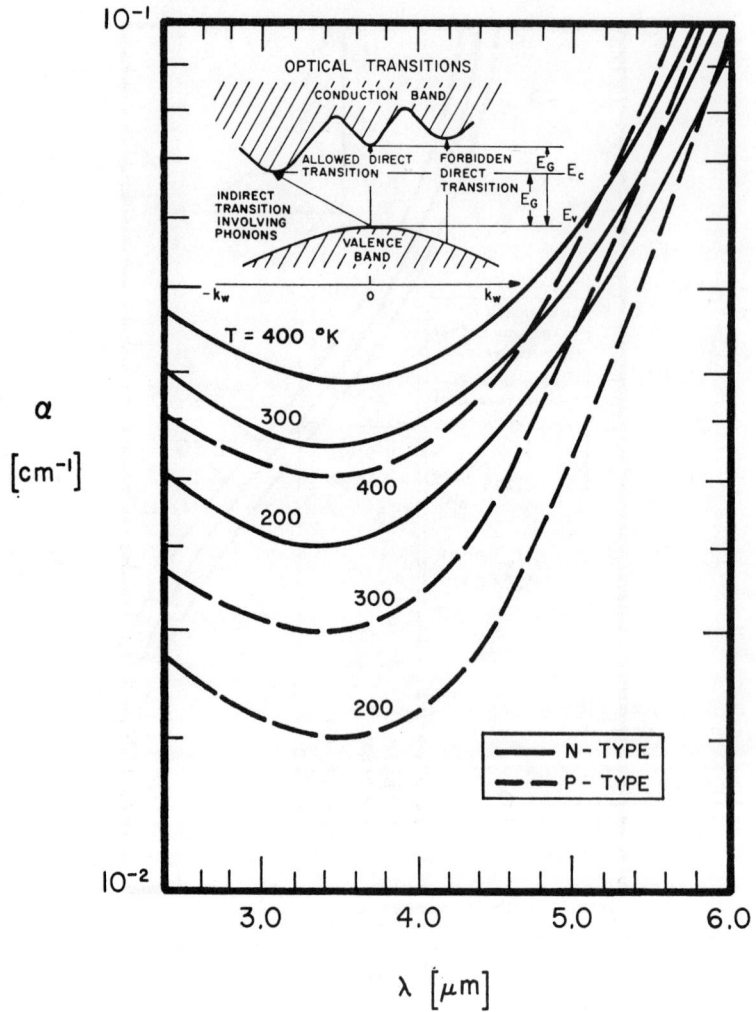

FIGURE 1.30

Absorption coefficient (α) vs. wavelength (λ) and temperature (T). Silicon. The inset shows allowed and forbidden optical transitions.

where α_{fc} is the free carrier absorption coefficient and α' takes into account factors other than free carriers.

In epitaxial structures, due to different carrier densities in substrate and epitaxial film, changes in optical characteristics occur during the transition of light from one region to the other resulting in partial light reflection and optical interference.

TABLE 1.33
Optical and Dielectric Properties of Selected Semiconductors

Semi-conductor	$E_{G, opt}$ [eV]	λ_E [μm]	n^*	ε_s	L_D [Å]	α[cm^{-1}] 0.5 μm	0.8 μm	10 μm	a_r 0.8 μm
Si	1.12	1.03	3.42	11.7	2.32	$1.1 \cdot 10^4$	$1.0 \cdot 10^3$	1.0	0.31
Ge	0.67	1.85	4.00	16.0	0.07			10^{-3}	0.36
AlAs	2.20	0.56	2.92	8.5					
AlP	3.00	0.41	3.41	11.6					
AlSb	1.60	0.77	3.18	10.1	118.0			0.5	
GaAs	1.43	0.86	3.30	10.4	28.80	$5.0 \cdot 10^5$	$8.0 \cdot 10^3$	1	0.29
GaP	2.24	0.55	2.91	8.5	0.001	$1.0 \cdot 10^3$	1		
GaSb	0.70	1.77	3.74	14.0	0.44		$2.0 \cdot 10^3$		
InAs	0.36	3.44	3.42	11.7	0.01		$5.0 \cdot 10^3$		
InP	1.26	0.98	3.26	10.3	0.03	10^6	$2.0 \cdot 10^4$		0.28
InSb	0.18	6.86	3.96	15.6	0.003			30.0	

Optical and dielectric properties of representative near-intrinsic semiconductors at 300°K are given in Table 1.33. For comparison, the dielectric constants of some other materials at 300°K and 10^6 Hz are given in Table 1.34.

TABLE 1.34
Dielectric Constants and Refractive Indices of Selected Materials

Material	Dielectric constant	Refractive index
Air	1.0	1.00
Teflon	2.1	1.45
SiO_2	3.9	1.98
Si_3N_4	8.0	2.82
Al_2O_3	8.8	2.97
H_2O	79.0	8.9
TiO_2	100.0	10.0
$BaTiO_3$	1140.0	33.8

The threshold of continuous optical absorption (optical absorption edge), i.e., the wavelength below which strong optical absorption occurs, is a function of the energy gap of the semiconductor. Two types of absorption are usually distinguished.

(a) Direct process:

In this case a photon is absorbed by the semiconductor and an electron and a hole are created. The absorption spectrum displays a strong increase at a wavelength λ_E which is called the absorption edge and which is given by

$$h\nu = E_G, \quad (1.117)$$

i.e.,

$$\lambda_E = ch/E_G. \quad (1.118)$$

Since $c = 3.0 \cdot 10^{10}$ cm/sec and $h = 4.14 \cdot 10^{-15}$ eV sec,

$$\lambda_E = 1.237/E_G \quad (1.119)$$

if E_G is expressed in eV and λ_E in μm.

(b) Indirect process:

In this case a photon is absorbed by the semiconductor and an electron and a hole are created, but, due to interaction between photon and lattice, a phonon is also involved. A strong absorption increase is observed at wavelength λ_E which is given by

$$h\nu = E_G + h\nu_{ph} \quad (1.120)$$

i.e.,

$$\lambda_E = c/(E_G/h + \nu_{ph})$$
$$= 1.237/(E_G + 4.14 \cdot 10^{-15}\nu_{ph}) \quad (1.121)$$

if E_G is expressed in eV, ν_{ph} in Hz, and λ_E in μm.

In the above equations

$$h\nu_{ph} = \text{phonon energy}$$
$$\nu_{ph} = \text{phonon frequency}.$$

The phonon energy is typically

$$h\nu_{ph} \approx 0.01 \text{ to } 0.03 \text{ eV} \ll E_G$$

so that in most cases the absorption edges for a direct and an indirect process are identical, provided that the energy gap is the same. If the phonon energy is not negligible compared to the energy gap of the semiconductor, then λ_E in the indirect process is shifted to a smaller wavelength compared to the direct process. The absorption edge depends upon temperature, pressure, magnetic field, and impurity concentration.

The absorption spectrum of a semiconductor can be divided into three ranges with respect to the band edge wavelength λ_E.

(a) $\lambda < \lambda_E$:
   Absorption is mainly due to carriers being raised from the valence band to the conduction band.
(b) $\lambda \approx \lambda_E$:
   Range of direct transition.
(c) $\lambda > \lambda_E$:
   Absorption is mainly due to free carriers, lattice impurities, defects, carriers associated with impurities and defects.

In a degenerate semiconductor the Fermi level lies within the conduction or valence band and unoccupied states exist only above an energy $(E_F - 4kT)$ for an n-type semiconductor or below an energy $(E_F + 4kT)$ for a p-type semiconductor. Direct optical transitions are then possible (for an n-type semiconductor) only above $(E_F - 4kT - E_G)$, i.e., at wavelengths for which $h\nu \geq E_{G,\text{opt}}$. $E_{G,\text{opt}}$ is called the optical energy gap. For a degenerate n-type semiconductor

$$E_{G,\text{opt}} = E_G + (E_F - 4kT - E_c)(1 + m_n/m_p) \qquad (1.122)$$

where m_n, m_p are the effective masses of electrons and holes, respectively. This effect is important in semiconductors with small effective mass of the charge carriers, e.g., InSb, since then the density of states is small and degeneracy occurs even at small impurity concentration and a displacement of the absorption edge toward shorter wavelengths is observed with increasing impurity concentration. For most semiconductors of low impurity concentration the optical energy gap is approximately equal to E_G.

2. FREE CARRIER ABSORPTION

Free carrier absorption is the absorption of infrared energy by conduction electrons or holes of a semiconductor. It is due to transitions of free electrons or holes between states in the same band or between different bands. Transitions within the same band (intraband absorption) cause absorption edges and, if the transitions are confined to a narrow spectral region, absorption peaks. Transitions between different bands (interband absorption) depend upon the density of free carriers and increase with increasing carrier density.

The magnitude of free carrier absorption depends upon the number (n and p) and mobility (μ_n and μ_p) of free carriers available for absorption and the wavelength of light used (λ).

$$\alpha_{fc} = \frac{\lambda^2 q^3}{4\pi^2 c^3 n^* \varepsilon_o}\left(\frac{n}{m_n^2 \mu_n} + \frac{p}{m_p^2 \mu_p}\right). \qquad (1.123)$$

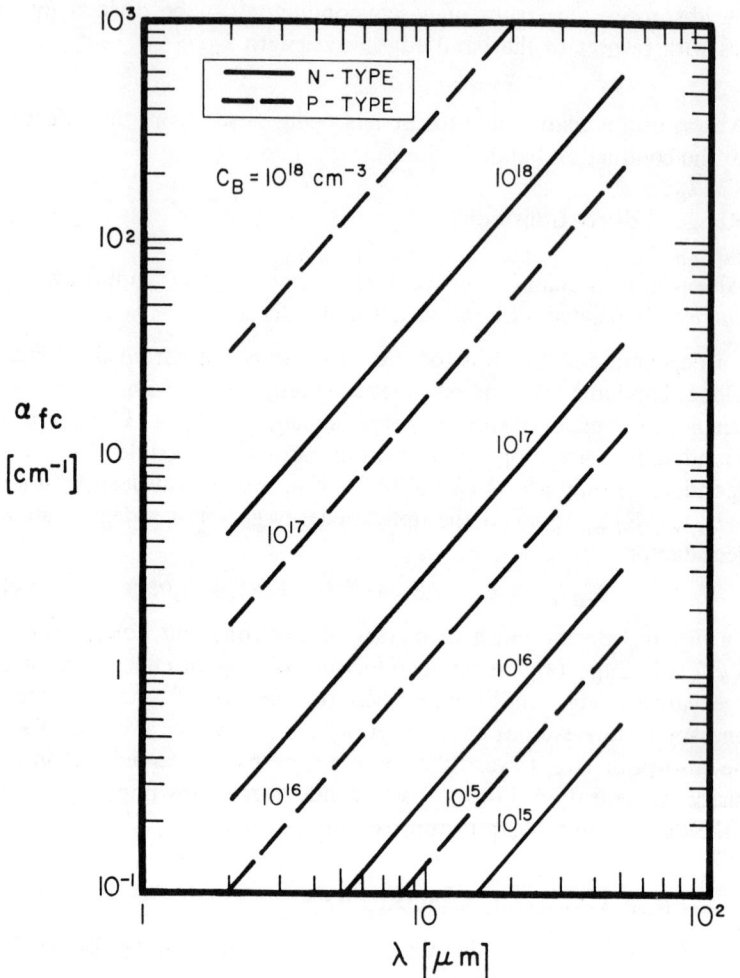

FIGURE 1.31

Free carrier absorption coefficient (α_{fc}) vs. wavelength (λ) and semiconductor impurity concentration. Silicon, 300°K.

In the infrared the contribution of free carriers to the dielectric constant of a nondegenerate semiconductor of impurity concentration C_B is

$$\varepsilon_{fc} = -q^2 C_B \lambda^2 / 4\pi^2 \varepsilon_o c^2 m_{n,p} \qquad (1.124)$$

where $m_{n,p}$ is the average effective mass of electrons and holes.

3. REFRACTIVE INDEX AND DIELECTRIC CONSTANT

In a single-carrier n-type or p-type semiconductor respectively:

$$n^{*2} - \delta^2 - \varepsilon_s = \frac{nq^2}{m_n \varepsilon_o} \frac{1}{4\pi^2/\lambda^2 + q^2/m_n^2 \mu_n^2} = A_{\alpha n} \qquad (1.125a)$$

or

$$n^{*2} - \delta^2 - \varepsilon_s = \frac{pq^2}{m_p \varepsilon_o} \frac{1}{4\pi^2/\lambda^2 + q^2/m_p^2 \mu_p^2} = A_{\alpha p} \qquad (1.125b)$$

The refractive index is a function of the wavelength:

(a) Short wavelength, i.e., at $\lambda < (2\pi/q)(m_n \varepsilon_o/n)^{1/2}$:
In this case absorption can be neglected and

$$n^{*2} = \varepsilon_s, \qquad (1.126)$$

i.e., the refractive index is equal to the square root of the dielectric constant (if n^* and ε_s are measured at the same frequency).

(b) Large wavelength, i.e., at $\lambda > (2\pi/q)(m_n \varepsilon_o/n)^{1/2}$:
In this case

$$n^{*2} = \varepsilon_s + \delta^2 + A_\alpha \qquad (1.127)$$

where for an n-type semiconductor:

$$A_{\alpha n} = \frac{nq^2}{m_n \varepsilon_o} \frac{1}{4\pi^2/\lambda^2 + q^2/m_n^2 \mu_n^2}. \qquad (1.128)$$

4. OPTICAL PENETRATION DEPTH

Penetration depth (x_l) is defined as twice the distance from the semiconductor surface at which the surface light intensity (A_o) has fallen to $1/e$ of that value; i.e., at $x = x_l$:

$$A(x_l) = 0.135 A_o. \qquad (1.129)$$

Semiconductor transparency (see Figure 1.32) at a given wavelength is a function of the absorption coefficient. Crystals which are electrical insulators are usually transparent in the visible range, i.e., $\alpha \ll 1$. Transparency of a crystal requires the absence of strong electronic or vibronic transitions in the

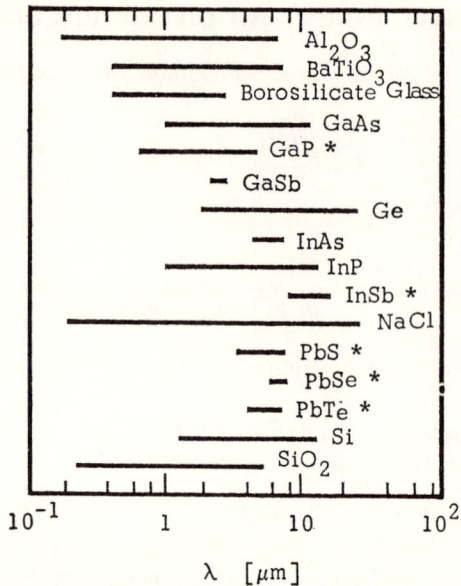

FIGURE 1.32

Regions of transparency for selected materials. The cut-off wavelength is that at which a 2-mm thick sample has 10% transmission at 300°K. The materials marked with an asterisk have less than 10% transmission throughout. The higher the wavelength of transparency, the lower usually the binding forces between atoms and therefore the higher the solubility in water.

visible range (these transitions correspond to an energy $E_G = 1.7$ to 3.5 eV within the visible range). If the crystal displays a certain color (=wavelength at which it is opaque) it has an electric dipole electronic transition corresponding to this wavelength. Silicon has a metallic appearance because its energy gap corresponds to a wavelength which is in the infrared; at wavelengths shorter than λ_E it is opaque since there electronic transitions from valence to conduction band will take place.

Radiation damage in a semiconductor (e.g., ion bombardment, ultraviolet light, etc.) can cause a color change due to enhanced transition; the introduction of impurity levels within the energy gap may also result in a change of the semiconductor color.

In the visible range the absorption coefficient of silicon decreases and the penetration depth increases by approximately one order of magnitude if the light color is changed from violet to red.

1-8 OPTICAL AND DIELECTRIC CHARACTERISTICS

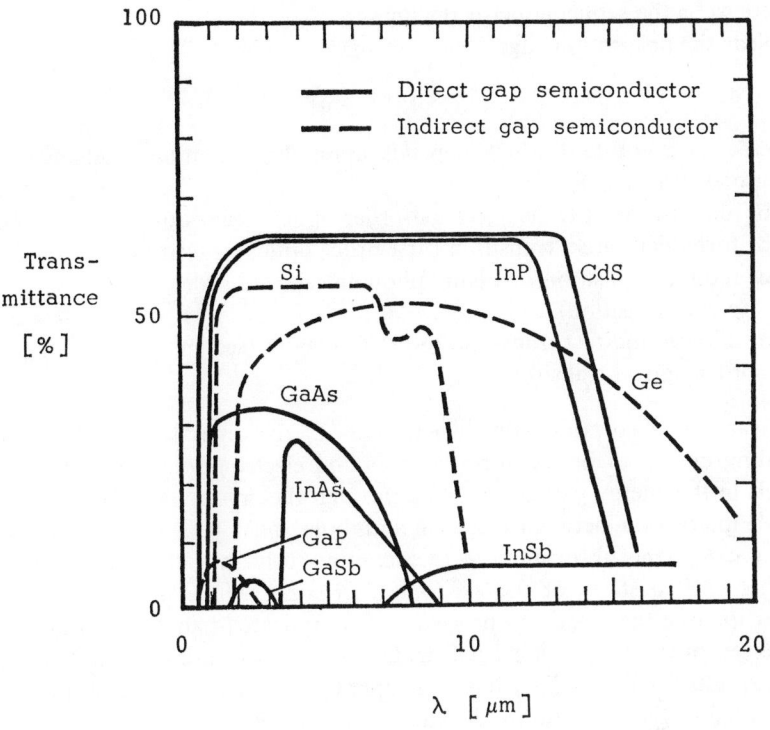

FIGURE 1.33

Optical transparency of selected semiconductors as function of wavelength (λ). Direct gap semiconductors are: CdS, GaAs, GaSb, InAs, InP, InSb. Indirect gap semiconductors are: Si, Ge, AlAs, AlP, AlSb, GaP.

5. TRANSMISSION AND REFLECTION

Free carrier absorption is one means of determining the energy gap of a semiconductor. Measurement of the transmission coefficient a_t and of the reflection coefficient a_r allows the determination of refractive index (n^*) and absorption coefficient (α) which in turn are related to the transition energy between bands. For normal incidence

$$a_t = \frac{(1 - a_r^2) \exp(-4\pi x_w/\lambda)}{1 - a_r^2 \exp(-8\pi x_w/\lambda)}, \quad (1.130)$$

$$a_r = \frac{(1 - n^*)^2 + (\alpha\lambda/4\pi)^2}{(1 + n^*)^2 + (\alpha\lambda/4\pi)^2}, \quad (1.131)$$

where x_w is the semiconductor thickness.

Near the absorption edge

$$\alpha \approx (h\nu - E_G)^j$$

where j is a constant which depends upon the transition mechanism (see inset of Figure 1.29);

for allowed direct transition (first-order optical transition) $j = 1/2$

for forbidden direct transition (first-order optical transition) $j = 3/2$

for indirect transition involving phonons (second-order optical transition) $j = 2$

for allowed indirect transition to exciton states (second-order optical transition) $j = 1/2$

Near the absorption edge where $h\nu - E_G$ becomes comparable to the binding energy of an exciton (i.e., a bound electron-hole pair with energy levels in the energy gap and moving through the crystal as a unit), the Coulomb interaction between electron and hole must be taken into account. If $h\nu < E_G$, the absorption merges continuously into that caused by the higher excited states of the exciton. If $h\nu \gg E_G$, higher energy bands will participate in the transition process and complicated band structures will be reflected in the absorption coefficients. The temperature dependence of the absorption coefficient (i.e., higher temperature results in larger values of α) is related to the temperature dependence of the energy gap.

6. DEBYE LENGTH

The Debye length (L_D) is the radius of a sphere around an impurity ion in which electrical neutrality is maintained; i.e., it is the distance over which small variations in potential (measured in eV) become insignificant in a semiconductor of uniform impurity concentration and uniform bulk characteristics. This definition is analogous to the Debye-Hückel theory of electrolytes.

The sphere surrounding an impurity ion causes, due to a relaxation effect, a velocity reduction of mobile charges in an electric field, i.e., it affects carrier mobility and, in turn, semiconductor resistivity. Consequently, the ion plasma can be divided into microcells of radius L_D which, in a first-order approximation, do not affect each other electrically and which obey statistical laws. In place of the mean free path, a relaxation length s_r may be introduced which is the distance an electron travels by interaction with the microcell if it experiences a thermal energy variation of magnitude kT.

$$s_r = (8/3)(2\pi\varepsilon_o)^2(kT/q^2)^2/[C_B \ln(1.7\pi\varepsilon_o kT/q^2 C_B^{1/3})]. \quad (1.132)$$

1-8 OPTICAL AND DIELECTRIC CHARACTERISTICS

A general expression for L_D for an arbitrary semiconductor is

$$L_D = \left(\frac{\varepsilon_s \varepsilon_o kT}{2q^2 C_B \exp(qV/2kT)}\right)^{1/2} \quad (1.133)$$

At small voltage, i.e., if

$$\exp(qV/2kT) \approx 1,$$

this can be simplified as follows.

(a) Extrinsic semiconductor:

$$L_D = (\varepsilon_s \varepsilon_o kT/2q^2 C_B)^{1/2} \quad (1.134)$$

(b) Intrinsic semiconductor:

$$L_D = (\varepsilon_s \varepsilon_o kT/2q^2 n_i)^{1/2}. \quad (1.135)$$

In the above expressions it is assumed that the semiconductor is nondegenerate, i.e., that the carrier concentration is equal to the semiconductor impurity concentration, C_B.

The Debye length is related to the radius r_F of the Fermi sphere (which, in k_w-space, is the surface bounding the occupied region; r_F is the radius which corresponds to the highest occupied energy level).

$$L_D = 2^{1/3} r_F \approx 1.26 r_F.$$

7. DIELECTRIC CONSTANT AND RELAXATION TIME

The Debye length can also be expressed by a dielectric relaxation time as follows:

$$L_D \approx (D_c \tau_d)^{1/2}, \quad (1.136)$$

where D_c is the majority carrier diffusion coefficient. The relaxation time is

$$\tau_d - \varepsilon_s \varepsilon_o \rho, \quad (1.137)$$

where ρ is the sample resistivity. For example, for silicon of resistivity $\rho = 1$ Ωcm the dielectric relaxation time

$$\tau_d = 1 \text{ psec}.$$

In a polar solid the portion of the dielectric constant resulting from dipole orientation decreases on cooling in an irregular fashion, since the dipoles do not rotate with perfect freedom but jump from one to another of a few permitted orientations. If an electric field is applied externally, the two

116 GENERAL PROPERTIES OF SEMICONDUCTORS

energy minima are no longer equivalent but transitions will favor the population of one. Thus, the dielectric relaxation time can also be expressed by

$$\tau_d = (1/2\pi v_o) \exp(q\phi_b'/kT) \tag{1.138}$$

which is the rate at which jumping of the polar dipoles of a dielectric occurs. The intrinsic frequency of jumping related to the upper limit of the lattice vibrational spectrum is v_o, i.e.,

$$v_o = kT_D/h. \tag{1.139}$$

Usually v_o is in the range 10^{11} to 10^{13} Hz; for silicon $v_o = 1.39 \cdot 10^{13}$ Hz. The barrier height which must be surmounted in jumping between two equilibrium positions is $\phi_{b'}$.

If the electric field is applied at frequency ω (which is large enough to affect the jumping response of permanent dipoles, i.e., it is in the order of $1/\tau_d$, but small enough that the induced ionic and electronic polarization can follow the field), then the complex permittivity (dielectric constant)

$$\begin{aligned}\varepsilon &= \varepsilon_1 + i\varepsilon_2 \\ &= a_\varepsilon + b_\varepsilon/(1 - i\omega\tau_d)\end{aligned} \tag{1.140}$$

where the real and imaginary parts of the dielectric constant

$$\varepsilon_1 = a_\varepsilon + b_\varepsilon/(1 + \omega^2\tau_d^2) \tag{1.141a}$$

and

$$\varepsilon_2 = b_\varepsilon \omega^3 \tau_d^3. \tag{1.141b}$$

The real part of the dielectric constant displays a minimum at a frequency $1/\tau_d$. In contrast to the relaxation type response of permanent dipoles, the dielectric response of induced dipoles is a resonant phenomenon.

8. PHOTOCONDUCTIVITY

Photoconductivity is the electrical conductivity of semiconductor or insulator resulting from illumination by radiant energy. Most of the nonmetallic elements having a refractive index greater than about 2 (e.g., B, C, Si, Ge, P, As, Se, Te, etc.) and a number of inorganic compounds (e.g., the alkali halides, the sulfides, selenides, and tellurides of zinc, cadmium, and mercury, silver and thallous halides, oxides, including Cu_2O, MgO, BaO, ZnO, and others, and the semiconducting compounds such as InSb, GaAs, InSe, and InTe) exhibit photoconductivity.

Photoconductivity can be explained by two models depending upon whether the semiconductor is single-crystalline or polycrystalline.

1–8 OPTICAL AND DIELECTRIC CHARACTERISTICS

(a) Recombination model:

In this model, which applies to crystalline semiconductors, it is assumed that changes in conductivity on illumination result from a change in the number of conducting electrons or holes per unit volume. In a semiconducting or insulating solid in thermal equilibrium, some small fraction of the electrons is in the conduction state due to thermal excitation. The fraction of electrons in the conduction state depends on the temperature of the specimen, the energy gap, and the density and nature of imperfections within the crystal. These electrons are responsible for the "dark" conductivity of the material, that is, the conductivity observed in the absence of illumination. Absorption of quanta whose energy is greater than the width of the energy gap produces free holes and electrons which increase the observed conductivity. The magnitude of the increase in conductivity depends on the rate of generation of charge carriers and the time which these carriers spend in the conduction state; this time depends in turn on the recombination processes.

The concept of hole or electron traps is important in the recombination theory of photoconductivity. Because of the presence of impurities and other lattice imperfections localized levels having energies in the energy gap exist in the crystal. Most recombination occurs through these recombination centers or traps. Only at high carrier densities and therefore at high intensities of illumination does the direct recombination process compete with trapping. The photoconductivity in equilibrium will be determined by a balance between the processes which create charge carriers and those by which they disappear. The generation processes by which carriers are produced include:

— generation of hole-electron pairs by thermal or radiant energy;
— creation of a free electron by thermal or radiant energy, leaving a trapped hole behind;
— creation of a free hole by thermal or radiant energy, leaving a trapped electron behind;
— release of a trapped electron;
— thermal release of a trapped hole.

The recombination processes by which carriers disappear include:

— direct recombination of a free electron and a free hole;
— trapping of a free electron;
— trapping of a free hole;
— recombination of a free electron and a trapped hole;
— recombination of a free hole and a trapped electron.

For a particular case some of these processes may be neglible. The rate-determining coefficients of some of these processes depend on the energy distribution of the traps.

(b) Barrier model:

In this model, which applies to polycrystalline semiconductors which possess a high surface state density, it is assumed that illumination produces little or no change in the density of charge carriers but that the effective mobility of such carriers is increased. It is also assumed that a large number of surface states are produced along the crystalline boundaries by, for instance, oxidation of these boundaries. These surface states capture electrons from the interior of the single-crystalline domains and produce space charge barriers. Illumination reduces the number of electrons in the surface states and thereby lowers the barrier height between domains, which may then be observed as an increase in conductivity. This model predicts that the photosensitivity (i.e., the fractional change in conductivity per unit change in light intensity) is independent of light intensity and proportional to temperature; this prediction does not agree with the recombination model.

The two models of photoconductivity may be complementary in that the observed photoconduction in a real polycrystalline material may be due to a combination of both processes. Both models predict that a sensitive photoconductor will have a long time constant both in response and in decay. The time constant of a photoconductor depends upon the purity and crystalline perfection of the material and on temperature. The response time and decay time may be different. Many photoconductors have time constants in the range 10^{-12} to 1 second.

9. PHOTOVOLTAIC EFFECT

The photovoltaic effect is a process by which a voltage is produced at the junction of two different materials (e.g., a metal-semiconductor contact or a p-n junction) through an incident photon flux. In a p-n junction photons of sufficient energy incident upon the semiconductor produce hole-electron pairs. Some of these holes and electrons diffuse toward the junction and, if they are created sufficiently close to it, have a high probability of reaching it before they recombine. At the junction they are separated by the barrier. If they are created at the p-side of the junction and diffuse toward the junction, the electrons will be swept across the junction into the n-side whereas the holes will be blocked by the barrier. If the hole-electron pairs diffuse in from the n-side of the junction, the holes are swept into the p-side by the barrier while the electrons are blocked.

The carriers which are swept across the barrier are the minority carriers in the region in which they are generated and their flow through the barrier constitutes reverse current. If there are no exterior connections to the device (open circuit conditions) the charge which accumulates on either side of the junction reduces the barrier height. This reduction causes the forward current to increase. The forward current results from the passage of thermally produced majority carriers across the barrier. An equilibrium condition is quickly established which results in an open circuit voltage intermediate between zero and the full barrier voltage. If the illumination is sufficiently intense to annihilate the barrier, the photovoltage has reached its saturation value which is equal in magnitude to the barrier voltage. Usually, however, the intensity of illumination is considerably less than the saturation value and then the open circuit photovoltage, V_{th}, is

$$V_{th} = -(kT/q)\ln(1 + G_{ph}/G_{th})$$

where G_{ph} = rate of generation of electron-hole pairs by photons
G_{th} = rate of generation of electron-hole pairs by thermal excitation.

In order to obtain a high photovoltaic efficiency the following requirements have to be met:

(a) The energy per photon has to be slightly greater than the band gap energy.
(b) The impurity concentration has to be high and the temperature low. These conditions insure that the potential barrier height of the junction is nearly equal to the semiconductor energy gap.
(c) The device area has to be small with respect to the diffusion lengths of the minority carriers, i.e., the average distance through which they diffuse before recombining, but long enough so that the incident radiation will be absorbed.
(d) The surface reflectivity has to be small.

10. REFERENCES

(1) K. J. Schmidt-Tiedemann, *Z. Naturforschg.*, **16a**, 639 (1961).
(2) W. Paul and H. Brooks, *Progress in Semiconductors*, Vol. 7, John Wiley and Sons, New York, 1963.
(3) R. B. Adler et al., *Introduction to Semiconductor Physics*, John Wiley and Sons, New York, 1964.
(4) S. Larach, Editor, *Photoelectronic Materials and Devices*, D. Van Nostrand Company, Princeton, N.J., 1965.
(5) A. R. Hilton and C. E. Jones, *J. El. Chem. Soc.*, **113**, 472 (1966).
(6) R. K. Willardson and A. C. Beer, Editors, *Semiconductors and Semimetals*, Vol. 3, Academic Press, New York, 1967.
(7) S. M. Sze, *Physics of Semiconductor Devices*, John Wiley and Sons, New York, 1969.
(8) J. S. Blakemore, *Solid State Physics*, W. B. Saunders Co., Philadelphia, 1969.

1-9 Stress in a Semiconductor

1. EFFECT OF STRESS ON SEMICONDUCTOR PROPERTIES

If an ionic crystal (that is, one which has a polar axis, i.e., where not each lattice constituent is a center of symmetry) is elastically compressed in certain directions it will become polarized or an existing electric polarization will be altered. As a consequence, dipole moments in the crystal bulk will occur or existing dipole moments will vary, resulting in surface charges, an effect which is called piezoelectricity; the associated resistance is called piezoresistance of the crystal. Similarly, an electric field applied to the crystal will result in strain in the crystal. The piezoelectric effect is reversible, i.e., a volume expansion of the crystal is observed if the surface is charged (ultrasound generator). The piezoresistive coefficient of semiconductors is usually significantly higher than that of metals.

If uniaxial stress is applied to a semiconductor, the bands are altered in such a way that points lying symmetrically in the Brillouin zone are displaced by different amounts of energy, resulting in changes in the electrical properties. The effects of stress on the semiconductor properties differ for isotropic and anisotropic conduction bands.

(a) Isotropic conduction band:
   In this case the change in band structure due to uniaxial stress is insignificant. The energy surfaces are deformed, but only a shift of one band edge relative to another will have a noticeable effect.

(b) Anisotropic conduction band:
   In this case there is a significant rearrangement of charge carriers within the conduction band. Depending upon the direction of the stress, the band edges of the individual minima will be moved by different amounts of energy and will, therefore, be occupied by different numbers of carriers. In a cubic crystal with minima lying at the same energy, all minima contribute to anisotropic parts of the conductivity; for no pressure these add up to an isotropic conductivity, i.e., there will be a pressure- and orientation-dependent anisotropy of the conductivity (piezoresistance).

Changes of the atomic spacing (lattice constant) will affect the semiconductor band structure. The changes in energy band boundaries correspond to variations in carrier potential energy (i.e., the deformation potential); variations in carrier kinetic energy result in reflection coefficients at the transition boundaries.

The deformation potential is defined as the displacement of the edge of the conduction band for electrons or of the edge of the valence band for holes per unit dilatation of the lattice. Anisotropic stress in a semiconductor causes an anisotropic change in carrier mobility, resulting in an anisotropic resistivity change. Tension in a valley direction causes electrons to be removed from that valley and transferred to valleys that are normal to the tension direction; hence it causes the resistance to increase in the tension direction and to decrease in the direction normal to the tension. Compression has the opposite effect of tension.

The resistivity change ($\Delta\rho$) observed as a result of applied longitudinal stress (σ_l^*) is

$$\Delta\rho/\rho_o = \pi_l \sigma_l^* \qquad (1.142)$$

where π_l is the piezoresistive coefficient which depends upon the crystallographic orientation. Generally, values of π_l are as follows.

Direction	π_l [cm²/dyne]
$\langle 100 \rangle$	π_{11}
$\langle 111 \rangle$	$(1/3)\pi_{11} + (2/3)(\pi_{12} + \pi_{44})$
$\langle 110 \rangle$	$(1/2)(-\pi_{11} + \pi_{12} + \pi_{44})$

The tensor components π_{ij} are given in the inset of Figure 1.34 as a function of semiconductor doping level at 300°K; an increase in temperature results in a slight decrease of π_{ij}. Piezoelectricity occurs only in ionic crystals (or in crystals which have at least some ionic character), whereas dipole moments are induced in an electric field in every dielectric material.

To be distinguished from the effect of stress on a bulk semiconductor, whose piezoresistance is due to majority carriers, is the effect of stress on a p-n junction where stress results in variations in minority carrier density. Changes in conduction and valence band energy levels with stress cause exponential changes in minority carrier density at the edges of the junction depletion region on both sides of the metallurgical junction. The minority carrier density variations as a result of applied stress are shown in Figure 1.35 for different crystallographic orientations.

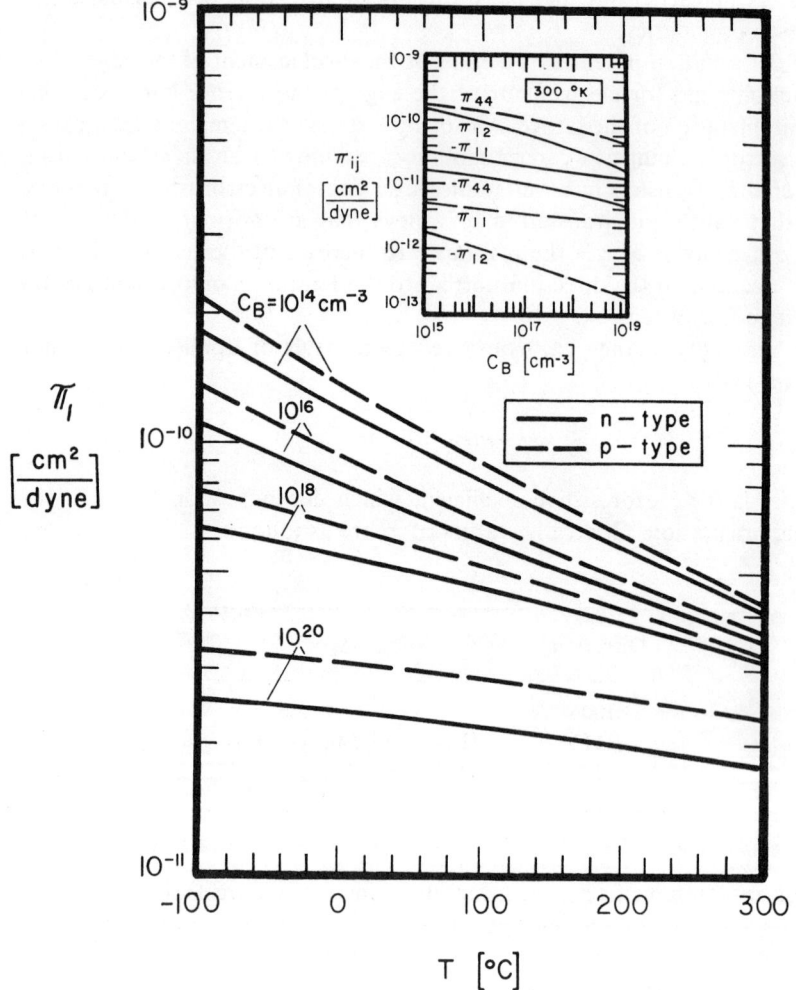

FIGURE 1.34

Piezoresistive coefficient (π_l) of silicon vs. temperature (T) and background impurity concentration (C_B). The inset shows the variation of the tensor components (π_{ij}) with silicon impurity concentration. Experimental data on silicon of $<110>$ orientation.

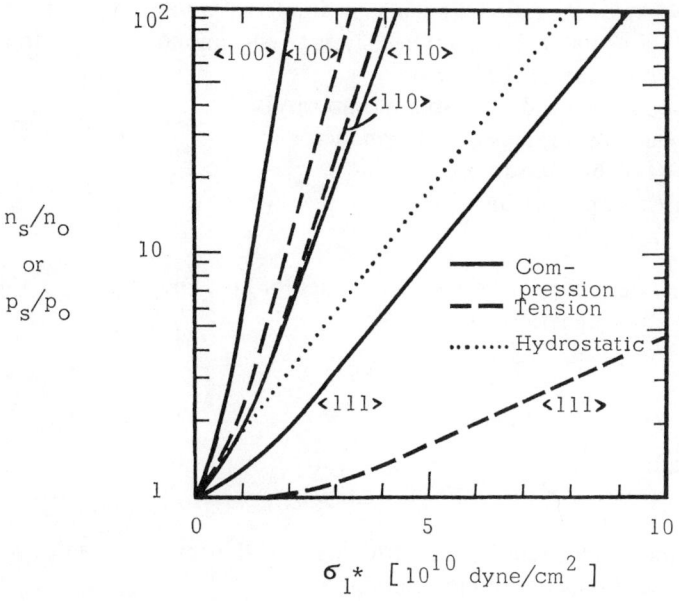

FIGURE 1.35
Ratio of minority carrier density under stress (n_s or p_s) to minority carrier density in the absence of stress (n_o or p_o) vs. applied stress (σ_i^*) for silicon of various crystallographic orientations at 300°K.

2. STRESS OWING TO THE PRESENCE OF IMPURITIES

Impurity atoms in a semiconductor may generate localized stress due to a mismatch of atomic radii of host and impurity, i.e., the generated stress is a function of the misfit factor ε_i^*. Its magnitude is

$$\sigma_i^* = [1 - (r_i/r_a)^3](C_S/C_L)Y_l/(1 - P)$$
$$= [1 - (1 - \varepsilon_i^*)^3](C_S/C_L)Y_l/1 - P). \tag{1.143}$$

This assumes that the impurity atoms enter the crystal lattice substitutionally. In the case of interstitial impurity migration the generated stress is a function of the ratio of the radius of the impurity atom to the radius of the voids between the lattice atoms.

In the above equation the following terms have been used.

C_S = impurity concentration at the semiconductor surface
C_L = density of semiconductor lattice atom sites. $C_L = Ld_s/w_{at}$
Y_l = Young's modulus

P = Poisson's ratio (defined as ratio of the amount of strain perpendicular to the applied stress to the amount of strain in the direction of the stress)
r_i = covalent radius of solute (impurity)
r_a = covalent radius of semiconductor
L = Loschmidt number.

Three cases have to distinguished:

(a) $r_i = r_a$:
In this case, because of a perfect match of host and impurity radii, there is no stress generated, i.e.,

$$\sigma_l^* = 0.$$

(b) $r_i < r_a$:
In this case

$$\sigma_l^* > 0,$$

i.e., localized stress is generated because the impurity radius is larger than the radius of the lattice atoms.

(c) $r_i > r_a$:
In this case

$$\sigma_l^* < 0,$$

i.e., the semiconductor is under localized compressional forces because the impurity radius is smaller than the radius of the lattice atoms.

The stress generated within the semiconductor lattice due to the presence of impurity atoms may have an effect on the motion of a dislocation line present in the lattice, as discussed below.

Misfit dislocation networks may occur in a semiconductor as a result of high impurity concentration. If phosphorus and boron are diffused into silicon at concentrations close to the maximum solid solubility, such dislocation networks have been observed. Some other impurities do not display this behavior; arsenic, for example, produces Frank hexagonal loops on (111) planes parallel to the semiconductor surface and stacking faults on the inclined (111) planes.

The energy of a Frank hexagonal loop is

$$E_{aF}^* = (1/2\pi\sqrt{6})a^2 l_H [Y_l/(1-P)] \ln (2l_H/a\sqrt{3}) + (\sqrt{6}/4) E_s l_H^2 \quad (1.144)$$

whereas the energy of a prismatic hexagonal loop is

$$E_{aP}^* = (3/4\pi 6^{1/2}) a^2 l_H [Y_l/(1-P)] \ln (2l_H/a\sqrt{6}). \quad (1.145)$$

In these equations
E_S = extrinsic stacking fault energy; for silicon: $E_S \approx 3.7 \cdot 10^{-3}$ eV/Å²
l_H = length of a side of a hexagon.

3. VELOCITY OF A DISLOCATION LINE UNDER STRESS

A dislocation line pinned between two points in a crystal lattice will move under the influence of stress. Its velocity is

$$v_{dis} = v_o a \exp(-E_{dis}/kT) \sinh(\sigma_l^* a^2 l_{dis}/kT). \qquad (1.146)$$

where v_o = frequency of lattice vibrations
a = lattice constant
σ_l^* = applied stress
E_{dis} = activation energy for the migration of a dislocation; for silicon:
$E_{dis} \approx 2.2$ eV = $3.5 \cdot 10^{-12}$ dyne cm
l_{dis} = length of the dislocation line between two pinning points.
The above expressions can be simplified if

$$\sigma_l^* \gg kT/a^2 l_{dis}.$$

In this case

$$v_{dis} = (v_o a/2) \exp[(\sigma_l^* a^2 l_{dis} - E_{dis})/kT]. \qquad (1.147)$$

For silicon at 1200°C

$$kT/a^2 = 4.3 \cdot 10^{13} \text{ eV/cm}^2,$$

i.e., the above simplification can be made if

$$\sigma_l^* l_{dis} \gg 4.3 \cdot 10^{13} \text{ eV/cm}^2 = 69 \text{ dyne/cm}.$$

The dislocation velocity decreases approximately exponentially with applied stress and dislocation length; an increase in either σ_l^* or l_{dis} has the same effect. The dislocation velocity increases also very rapidly with increasing lattice constant. If crystal imperfections are encountered by the dislocation during its travel through the crystal, the dislocation velocity is decreased because more energy is required to overcome this barrier. The distance over which a dislocation accelerates before encountering imperfections is usually small compared to the distances involved in making an experimental measurement of dislocation velocity.

Three stress conditions can be distinguished if $\sigma_l^* \gg kT/a^2 l_{dis}$:

(a) $\sigma_l^* = E_{dis}/a^2 l_{dis}$:
In this case

$$v_{dis} = v_o a/2.$$

For silicon (where $v_o = 1.39 \cdot 10^{13}$ Hz, $a = 5.43$ Å, $E_{dis} = 2.2$ eV)

$$v_{dis} = 3.8 \cdot 10^5 \text{ cm/sec};$$

the phonon velocity, on the other hand, which is the maximum velocity the dislocation line can attain

$$c_{ac} = 6.6 \cdot 10^5 \text{ cm/sec}.$$

For silicon this corresponds to the condition

$$\sigma_l^* = E_{dis}/a^2 l_{dis} = 1.1 \cdot 10^9 \text{ dyne/cm}^2$$

if the length of the dislocation line is assumed to be 100 Å.

(b) $\sigma_l^* \gg E_{dis}/a^2 l_{dis}$:

In this case

$$v_{dis} = (v_o a/2) \exp(\sigma_l^* a^2 l_{dis}/kT).$$

As in the first case, the dislocation velocity cannot exceed the phonon velocity, i.e., under all circumstances

$$v_{dis} \leq c_{ac}.$$

(c) $\sigma_l^* \ll E_{dis}/a^2 l_{dis}$:

In this case

$$v_{dis} = (v_o a/2) \exp(-E_{dis}/kT),$$

i.e., the dislocation velocity is independent of the stress. In silicon at 25°C

$$v_{dis} = 3.0 \cdot 10^{-32} \text{ cm/sec} \approx 10^{15} \text{ Å/year},$$

i.e., the dislocation line will not move; and at 1200°C

$$v_{dis} = 1.0 \cdot 10^{-2} \text{ cm/sec} = 100 \text{ Å/sec}.$$

In the neighborhood of an edge dislocation the maximum binding energy of an impurity atom of radius r_i in a semiconductor lattice of atomic radius r_a is, assuming a misfit factor $\varepsilon_i^* < 1$,

$$E_B = 4 Y_B \varepsilon_i^* r_a^3 d \sin \alpha^* / r_o^*$$
$$\approx 7(d \varepsilon_i^* / r_o^*) \text{ [eV]} \tag{1.148}$$

where Y_B = bulk modulus
ε_i^* = misfit factor; $\varepsilon_i^* = (r_i - r_a)/r_a$ or $r_i = r_a(1 - \varepsilon_i^*)$; for self-diffusion: $\varepsilon_i^* = 0$
d = apparent lattice spacing (i.e., the separation between two adjacent parallel planes)

$$d = a/(k_1^2 + k_2^2 + k_3^2)^{1/2}$$

k_i = Miller indices
α^* = angle between the dislocation core and the glide plane
r_o^* = distance between the dislocation core and the edge of the extra half-plane.

4. ELASTICITY OF A SEMICONDUCTOR

At frequencies below 10^{11} Hz (corresponding to elastic waves longer than 10^{-6} cm) a crystal can be viewed as a homogeneous medium rather than as a periodic array of atoms if the elastic properties of the crystal are considered. Any body reacts to applied pressure by a volume change. In most solids, including all semiconductors, this results in a counterforce which tends to reduce the deformation, i.e., they are elastic. The elastic constants (stiffness constants) describe the elastic properties of a solid.

Generally, for a solid there are 36 elastic constants; however, because of $c_{ik} = c_{ki}$, this can be reduced to 24 constants which for cubic crystals reduces further to three constants: c_{11}, c_{12}, c_{44}. These are related by

$$c_{11} = c_{12} + 2c_{44},$$

i.e., a cubic crystal has two independent elastic constants. These three constants describe volume compressibility (κ_v^*), bulk modulus (Y_B), Young's modulus (Y_l), Poisson's ratio (P), the velocity of elastic waves (v_l and v_t) and other mechanical properties of semiconductors.

For a cubic crystal

$$\kappa_v^* = 3/(c_{11} + 2c_{12}) = 1/Y_B \tag{1.149}$$

$$Y_B = (c_{11} + 2c_{12})/3 \tag{1.150}$$

$$Y_l = (c_{11} - c_{12})(c_{11} + 2c_{12})/(c_{11} + c_{12}) \tag{1.151}$$

$$P = -2c_{12}/(2c_{11} + c_{12}) \tag{1.152}$$

$$v_l = (c_{11}/d_s)^{1/2} \quad \text{(longitudinal wave)} \tag{1.153}$$

$$v_t = [(c_{11} - c_{12})/2d_s]^{1/2} \quad \text{(transverse wave)} \tag{1.154}$$

where p = applied pressure

d_s = density; $d_s = C_L w_{at}/L$.

The three elastic constants decrease with increasing temperature. Elastic properties of selected materials at 300°K are given in Tables 1.35 and 1.36. Table 1.35 gives also the average atomic weight (w_{at}) and the density of lattice sites (C_L).

TABLE 1.35
Elastic Constants, Density, and Atomic Weight of Selected Materials (300°K)

Material	c_{11}	c_{12}	c_{44}	d_s	w_{at}	C_L
	[10^{11} dyne/cm²]			[g/cm³]		[10^{22} cm⁻³]
C	107.60	12.50	57.60	3.52	12.01	17.60
Si	16.74	6.52	7.96	2.33	28.09	5.00
Ge	12.90	4.80	6.70	5.36	72.59	4.45
AlSb	8.94	4.43	4.16	4.26	74.37	3.46
GaAs	11.81	5.32	5.94	5.32	72.32	4.44
GaSb	8.85	4.04	4.33	5.60	90.74	3.72
InAs	8.33	4.53	3.96	5.66	94.87	3.60
InSb	6.70	3.65	3.02	5.77	118.29	2.94
NaCl	4.87	1.24	1.26	2.16	29.72	3.95
Al	10.68	6.07	2.82	2.73	26.98	6.10
K	0.37	0.31	0.19	0.86	39.10	1.33
BaF₂	8.91	4.00	2.54	4.89	58.45	5.04

TABLE 1.36
Elastic Properties of Selected Materials (300°K) (Calculated from elastic constants of Table 1.35)

Material	κ_v^* [10^{-11} cm²/dyne]	Y_B [10^{11} dyne/cm²]	Y_t [10^{11} dyne/cm²]	$v_l(\langle 100 \rangle)$ [10^5 cm/sec]	$v_t(\langle 100 \rangle)$ [10^5 cm/sec]	$-P$
C	0.02	44.20	105.00	17.50	12.80	0.20
Si	0.10	9.93	13.10	8.47	5.85	0.33
Ge	0.13	7.50	10.30	4.90	3.54	0.30
AlSb	0.17	5.94	6.01	4.59	3.13	0.40
GaAs	0.13	7.49	8.50	4.72	3.20	0.37
GaSb	0.18	5.65	6.33	3.98	2.78	0.37
InAs	0.17	5.79	5.13	3.84	2.65	0.48
InSb	0.21	4.66	4.12	3.41	2.28	0.43
NaCl	0.42	2.38	4.24	4.76	2.41	0.23
Al	0.13	7.61	6.29	6.26	3.25	0.44
K	3.03	0.33	8.75	2.08	1.48	0.59
BaF₂	0.18	5.64	6.44	4.27	2.28	0.37

5. ACOUSTOELECTRIC EFFECT

The acoustoelectric effect is the generation of an electric current in a crystal by a traveling longitudinal acoustic wave. A traveling longitudinal acoustic wave interacts with the free electrons in a semiconductor and transfers to the electrons momentum in the direction of propagation of the wave. The electrons transfer a portion of this momentum to the thermal lattice vibrations but are left with a net momentum in the direction of propagation. If the material is electrically insulated, charges will accumulate at its boundaries and create an electric field which counterbalances the effect of the acoustic wave. The magnitude of the effect can be measured by the potential difference between the extremities of the material. In semiconductors the acoustoelectric effect is about 10^3 to 10^4 times as large as in metals. The effect should not be observed in intrinsic semiconductors unless an electric field is applied; however, if a field is applied, an acoustoelectric current is to be expected even in an intrinsic semiconductor. For a transverse acoustic wave traveling in the < 100 > direction and polarized in the < 010 > direction, the acoustoelectric field is

$$E_{ae} = (6\pi^2/q)(q_{ac}^2 \tau_{iv} S_{ac})/(\lambda_{ac}^2 kT)$$

where q_{ac} = acoustic charge of the carriers (the acoustic charge is proportional to the deformation potential; the potential energy of a particle is proportional to the dilation of the crystal)

τ_{iv} = intervalley scattering time

S_{ac} = acoustic power density

λ_{ac} = acoustic wavelength.

6. REFERENCES

(1) W. Shockley, *Electrons and Holes in Semiconductors*, D. Van Nostrand, New York, 1950.
(2) C. S. Smith, *Phys. Rev.*, **94**, 42 (1954).
(3) J. J. Wortman et al., *J. Appl. Phys.*, **35**, 2122 (1964).
(4) H. J. McSkimin and P. Andreatch, Jr., *J. Appl. Phys.*, **35**, 2161 (1964).
(5) O. Madelung, *Physics of III–V Compounds*, John Wiley and Sons, New York, 1964.
(6) W. R. Runyan, *Silicon Semiconductor Technology*, McGraw-Hill Book Co., New York, 1965.
(7) J. E. Lawrence, *J, El. Chem. Soc.*, **113**, 819 (1966).
(8) L. K. Russell and W. H. Legat, *J. El. Chem. Soc.*, **114**, 277 (1967).
(9) N. D. Thai, *Solid-State Electronics*, **13**, 165 (1970).
(10) S. Dash and M. L. Joshi, *IBM J. Res. Develop.*, **14**, 453 (1970).

1-10 Etching

1. ETCHING MECHANISMS

Chemical etching of a semiconductor surface is the result of the dissolution of the semiconductor in an oxidizing electrolyte solution. The etching process is determined by the nature of the semiconductor, the presence or absence of lattice defects and the crystal orientation, type and concentration of impurity atoms, and the flow of the etchant solution over the semiconductor surface.

In an elemental semiconductor, the dissolution process involves the oxidation of the semiconductor from its zero oxidation state to a higher state which is either soluble in the electrolyte or forms insoluble products away from the surface. In a compound semiconductor, the same conditions apply but several oxidation reactions may occur simultaneously.

Chemical etching of a semiconductor usually proceeds in two steps: an oxidation-reduction step (usually facilitated by HNO_3 for Si or by H_2O_2 for Ge); and the dissolution of oxidation products in the form of complex soluble ions (usually facilitated by HF for Si and Ge). The oxidation step is an anodic reaction while the reduction step is a cathodic reaction; the actual etching is a result of corrosion currents of significant magnitude due to the anodic and cathodic reactions. Uniform etching occurs if everywhere both reactions take place over approximately equal amounts of time; selective etching occurs if anodic and cathodic reactions take place over widely differing times. During an intermediate step in the etching of silicon H_2SiF_6 is formed which acts as a complexing anion to remove the intermediate oxide. All of these processes usually occur within a single etchant.

The degree of selectivity and the etch rate itself are a function of crystallographic orientation, surface and bulk defects, etchant composition, temperature, and the hydrodynamics of the semiconductor-etchant interface. The overall reaction taking place during the etching of silicon is

$$3Si + 4HNO_3 + 18HF \longrightarrow 3H_2SiF_6 + 8H_2O + 4NO + 3n + 3p$$

where n and p are the number of electrons and holes generated by the etching process which move to conduction or valence band. The etching process may be reaction-rate-limited (etch rate is a function of the reaction rate) or diffusion-limited (etch rate is a function of how fast the etchant can diffuse through a surface layer). In both cases an etchant movement (rotation) will enhance the etch rate; this will also prevent the generation of localized temperature increase. Usually HNO_3-HF etching of silicon is limited by the reaction rate.

The reactions taking place during the etching of germanium are

$$Ge + O_2 \longrightarrow GeO_2$$
$$GeO_2 + 6HF \longrightarrow GeF_6 + 2H_2O + 2H$$
$$Ge + 3H_2O + 2p \longrightarrow H_2GeO_3 + 4H^+ + 2n$$

2. ETCH PROCESSES

Etching of semiconductors can be divided into two types: chemical etching and vapor etching.

(a) Chemical etching:
Most chemical etches contain HF as the complexing agent and HNO_3 as the oxidizing agent. Some etches act preferentially, others nonpreferentially; however, some etches can act both ways depending upon relative HF/HNO_3 proportions, temperature, and other conditions. Nonpreferential (isotropic) etches are used to remove surface contaminations, to remove work-damaged material (polishing), and to control thickness. Preferential (anisotropic) etches are used to determine crystal orientation, to expose crystal imperfections, and to facilitate dielectric component isolation.

(b) Vapor etching:
This is a high-temperature process and usually involves the oxidation or another conversion of a surface layer into a silicon compound and its subsequent removal. It avoids contamination of the surface by the reagents employed in chemical etches. It is used to expose a fresh silicon surface immediately prior to metal evaporation or deposition of an epitaxial layer. Most commonly used are O_2, H_2O, and HCl vapor etching processes.

3. ETCII RATE

The etching of a semiconductor surface is composed of two processes:

(a) transport of etchant and reaction products to and from the surface;
(b) surface reaction itself.

If the first of these limits the etch process, then the etch reaction is diffusion-limited; if the second limits the etch process, then the etch reaction is reaction-rate-limited.

132 GENERAL PROPERTIES OF SEMICONDUCTORS

The following factors affect the etch rate:

(a) Temperature:
Since the surface reaction rate depends exponentially upon temperature, rapid etchant movement (e.g., rotation) prevents localized temperature increases and results in a more uniform etching process.

(b) Surface damage:
Surface damage may lead to a larger effective surface area and hence to an increased etch rate. Mechanical etching produces more surface damage than chemical etching and, as a consequence, results in a higher etch rate.

(c) Surface films:
Films resulting from insoluble reaction products reduce the etch rate, e.g., a high HNO_3 concentration used to etch silicon may produce oxidation products which are generated at a faster rate than they are dissolved in HF.

(d) Crystal defects:
Defects, e.g., dislocations, are usually associated with the segregation of impurities and can lead to an increased etch rate in their vicinity.

(e) Impurities in the etchant:
Due to their influence on the absorption processes of the semiconductor surface, contamination of the etchant may increase or reduce the etch rate, depending upon the nature of the impurities.

Anisotropic etching is the preferential etching along a crystallographic plane. Etchants containing KOH or NaOH are frequently used to etch preferentially $\langle 100 \rangle$ silicon (see inset of Figure 1.37). The anisotropic etch rate is considerably lower than the isotropic etch rate.

TABLE 1.37
Etchants Used for Nonpreferential Isotropic Etching of Selected Semiconductors

Semi-conductor	Etchant composition	Temperature (T) [°C]	Etch rate (r_e) [μm/min]
Si	8HF:1 HNO_3:1 H_2O (by volume)	25	20.0
Ge	1HF:1 H_2O_2:4H_2O (by volume)	25	0.5
GaAs	99 MeOH:1 Br_2 (by weight);	60	5.0
	5NaOCl:95 H_2O (by weight);	25	1.0
	5H_2SO_4:1H_2O_2:1H_2O (by volume)	50	2.0
GaP	3HCl:1HNO_3 (by volume)	0	5.0

A condensed summary of etchant characteristics is given in Table 1.37. Some mechanical and chemical surface polishing techniques are summarized for silicon in Table 1.38. The etch rate of silicon is given in Figures 1.36 and 1.37 for specific etchant compositions.

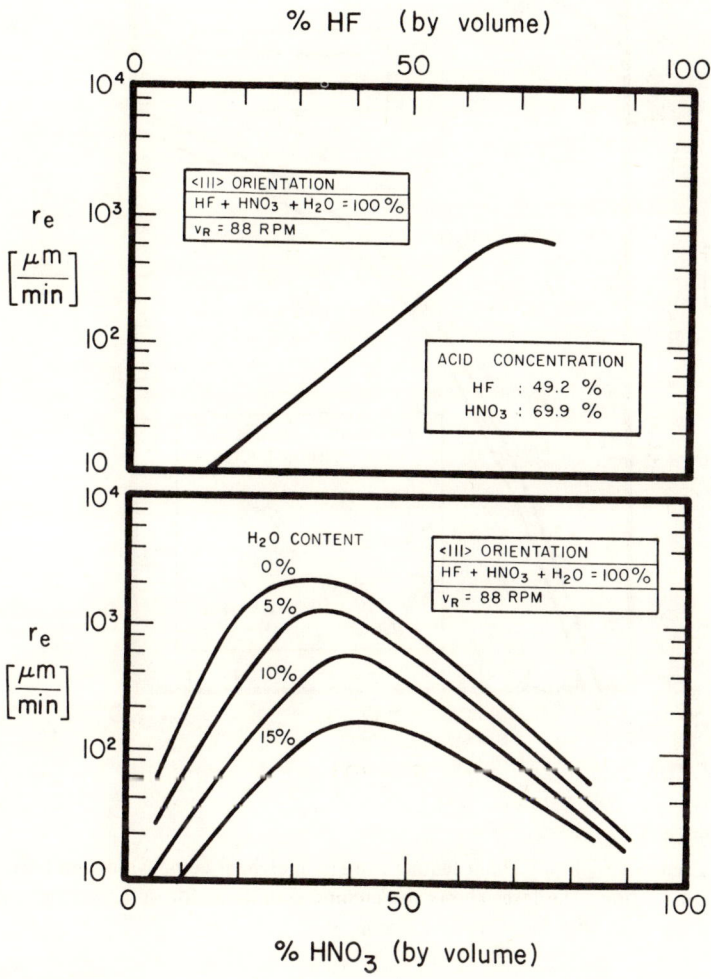

FIGURE 1.36

Etch rate (r_e) vs. etchant composition. Silicon, 300°K.

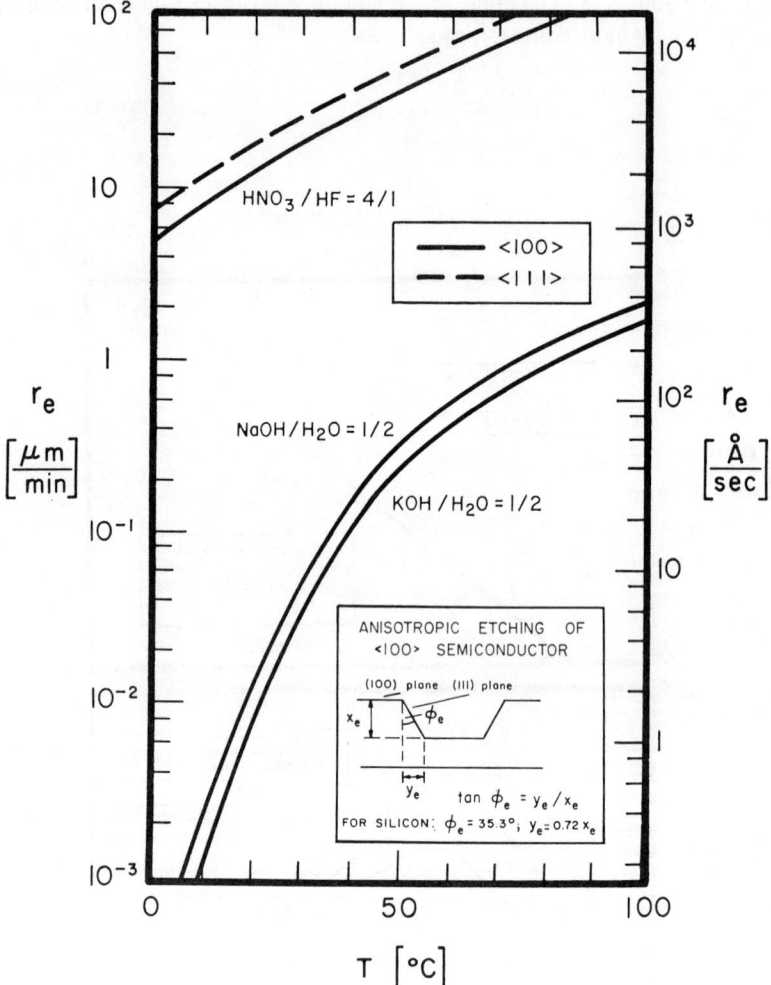

FIGURE 1.37

Etch rate (r_e) vs. etchant temperature (T), etchant composition, and crystal orientation. The inset shows the etching conditions for an anisotropic process, especially for $<100>$ silicon.

TABLE 1.38

Silicon Surface Polishing Techniques

Polishing technique	Removal rate [μm/min]	Surface damage	Pressure requirement	Remarks
Mechanical				
1 μm diamond grid	< 2.5	Heavy	High	
0.1 μm diamond grid	< 0.3	Heavy	High	
0.1 μm Al$_2$O$_3$	≈ 0.3	Heavy	High	
Chemical				
HNO$_3$-CH$_3$COOH-HF	> 7.5	Light	None	Intrinsically clean; details are sacrificed
Chemical-mechanical				
Rare-earth oxide	< 2.5	Heavy	High	Optical type polishing
ZrO$_2$	≈ 3.3	Light	High	
SiO$_2$	< 0.8	Light	High	Difficult to clean
Cupric-ion	< 8.5	Light	Light	Intrinsically clean, but residual Cu difficult to remove

4. REFERENCES

(1) D. L. Klein and D. J. D'Stefan, *J. El. Chem. Soc.*, **109**, 37 (1962).
(2) P. J. Holmes, Editor, *The Electrochemistry of Semiconductors*, Academic Press, New York, 1962.
(3) B. N. Padnos, *RTI Report ASD-TDR-63-316*, Vol. X, Nov. 1965.
(4) W. C. Rosvold et al., *IEEE Transact. Electr. Dev.*, **ED-15**, 640 (1968).
(5) E. Mendel and K.-H. Yang, *Proc. IEEE*, **57**, 1476 (1969).

1-11 Examples

PROBLEMS AND SOLUTIONS

1. **Problem:** Given a silicon sample doped with phosphorus to a uniform concentration $N_D = 10^{17}$ cm^{-3}. Determine the position of the Fermi level with respect to conduction band edge and donor level at 25°C.

   Solution: The activation energy of the impurity is
   $$E_{aD} = E_c - E_D = 0.05 \text{ eV}.$$
   Hence
   $$(E_c - E_D)/kT = 1.9,$$
   $$N_D T^{-3/2} = 10^{17} \cdot 300^{-3/2}$$
   $$= 1.9 \cdot 10^{13} \text{ cm}^{-3} \text{ °K}^{-3/2}$$
   $$\eta_n = (E_F - E_c)/kT = -5.6$$
   $$E_F - E_c = 0.15 \text{ eV}.$$
   This means that the Fermi level is 0.15 eV below the conduction band and 0.10 eV below the donor level.

2. **Problem:** Determine the contribution of gold and phosphorus impurity atoms (assumed to be donors) of density $N_D = 10^{17}$ cm^{-3} in otherwise intrinsic silicon to the electron density at 25°C.

   Solution: (a) Gold:
   Since in this case
   $$N_D \gg (1/4) N_c \exp\left[-(E_c - E_D)/kT\right],$$
   the electron concentration will be determined by
   $$n = (N_c N_D)^{1/2} \exp\left[-(E_c - E_D)/2kT\right].$$
   With
   $$E_c - E_D = 0.76 \text{ eV}, \, kT = 0.025 \text{ eV}, \, N_c = 2.8 \cdot 10^{19} \text{ cm}^{-3};$$
   the electron density
   $$n = 4.5 \cdot 10^{11} \text{ cm}^{-3};$$
   i.e., only about five out of one million gold atoms will contribute an electron to the number of free carriers.

(b) Phosphorus:
In this case

$$N_D \ll (1/4)N_c \exp[-(E_c - E_D)/kT] = 9 \cdot 10^{17} \text{ cm}^{-3};$$

therefore, all phosphorus atoms are ionized and the number of free electrons is equal to the number of impurity atoms, i.e.,

$$n = 10^{17} \text{ cm}^{-3}.$$

3. **Problem:** Compare the optical attenuation in an intrinsic silicon sample of 200 μm thickness at $\lambda = 0.6$ μm and $\lambda = 3.5$ μm.

   Solution: (a) $\lambda = 0.6$ μm

   $$\alpha = 4 \cdot 10^3 \text{ cm}^{-1}, \quad x_l = 2.5 \text{ μm};$$
   $$A(x)/A_o \approx 0,$$

   i.e., the sample is opaque.
   (b) $\lambda = 3.5$ μm

   $$\alpha = 3.1 \cdot 10^{-2} \text{ cm}^{-1};$$
   $$A(x)/A_o \approx 0.99997,$$

   i.e., the attenuation is less than 0.003 % and the sample is transparent.

4. **Problem:** Determine the stress introduced in a silicon crystal due to the diffusion of boron if the surface impurity concentration equals the maximum solid solubility.

   Solution: For boron

   $$r_i = 0.98 \text{ Å}, \quad C_S = C_{B\,\max} = 5 \cdot 10^{20} \text{ cm}^{-3}.$$

   Hence, the maximum internal stress generated by boron atoms in silicon

   $$\sigma_l^* \approx 10^{10} \text{ dyne/cm}^2.$$

5. **Problem:** Determine the velocity of a dislocation loop of length $l_{\text{dis}} = 0.1$ μm in silicon at 25°C under applied stresses of $\sigma_l^* = 10^7, 10^8,$ and 10^9 dyne/cm².

   Solution: Since

   $$\sigma_l^* a^2 l_{\text{dis}} = 2.9 \cdot 10^{-10} \text{ dyne cm},$$
   $$kT = 4.14 \cdot 10^{-14} \text{ dyne cm},$$
   $$\sigma_l^* a^2 l_{\text{dis}} \gg kT;$$

   generally

   $$v_{\text{dis}} = 2.7 \cdot 10^5 \exp(7.1 \cdot 10^{-7} \sigma_l^* - 85).$$

Specifically for

(a) $$\sigma_l^* = 10^7 \text{ dyne/cm}^2:$$
$$v_{\text{dis}(7)} = 1.9 \cdot 10^{-8} \text{ Å/day}.$$

(b) $$\sigma_l^* = 10^8 \text{ dyne/cm}^2:$$
$$v_{\text{dis}(8)} = 680 \text{ cm/sec}.$$

(c) $$\sigma_l^* = 10^9 \text{ dyne/cm}^2:$$
$$v_{\text{dis}(9)} \gg v_{\text{dis}(8)}.$$

This means that the dislocation velocity increases very rapidly with stress level until it reaches the phonon velocity in the crystal which is reached slightly above $\sigma_l^* = 10^8$ dyne/cm^2.

PROBLEMS FOR WHICH A SOLUTION IS NOT GIVEN

1. Compare the energy band gaps, the wavelengths of the absorption edge, and the optical penetration depth at the absorption edge of Si, Ge, and GaAs at 25°C.

2. Given two n-type silicon samples of resistivity $\rho_o = 1.0 \, \Omega$ cm, one of $\langle 100 \rangle$ and one of $\langle 111 \rangle$ orientation. Determine the relative resistivity change under the same stress σ_l^* in both samples at 25°C.

3. Determine the increase in minority carrier density if an n-type silicon sample is compressed in the $\langle 100 \rangle$ direction by a stress $\sigma_l^* = 10^{10}$ dyne/cm^2. Discuss the increase if the stress is raised to $2 \cdot 10^{10}$ dyne/cm^2.

2
IMPURITIES IN SEMICONDUCTORS AND EPITAXIAL GROWTH

2–1 Diffusion
2–2 Impurity Concentration Profile
2–3 Properties of Impurities Used in Silicon Technology
2–4 Gold in Silicon
2–5 Ion Implantation
2–6 Energy Levels of Impurities
2–7 Growth of Epitaxial Film
2–8 Phase Diagrams
2–9 Examples

The addition of impurities to an intrinsic semiconductor affects most of its electrical properties, e.g., carrier generation and mobility, resistivity, conduction type, and also most of its other properties. The introduction of impurities can be achieved by any of four basic methods:

(a) addition of elements in the melt during crystal growth,
(b) addition of elements during epitaxial deposition of a semiconductor film,
(c) solid-state diffusion of elements from the surface or a suitable interface,
(d) ion implantation of elements into the semiconductor at relatively low temperature.

The first two methods result in uniform doping of the semiconductor, whereas the last two methods result in an impurity concentration gradient with the highest impurity concentration at the source (i.e., at the surface) in the case of diffusion, and within the semiconductor (i.e., below the surface) in the case of ion implantation. In the first three methods impurity introduction is achieved at a temperature close to or above the melting point of the semiconductor. In the last method it may be achieved at any temperature at which the semiconductor is in its solid state, i.e., even at or below room temperature; in this case, the damage to the crystal as a result of the implantation is, among others, dependent upon the implantation temperature. Selective diffusion of impurities into a semiconductor is one of the most important methods of affecting the carrier density in a semiconductor.

Diffusion of impurity atoms and charge carriers and transfer of heat proceed by analogous processes. Impurity and thermal diffusion can be considered as a return to the state of equilibrium; they take place wherever there exists a concentration gradient or a thermal gradient. The transfer of heat is due to random molecular motions, whereas the transfer of impurity atoms is due to random motions of the crystal lattice. Because of the analogy between heat transfer and matter transport the equations of heat conduction derived by Fourier can be adopted to diffusion on a quantitative basis, as first suggested by Fick. The rate of impurity transfer per unit area (flux) depends upon the impurity gradient ($\partial C/\partial x$). The coefficient of proportionality is the diffusion coefficient (D); in heat conduction problems it corresponds to thermal conductivity.

The impurity flux depends also upon a drift term which is proportional to the impurity concentration (C), the built-in electric field (E), and the impurity mobility (μ); diffusion coefficient and mobility of the impurity ions are related by $D = kT\,\mu/q$. Therefore the total impurity flux in the one-dimensional case

$$F(x) = F_{\text{drift}} + F_{\text{diffusion}} = D\left(\frac{q}{kT}EC(x,t) - \frac{\partial C(x,t)}{\partial x}\right) \qquad (2.1)$$

and the one-dimensional transport equation

$$\frac{\partial C(x,t)}{\partial t} = -\frac{\partial F}{\partial x} = D\left(\frac{\partial^2 C(x,t)}{\partial x^2} - E\frac{q}{kT}\frac{\partial C(x,t)}{\partial x}\right). \quad (2.2)$$

For negligible built-in electric field (i.e., $E = 0$), these equations reduce to

$$F(x) = -D\frac{\partial C}{\partial x} \quad \text{and} \quad \frac{\partial C}{\partial t} = D\frac{\partial^2 C}{\partial x^2}. \quad (2.3)$$

Two boundary conditions are important. Both of them are usually employed successively in the formation of diffused layers. In the first case, the impurity concentration at the semiconductor surface is ideally kept constant ("infinite source") while the initial impurity concentration is negligible elsewhere within the semiconductor. This process is referred to as predeposition or constant-source diffusion. In the second case, the total impurity content of the semiconductor is ideally kept constant, i.e., the thin predeposited layer is approximated by a plane of infinitely small thickness which is diffused into the wafer. This process, which is a redistribution of impurities within a surface layer, is referred to as drive-in diffusion or diffusion from an instantaneous source.

The predeposition process is best described by a complementary error function impurity distribution. With high impurity concentrations or short diffusions deviations occur because the diffusion coefficient is a function of the concentration of the diffusing impurity and of the already present impurities, and because it takes a finite time for the surface concentration to reach its equilibrium value and for the semiconductor to reach the final diffusion temperature.

The drive-in diffusion process is best described by a Gaussian distribution. Certain impurities (e.g., gold in silicon) show deviations from a Gaussian distribution. Frequently the final impurity distribution can best be represented by a weighted combination of erfc and Gaussian distributions.

The drive-in diffusion is normally accompanied by a thermal oxidation of the semiconductor surface. During this oxidation the semiconductor-oxide interface moves into the semiconductor while the oxide-ambient interface moves into the ambient. This leads to a redistribution of impurities at the resulting semiconductor-oxide interface and is associated with a volume change of the surface layer resulting in stresses at the semiconductor-oxide interface.

Selective diffusion in certain areas of the semiconductor surface is accomplished by masking of the remaining areas against these impurities. Masks usually employed in semiconductor technology are either SiO_2 or deposited layers of Si_3N_4, Al_2O_3, or others, or in low-temperature processes certain metals. The diffusion profile in the proximity of a mask edge is disturbed by fringing effects which are important if the mask opening dimensions are comparable to the fringing dimensions.

Diffusion also takes place along the mask-semiconductor interface (lateral diffusion). Usually its depth is less than the corresponding diffusion depth normal to the semiconductor surface. The ratio of diffusion depths along and normal to the semiconductor-mask interface is also a function of the stresses at the semiconductor surface and of the width of the mask opening.

During epitaxial growth a crystalline film is deposited on a substrate of similar crystal structure. Conductivity type and impurity concentration of film and substrate away from the interface are not related and, within limits, may be arbitrary.

The growth of epitaxial silicon films involves a gaseous atmosphere. In a direct process (e.g., evaporation, sputtering, sublimation) silicon is transferred from source to substrate directly without an intermediate chemical reaction; epitaxial growth occurs because of good crystal lattice matching and of the influence of the substrate on the nucleus. In an indirect process (e.g., hydrogen reduction of $SiCl_4$, $SiBr_4$, $SiHCl_3$, all at elevated temperature) silicon atoms are deposited by the decomposition of a vapor of a silicon compound at the substrate surface. A reversal of this process is HCl etching of a silicon surface under conditions similar to those of epitaxial growth, the difference being a variation in $SiCl_4$ mole fraction.

The most frequently used methods of epitaxial deposition of silicon involve either $SiCl_4$ (silicon tetrachloride), $SiHCl_3$ (trichlorosilane) or SiH_4 (silane). In the first two methods the deposition follows the reactions

$$SiCl_4 + 2H_2 \longrightarrow Si + 4HCl$$
$$SiHCl_3 + H_2 \longrightarrow Si + 3HCl.$$

The third method utilizes the pyrolytic decomposition of silane

$$SiH_4 \longrightarrow Si + 2H_2.$$

Deposition rates are in the order of 1 $\mu m/min$.

The deposition of a doped silicon film is accomplished by the addition of B_2H_6 (diborane), PH_3 (phosphine), or AsH_3 (arsine) to the gaseous atmosphere.

The epitaxial deposition results in a redistribution of impurities at the substrate-film interface which depends upon the impurity concentrations of substrate and film and takes place over a distance of less than $2\sqrt{Dt}$ from the original substrate surface. The net impurity concentration at the metallurgical interface between substrate and film remains essentially constant during this redistribution and is approximately half of the concentration on the higher-doped side.

Ion implantation is closely linked to target and projectile characteristics. Although ion implantation will result in a graded impurity profile due to the random distribution of ions as a result of two basic collision mechanisms, the highest impurity concentration will usually not be at the semiconductor surface but within the semiconductor. Impurity gradients and achievable impurity concentrations are similar to those of diffused layers; achievable junction depths are frequently substantially less. Potential advantages of ion implantation over other doping methods are: the ability to introduce into a variety of substrates precise amounts of nearly any impurity desired; the ability to control doping profiles in three dimensions by modulating energy, current, and position of the ion beam; and the possibility of avoiding problems associated with high-temperature diffusion processes.

GENERAL REFERENCES

(1) W. Jost, *Diffusion in Solids, Liquids and Gases*, Academic Press, New York, 1963.
(2) B. I. Boltaks, *Diffusion on Semiconductors*, Academic Press, New York, 1963.
(3) J. Crank, *The Mathematics of Diffusion*, Clarendon Press, Oxford, 1964.
(4) L. Eriksson et al., *J. Appl. Phys.*, **40**, 842 (1969).

2–1 Diffusion

1. SOLID-STATE DIFFUSION

Contrary to the diffusion of foreign atoms in a gas or in a liquid, solid-state diffusion in a crystal takes place in jumps whose frequency is a function of the frequency of lattice vibrations and of the number of available lattice positions.

In the absence of a concentration or temperature gradient or an external electric field, solid-state diffusion occurs randomly in three dimensions and no net transport of impurities is involved.

A net transport of impurities will occur, however, in the presence of an impurity gradient so that the impurity flux is proportional to the gradient. The proportionality factor, i.e., the diffusion coefficient, is, in turn, a function of the impurity density as a result of the electric field arising from the fact that the rapidly diffusing electrons and holes, compared to the impurity atoms, come to equilibrium instantaneously. If, therefore, the density of diffusing impurity atoms approaches or exceeds the density of intrinsic carriers at the diffusion temperature, the impurity diffusion coefficient will be enhanced.

For only one diffusing impurity in the one-dimensional case impurity flux and diffusion coefficient are (assuming small electric field, i.e., $E \ll kT/qa$)

$$F(x) = -D_s(\partial C/\partial x) + \mu E C(x) \qquad (2.4)$$

$$D_s = D_i[1 + qEC(x)/(kT\, \partial C/\partial x)]$$
$$= D_i[1 + C(x)/\sqrt{C(x)^2 + 4n_i^2}]. \qquad (2.5)$$

These equations apply to both donor- and acceptor-type impurities since the fields are always in such a direction as to retard the free carriers; i.e., these fields always aid the oppositely charged impurities. The quantity a is the lattice constant of the semiconductor host.

Three diffusion mechanisms are of importance.

(a) Interstitial diffusion:

In this case impurity atoms move through the semiconductor by jumping from one interstitial site to the next; the presence of vacancies is not required. Interstitially located impurity atoms jump from one void to the next by squeezing through the constrictions between the voids. The interaction (activation) energy between the impurity atom and the center between the two interstitial sites is E_{aI}^*. For crystals of the diamond and zincblende structure where the tetrahedral radius is

$$r_a = (\sqrt{3}/8)a, \qquad (2.6)$$

the diameter of the interstitial voids is

$$d_{\text{void}} = (\sqrt{3}/8)a \qquad (2.7)$$

and the diameter of the constrictions between the voids is

$$d_{\text{constr}} = (3/16)a. \qquad (2.8)$$

This assumes the validity of the hard-sphere model of the lattice atoms; in this model $\pi\sqrt{3}/16 = 34\%$ of the intertwined face-centered cubic lattice (which consists of two interpenetrating face-centered cubic lattices, with one atom of the second sublattice located at one-fourth of the distance along a major diagonal of the first sublattice) is occupied by atoms, compared to 74% occupation in the simple face-centered cubic lattice.

These diameters are given for several semiconductors in Table 2.1. Radii of some impurity atoms of interest are given for comparison in Table 2.12. The frequency of jumping, i.e., the frequency with which thermal energy fluctuations occur with sufficiently large magnitude to overcome the

TABLE 2.1

Diameter of Voids in the Lattice and of the Constrictions between Voids

Semi-conductor	a [Å]	r_a [Å]	d_{void} [Å]	d_{constr} [Å]
Si	5.43	1.18	1.18	1.02
Ge	5.66	1.23	1.23	1.06
AlAs	5.66	1.23	1.23	1.06
AlP	5.46	1.19	1.19	1.02
AlSb	6.14	1.33	1.33	1.15
GaAs	5.65	1.22	1.22	1.06
GaP	5.45	1.18	1.18	1.02
GaSb	6.10	1.32	1.32	1.14
InAs	6.06	1.31	1.31	1.13
InP	5.07	1.10	1.10	0.95
InSb	6.48	1.41	1.41	1.21

potential barrier, is for crystals of the diamond and zincblende structure

$$v_I = 4v_o \exp(-E^*_{aI}/kT) \tag{2.9}$$

where v_o is the frequency of lattice vibrations which is related to the Debye temperature by

$$v_o = T_D k/h. \tag{2.10}$$

The activation energy is in the order of

$$E^*_{aI} \approx 1 \text{ eV}.$$

The activation energy is the interaction energy for the jumping process and is to be distinguished from the ionization energy of an impurity atom involving a donor or acceptor level. Most interstitial impurities are characterized by deep energy levels.

The effective size of the interstitial species is indicative of the energy of closed shell atoms or ions and thus their solubility in such sites. The repulsive energy between host atom and interstitial ion (i.e., the activation energy for interstitial migration) also affects the diffusion coefficient of an interstitial solute.

The activation energy for the motion of a singly ionized interstitial atom is

$$E^*_{aI} = |E_{P(H)} - E_{P(T)} - (E_{R(H)} - E_{R(T)})| \tag{2.11}$$

where E_P and E_R are the polarization and repulsive energies and the subscripts H and T denote the interstitial ion located at the hexagonal and tetrahedral interstitial sites.

In silicon for singly ionized interstitial atoms

$$E_{P(H)} - E_{P(T)} = 0.75 \text{ eV}.$$

$E_{R(H)} - E_{R(T)} = 0.47$ eV for Au in Si, 0.11 eV for Cu in Si, 0.023 eV for Li in Si; for Au in Si: $E^*_{aI} = 0.28$ eV.

(b) Substitutional diffusion:

In this case impurity atoms move through the semiconductor by jumping from one lattice site to the next; this requires the presence of adjacent vacancies, e.g., by thermal fluctuations. Since the number of vacancies in a crystal is relatively small, substitutional diffusion takes place at a slower rate than interstitial diffusion; i.e., the diffusion coefficient of substitutional impurities is usually smaller than that of interstitial impurities. Similar to interstitial jumps, the frequency of jumping of an impurity atom from one substitutional site to the next is for crystals of the diamond and zincblende structure

$$v_S = 4v_o \exp(-E^*_{aS}/kT) \quad (2.12)$$

where the activation energy

$$E^*_{aS} = E_p + E_s \quad (2.13)$$

i.e., it is the sum of the height of the potential barrier (E_p) and the energy of forming a Schottky defect (E_s). Experimentally observed values are

$$E^*_{aS} \approx 3 \text{ to } 4 \text{ eV}.$$

Most substitutional impurities are characterized by shallow energy levels. In Group IV semiconductors the predominant diffusion mechanism of Group III and Group V elements is by substitutional migration.

Jumping of the impurity ions in the case of substitutional diffusion takes place between tetrahedral sites of spacing d which is related to the lattice constant a and the Miller indices k_i by

$$d = a/(k_1^2 + k_2^2 + k_3^2)^{1/2}. \quad (2.14)$$

For example, for the $\langle 111 \rangle$ plane $k_1 = 1, k_2 = 1, k_3 = 0$. The $\langle 111 \rangle$ plane has the highest packing density, as seen in the case of silicon:

$$d_{\langle 100 \rangle} = a = 5.42 \text{ Å}$$
$$d_{\langle 110 \rangle} = a/\sqrt{2} = 3.84 \text{ Å}$$
$$d_{\langle 111 \rangle} = a/\sqrt{3} = 3.14 \text{ Å}.$$

(c) Interchange diffusion:
In this case two or more atoms (lattice or impurity atoms) exchange sites and contribute to the diffusion. This mechanism is one of very small probability and is characterized by a very small diffusion coefficient.

All three of these diffusion mechanisms may take place simultaneously; however, the first two are most important. Elements of Groups III, IV, and V diffuse in Group IV semiconductors through a vacancy and interchange mechanism, i.e., basically substitutionally. For impurities of other Groups, e.g., Au, Cu, and Li, where the ionic size accounts for the diffusion mechanism, the interstitial position is energetically more favorable than the substitutional position.

The physical meaning of the activation energy is as follows: For interstitial diffusion the activation energy corresponds to the energy required to move the impurity atom from one interstitial site to the next one. For substitutional diffusion it corresponds to the energy required to form a vacancy in the semiconductor rather than to the energy required to move the impurity. The substitutional activation energy is approximately three to four times the energy gap of the semiconductor since semiconductor bonds must be broken to form a vacancy.

The diffusion coefficient D_s is a function of the frequency (v^*) of jumping of impurity atoms from one site to the next and of the availability of interstitial or substitutional lattice positions. Since the number of available lattice positions is

$$N_v = C_L \exp(-E_a^*/kT), \quad (2.15)$$

where E_a^* is the energy required for a jump and C_L is the density of lattice constituents, the exponential term $\exp(-E_a^*/kT)$ gives the fractional density of available lattice positions. Due to the dependence of D_s on the apparent lattice spacing d, the diffusion coefficient is also a function of the crystallographic orientation. It is related to jumping frequency and apparent lattice spacing by

$$D_s = v^* d^2/6 \quad (2.16)$$

where v^* is synonymous with either v_I or v_S, depending upon the diffusion mechanism involved. The factor of 1/6 results from directional averaging. This equation simplifies the true diffusion conditions since it ignores the fact that environments may vary greatly from site to site. Nevertheless, for most practical purposes, the above equation is a useful approximation.

For interstitial diffusion

$$D_I = (2/3)v_o d^2 \exp(-E_{aI}^*/kT) = D_\infty \exp(-E_{aI}^*/kT) \quad (2.17)$$

and for substitutional diffusion

$$D_S = (2/3)v_o d^2 \exp(-E_{aS}^*/kT) = D_\infty \exp(-E_{aS}^*/kT). \quad (2.18)$$

The term

$$D_\infty = (2/3)v_o d^2 = (2/3)(k/h)T_D d^2 \quad (2.19)$$

is the asymptotic value of the diffusion coefficient at infinite temperature ($1/T = 0$) and is independent of the diffusion mechanism. The temperature dependence of D_I and D_S is due to the exponential terms.

A numerical calculation will illustrate this. If it is assumed that for impurity atoms in silicon

$$E_{aI}^* = 1.0 \text{ eV} \quad \text{or} \quad E_{aS}^* = 3.5 \text{ eV},$$

then

$$\text{at } 25°C: v_I = 1.1 \cdot 10^{-5} \text{ Hz},$$
$$v_S = 7.6 \cdot 10^{-48} \text{ Hz};$$
$$\text{at } 1000°C: v_I = 1.4 \cdot 10^{10} \text{ Hz},$$
$$v_S = 1.3 \cdot 10 \quad \text{Hz}.$$

This means that at 25°C an impurity atom in silicon will jump interstitially from one site to the next on the average about once a day or substitutionally about once every $4 \cdot 10^{39}$ years; at 1000°C it will jump interstitially from one site to the next on the average about once every $7.4 \cdot 10^{-11}$ sec or substitutionally once every $8.0 \cdot 10^{-2}$ sec.

Theoretically expected diffusion coefficients for silicon of $\langle 111 \rangle$ crystal orientation based on the above assumptions are as follows:

$$\text{at } 25°C \quad D_I = 1.9 \cdot 10^{-21} \text{ cm}^2/\text{sec},$$
$$D_S = 1.2 \cdot 10^{-63} \text{ cm}^2/\text{sec};$$
$$\text{at } 1000°C \quad D_I = 2.2 \cdot 10^{-6} \text{ cm}^2/\text{sec},$$
$$D_S = 2.1 \cdot 10^{-15} \text{ cm}^2/\text{sec}.$$

These diffusion coefficients agree well with experimentally observed values. The corresponding diffusion length (which gives an indication of the progression of the diffusion front) is for the two mechanisms for one year at 25°C:

$$L_I = (D_I t)^{1/2} = 2.45 \cdot 10^{-7} \text{ cm} = 24.5 \text{ Å}$$
$$L_S = (D_S t)^{1/2} = 1.67 \cdot 10^{-28} \text{ cm}.$$

Table 2.2 gives calculated values of the fractional density of available lattice positions as a function of activation energy and diffusion temperature for arbitrary semiconductor and interstitial or substitutional diffusion mechanism. For a given semiconductor the fractional density of available lattice positions is

$$f_L = N_v/C_L = D_s/D_\infty = v^*/v_o = \exp(-E_a^*/kT). \qquad (2.20)$$

For a given semiconductor all curves showing the diffusion coefficient D_s of all impurities (both interstitial and substitutional) vs. inverse temperature should form a family of straight lines which intersect at the same point (namely, $D_s = D_\infty$ and $1/T = 0$) and which should differ only in their slopes which are a function of the activation energy alone. The apparent diffusion coefficient (D_∞) is expected to be independent of the impurity species; in

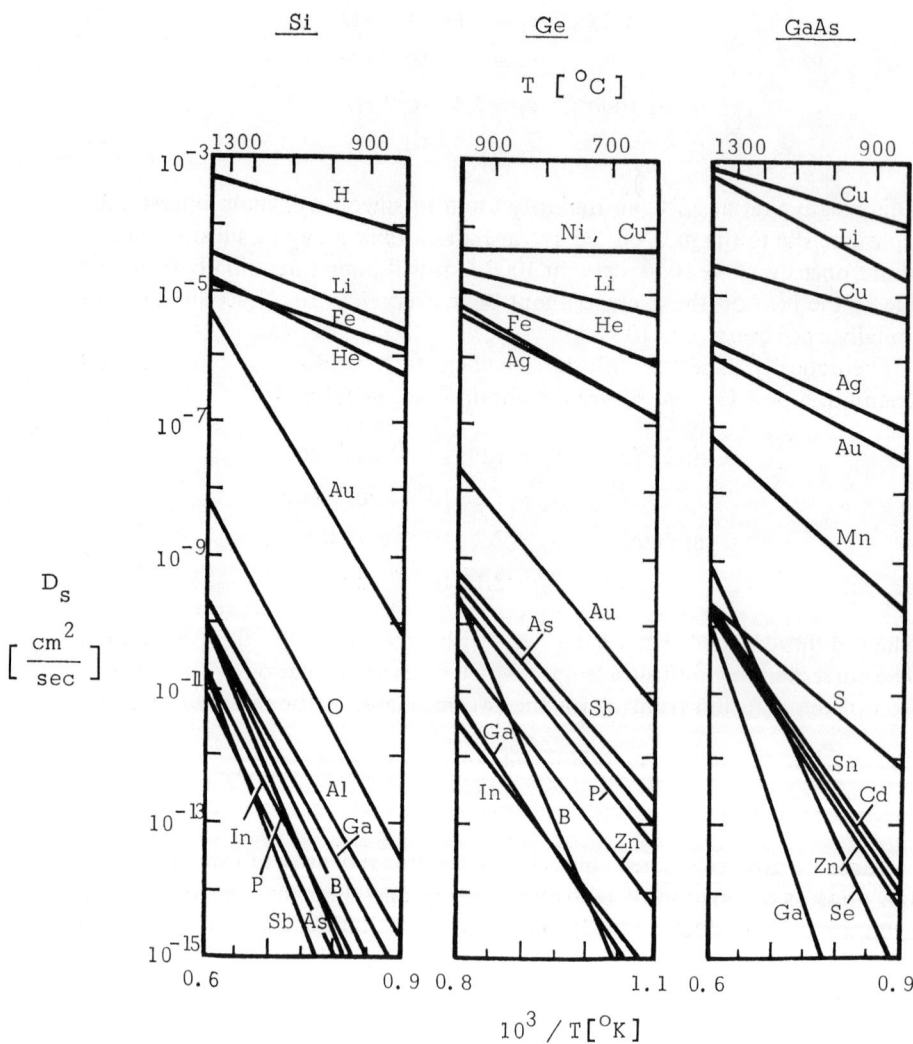

FIGURE 2.1
Diffusion coefficients (D_s) of impurities in near-intrinsic Si, Ge, and GaAs vs. temperature (T).

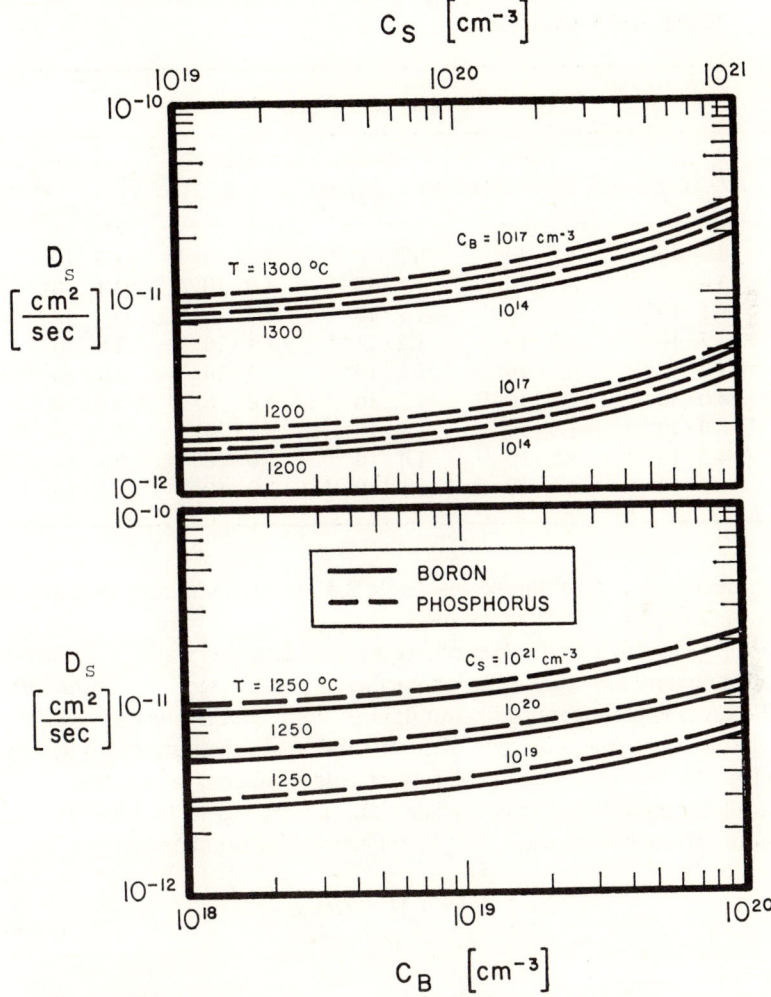

FIGURE 2.2

Enhancement of impurity diffusion coefficient (D_s) vs. background impurity concentration (C_B) and surface concentration of in-diffusing impurity (C_S) for boron and phosphorus in silicon.

TABLE 2.2

Fractional Density of Available Lattice Positions as a Function of Activation Energy and Temperature

E_a^* [eV]	Fractional density of lattice positions (f_L)				
	25°C $kT = 0.026$ eV	800°C $kT = 0.093$ eV	1000°C $kT = 0.110$ eV	1200°C $kT = 0.127$ eV	1400°C $kT = 0.144$ eV
0.3	$8.3 \cdot 10^{-6}$	$4.0 \cdot 10^{-2}$	$6.5 \cdot 10^{-2}$	$9.4 \cdot 10^{-2}$	$1.3 \cdot 10^{-1}$
0.5	$3.4 \cdot 10^{-9}$	$4.7 \cdot 10^{-3}$	$1.1 \cdot 10^{-2}$	$1.9 \cdot 10^{-2}$	$3.1 \cdot 10^{-2}$
1.0	$1.1 \cdot 10^{-17}$	$2.1 \cdot 10^{-5}$	$1.1 \cdot 10^{-4}$	$3.7 \cdot 10^{-4}$	$9.7 \cdot 10^{-4}$
1.5	$3.7 \cdot 10^{-26}$	$1.0 \cdot 10^{-7}$	$1.2 \cdot 10^{-6}$	$1.4 \cdot 10^{-5}$	$3.0 \cdot 10^{-5}$
2.0	$1.2 \cdot 10^{-34}$	$4.5 \cdot 10^{-10}$	$1.2 \cdot 10^{-8}$	$1.4 \cdot 10^{-7}$	$1.1 \cdot 10^{-6}$
2.5	$4.0 \cdot 10^{-43}$	$2.1 \cdot 10^{-12}$	$1.3 \cdot 10^{-10}$	$2.8 \cdot 10^{-9}$	$2.9 \cdot 10^{-8}$
3.0	$6.0 \cdot 10^{-51}$	$1.0 \cdot 10^{-14}$	$1.4 \cdot 10^{-12}$	$5.3 \cdot 10^{-11}$	$8.9 \cdot 10^{-10}$
3.5	$4.5 \cdot 10^{-60}$	$4.5 \cdot 10^{-17}$	$1.4 \cdot 10^{-14}$	$1.0 \cdot 10^{-12}$	$2.8 \cdot 10^{-11}$
4.0	$1.5 \cdot 10^{-68}$	$2.1 \cdot 10^{-19}$	$1.5 \cdot 10^{-16}$	$2.0 \cdot 10^{-14}$	$8.7 \cdot 10^{-13}$

practice, deviations are observed (as Table 2.4 indicates), mainly because of measurement errors.

Table 2.3 gives calculated values of frequency of lattice vibrations, apparent lattice spacing, and apparent diffusion coefficient for selected semiconductors. Table 2.4 gives experimental diffusion data of impurities in silicon.

For the simultaneous presence of interstitial and substitutional diffusion mechanisms (as, for example, in the case of gold in silicon), a weighted combination of the two diffusion coefficients D_I and D_S has to be used. If f_I is the fraction of interstitial sites, then the effective diffusion coefficient

$$D_{\text{eff}} = f_I D_I + (1 - f_I) D_S$$
$$= f_I (D_I - D_S) + D_S. \tag{2.21}$$

The fraction f_I is slightly concentration-dependent. For gold in silicon $f_I \approx 0.9$.

In a similar manner the diffusion coefficient is affected by the presence of dislocations. If D_{dis} is the diffusion coefficient along a dislocation, D_s is the diffusion coefficient in defect-free material, and f_d is the fractional dislocation density, then the effective diffusion coefficient in the semiconductor

$$D_{\text{eff}} = f_d D_{\text{dis}} + (1 - f_d) D_s$$
$$= f_d (D_{\text{dis}} - D_s) + D_s. \tag{2.22}$$

TABLE 2.3

Frequency of Lattice Vibrations, Apparent Lattice Spacing, and Apparent Diffusion Coefficient at Infinite Temperature

Semi-conductor	ν_o [10^{13} Hz]	d [Å] ⟨100⟩	d [Å] ⟨110⟩	d [Å] ⟨111⟩	D_∞ [10^{-3} cm²/sec] ⟨111⟩
Si	1.37	5.43	3.84	3.14	9.03
Ge	0.79	5.66	4.00	3.27	5.61
AlAs	0.87	5.66	4.00	3.27	6.21
AlP	1.23	5.46	3.86	3.15	8.15
AlSb	0.61	6.14	4.33	3.54	5.09
GaAs	0.72	5.65	3.99	3.26	5.11
GaP	0.91	5.45	3.85	3.14	5.97
GaSb	0.54	6.10	4.31	3.52	4.46
InAs	0.52	6.06	4.28	3.50	4.25
InP	0.67	5.87	4.15	3.39	5.14
InSb	0.42	6.48	4.58	3.74	3.91

TABLE 2.4

Diffusion Characteristics of Impurities in Silicon

Impurity	Dominant diffusion mechanism[a]	D_∞ [cm²/sec]	E_a^* [eV]
Ag	i	0.002	1.60
Al	s	4.80	3.36
As	s	68.60	4.23
Au	i	0.001	1.12
B	s	25.00	3.51
Fe	i	0.006	0.87
H	i	0.01	0.48
He	i	0.11	1.26
O	s	135.00	3.50
P	s	10.50	3.69
Sb	s	12.90	3.98

[a] i = interstitial, s = substitutional

2. IMPURITIES IN SEMICONDUCTORS

Impurities in semiconductors may act as acceptors (A) or donors (D) or their energy level within the energy gap of the semi-conductor may be so far from either band edge (deep-lying impurity) that they behave like electrically neutral (N) atoms, although they may be active carrier recombination centers. Impurities may migrate through the semiconductor either interstitially (i) or substitutionally (s). The electrical behavior is given for a number of impurities in selected semiconductors in Table 2.5. Tables 2.6 and 2.7 list values of the diffusion coefficient at infinite temperature (D_∞) and of the activation energy of the diffusion process (E_a^*), both applicable to the Arrhenius plot of the diffusion coefficient D_s at finite temperature T:

$$D_s = D_\infty \exp(-E_a^*/kT). \qquad (2.23)$$

Data given in Tables 2.5, 2.6, and 2.7 are based on measurements. Blank spaces indicate that either no data exist for the particular semiconductor or impurity or that they do not apply.

TABLE 2.5
Diffusion Mechanisms of Acceptors and Donors in Selected Semiconductors*

Semi-conductor	Impurity									
	Ag	Al	Au	Cd	Cu	In	S	Se	Te	Zn
Si	A(i) D(i)	A	N		A D	A	D			A(i)
GaAs				A	A		D	D	D	A(s) D(i)
GaP							D	D	A	A
GaSb				A	A(s) D(i)	A		D	D	A
InAs	D		D	A	D(i)		D	D	D	A
InSb	A	A	A	A	A		D	D	D	A
CdS					D	A(s)				
CdSe	A				A					
CdTe			A			D				
SiC		A								
ZnSe		A			A(i) D(s)					
ZnTe	A				A	D				

* Symbols are explained in the text.

TABLE 2.6
Apparent Diffusion Coefficient at Infinite Temperature (D_∞, in cm^2/sec) for Various Impurities

Semi-conductor	Impurity									
	Ag	Al	Au	Cd	Cu	In	S	Se	Te	Zn
Si	$2.0 \cdot 10^{-3}$	4.8	$1.0 \cdot 10^{-3}$		$4.0 \cdot 10^{-2}$	16.5	0.92			$6.1 \cdot 10^{-7}$
GaAs	$2.5 \cdot 10^{-3}$		$1.0 \cdot 10^{-3}$	$5.0 \cdot 10^{-2}$	$3.0 \cdot 10^{-2}$		$2.6 \cdot 10^{-5}$	$3.0 \cdot 10^{3}$		
GaP										1.0
GaSb					$1.2 \cdot 10^{-7}$				$3.8 \cdot 10^{-4}$	
InAs	$7.3 \cdot 10^{-4}$		$5.8 \cdot 10^{-3}$	$7.4 \cdot 10^{-4}$	$3.6 \cdot 10^{-3}$		6.8	13.0	$3.4 \cdot 10^{-5}$	
InSb			$7.0 \cdot 10^{-4}$	$1.0 \cdot 10^{-5}$	$3.0 \cdot 10^{-5}$	$5.0 \cdot 10^{-2}$			$1.7 \cdot 10^{-7}$	
CdS	25			3.4	$1.5 \cdot 10^{-3}$					
CdSe										$2.6 \cdot 10^{-3}$
CdTe			67			0.41				
SiC		0.20								
ZnSe					$1.7 \cdot 10^{-5}$					

156 IMPURITIES IN SEMICONDUCTORS

TABLE 2.7

Activation Energy (E_a^*, in eV) of Various Impurities in Selected Semiconductors

Semi-conductor	Impurity									
	Ag	Al	Au	Cd	Cu	In	S	Se	Te	Zn
Si	1.6	3.4	1.1		1.0	3.9	2.2			≪ kT
GaAs	1.5		1.0	2.6	0.5		1.9	4.2		
GaP										2.1
GaSb					0.5				1.2	
InAs	0.3		0.7	1.2	0.5		2.2	2.2	1.3	1.0
InSb				0.3	1.1	0.4	1.8		0.6	
CdS	1.2			2.0	0.8					
CdSe										1.6
CdTe			2.0			1.6				
SiC		4.9								
ZnSe					0.6					
ZnTe										

3. MAXIMUM SOLID SOLUBILITY

The maximum solid solubility of an impurity species in a semiconductor is the maximum density of impurity atoms, the solid host crystal can accomodate at a given temperature. The maximum solubility is a function of the mechanism by which the impurity atoms are incorporated in the crystal.

The theoretical maximum solid solubility relative to the lattice density C_L is given by the fractional density of available lattice sites (f_L):

$$C_{B\,max} = f_L C_L = C_L \exp(-E_a^*/kT) = C_L(D_s/D_\infty). \qquad (2.24)$$

Particularly for substitutional impurities this is in disagreement with experimentally observed solid solubilities.

The solid solubility increases with increasing temperature; empirically a decrease in solubility has been observed close to the semiconductor melting point. Temperature reduction after diffusion to ambient temperature will normally not affect the impurity density present in the lattice, as there is usually no impurity redistribution and precipitation observed during the temperature reduction.

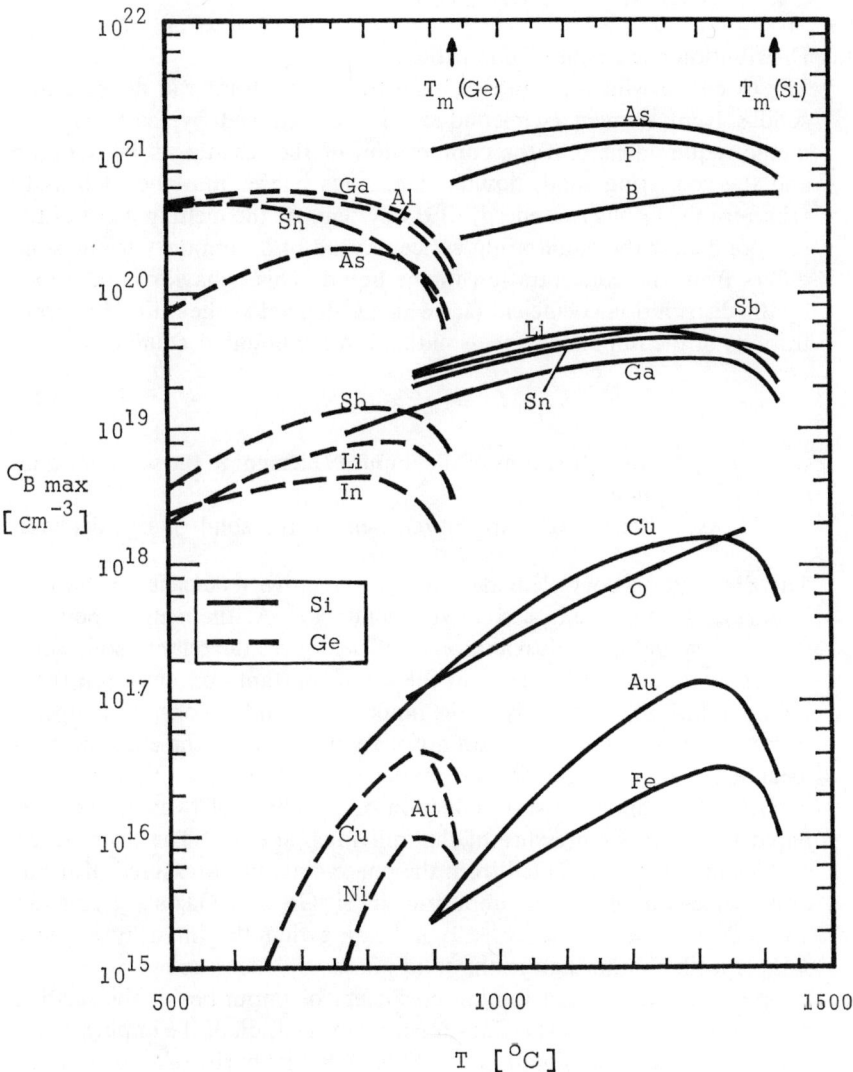

FIGURE 2.3

Maximum solid solubility ($C_{B\,max}$) of impurities in silicon and germanium vs. temperature (T). The melting points (T_m) of the semiconductors are indicated.

4. LIQUID SEMICONDUCTORS

(a) Distribution coefficient of impurities:
The extent to which a solute element (impurity atom) will dissolve in a solid solvent element (semiconductor) is determined by the thermodynamic requirement that the composition of the resulting solid solution and the coexisting solid, liquid, or gaseous phases must be such as to minimize the Gibbs free energy of the system. At the melting point of the semiconductor the equilibrium concentration of an impurity in the solid differs from the concentration in the liquid. This behavior is described by the distribution coefficient (k_o) which is defined as the ratio of concentrations of the impurity in the solid and in the liquid at equilibrium:

$$k_o = f_s/f_l \approx N_s/N_l \qquad (2.25)$$

where f_s, f_l = atom fractions of the impurity element in the solid or liquid phase
N_s, N_l = density of impurity element in the solid or liquid phase.

The distribution coefficient describes the relative tendencies of various impurities to dissolve in a solid semiconductor. At the melting point of the semiconductor the distribution coefficient gives the relative solid solubilities of the impurities, each at the same constant concentration (near infinite dilution) of impurity in the liquid phase and at constant temperature. The values given for k_o do not take into account the effect of non-ideal liquid solution behavior, i.e., of departures from Raoult's law (which describes the vapor pressure reduction of a mixture of two species compared to the vapor pressure of the individual species). The distribution coefficient can be calculated from the phase diagram. Measured distribution coefficients of various impurities in Si, Ge, and GaAs are given in Table 2.8; in most cases $k_o < 1$, indicating that the impurity is more soluble in the liquid than in the solid phase.

Figure 2.4 gives the distribution coefficient of impurities at the melting point of silicon ($T_m = 1417°C$) as function of the radii of the impurity ions and of the heat of sublimation at 300°K. For silicon the relative order of k_o of elements from different groups of the Periodic System is roughly IV > V > III > I; the distribution coefficient within a group generally decreases with increasing atomic radius. Gold shows a slight deviation because interstitial and substitutional diffusion take place simultaneously. The heat of sublimation of the impurity element to the monomeric vapor species (not to the equilibrium vapor) is related to the distribution coefficient; the relationship is only qualitative.

TABLE 2.8
Distribution Coefficients of Impurities in Si, Ge, and GaAs

Impurity	Distribution coefficient (k_o)		
	Si	Ge	GaAs
Al	0.002	0.07	2.5
As	0.30	0.02	
B	0.85	20.0	0.10
Bi	$7 \cdot 10^{-4}$	$5 \cdot 10^{-5}$	
Cd	$1 \cdot 10^{-8}$	$1 \cdot 10^{-5}$	0.20
Cu	$4 \cdot 10^{-4}$	$2 \cdot 10^{-5}$	0.002
Fe	$8 \cdot 10^{-6}$	$3 \cdot 10^{-5}$	0.003
Ga	0.01	0.09	
Ge	0.33	1	0.02
In	$4 \cdot 10^{-4}$	0.001	
Li	0.01	0.002	
O	0.50		
P	0.35	0.08	1.0
S	10^{-5}	10^{-6}	0.50
Sb	0.02	0.003	
Se		10^{-6}	0.50
Si	1	5.5	0.10
Sn	0.02	0.02	
Te		10^{-6}	0.06
Zn	10^{-5}	$4 \cdot 10^{-4}$	0.42

(b) Pseudodiffusion coefficient of impurities:
Figure 2.5 shows the empirically determined dependence of the ratio of the pseudodiffusion coefficient in the liquid (D_P) to the diffusion coefficient of a substitutional impurity at the melting point (D_{Sm}) upon the distribution coefficient. The pseudodiffusion coefficient is defined as

$$D_P = D_\infty \exp(-E'_a/kT)$$
$$= (A_p/w_{at}^{1/2}) \exp[-(H_E/RT)(r_i/r_a)^3] \qquad (2.26)$$

where A_p = constant for the solvent atom (semiconductor)
H_E = heat of evaporation of the solvent
R = Avogadro's constant
r_i, r_a = radius of the fully ionized solute ion (impurity) or tetrahedral radius of the solvent atom
w_{at} = atomic weight of the impurity.

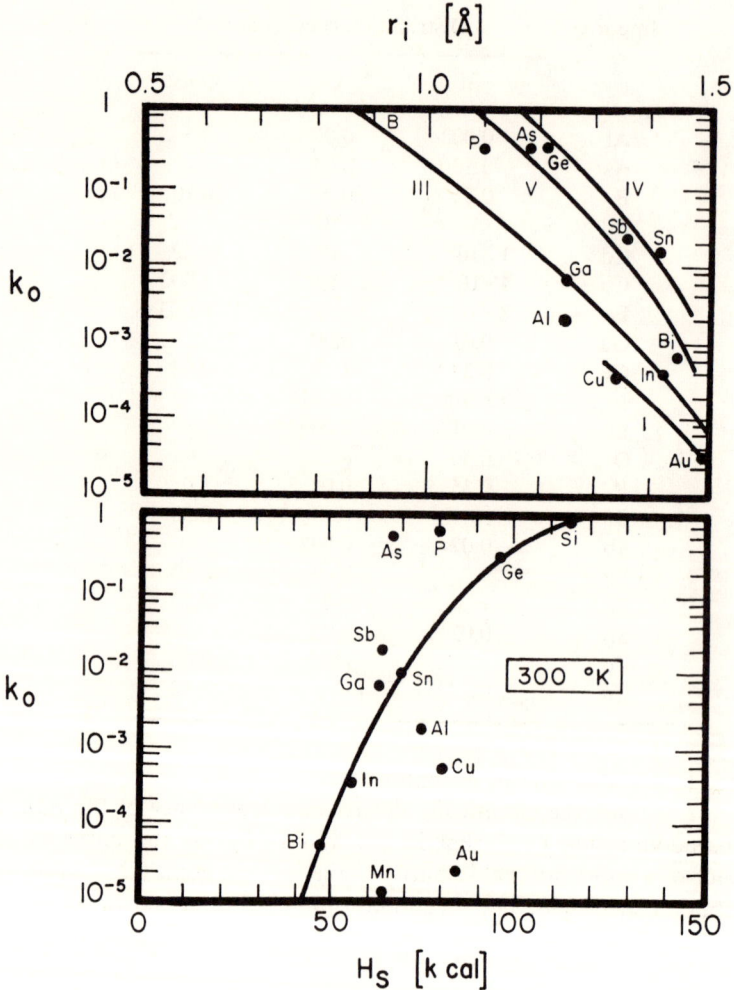

FIGURE 2.4

Impurity distribution coefficient (k_o) in liquid silicon vs. radius of impurity atom (r_i) and heat of sublimation (H_S) for various impurities.

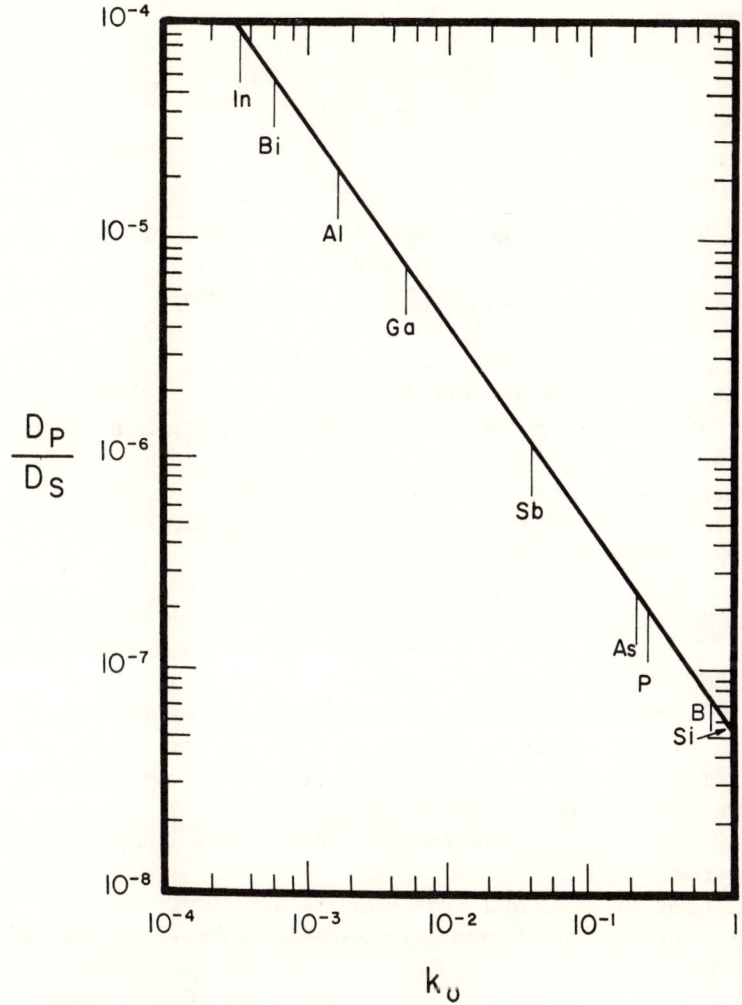

FIGURE 2.5
Ratio of pseudodiffusion coefficient (D_P) of impurities in liquid silicon to diffusion coefficient of impurities at the melting point of silicon (D_{Sm}) vs. impurity distribution coefficient (k_o) in liquid silicon.

The first term in the above equation represents the pseudodiffusion coefficient at infinite temperature and the exponential term contains the activation energy:

$$D_\infty = A_p / w_{at}^{1/2} \qquad (2.27)$$

$$E'_a = (H_E k / R)(r_i / r_a)^3. \qquad (2.28)$$

5. REFERENCES

(1) C. S. Fuller and J. A. Ditzenberger, *J. Applied Physics*, **27**, 544 (1956).
(2) F. A. Trumbore, *BSTJ*, **39**, 205 (1960).
(3) E. Tannenbaum, *Solid-State Electronics*, **2**, 123 (1961).
(4) J. C. Irvin, *BSTJ*, **41**, 387 (1962).
(5) J. C. Brice, *Solid-State Electronics*, **6**, 673 (1963).
(6) A. M. Smith, *RTI Report ASD-TDR-63-316*, Vol. IV, February, 1964.
(7) D. L. Kendall, Report No. 65-29, Stanford University, Dept. Material Science, 1965.
(8) F. Gereth et al., *J. El. Chem. Soc.*, **112**, 323 (1965).
(9) O. Leistiko et al., *IEEE Transact. Electr. Dev.*, ED-**12**, 248 (1965).
(10) K. H. Nicholas, *Solid-State Electronics*, **9**, 35 (1966).
(11) M. F. Millea, *J. Phys. Chem. Solids*, **27**, 315 (1966).
(12) R. A. Evans and R. P. Donovan, *Solid-State Electronics*, **10**, 155 (1967).
(13) S. K. Ghandhi, *The Theory and Practice of Microelectronics*, John Wiley and Sons, New York, 1968.
(14) D. W. Yarbrough, *Solid State Technology*, **11**, 23 (Nov. 1968).
(15) D. L. Kendall and D. B. DeVries, *First International Symposium on Silicon Materials, Science and Technology*, Abstract No. 296, The Electrochemical Society, Inc., New York, May 5–9, 1969.
(16) S. M. Sze, *Physics of Semiconductor Devices*, John Wiley and Sons, New York, 1969.
(17) T. C. Chan and C. C. Mai, *Proc. IEEE*, **58**, 588 (1970).
(18) L. P. Hunter, Editor, *Handbook of Semiconductor Electronics*, McGraw-Hill Book Co., New York, 1970.

2-2 Impurity Concentration Profile

1. MATHEMATICAL DESCRIPTION OF IMPURITY PROFILES

The most important expressions that describe frequently used impurity profiles in a semiconductor are given in the following. It is assumed that diffusion takes place into a semiconductor of background concentration C_B which is of the opposite conductivity type as the in-diffusing species. In this case a *p-n* junction is formed where the concentration of the in-diffusing species equals that of the background concentration. Except in the proximity of the junction, the background concentration is negligible compared to the concentration of the in-diffusing impurities. It takes a finite time for the surface impurity concentration C_S to reach its steady-state value, as shown below, which is of significant influence for very short diffusion times.

(a) Complementary error function distribution (C_S = constant):

$$C(x, t) = C_S \, \text{erfc} \, (x/2\sqrt{Dt}) - C_B \quad (2.29)$$
$$C(x) = C_S \, \text{erfc} \, [(x/x_j)/\text{erfc} \, (C_B/C_S)] - C_B \quad (2.30)$$
$$C_S = (\pi/2)Q(t)/\sqrt{Dt} = \text{constant} \quad (2.31)$$
$$Q(t) = (2/\sqrt{\pi})\sqrt{Dt} \, C_S \quad (2.32)$$
$$dC(x, t)/dx = -C_S/\sqrt{\pi}\sqrt{Dt} \, \exp(-x^2/4 \, Dt)$$
$$\approx -(x/2 \, Dt)C(x, t)$$
$$\approx -2(C_B/x_j) \ln (C_S/C_B) \quad (2.33)$$

(b) Gaussian distribution (Q = constant):

$$C(x, t) = C_S \exp[-(x/2\sqrt{Dt})^2] - C_B$$
$$= C_S \exp(-x^2/4Dt) - C_B \quad (2.34)$$
$$C(x) = C_S \exp[(x/x_j)^2 \ln (C_B/C_S)] \quad (2.35)$$
$$C_S(t) = (1/\sqrt{\pi})Q/\sqrt{Dt} \quad (2.36)$$
$$dC(x, t)/dx = -(x/2Dt)C(x, t)$$
$$= -2(C_B/x_j) \ln (C_S/C_B) \quad (2.37)$$

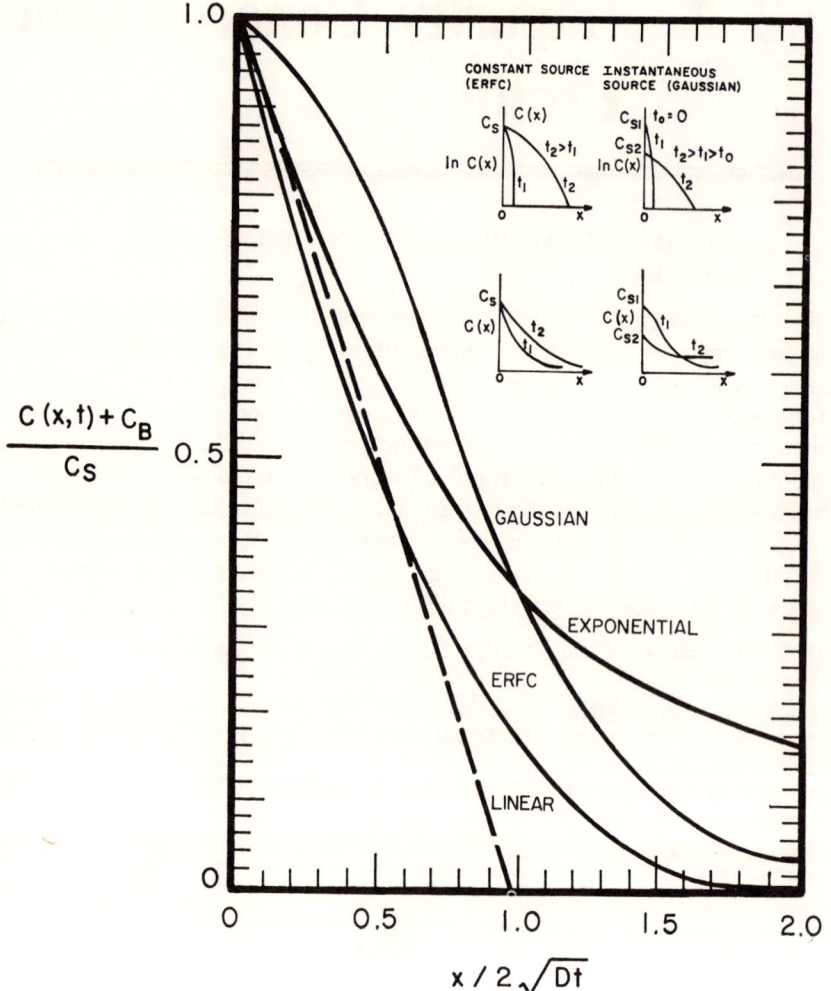

FIGURE 2.6

Ratio of impurity concentration at distance x from semiconductor surface ($C(x)$) to surface impurity concentration (C_S) vs. normalized distance from surface. The inset shows the variation of impurity profiles with diffusion time for the two major types of diffusions (erfc and Gaussian distributions) on logarithmic and linear scales.

2-2 IMPURITY CONCENTRATION PROFILE

(c) Exponential distribution:

$$C(x, t) = C_S \exp(-x/2\sqrt{Dt}) - C_B \qquad (2.38)$$
$$C(x) = C_S \exp[(x/x_j) \ln C_B/C_S] \qquad (2.39)$$
$$dC(x, t)/dx = -(C_B/x_j) \ln(C_S/C_B) \qquad (2.40)$$

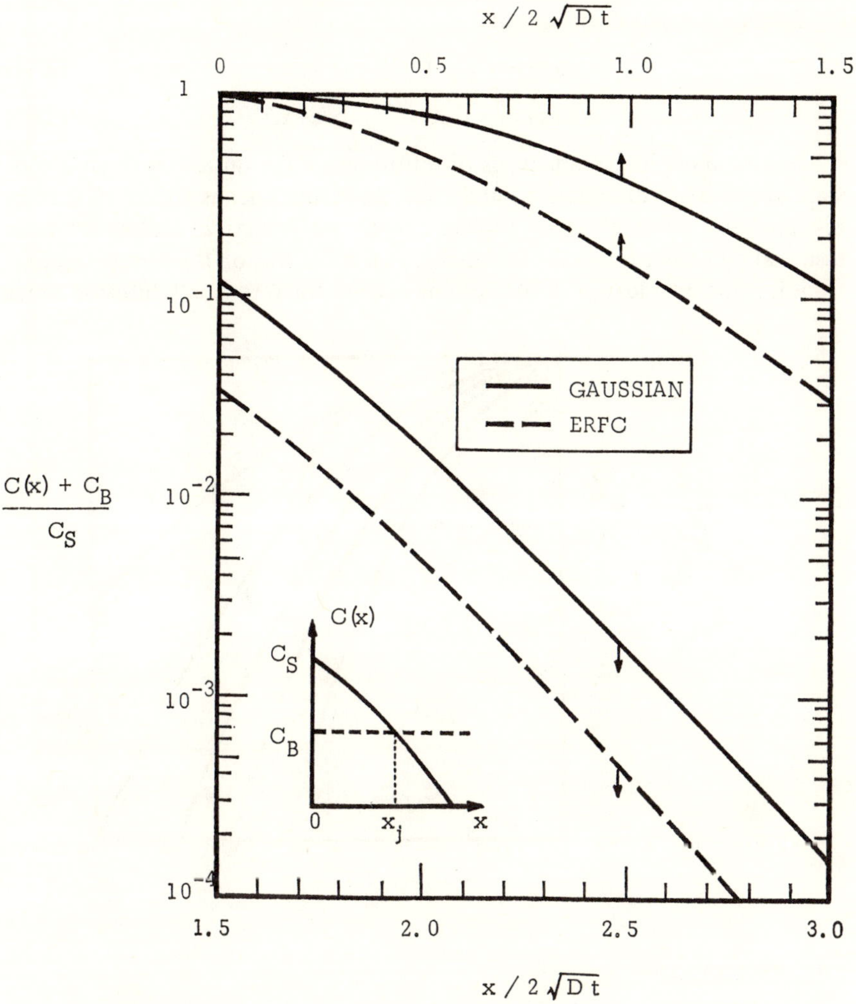

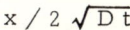

FIGURE 2.7

Ratio of impurity concentration ($C(x)$) at distance x from semiconductor surface ($x = 0$) to impurity concentration at surface (C_S) vs. distance from surface (x) and diffusion length ($\sqrt{Dt}$).

166 IMPURITIES IN SEMICONDUCTORS

(d) Linear distribution:

$$C(x, t) = C_S(1 - x/2\sqrt{Dt}) - C_B \quad (2.41)$$

$$C(x) = (dC/dx)(x - x_j) - C_B, \quad x \approx x_j \quad (2.42)$$

$$dC(x, t)/dx = -C_S/x_j \quad (2.43)$$

(e) Step (box-type) distribution:

$$C(x) = C_S - C_B, \quad x < x_j \quad (2.44)$$

$$C(x) = -C_B, \quad x > x_j \quad (2.45)$$

The two most important types of diffusion are the diffusions from a constant source (erfc distribution) and from an instantaneous source (Gaussian distribution). Both of these diffusion types are similar in character; their essential difference lies in the fact that for an erfc diffusion the surface concentration is independent of diffusion time (except for very small diffusion times

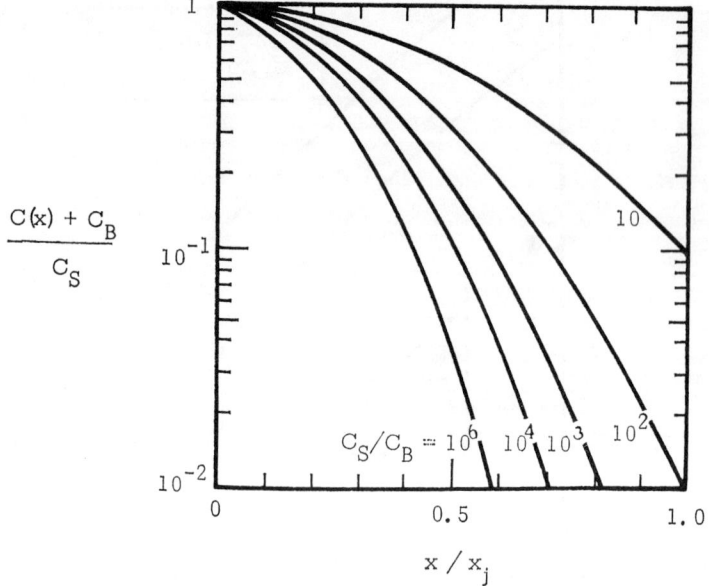

FIGURE 2.8

Impurity concentration ($C(x)$) normalized with respect to surface impurity concentration (C_S) vs. distance (x) from semiconductor surface normalized with respect to junction depth (x_j), at which $C(x) = C_B$, for various ratios of surface impurity concentration to background impurity concentration. Gaussian impurity distribution.

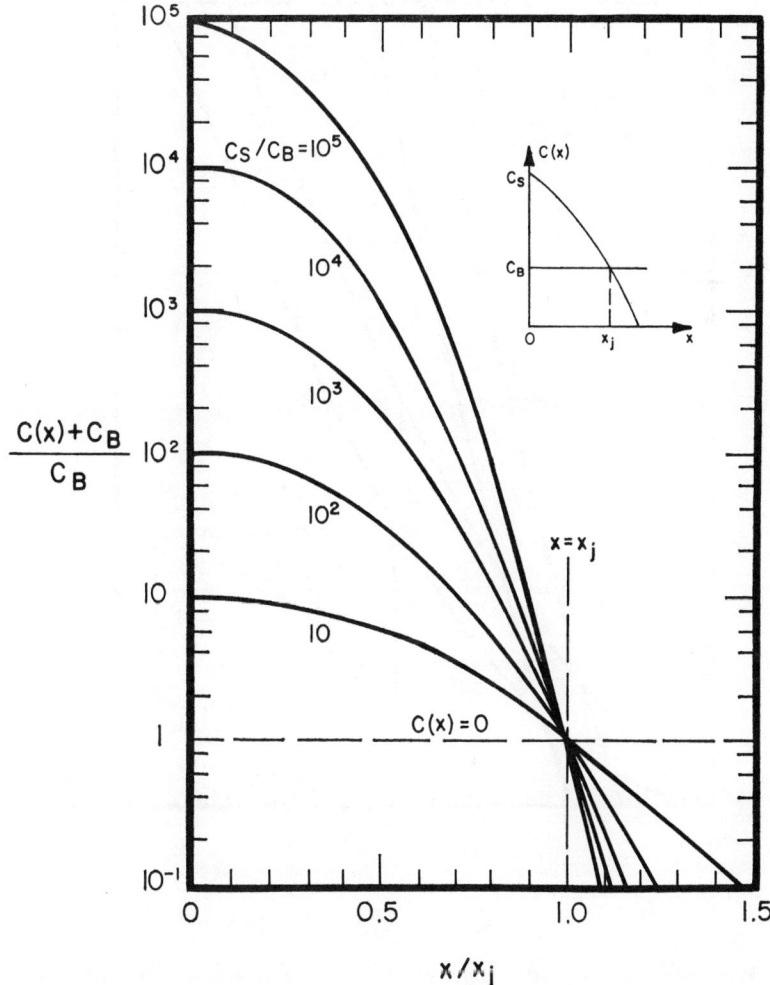

FIGURE 2.9

Impurity concentration ($C(x)$) at distance x normalized with respect to background impurity concentration (C_B) vs. distance from semiconductor surface (x) normalized with respect to junction depth (x_j) for various ratios of surface impurity concentration (C_S) to background impurity concentration. At the junction $C(x) = C_B$.

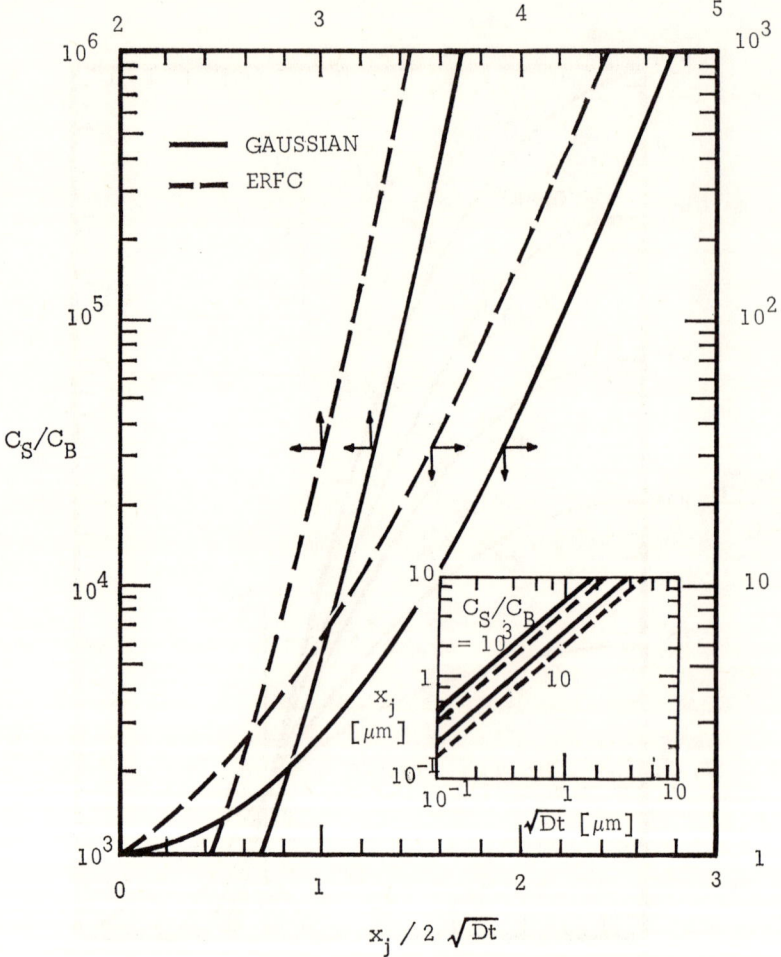

FIGURE 2.10

Ratio of surface impurity concentration (C_S) to background impurity concentration (C_B) vs. junction depth (x_j) normalized with respect to diffusion length ($\sqrt{Dt}$). The difference between complementary error function (erfc) and Gaussian distributions is insignificant for shallow junctions ($x_j \approx 0$); the ratio of erfc to Gaussian values of C_S/C_B is constant (≈ 2) for $x_j/\sqrt{Dt} > 2$.

The inset shows the relationship between junction depth and diffusion length for two ratios C_S/C_B.

because of the finite time required for the surface concentration to reach its final steady-state value), whereas for a Gaussian diffusion surface concentration decreases with increasing time (however, the total impurity charge, Q, remains constant except that during a simultaneous oxidation of silicon certain impurities are rejected or taken up by the oxide which affects the effective Q depending upon the segregation coefficient).

If the diffusion of an impurity is carried out successively at two different temperatures corresponding to diffusion coefficients D_1 and D_2 and to diffusion times t_1 and t_2, then the total effective diffusion length

$$(Dt)^{1/2} = (D_1 t_1 + D_2 t_2)^{1/2}. \tag{2.46}$$

2. CONCENTRATION DEPENDENCE OF DIFFUSION COEFFICIENT

In special cases the diffusion coefficient depends upon the background impurity concentration C_B and the surface concentration of the in-diffusing species, C_S. A simple dependence of D upon impurity concentration can be given by

$$D = D_i (C_B/C_S)^\Theta \tag{2.47}$$

where D_i is the impurity diffusion coefficient in the intrinsic semiconductor (as usually quoted) and Θ is a positive parameter describing the impurity concentration dependence. The complementary error function corresponds to the condition $\Theta = 0$ for which $D = D_i$. Increasing concentration dependence (increasing value of Θ) leads to a steeper impurity profile and to more significant deviations from the normal erfc-type distribution.

In practical applications the concentration dependence of the diffusion coefficient becomes important mainly at impurity concentrations greater than a characteristic concentration C_{crit}.

$$C_{\text{crit}} = C_L [6(Y_B/Y_l)(1-P)^3 \, d^3 N_v]^{1/2}/[1-(1-\varepsilon_i^*)^3] \tag{2.48}$$

where C_L = density of lattice atoms
N_v = density of available lattice positions

$$N_v = C_L \exp(-E_a^*/kT)$$

Y_B, Y_l = bulk and Young's modulus, respectively
d = apparent lattice spacing
P = Poisson's ratio
ε_i = misfit factor; $\varepsilon_i^* = 1 - r_i/r_a$
r_i, r_a = atomic radius of impurity and lattice atoms, respectively.

For example, the characteristic impurity concentration for
(a) boron in silicon:
$$\text{at } 900°C: C_{\text{crit}} = 2.5 \cdot 10^{19} \text{ cm}^{-3},$$
$$\text{at } 1200°C: C_{\text{crit}} = 2.5 \cdot 10^{20} \text{ cm}^{-3};$$
(b) phosphorus in silicon:
$$\text{at } 900°C: C_{\text{crit}} = 1.0 \cdot 10^{20} \text{ cm}^{-3},$$
$$\text{at } 1200°C: C_{\text{crit}} = 1.0 \cdot 10^{21} \text{ cm}^{-3}.$$

The characteristic impurity concentration increases with increasing diffusion temperature.

If the surface concentration of the in-diffusing impurity is greater than the critical concentration, i.e., if $C_S > C_{\text{crit}}$, then in practice the diffusion enhancement depends on the ratio $(C_S/C_{\text{crit}})^2$ and deviations from the conventional Arrhenius plot are observed.

The effective concentration-dependent diffusion coefficient $D(C_S)$ of a substitutional impurity is

$$D(C_S) = D_i c_1 c_2 \tag{2.49}$$

where $c_1 = (C_S/C_{\text{crit}})^2 + 1$ (2.50)

$c_2 = C_S/(C_S^2 + 4n_i^2)^{1/2} + 1.$ (2.51)

The critical impurity concentration is usually close (within an order of magnitude) to the maximum solid solubility of the impurity at a given temperature. Near the maximum solubility the effective diffusion coefficient

$$D_{\text{eff}} = D_i \frac{C_{B\,\text{max}} + C(x)}{C_{B\,\text{max}} - C(x)}. \tag{2.52}$$

3. TWO-STEP DIFFUSION

In a two-step process (i.e., an erfc-type predeposition at low temperature followed by a Gaussian-type drive-in diffusion at high temperature) the resulting impurity profile lies between the corresponding erfc and Gaussian profiles and is a function of the diffusion conditions. A more erfc-like distribution (subscript p) can be expected if $D_p t_p \gg D_d t_d$ and a more Gaussian-like distribution (subscript d) if $D_p t_p \ll D_d t_d$; the actual impurity profile can be described by

$$[C(x, t_p, t_d) + C_B]/C_{Sd} = \tan D_1 \int_{u=0}^{D_1} \exp\left[-(x^2/4D_2)(1 + u^2)\right]/(1 + u^2)\, du \tag{2.53}$$

where $C_{Sd} = (2/\pi) C_{Sp}/\tan D_1$;
$$D_1 = (D_p t_p / D_d t_d)^{1/2}, \qquad D_2 = D_p t_p + D_d t_d.$$

2-2 IMPURITY CONCENTRATION PROFILE

C_{Sd} is the final surface concentration after the drive-in diffusion and C_{Sp} the surface concentration after the first diffusion. Various profiles are shown in Figure 2.11 for two-step diffusions together with values of the integral as function of $(D_p t_p / D_d t_d)^{1/2}$ and $x/2(D_p t_p + D_d t_d)^{1/2}$.

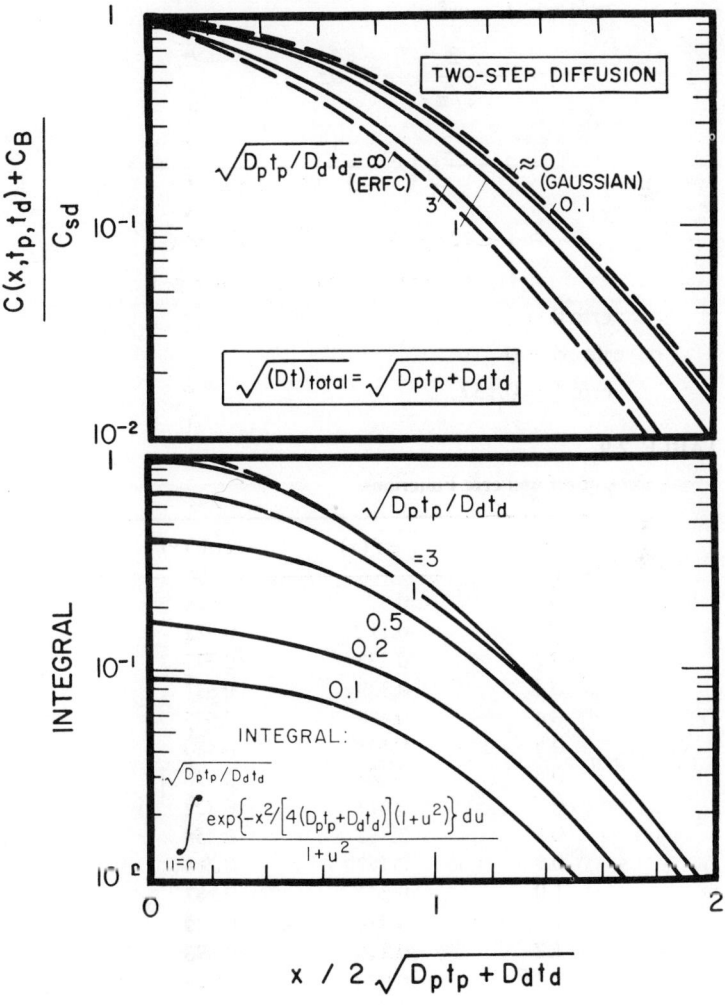

FIGURE 2.11

Impurity concentration profile in a semiconductor after a two-step diffusion vs. normalized distance from semiconductor surface. The lower part of this figure shows the integral used in the calculation of the impurity profile (see text).

4. THE COMPLEMENTARY ERROR FUNCTION

The most important mathematical properties of the complementary error function (erfc) which play a role in the description of diffusion characteristics are as follows.

$$\text{erfc } x = 1 - \text{erf } x$$

$$= 1 - \frac{2}{\sqrt{\pi}} \int_0^x e^{-z^2} dz$$

$$= 1 - \frac{2}{\sqrt{\pi}} \left(x - \frac{x^3}{3 \cdot 1!} + \frac{x^5}{5 \cdot 2!} - + \cdots \right)$$

$$\approx 1 - (2/\sqrt{\pi})x \qquad \text{for } x \ll 1$$

$$\approx (1/\sqrt{\pi}) \exp(-x^2)/x \qquad \text{for } x \gg 1$$

erfc 0 = 1
erfc ∞ = 0
erfc 1 = 0.157.

TABLE 2.9

Values of erf and erfc Functions

x	erf (x)	erfc (x)
0	0	1.000
0.1	0.112	0.888
0.2	0.223	0.777
0.3	0.329	0.671
0.4	0.428	0.572
0.5	0.521	0.480
0.6	0.604	0.396
0.7	0.678	0.322
0.8	0.742	0.258
0.9	0.797	0.203
1.0	0.843	0.157
1.1	0.880	0.120
1.2	0.910	0.090
1.3	0.934	0.066
1.4	0.952	0.048
1.5	0.966	0.034
2.0	0.995	0.005
2.5	0.99959	0.00041
3.0	0.99998	0.00002

2-2 IMPURITY CONCENTRATION PROFILE

Numerical values of the error function (erf) and of the complementary error function (erfc) are given in Table 2.9 for the most often used range of x.

5. IMPURITY DOSAGE AND IMPURITY GRADIENT

(a) Impurity dosage (charge):
The quantity Q is the total number of impurity atoms per unit area of the semiconductor and is defined as

$$Q(t) = \int_0^\infty C(x, t)\, dx. \qquad (2.54)$$

It is the area under one of the distribution curves of Figure 2.6. If this area is approximated by a triangle of height C_S and base $2\sqrt{Dt}$, the area under one of the curves is $C_S\sqrt{Dt} = Q$.

(b) Impurity gradient:
The differential $dC(x)/dx$ is the impurity gradient and is usually defined at $x = x_j$, i.e., at the p-n junction if the diffusion has taken place into a semiconductor of opposite conductivity type. This impurity gradient gives the slope of the impurity profile and is as follows.

(i) *erfc distribution*:

$$dC/dx = (C_B/x_j)\exp[-\mathrm{erfc}^{-2}(C_B/C_S)]/\mathrm{erfc}(C_B/C_S)$$
$$\approx 2(C_B/x_j)\ln(C_S/C_B) \qquad (2.55)$$

(ii) *Gaussian distribution*:

$$dC/dx = 2(C_B/x_j)\ln(C_S/C_B) \qquad (2.56)$$

(iii) *Exponential distribution*:

$$dC/dx = (C_B/x_j)\ln(C_S/C_B) \qquad (2.57)$$

The difference between complementary error function and Gaussian distributions is negligible. To find dC/dx of an exponential distribution, dC/dx of Figure 2.14 has to be divided by two.

The impurity gradient at a p-n junction is temperature-sensitive. At temperature T the gradient is related to the gradient at 300°K by the correction factor a_T (see inset of Figure 2.14):

$$dC/dx\,|_T = dC/dx\,|_{300°K} \cdot a_T. \qquad (2.58)$$

(c) Impurity dosage and impurity gradient:
Impurity dosage (Q) and impurity concentration gradient (dC/dx) are

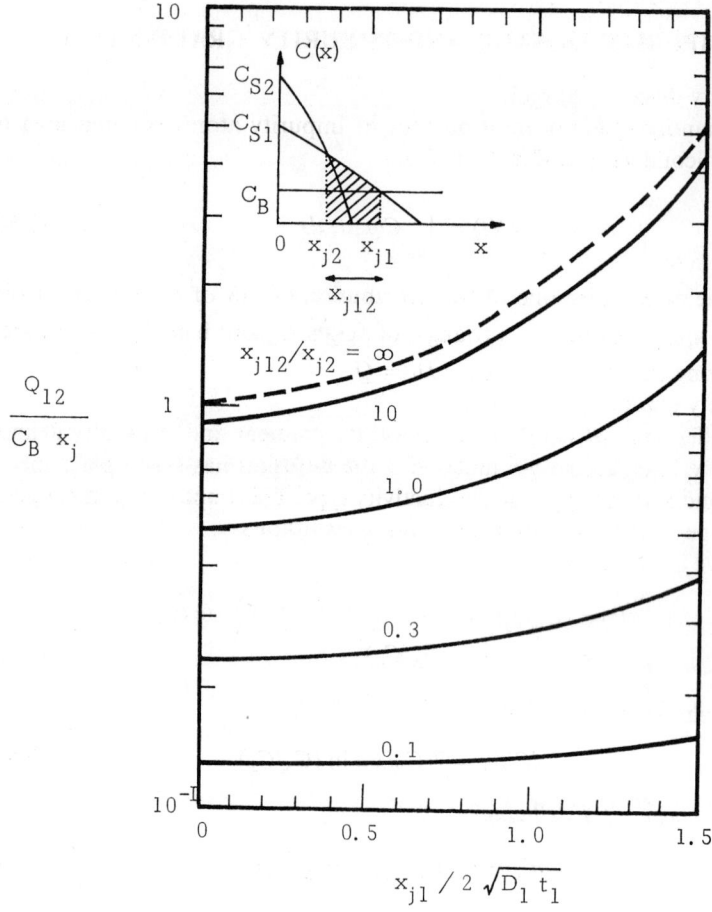

FIGURE 2.12

Impurity dosage (Q_{12}) of a buried impurity layer framed by two Gaussian distributions vs. normalized junction depth (x_{j1}) of the deeper of the two junctions. If $x_{j2} = 0$, then the impurity layer extends to the surface (dashed curve). The inset defines impurity concentrations and junction depths. The layer thickness is $x_{j12} = x_{j1} - x_{j2}$.

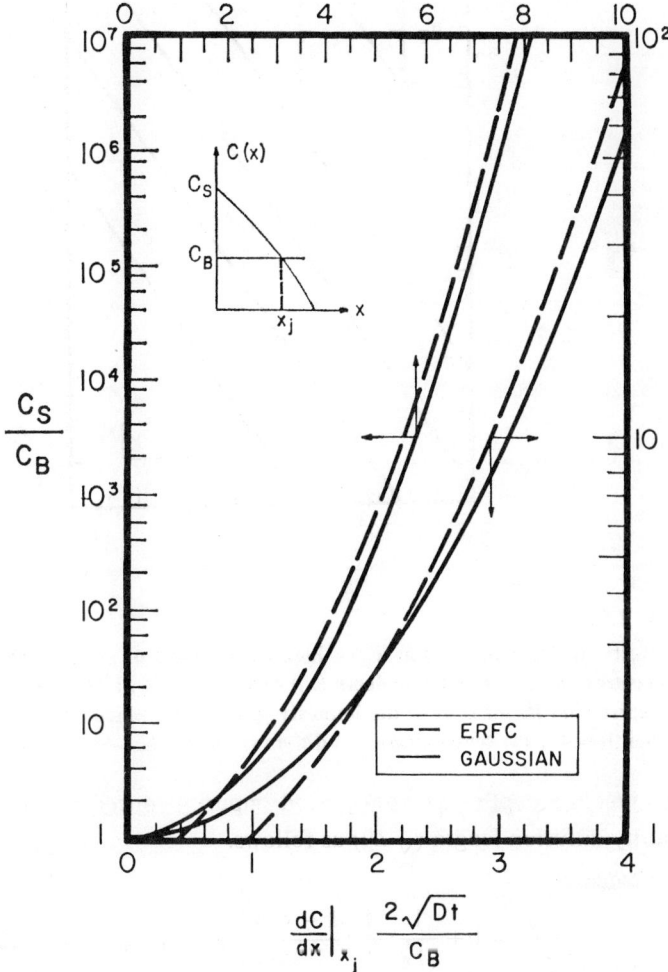

FIGURE 2.13
Ratio of surface impurity concentration (C_S) to background impurity concentration (C_B) vs. impurity concentration gradient (dC/dx) at the junction, diffusion length ($\sqrt{Dt}$), and background impurity concentration.

176 IMPURITIES IN SEMICONDUCTORS

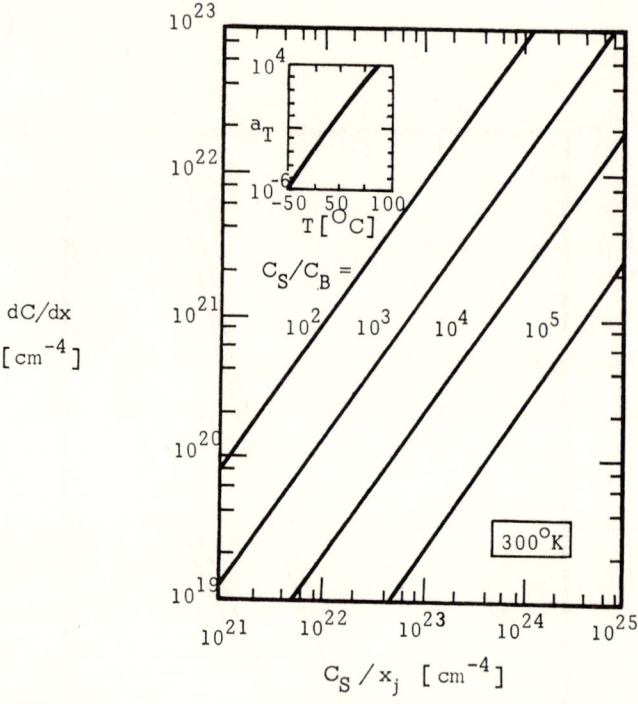

FIGURE 2.14

Impurity gradient (dC/dx) at a *p-n* junction vs. ratio of surface impurity concentration (C_S) to junction depth (x_J) and ratio C_S/C_B. The inset shows the correction factor (a_T) of the impurity gradient by which dC/dx has to be multiplied if the temperature (T) differs from 300°K.

related to junction depth (x_j), background impurity concentration (C_B), and impurity diffusion length ($\sqrt{Dt}$) as follows.

(i) *erfc distribution*:

$$\frac{Q_j}{C_B x_j} = 1 + \frac{1}{\sqrt{\pi}} \cdot \frac{1}{x_N} \cdot \frac{1 - \exp(-x_N^2)}{\text{erfc}(x_N)} \qquad (2.59)$$

$$\frac{dC}{dx} \frac{x_j}{C_B} = \frac{2}{\sqrt{\pi}} \cdot x_N \cdot \frac{\exp(-x_N^2)}{\text{erfc}(x_N)} \qquad (2.60)$$

(ii) *Gaussian distribution*:

$$\frac{Q_j}{C_B x_j} = \frac{\sqrt{\pi}}{2} \cdot \frac{1}{x_N} \cdot \frac{\text{erfc}(x_N)}{\exp(-x_N^2)} \qquad (2.61)$$

$$\frac{dC}{dx} \cdot \frac{x_j}{C_B} = 2 x_N^2. \qquad (2.62)$$

6. ADJUSTMENT OF SURFACE CONCENTRATION (EXTERNAL RATE LIMITATION)

Previously given expressions for impurity profiles assume the existence of a steady-state surface impurity concentration. In practice, the establishment of the steady-state surface concentration from zero at the beginning of the diffusion to its final value when the steady state has been reached, will take a finite time. The adjustment of the surface concentration from its initial value at $t = 0$ (at which it is zero) to its final value (C_S) will take place in a time of the order

$$t \approx 100 \, D/h_S^2. \tag{2.63}$$

The gas-phase mass-transfer coefficient in terms of concentration in the semiconductor (h_S) is defined as

$$h_S = h_G/H_G kT \tag{2.64}$$

and has the dimension of a velocity; it is in the order of 10^{-6} cm/sec. The term h_G is the mass-transfer coefficient in terms of concentration in the ambient gas, and H_G is Henry's Law constant. The boundary conditions are:

$$C(0, 0) = 0,$$
$$C(0, \infty) = C_S.$$

The duration of the adjustment period in which the usually assumed boundary condition $C(0, t) = C_S$ will not hold is determined by the relative rates of

(a) transport of impurities from the bulk of the ambient gas to the semiconductor surface;
(b) solid-state diffusion from the surface to the bulk of the semiconductor.

The impurity concentration $C(x, t)$ at distance x from the semiconductor surface depends upon predeposition time t as follows:

$$\frac{C(x, t)}{C_S(0, \infty)} = \mathrm{erfc}\left(\frac{x}{2\sqrt{Dt}}\right)$$
$$- \exp\left[\left(\frac{h_S t}{Dt}\right)^2\right] \exp\left(\frac{h_S t}{\sqrt{Dt}} \frac{x}{2\sqrt{Dt}}\right) \mathrm{erfc}\left(\frac{h_S t}{\sqrt{Dt}} + \frac{x}{2\sqrt{Dt}}\right). \tag{2.65}$$

The impurity concentration $C(x, t)$ consists of two terms, the usual erfc term from which a corrective term is subtracted. The corrective term decreases with increasing time. At $x = 0$, i.e., at the semiconductor surface,

$$C_S(0, t)/C_S(0, \infty) = 1 - \exp[(h_S t/\sqrt{Dt})^2]\,\mathrm{erfc}(h_S t/\sqrt{Dt}). \tag{2.66}$$

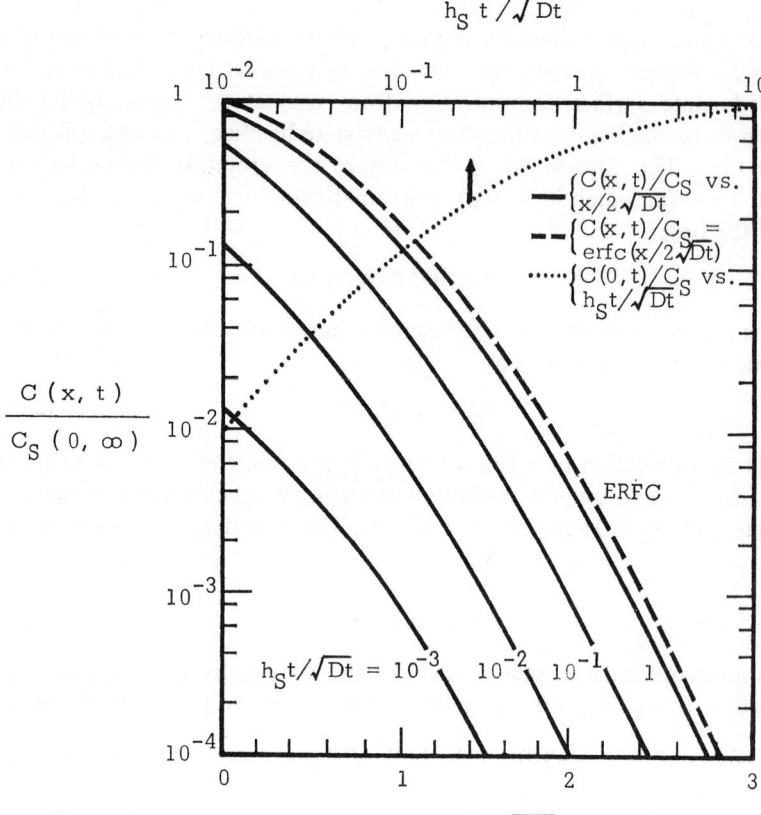

FIGURE 2.15

Impurity concentration profile vs. predeposition time (external rate limitation). This illustration shows the ratio of impurity concentration ($C(x, t)$) at distance x and time t to the final surface concentration at ($C_s(0, \infty)$) at $x = 0$ and $t = \infty$ vs. normalized distance from the semiconductor surface for various mass-transfer coefficients (h_S). Also shown is the variation of surface impurity concentration $C(0, t)$ relative to the final surface concentration with mass-transfer coefficient and time.

2-2 IMPURITY CONCENTRATION PROFILE

Two extreme cases can be distinguished at the surface:

(a) $t = 0$
In this case the corrective term is unity so that
$$C_S(0, t)/C_S(0, \infty) = 0.$$

(b) $t = \infty$
In this case the corrective term is insignificant so that
$$C(x, t)/C_S(0, \infty) = \text{erfc } (x/2\sqrt{Dt})$$
$$C_S(0, t)/C_S(0, \infty) = 1.$$

In practice this is achieved if
$$h_S t/\sqrt{Dt} > 10 \quad \text{or} \quad t > 100 D/h_S^2.$$

At this point the surface concentration has reached approximately 85% of its final value.

7. DIFFUSION FROM TWO SIDES OF A SEMICONDUCTOR

If simultaneous diffusions (with identical Dt) from constant sources (erfc-type diffusions) are made from each side of a semiconductor of thickness x_w and for duration t, then the total impurity distribution

$$C(x, t)/(C_{Sa} + C_{Sb}) = (1/2)\{1 - (4/\pi)[\exp(-\pi^2 Dt/x_w^2) \sin(\pi x/x_w)$$
$$+ (1/3) \exp(-9\pi^2 Dt/x_w^2) \sin(3\pi x/x_w) \quad (2.67)$$
$$+ (1/5) \exp(-25\pi^2 Dt/x_w^2) \sin(5\pi x/x_w) + \cdots]\}.$$

As $\pi^2 Dt/x_w$ becomes smaller (e.g., if the wafer thickness becomes larger), progressively more terms are required. For very small $\pi^2 Dt/x_w$ the above equation can be reduced to that of standard erfc-type diffusions, i.e., the diffusion fronts originating at the two opposite sides display no interaction. The two surfaces of the semiconductor are defined by $x/x_w = 0$ and $x/x_w = 1$, respectively.

8. IN-DIFFUSING AND OUT-DIFFUSING IMPURITIES

If two diffusions (one of an in-diffusing impurity whose source is at the semiconductor surface, one of an out-diffusing impurity whose source is underneath the surface, e.g., a buried subsurface layer) are made at times t_a and t_b, respectively, then the total impurity concentration at x is

$$\frac{C(x, t)}{C_{Sa} + C_{Sb}} = \frac{\text{erfc } (x/2\sqrt{D_b t_b})}{1 + C_{Sa}/C_{Sb}} - \frac{\text{erf } (x/2\sqrt{D_a t_a})}{1 + C_{Sb}/C_{Sa}} \quad (2.68)$$

The curves of Figure 2.16 are based on this equation.

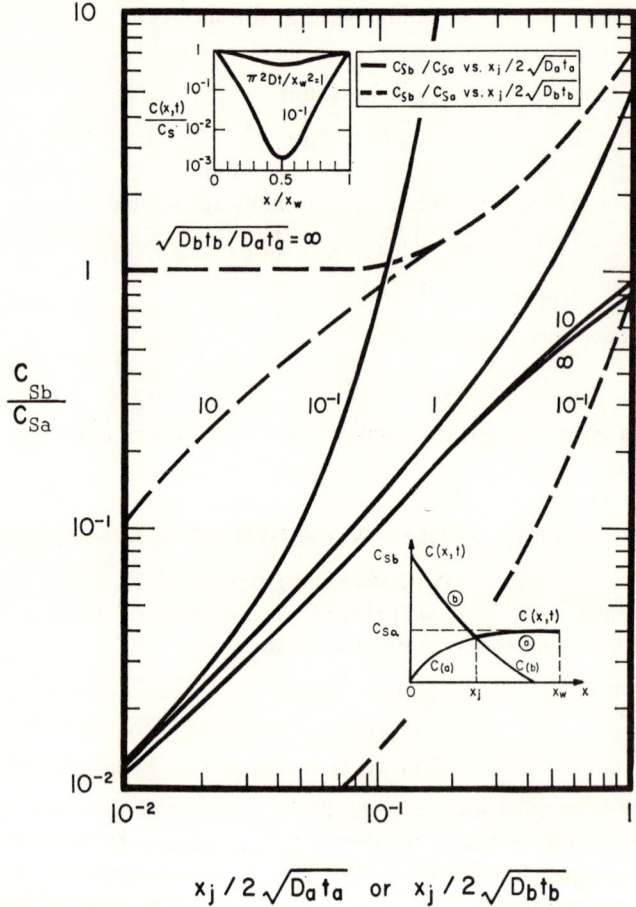

FIGURE 2.16

Ratio of surface concentration of in-diffusing impurity (C_{Sb}) to surface concentration of out-diffusing impurity (C_{Sa}) vs. normalized junction depth (x_j). For definition of terms see the lower inset. The upper inset shows total impurity concentration ($C(x)$) as function of distance from surface (x) thickness of semiconductor sample (x_w), and diffusion length ($\sqrt{Dt}$).

9. EFFECT OF BUILT-IN ELECTRIC FIELD ON IMPURITY DISTRIBUTION

The impurity distribution in a semiconductor during high-temperature diffusion is affected by the electric field which arises from the electric charge of the impurities and which occurs in the region where $C(x) > n_i$. This field is proportional to the impurity gradient and is therefore highest at the source, i.e., at the surface. The field enhances the motion of impurities which results in an effective diffusion coefficient ($D_{\text{eff}} > D$) which is composed of two terms:

(a) the conventional (low concentration) diffusion coefficient D which neglects the charge of impurities,
(b) an impurity concentration-dependent term (f_D) which takes the impurity charge into account (shown in the inset of Figure 2.18), i.e., the effective diffusion coefficient

$$D_{\text{eff}} = D\{1 + [1 + 4(n_i/C(x))^2]^{-1/2}\} \quad (2.69)$$

where $D \leq D_{\text{eff}} \leq 2D$.

As a result of this enhanced diffusion the true surface concentration (C_S^*) is usually smaller (sometimes by orders of magnitude) than the apparent surface concentration (C_S) obtained by neglecting the built-in field. At $C_S < n_i$ the true surface concentration approaches the apparent surface concentration asymptotically so that $C_S \approx C_S^*$. The intrinsic concentration n_i is taken at the diffusion temperature. The ratio $n_i/C(x)$ is temperature-dependent. Three cases are distinguished:

(a) $C(x) < n_i$
   In this case $D_{\text{eff}}/D = 1.0$, i.e., the built-in field is insignificant.
(b) $C(x) = n_i$
   In this case $D_{\text{eff}}/D = 1.45$, i.e., the built-in field has significantly increased the diffusion coefficient (by 45%).
(c) $C(x) \gg n_i$
   In this case $D_{\text{eff}}/D = 2.0$, i.e., the built in field may double the diffussion coefficient.

The ion flux into the semiconductor is

$$\begin{aligned} F &= -D\{1 + [1 + 4(n_i/C(x))^2]^{-1/2}\}(dC/dx) \\ &= -D_{\text{eff}}(dC/dx). \end{aligned} \quad (2.70)$$

The most significant influence of the built-in field is found near the semiconductor surface where the impurity concentration is highest; it is accompanied by a reduction of the impurity gradient at the surface.

Assumptions:

(a) The charge-independent diffusion coefficient D is independent of impurity concentration.
(b) The diffusion temperature is so high that at all times $n_i \gg C_B$ (i.e., the semiconductor is intrinsic at all times).

(c) Impurity concentration at $x \gg 0$ is negligible.
(d) All impurities are ionized.

The electric field resulting from the electric charge of the impurities is

$$E = -(kT/q)(1/C(x))[1 + 4(n_i/C(x))^2]^{-1/2}(dC/dx) \qquad (2.71)$$

If $C(x) > n_i$ then the electric field and the impurity gradient are proportional. The electric field in case of simultaneous diffusion of impurities of opposite type is

$$\begin{aligned} E = &-(kT/q)(1/|N_D - N_A|)[1 + 4n_i^2/(N_D - N_A)^2]^{-1/2} \cdot \\ &\cdot (dN_D/dx - dN_A/dx) \end{aligned} \qquad (2.72)$$

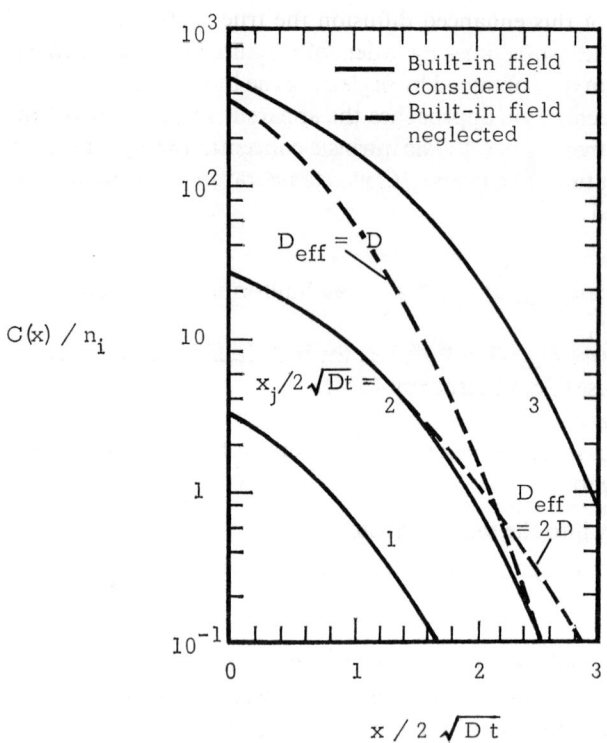

FIGURE 2.17

Ratio of impurity concentration ($C(x)$) at distance x from semiconductor surface to intrinsic carrier concentration (n_i) at diffusion temperature vs. normalized distance from surface (x) and normalized junction depth. The junction depth (x_j) is defined by $C(x) = n_i$.

where N_D and N_A are a function of x, and the maximum electric field at the junction in case of simultaneous diffusion

$$E_{max} = -(kT/q)(1/n_i)(dN_D/dx|_{pn} - dN_A/dx|_{pn}) \quad (2.73)$$

and in case of only one type of impurity

$$E_{max} = -(kT/q)(1/n_i)(dC/dx). \quad (2.74)$$

Figure 2.17 shows the ratio of impurity concentration ($C(x)$) to intrinsic carrier concentration (n_i) vs. distance from semiconductor surface, i.e., the

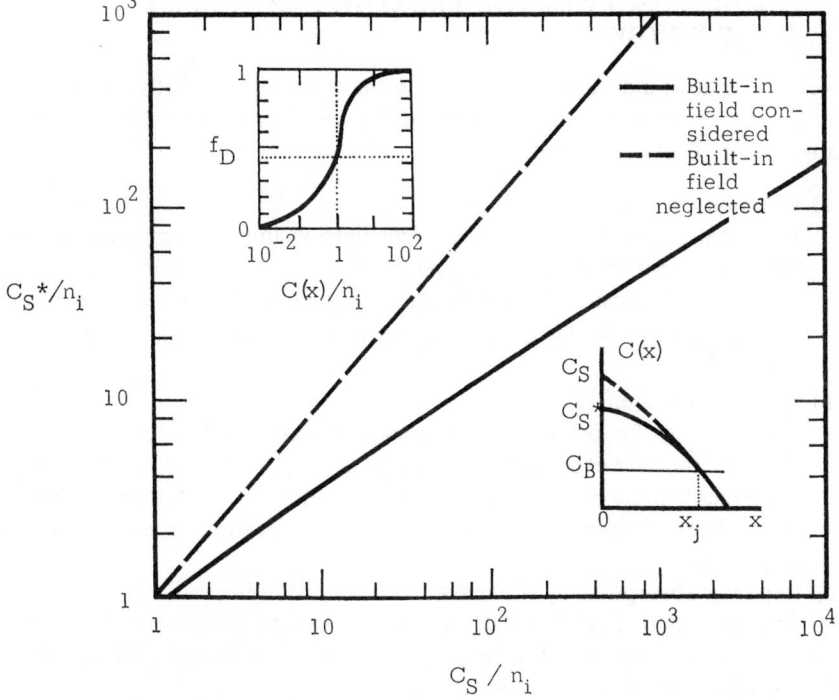

FIGURE 2.18

Actual surface impurity concentration (C_S^*) under consideration of the built-in electric field due to the charge of impurity ions vs. apparent surface concentration (C_S), given by erfc distribution, both normalized with respect to the intrinsic carrier concentration (n_i) at the diffusion temperature. The upper inset shows the charge-dependent term (f_D) of the effective diffusion coefficient which is given by

$$D_{eff} = D(1 + f_D) = D\{1 + [1 + 4(n_i/C(x))^2]^{-1/2}\}.$$

The lower inset defines true and apparent surface concentrations.

diffusion source (x). The solid lines represent the impurity distribution in the semiconductor under consideration of the built-in electric field due to the impurity charge. The dashed lines correspond to a complementary error function distribution neglecting the built-in field for the two extreme conditions $D_{\text{eff}} = D$ and $D_{\text{eff}} = 2D$ for $x_j/2\sqrt{Dt} = 2.0$.

Figure 2.18 shows the actual surface impurity concentration under consideration of the built-in field (C_S^*) vs. the apparent surface concentration, as given by the complementary error function (C_S).

10. DIFFUSION IN THE PROXIMITY OF A MASK WINDOW

Near the edge of a diffusion mask (i.e., in the proximity of a diffusion window in an SiO_2 layer on a semiconductor) deviations from ideal diffusion conditions occur. These may be due to

(a) Geometrical effects:
For small openings and in the immediate vicinity of any opening the ordinary one-dimensional solution to the diffusion equation is not valid and a two-dimensional solution must be applied. For window openings of less than $4\sqrt{Dt}$ the junction depth is a function of the width of the opening.

(b) Strain-induced effects:
Stresses may be present at the semiconductor-mask interface due to different thermal expansion coefficients of both materials or because of the larger volume of thermally grown SiO_2 compared to that of the underlying silicon. These stresses may result in acceleration or retardation of impurities.

(c) Impurity absorption by the mask:
The mask may deplete the dopant source by absorbing impurities. This may result in a reduced impurity concentration in the affected region.

(d) Out-diffusion from the mask:
If the mask contains impurities, then interaction between these impurities and the intentionally introduced impurities may be significant.

(e) Source edge effects:
If the thickness of the impurity source is different near the mask edge than elsewhere, then a different impurity profile is expected after diffusion because of a slower or faster depletion of the source.

Of these five causes of deviations of diffusion from ideal conditions, the geometrical effects are most important under ordinary circumstances. Geometrical effects can be divided into two categories:

2-2 IMPURITY CONCENTRATION PROFILE

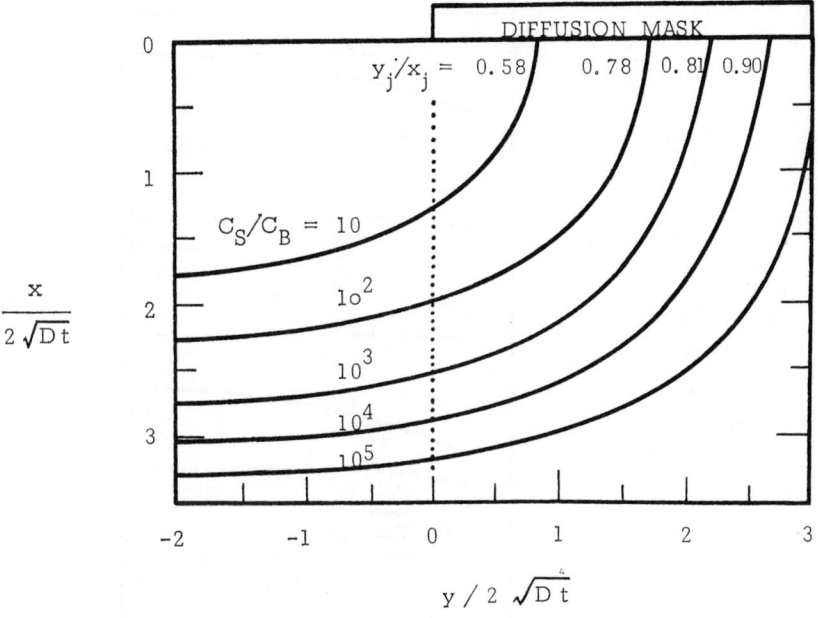

FIGURE 2.19

Normalized vertical junction depth (x_j) of a diffusion front at $y < 0$ and normalized horizontal junction depth (y_j) at $x = 0$ (representing contours of constant impurity concentration) for a given ratio C_S/C_B. Curves were calculated for Gaussian impurity distributions. The vertical junction depth is taken at $y = y_m/2$, the lateral junction depth at $x = 0$. The direction normal to the semiconductor surface is given as x and the direction parallel to the surface as y. The surface impurity concentration (C_S) is taken at $x = y = 0$. The ratio y_j/x_j given at the top of the illustration corresponds to values at the semiconductor surface. The impurity source is at $x = 0$ and $y < 0$.

(a) Effects due to the influence of the edge of the mask:
In this case it is assumed that the width of the mask opening is large, i.e.,

$$y_m > 8\sqrt{Dt}$$

(b) Effects due to the width of the mask opening:
These effects are significant if

$$y_m < 4\sqrt{Dt}$$

If it is assumed that diffusion from a semiconductor surface proceeds in a semicircle with the point source in the center, then disturbances of the diffusion profile occur in the proximity of an opening in an otherwise opaque

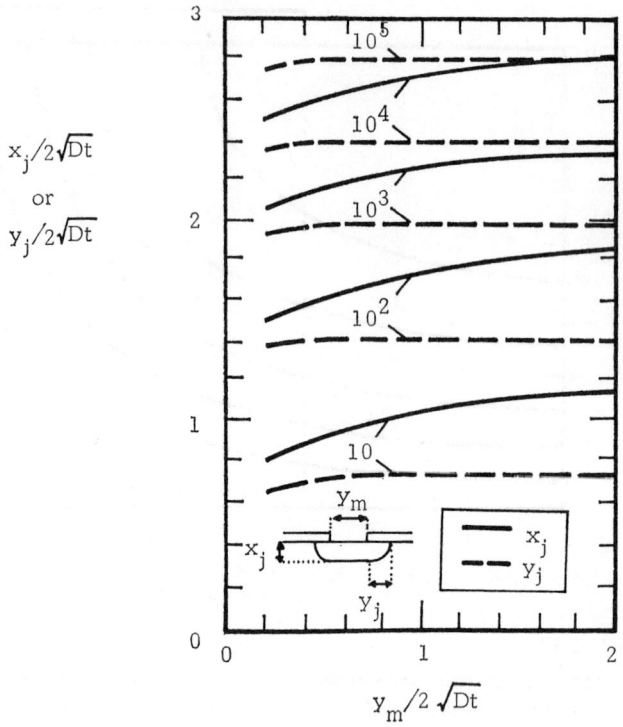

FIGURE 2.20
Normalized vertical and horizontal junction depths (x_j and y_j) vs. normalized width of mask opening (y_m) and ratio C_S/C_B. Gaussian distribution.

diffusion mask (e.g. SiO_2). In the case of a wide mask opening the vertical junction depth (normal to the semiconductor surface) is less than the lateral diffusion depth (parallel to the semiconductor surface); in the case of a narrow mask opening the lateral diffusion depth is less than the vertical diffusion depth.

The width of the diffusion mask window has significant influence on the vertical diffusion depth if $y_m < 4\sqrt{Dt}$. It results in a decrease of the vertical junction depth and in an increase in the ratio y_j/x_j which is shown in Figure 2.21. The ratio y_j/x_j is the ratio of lateral diffusion depth at the mask-semiconductor interface to the maximum vertical diffusion depth at the center of the mask opening. For example, if $C_S/C_B = 10^3$, then the lateral diffusion depth at the semiconductor surface is 81% of the vertical diffusion depth, and if $C_S/C_B = 10$, then the lateral diffusion depth is only 58% of the vertical diffusion depth, if a Gaussian impurity distribution is considered.

2-2 IMPURITY CONCENTRATION PROFILE

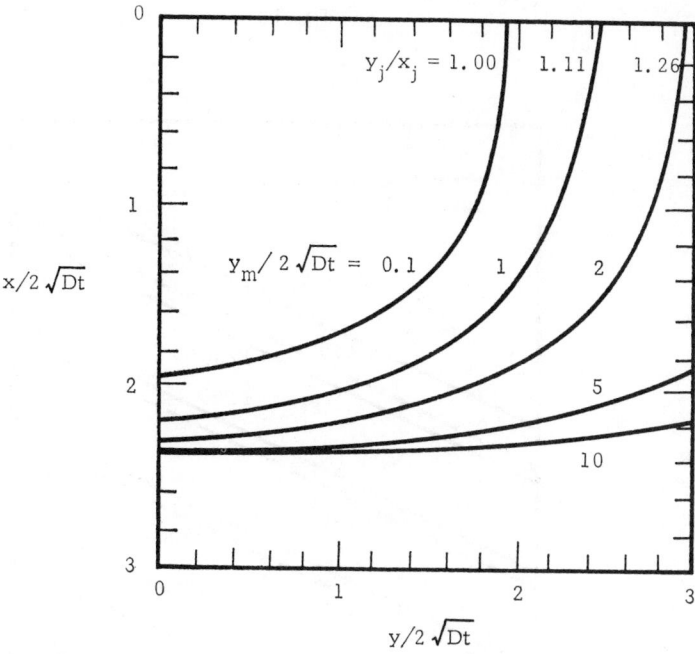

FIGURE 2.21

Normalized vertical junction depth (x_j) at $y \ll 0$ and normalized horizontal junction depth (y_j) at $x = 0$ and contours of constant impurity concentration for various mask opening widths (y_m). Gaussian impurity distribution, $C_S/C_B = 10^3$. The ratio y_j/x_j given at the top of the illustration corresponds to values at the semiconductor surface. The impurity source is at $x = 0$ and $y < 0$.

11. JUNCTION DEPTH

Junction depth is defined as the distance from a diffusion source (usually the semiconductor surface) at which the concentration of the in-diffusing species has fallen to the concentration of the already present species (usually referred to as background impurity concentration). In the case of a p-n junction, background species and in-diffusing species are of opposite conductivity type; at the junction the two impurity concentrations are equal and compensate each other, i.e., the net impurity concentration $|N_D - N_A| = 0$. Carrier mobility and related parameters, however, are affected by the sum of both scattering centers.

The curves of Figure 2.22 have been calculated for boron and phosphorus. For a Gaussian distribution at the junction

$$C(x)|_{x_j} = C_B = C_S \exp(-x_j^2/4Dt) \qquad (2.75a)$$

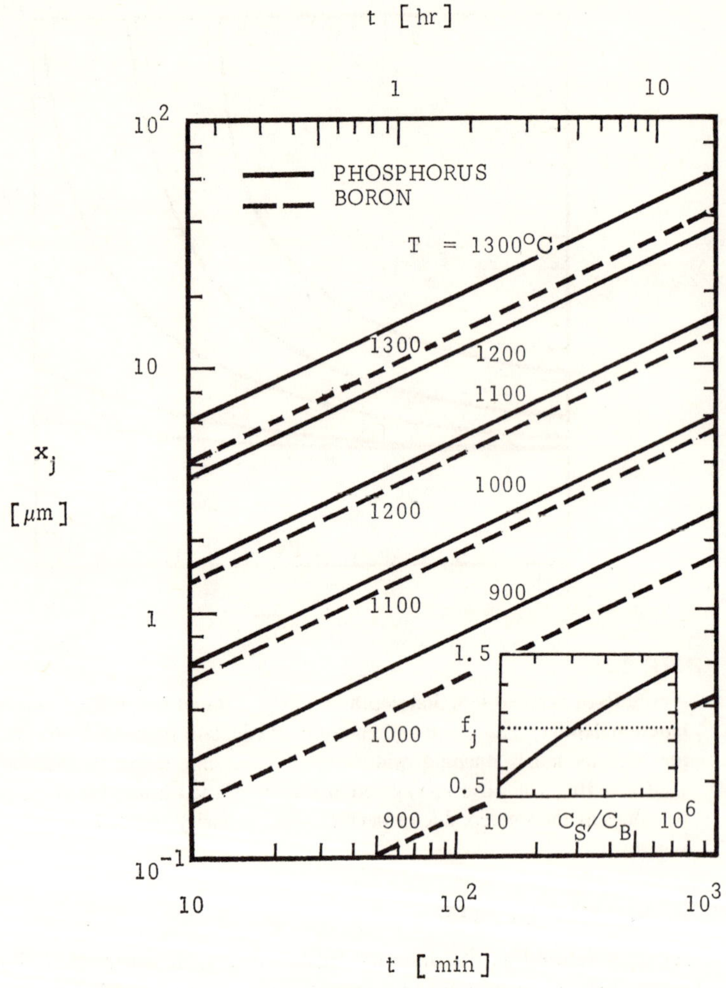

188 IMPURITIES IN SEMICONDUCTORS

FIGURE 2.22

Junction depth (x_j) vs. diffusion time (t) and diffusion temperature (T) for phosphorus and boron in silicon. Gaussian impurity distribution, $C_S/C_B = 10^3$. The inset shows the correction factor applicable to the junction depth if $C_S/C_B \neq 10^3$.

and for an erfc distribution at the junction

$$C(x)|_{x_j} = C_B = C_S \, \text{erfc} \, (x_j/2\sqrt{Dt}). \tag{2.75b}$$

Assumptions:

(a) The value of D is independent of impurity concentration.
(b) The built-in field due to impurity charge is neglected.

(c) The junction depth is a function of the ratio C_S/C_B but is independent of the individual values of C_S and C_B.

For ratios other than $C_S/C_B = 10^3$, the curves can be modified as follows. For example, for a Gaussian distribution:

$$x_j^2 = 4Dt \ln(C_S/C_B). \qquad (2.76)$$

Consequently, x_j as given in the illustration can be modified for $C_S/C_B \neq 10^3$ for both erfc and Gaussian impurity distribution by the following factors f_j which apply generally:

C_S/C_B	Multiply x_j by
10	0.58
10^2	0.82
10^3	1.00
10^4	1.15
10^5	1.29
10^6	1.41

The inset of Figure 2.22 displays f_j graphically.

12. LAYER RESISTIVITY (ρ)

(a) Diffused layer:

The sheet resistivity, $\rho_s(x)$, of a semi-infinite layer of bulk resistivity ρ and thickness $(x - x_j)$ is

$$1/\rho_s(x) = \int_x^{x_j} 1/\rho(x)\,dx. \qquad (2.77)$$

Differentiation leads to

$$\frac{1}{\rho} = -\frac{d(1/\rho_s)}{dx} = -\frac{1}{\rho_s} \cdot \frac{d \ln(1/\rho_s)}{dx} = \frac{1}{\rho_s} \cdot \frac{d \ln \rho_s}{dx} \qquad (2.78)$$

from which the resistivity of the diffused layer

$$\rho = \frac{0.4343 \rho_s}{d(\log \rho_s)/dx}. \qquad (2.79)$$

(b) Uniform layer:

The resistivity of a uniform layer is related to layer thickness (x_j) and four-point probe V/I by

$$\rho = 4.53 \cdot x_j(V/I) = 1/q\mu C_S, \qquad (2.80)$$

i.e., the impurity concentration of the layer

$$C_S = 1/4.53\, q\mu x_j(V/I). \qquad (2.81)$$

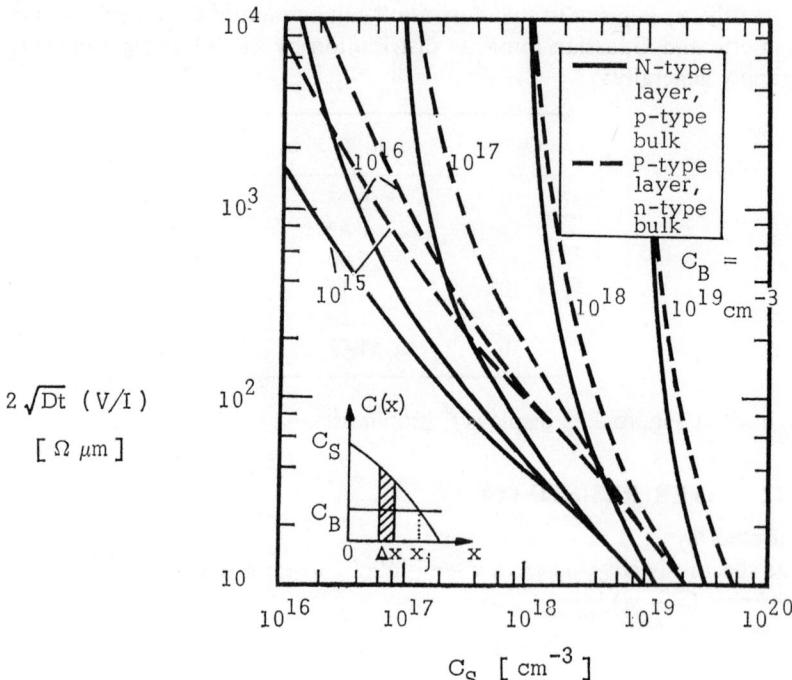

FIGURE 2.23

Product of impurity diffusion length ($\sqrt{Dt}$) and surface resistivity (V/I) vs. surface impurity concentration (C_S) and background impurity concentration (C_B). Gaussian impurity distribution.

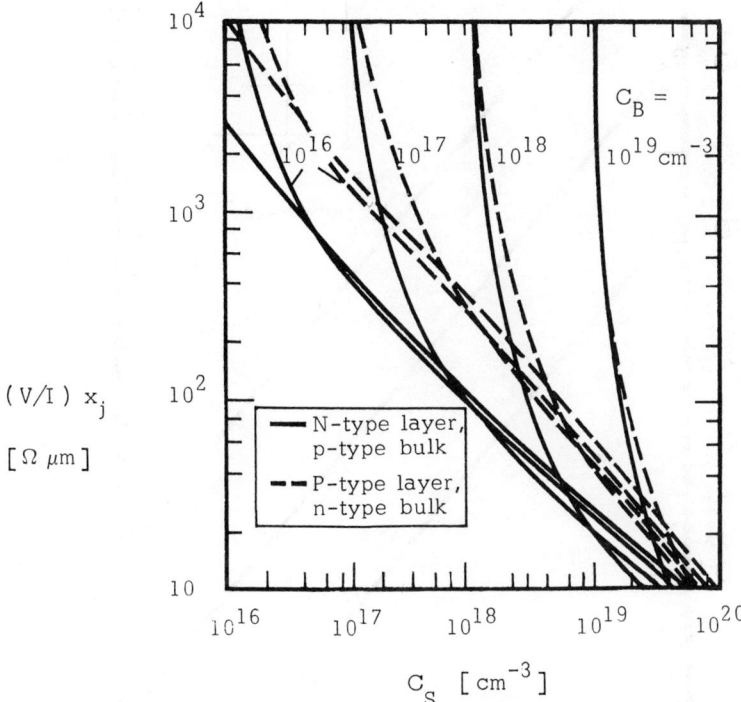

FIGURE 2.24
Product of surface resistivity (V/I) and junction depth (x_J) vs. surface impurity concentration (C_S) and background impurity concentration (C_B). Gaussian impurity distribution.

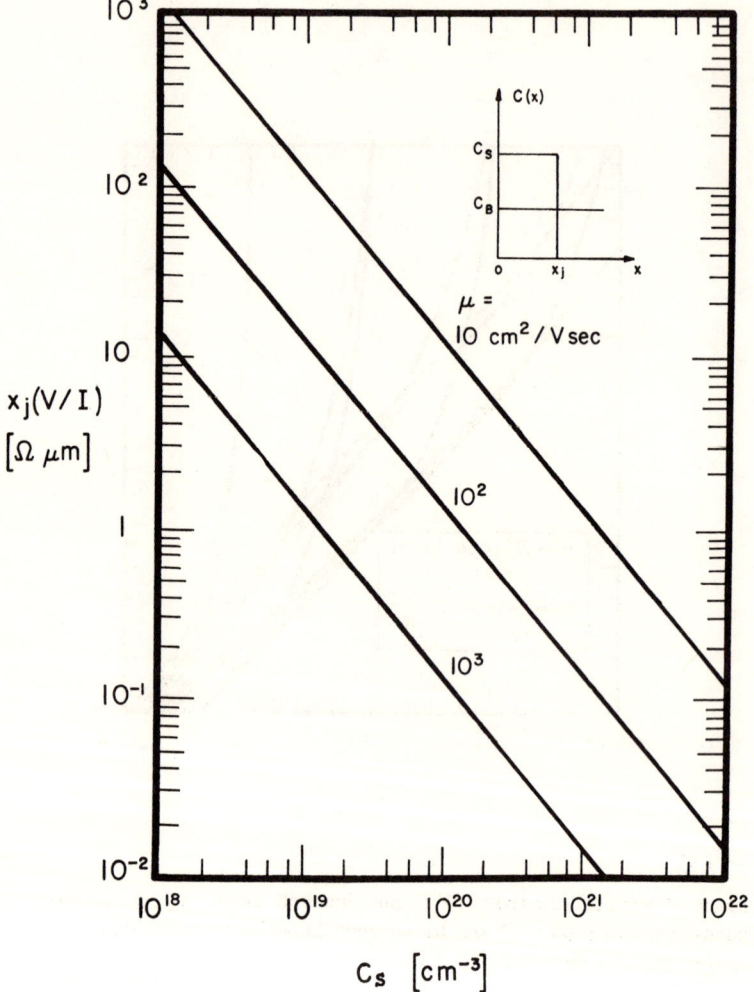

FIGURE 2.25

Product of junction depth (x_j) and surface resistance (V/I) vs. surface impurity concentration (C_S) and carrier mobility (μ) for box-type impurity profile.

13. SURFACE RESISTANCE (V/I)

The conductance per square ($g_\square$) of a diffused layer of thickness Δx and of conductivity σ is

$$g_\square = \sigma \Delta x. \tag{2.82}$$

For the entire diffused region

$$g_\square = \int_0^{x_j} \sigma(x)\, dx$$

$$= \int_0^{x_j} q\bar{\mu}[C(x)^2 - C_B^2]\, dx$$

$$= 1/[4.53(V/I)], \tag{2.83}$$

i.e.,

$$\frac{V}{I} = \frac{1}{4.53} \cdot \frac{1}{q} \cdot \frac{1}{\int_0^{x_j} [C(x)^2 - C_B^2]\, dx}. \tag{2.84}$$

Equation (2.84) forms the basis of all curves of Figures 2.23 through 2.25. The surface resistance (V/I) can be obtained from four-point probe measurements.

14. REFERENCES

(1) R. C. T. Smith, *Australian J. Phys.*, **6**, 127 (1953).
(2) H. S. Carslaw and J. C. Jaeger, *Conduction of Heat in Solids*, Oxford University Press, 1959.
(3) *Tables of the Error Function and its Derivative*, NBS Appl. Math. Series, 40 (1961).
(4) K. Lehovec and A. Slobodskoy, *Solid-State Electronics*, **3**, 45 (1961).
(5) J. C. Irvin, *BSTJ*, **41**, 387 (1962).
(6) W. Jost, *Diffusion in Solids, Liquids and Gases*, Academic Press, Inc,, New York, 1962.
(7) L. K. Monteith, *RTI Report ASD-TDR-63-316*, Vol. II, Oct. 1963.
(8) J. Crank, *The Mathematics of Diffusion*, Clarendon Press, Oxford, 1964.
(9) K. M. Busen and G. A. Shirn, *Solid-State Electronics*, **7**, 49 (1964).
(10) D. P. Kennedy and R. R. O'Brien, *IBM J. Res. Develop.*, **9**, 179 (1965).
(11) D. P. Kennedy and P. C. Murley, *IBM J. Res. Develop.*, **10**, 6 (1966).
(12) W. Kass and M. O'Keefe, *J. Appl. Phys.*, **37**, 2377 (1966).
(13) A. S. Grove, *Physics and Technology of Semiconductor Devices*, John Wiley and Sons, New York, 1967.

(14) W. Mönch, *Solid-State Electronics*, **10**, 745 (1967).
(15) S. M. Hu and S. Schmidt, *J. Appl. Phys.*, **39**, 4272 (1968).
(16) R. Berry, *Proc. IEEE*, **57**, 153 (1969).
(17) D. P. Kennedy, *Proc. IEEE*, **57**, 1202 (1969).
(18) B. Polata, private communication.
(19) N. D. Thai, *J. Appl. Phys.*, **41**, 2859 (1970).
(20) P. E. Bakeman, Jr., and J. M. Borrego, *J. El. Chem. Soc.*, **117**, 688 (1970).
(21) C. F. Gibbon et al., Abstract No. 176, 138th Nat. Mtg. Electrochemical Society, Atlantic City, Oct. 4–8, 1970.

2–3 Properties of Impurities Used in Silicon Technology

1. n-TYPE AND p-TYPE DIFFUSION SOURCES AND CHEMICAL REACTIONS

(a) Diffusion of boron in silicon:
Because of its superior diffusion characteristics, boron is almost exclusively used as p-type impurity (acceptor) in silicon; other p-type impurities have not achieved practical importance for the following reasons: Indium has a moderately deep-lying energy level ($E_a = 0.16$ eV) which makes it an inefficient acceptor. Aluminum combines with oxygen in the silicon crystal leading to anomalous effects. Gallium has an excessive diffusion coefficient in SiO_2 and cannot be used, therefore, for selective diffusion in silicon if SiO_2 is the masking barrier.

Boron, a substitutional impurity in silicon, has a misfit factor -0.25 in silicon which results in strain-induced effects on the silicon lattice and crystal damage if boron is present in amounts approaching the solid solubility (about 10^{20} cm^{-3}). Electrically active boron is found in concentrations of less than $5 \cdot 10^{19}$ cm^{-3}; boron concentrations in excess of that do not contribute to electric conductivity but increase crystal damage.
Predominant reactions during high-temperature diffusion:

(i) *Basic reaction of* B_2O_3 *with Si* (*delivering boron to the silicon surface*):

$$2B_2O_3 + 3Si \rightleftharpoons 4B + 3SiO_2.$$

(ii) *Preliminary reactions of original diffusion source with oxygen*:

$$4BBr_3 + 3O_2 \longrightarrow 2B_2O_3 + 6Br_2$$
$$4BCl_3 + 3O_2 \longrightarrow 2B_2O_3 + 6Cl_2$$
$$2(CH_3O)_3B + 9O_2 \longrightarrow B_2O_3 + 6CO_2 + 9H_2O$$
$$B_2H_6 + 3O_2 \longrightarrow B_2O_3 + 3H_2O.$$

The diffusion of boron into silicon is accompanied by the formation of SiO_2 at the silicon surface.

196 IMPURITIES IN SEMICONDUCTORS

(b) Diffusion of phosphorus in silicon:
The most common n-type impurity (donor) in silicon is phosphorus, whose diffusion coefficient is similar to that of boron. Other n-type impurities (mainly arsenic and antimony) have found extensive use mostly because of their small diffusion coefficients which makes them ideal under conditions where small out-diffusion is desirable. Phosphorus, a substitutional impurity in silicon, has a misfit factor -0.06 in silicon due to an only slightly smaller atomic radius compared to silicon so that strain-induced effects on the silicon lattice are insignificant. Electrically active phosphorus is found in concentrations of less than $5 \cdot 10^{19}$ cm^{-3}, as in the case of boron.

Predominant reactions during high-temperature diffusion:

(i) *Basic reaction of* P_2O_5 *with* Si (*delivering phosphorus to the silicon surface*):

$$2P_2O_5 + 5Si \rightleftarrows 4P + 5SiO_2.$$

(ii) *Preliminary reactions of original diffusion source with oxygen*:

$$4POCl_3 + 3O_2 \longrightarrow 2P_2O_5 + 6Cl_2$$
$$4PCl_3 + 5O_2 \longrightarrow 2P_2O_5 + 6Cl_2$$
$$2PH_3 + 4O_2 \longrightarrow P_2O_5 + 3H_2O.$$

The diffusion of phosphorus into silicon is accompanied by the formation of SiO_2 at the silicon surface.

2. BORON AND PHOSPHORUS DIFFUSION SYSTEMS

Tables 2.10 and 2.11 list the characteristics of various diffusion systems used for the diffusion of boron and phosphorus into silicon. Table 2.12 summarizes the properties of impurities used in silicon technology. The following comments apply to this table.

(a) The impurity radius within the silicon crystal differs from the radius of the atom of its own lattice because of a different internal field.
(b) The misfit factor (ε_i^*) describes the strain in the lattice introduced by impurity atoms; if r_a is the radius of the semiconductor host material

$$\varepsilon_i^* = \frac{r_i - r_a}{r_a} < 0: \quad r_i < r_a$$

$$\varepsilon_i^* = \frac{r_i - r_a}{r_a} > 0: \quad r_i > r_a$$

TABLE 2.10
Boron Diffusion Systems

Original impurity source	Achievable impurity concentration [cm^{-3}]	Advantages	Disadvantages
H$_3$BO$_3$ (boric acid; converts to B$_2$O$_3$ upon heating)	10^{17} to 10^{21}	Readily available	Tube contamination; difficult control
(CH$_3$O)$_3$B (methyl borate)	$> 10^{20}$	Easy operation	Restricted to high surface concentration
BBr$_3$ (boron tribromide)	10^{17} to 10^{21}	Clean system; good control of impurity concentration	
BCl$_3$ (boron trichloride)	10^{17} to 10^{21}	Same as BBr$_3$; easy operation	
B$_2$H$_6$ (diborane)	10^{17} to 10^{21}	Same as BBr$_3$	Toxic and explosive

TABLE 2.11
Phosphorus Diffusion Systems

Original impurity source	Achievable impurity concentration [cm^{-3}]	Advantages	Disadvantages
P (red phosphorus)	$< 10^{19}$	Low surface concentration possible	Control of impurity concentration marginal
P$_2$O$_5$ (phosphorus pentoxide)	$> 10^{20}$	High surface concentration possible	Sensitive to water vapor; tube contamination; restricted to high surface concentration
(NH$_4$)$_2$H$_2$PO$_4$ (ammonium diphosphate)	10^{17} to 10^{21}	Not dependent upon water vapor	Purification marginal
POCl$_3$ (phosphorus oxychloride)	10^{17} to 10^{21}	Clean system; good control over impurity concentration	
PCl$_3$ (phosphorus trichloride)	10^{17} to 10^{21}	Same as POCl$_3$	
PH$_3$ (phosphine)	10^{17} to 10^{21}	Same as POCl$_3$; good control; easy operation	Toxic

Summary of the Properties of Impurities in Silicon

Element	Element number	Donor (D) or acceptor (A)	Radius of impurity atom in Si crystal (a) r_i [Å]	Misfit factor (b), ϵ_i^*	Capture radius for thermal carriers, r_c [Å]	Maximum interaction energy, $E_{ia\,max}$ [eV]	Atomic weight w_{at}	Distribution coefficient, k_o	Melting point of impurity [°C]	Atomic density, C_L [atoms/g]	Heat of fusion, H_F [eV]	Partial molar heat of mixing Si in liquid solution $H_{Si(l)}$ [kcal/g]	Partial molar excess entropy of mixing Si in liquid solution, $S_{Si(l)}$ [cal/g°C]	Type of diffusion (c)	Common diffusion sources (d)
Ag	47	Deep-lying	1.52	0.23	21	0.70	107.88	$4 \cdot 10^{-4}$	961			0.12	−7.91	−7.63	i	
Al	13	A	1.26	0.08	7	0.23	26.97	$2 \cdot 10^{-3}$	659	$2.23 \cdot 10^{22}$	0.11	−4.14	−1.22	s	
As	33	D	1.18	0.01	8	0.03	74.91	0.30	817	$8.03 \cdot 10^{21}$	0.28	−49.99	−32.40	s	$As_2O_3(s)$
Au	79	Deep-lying	1.44	0.23	21	0.70	197.20	$3 \cdot 10^{-5}$	1063	$3.05 \cdot 10^{21}$	0.13	−19.54	−10.28	i	
B	5	A	1.50	0.28	26	0.85	10.82	0.80	2027	$5.55 \cdot 10^{22}$	0.23			s	$(CH_3O)_3B(g)$
			0.88	−0.25	24	0.75								s	$BCl_3(g)$, $BBr_3(l)$, $B_2O_3(s)$, $Bi_2O_3(s)$
Bi	83	D	1.45	0.23	9	0.28	209.00	$7 \cdot 10^{-4}$	271	$2.88 \cdot 10^{21}$	0.11	14.84	2.06	i	
Cu	29	Deep-lying	1.23	0.09	14	0.47	63.57		1083	$9.47 \cdot 10^{21}$	0.14	11.91	−7.19	s	
Fe	26	D	1.35	0.15			55.84	$8 \cdot 10^{-6}$	1539	$1.08 \cdot 10^{22}$	0.16			i	$FeCl_3(s)$
Ga	31	A	1.26	0.08	7	0.23	69.22	$8 \cdot 10^{-3}$	30	$8.69 \cdot 10^{21}$	0.06	3.25	0.83	s	$Ga_2O_3(g)$
In	49	A	1.26	0.08	7	0.23	114.76	$4 \cdot 10^{-4}$	156	$5.25 \cdot 10^{21}$	0.03			s	$In_2O_3(s)$
Li	3	D	1.44	0.23	21	0.70	6.94		181		0.03	11.45	3.37	s	$LiOH(s)$, LiOH-LiCl mixture (s)
Ni	28	Deep-lying	1.24	0.06	5	0.18	58.71		1455		0.18			i	
O	8		0.66	−0.44	17	0.57	16.00	0.50							$O_2(g)$
P	15	D	1.10	−0.06	5	0.18	30.98	0.35	597 (red)	$1.94 \cdot 10^{22}$		−53.24	−31.63	s	$POCl_3(l)$, $PCl_3(l)$, $P_2O_5(l)$
Sb	51	D	1.36	0.15	15	0.49	121.76	0.02	630	$4.94 \cdot 10^{21}$	0.21	3.29	−1.61	s	$Sb_2O_4(g)$, $Sb_3Cl_5(l)$, $Sb_2O_3(s)$
Si	14		1.17	0			28.06	1.00	1410	$5.00 \cdot 10^{22}$	0.53				

For self-diffusion $\varepsilon_i^* = 0$, i.e., $r_a = r_i$.
(c) i = interstitial, s = substantial.
(d) At temperatures normally employed the impurity sources are either gas (g), liquid (l), or solid (s).

3. REFERENCES

(1) C. D. Thurmond and M. Kowalchik, *BSTJ*, **39**, 169 (1960).
(2) F. A. Trumbore, *BSTJ*, **39**, 205 (1960).
(3) R. Bullough and R. C. Newman, *Progress in Semiconductors*, John Wiley and Sons, New York, 1963.
(4) A. M. Smith and R. P. Donovan, paper presented at the Third Annual Microelectronics Symposium, St. Louis, April 1964.
(5) S. K. Ghandhi, *The Theory and Practice of Microelectronics*, John Wiley and Sons, New York, 1968.

2–4 Gold in Silicon

1. GOLD DIFFUSION MECHANISMS

The diffusion of gold in silicon is affected mainly by two simultaneously occurring mechanisms. The dominant mechanism is by interstitial movement of gold atoms associated with a high interstitial diffusion coefficient; a small fraction (about 10%) of the gold atoms moves by substitutional diffusion associated with a small diffusion coefficient. The ratio of interstitial to substitutional diffusion coefficients is a function of temperature; it is usually several orders of magnitude greater than unity.

Since interstitially located gold atoms can become substitutional depending upon the availability of substitutional sites within their capture range, the diffusion mechanism is quite complex. In a region of high defect concentration (dislocations) the equilibrium between substitutional and interstitial sites is disturbed, resulting in a higher overall diffusion coefficient. In a defect-free semiconductor the vacant lattice sites are provided from the crystal surface; the corresponding gold diffusion coefficient is relatively small since it depends upon a dissociative mechanism in which gold atoms move from substitutional to interstitial sites leaving behind vacancies.

The true diffusion coefficient of gold in silicon is a composite given by the various mechanisms and depends upon defect and impurity density, temperature, and oxygen content. If D_I and D_S are the interstitial and substitutional diffusion coefficients of gold in silicon, then the effective diffusion coefficient which takes both diffusion mechanisms proportionally into account is approximately

$$D_{\text{eff}} \approx 0.9 D_I + 0.1 D_S. \tag{2.85}$$

2. SOLUBILITY ENHANCEMENT

The maximum solid solubility of gold in silicon in saturation is determined by the availability of interstitial and substitutional lattice sites. The number of substitutional sites, however, is usually larger than the number of interstitial sites.

The solid solubility is the total density of atoms in all three charge states (negative, neutral, positive). It is enhanced by the presence of impurity atoms other than gold (of density C_B) in the silicon lattice. The total solubility of gold, an amphoteric impurity, is theoretically

$$C_{\text{Au}} = C_{\text{Au}}^{(-)} + C_{\text{Au}}^{(0)} + C_{\text{Au}}^{(+)} \tag{2.86}$$

with the individual contributions

$$C_{Au}{}^{(-)} = C_{Aui}{}^{(-)}[(C_B + \sqrt{C_B{}^2 + 4n_i{}^2})/2n_i] \geq C_{Aui}{}^{(-)} \quad (2.87)$$
$$C_{Au}{}^{(0)} = C_{Aui}{}^{(0)} \quad (2.88)$$
$$C_{Au}{}^{(+)} = C_{Aui}{}^{(+)}[2n_i/(C_B + \sqrt{C_B{}^2 + 4n_i{}^2})] \leq C_{Aui}{}^{(+)}. \quad (2.89)$$

The subscript i refers to the solubility in the intrinsic semiconductor (where $C_B \leq n_i$) and n_i is the intrinsic carrier concentration at the gold diffusion temperature.

It is assumed that always $C_B \gg C_{Au}$ and that all impurities are ionized. In n-type silicon, the solubility enhancement is due to the gold acceptor level; in p-type silicon, a less significant effect has been observed. From the above expressions it follows that $C_{Au}{}^{(-)}$ increases and $C_{Au}{}^{(+)}$ decreases with increasing background concentration C_B, whereas $C_{Au}{}^{(0)}$ remains unaffected. The ratio of densities of gold atoms in the three charge states is

$$C_{Au}{}^{(-)} : C_{Au}{}^{(0)} : C_{Au}{}^{(+)} = \exp\left[(E_F - E_A)/kT_{Au}\right] : 1 : \exp\left[(E_D - E_F)/kT_{Au}\right]. \quad (2.90)$$

The maximum solid solubility of gold in silicon as a function of background impurity concentration is

$$\begin{aligned}C_{Au}(C_B) &= C_{Aui} \exp\left[(E_F - E_i)/kT_{Au}\right] \\ &= C_{Aui} + C_{Au}{}^{(-)}[(C_B - 2n_i + \sqrt{C_B{}^2 + 4n_i{}^2})]/2n_i \end{aligned} \quad (2.91)$$

where $C_{Au\,i}$ is the total solubility of gold in intrinsic silicon, E_F and E_i are the Fermi levels in extrinsic and intrinsic silicon, and T_{Au} is the temperature at which gold is diffused into silicon.

The solubility of gold in silicon is enhanced whenever the impurity density C_B exceeds the intrinsic carrier density n_i at the diffusion temperature (dotted curve in Figure 2.26). Figure 2.26 shows experimental data of solubility of gold in silicon. The background concentration above which a difference between n-type and p-type silicon is observed shifts to higher values of C_B as the diffusion temperature is increased; e.g., for $T_{Au} = 800°C$ this separation occurs at $C_B \approx 10^{18}$ cm^{-3}, for $T_{Au} = 1200°C$ at $C_B \approx 10^{19}$ cm^{-3}. Within the range of this illustration the difference between C_{Au} of n-type and p-type silicon is less than 50%. No difference is observed at 900°C.

If an amphoteric impurity (e.g., gold in silicon) is diffused to saturation, neutral and charged impurity atoms display a different behavior.

(a) Neutral atoms:
   The distribution of neutral atoms in a semiconductor saturated with gold is uniform throughout the semiconductor regardless of local variations in background impurity concentration since only diffusion and not an electric field affects their migration. The density of neutral gold atoms is, therefore, independent of the local Fermi level.

(b) Charged atoms:

The density of charged impurity atoms is not uniform throughout the semiconductor if there are variations in the background concentration. The density of charged gold atoms depends, therefore, on the local Fermi level wherever their density is sufficiently large to affect the position of the Fermi level.

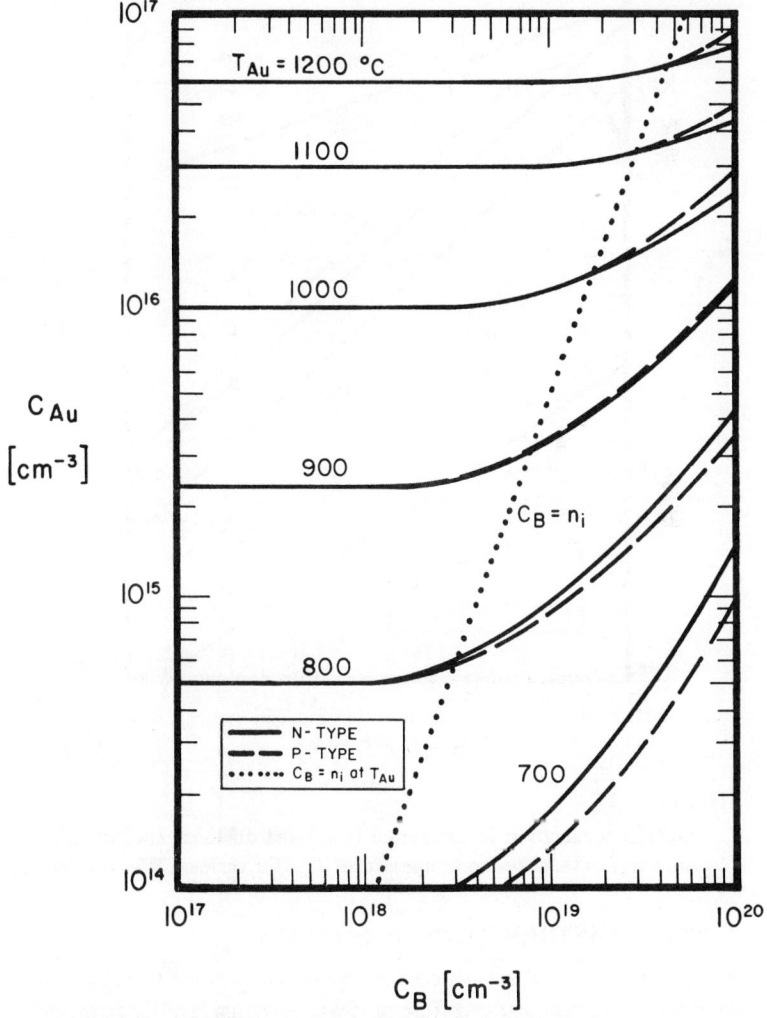

FIGURE 2.26

Maximum concentration (C_{Au}) of gold in silicon vs. gold diffusion temperature (T_{Au}) and background impurity concentration (C_B).

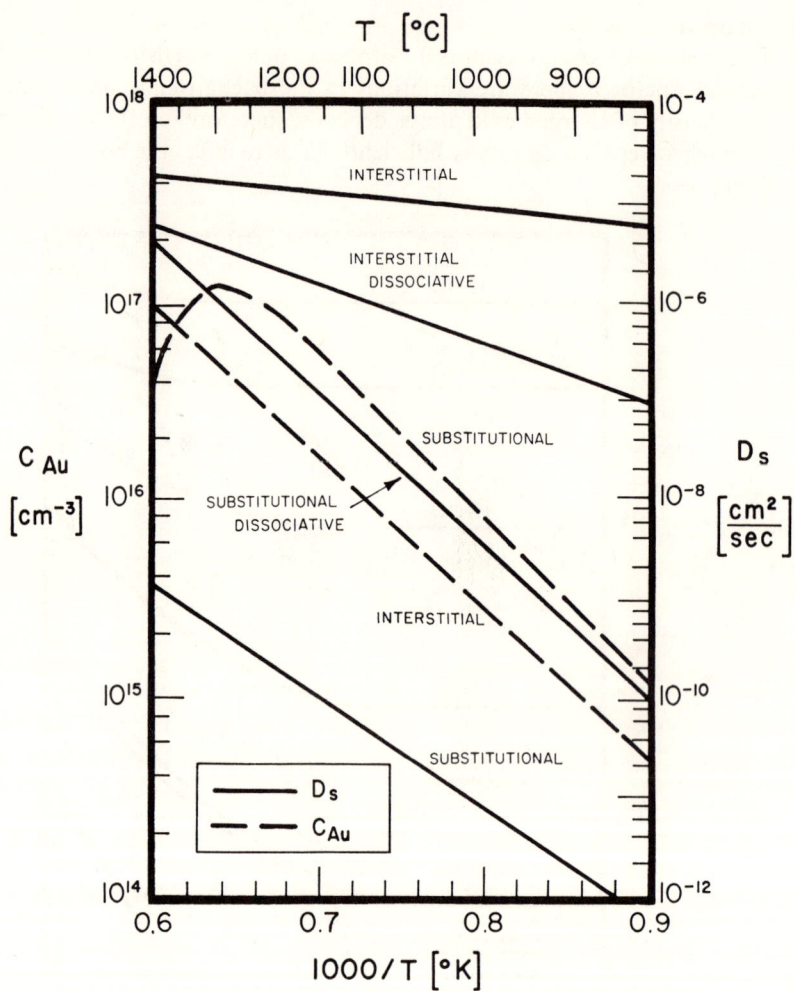

FIGURE 2.27

Gold concentration in saturation (C_{Au}) and diffusion coefficient of gold in silicon (D_s) vs. diffusion temperature (T_{Au}) for various diffusion mechanisms.

3. GOLD DISTRIBUTION IN SILICON

If the silicon lattice is not saturated with gold atoms, then the distribution of gold after a high-temperature diffusion deviates considerably from erfc- and Gaussian-type impurity distributions normally observed with other impurities. This is partially due to the complex diffusion mechanism found in the migration of gold in silicon. Figure 2.28 compares the empirically determined concentration profile of gold as established at a given temperature with the theoretical impurity profiles of other elements. The profiles for gold were established in dislocation-free n-type and p-type silicon of $\langle 111 \rangle$ orientation at temperatures between 800 and 1100°C. The true diffusion coefficient (D_s) and the effective diffusion or permeation coefficient (K) are shown in the inset as function of gold diffusion temperature.

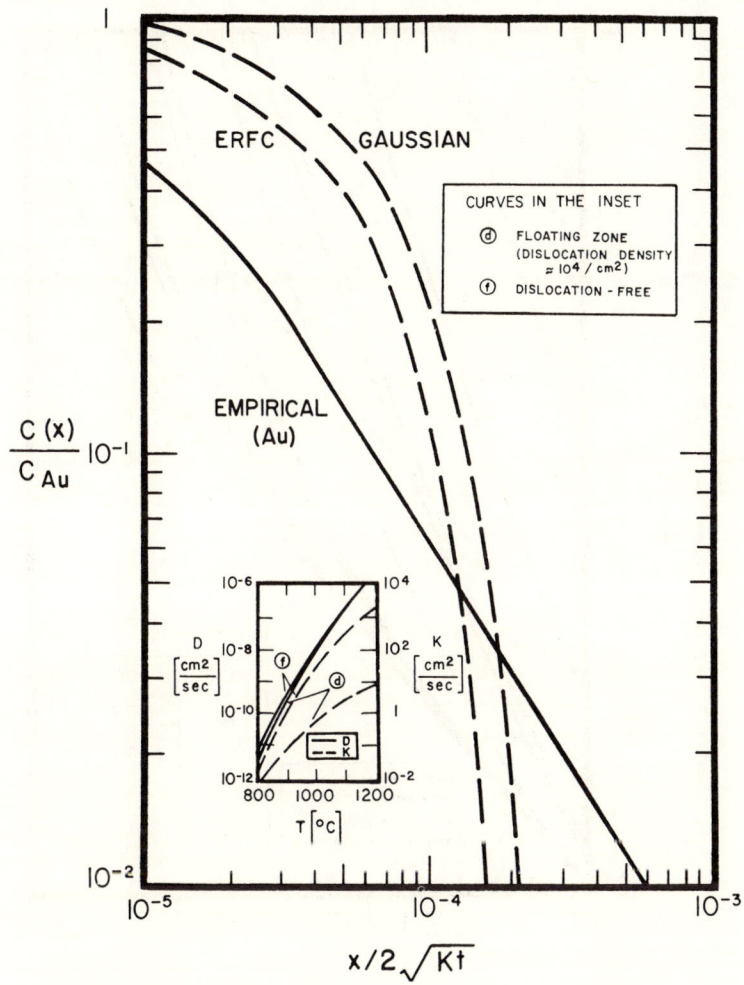

FIGURE 2.28

Ratio of gold concentration ($C(x)$) at distance x from silicon surface to gold concentration in saturation (C_{Au}) vs. normalized distance from surface. For comparison, erfc- and Gaussian-type impurity distributions are given. The inset shows the variation of true and effective diffusion coefficients (D_s and K) vs. gold diffusion temperature (T_{Au}).

4. RESISTIVITY INCREASE OWING TO DEEP-LYING IMPURITIES

The presence of deep-lying impurities in a semiconductor results in a reduction of the density of free carriers. Each atom having a deep-lying energy level removes one majority carrier from the conduction band of an *n*-type

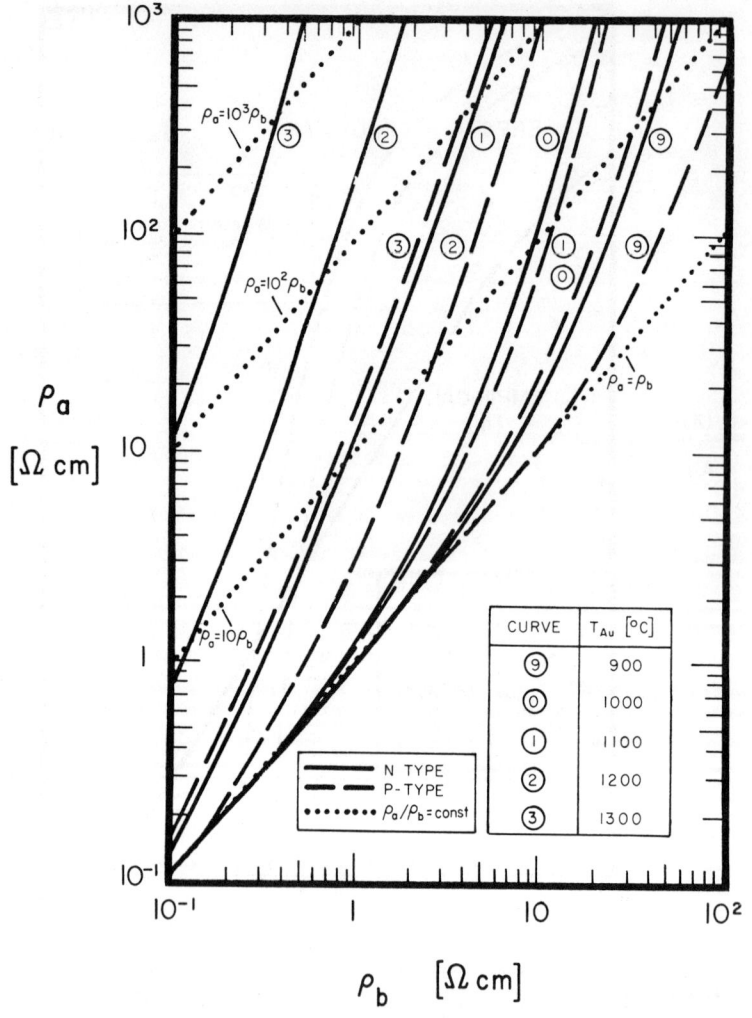

FIGURE 2.29

Silicon resistivity after saturating gold diffusion (ρ_a) vs. resistivity before diffusion (ρ_b) for various gold diffusion temperatures (T_{Au}). The dotted curves correspond to theoretical values of $\rho_a/\rho_b = $ constant.

semiconductor or from the valence band of a *p*-type semiconductor. Therefore the available carrier density in an *n*-type semiconductor:

$$n \approx N_D - N_t \tag{2.92a}$$

and in a *p*-type semiconductor:

$$p \approx N_A - N_t, \tag{2.92b}$$

where N_t is the density of recombination centers, i.e., the density of deep-lying impurity atoms.

In a nondegenerate semiconductor in the absence of recombination centers the carrier density is equal to the density of impurity atoms. The carrier

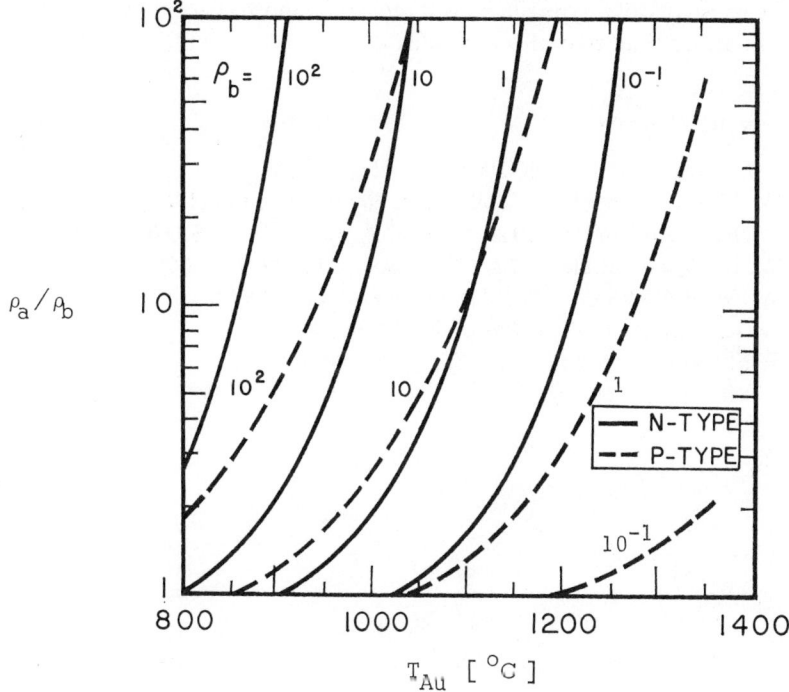

FIGURE 2.30

Increase in semiconductor resistivity over its value prior to gold diffusion (ρ_b) vs. gold diffusion temperature (T_{Au}), assuming semiconductor saturation with gold. Near-intrinsic silicon. The resistivity after saturating gold diffusion is ρ_a. Curves are based on experimental data on 300 μm thick *n*-type and *p*-type wafers into which gold was diffused for more than 24 hours. Saturation was achieved when the gold concentration had reached the maximum solid solubility at the given temperature.

removal is, therefore, within limits, equivalent to a lighter impurity doping. This, in turn, leads to an increased semiconductor resistivity in the presence of deep-lying impurity atoms. If $N_t \ll C_B$, the resistivity is not significantly affected by carrier removal.

In silicon, gold is a deep-lying impurity, since its two energy levels are close to the center of the band gap. As a consequence of its supply of recombination centers, it reduces the average carrier lifetime.

Due to the increased maximum solid solubility of gold in silicon at higher temperature, the resistivity change is most pronounced at high diffusion temperature. A high initial semiconductor resistivity requires fewer recombination centers for compensation and consequently is more sensitive to gold doping, resulting in a greater variation of resistivity.

The resistivity of n-type silicon is significantly more affected by the presence of gold atoms than that of p-type silicon.

5. REFERENCES

(1) J. D. Struthers, *J. Appl. Phys.*, **27**, 1560 (1956).
(2) W. Shockley and J. L. Moll, *Phys. Rev.*, **119**, 1480 (1960).
(3) W. R. Wilcox and T. J. LaChapelle, *J. Appl. Phys.*, **35**, 240 (1964).
(4) W. R. Wilcox et al., *J. El. Chem. Soc.*, **111**, 1377 (1964).
(5) W. M. Bullis, *Solid-State Electronics*, **9**, 143 (1966).
(6) D. R. Collins, *J. Appl. Phys.*, **39**, 4133 (1968).
(7) S. F. Cagnina, *J. El. Chem. Soc.*, **116**, 498 (1969).

2-5 Ion Implantation

1. GENERAL CHARACTERISTICS

Ion implantation is an alternate method to solid-state diffusion for introducing impurities into a semiconductor. The penetration depth is a function of the original kinetic energy of the projectile (i.e., the impurity ion) and the properties of the target matter (i.e., the semiconductor) and is determined by the interaction between projectile and target. In a first-order approximation it is not a function of temperature and time. The kinetic energy of the projectile is a function of its mass.

Target properties of importance are the mass of the target atoms, lattice spacing, and atomic arrangement (i.e., the apparent behavior of the target as a crystal or an amorphous material, depending upon the incident angle). Because of enhanced reproducibility, ions are normally implanted at an angle with respect to the crystal orientation such that the semiconductor appears to be amorphous, i.e., no projectile channeling is assumed.

Ion bombardment of a crystalline semiconductor usually results in crystal damage which degrades the semiconductor characteristics and which can be partially removed in a high-temperature annealing treatment. The amount of damage depends upon the properties of projectile and target, the ion energy, the peak impurity concentration, and ambient conditions. It is heaviest in the region of nuclear scattering and substantially less in the region of electronic scattering, usually resulting in a subsurface layer of high crystal damage.

The impurity profiles obtained by solid-state diffusion and by ion implantation are different. Solid-state diffusion generally yields impurity concentration profiles ranging from complementary error function to Gaussian distributions depending on the boundary conditions. The concentration profile in a solid target obtained through ion implantation consists of an impurity concentration peak at the most probable penetration depth (x_P) and a long tail representing components of deep penetration ($\geq 5x_P$). The deep penetration is attributed to "channeling" which occurs readily in crystal lattices with an open structure; most semiconductors have an open structure so that channeling effects are very pronounced. In the channeling event, an atom with high energy can readily move between atoms aligned in the close-packed directions. Because the injected ion makes only a glancing collision with the channel wall, the total energy lost per unit distance traveled is small and deep penetration occurs.

210 IMPURITIES IN SEMICONDUCTORS

The slowing down of energetic ions depends strongly on the crystallographic orientation of the lattice with respect to the incident beam. Even for the least penetrated directions, the observed depth is about three times the value expected in the absence of crystal defects. A drastic change in penetration depth occurs when the direction of the beam is misaligned with that of the crystal (Table 2.13). A reduction of more than one order of magnitude in penetration depth is found when the crystal is tilted by 7° from the $\langle 110 \rangle$ orientation. The penetration depth increases with the incident ion energy. However, the penetration depth is reduced with an increase in ion flux. The penetration depth of incident ions in an amorphous target is much smaller than in a crystalline target of the same material. The deep channeling tail is absent in amorphous targets.

TABLE 2.13

Critical Angle Below Which Channeling Occurs for Selected Ions in Silicon as a Function of Target Orientation. $E_{I_o} = 50$ keV*

Ion	Critical angle (deg)		
	$\langle 100 \rangle$	$\langle 110 \rangle$	$\langle 111 \rangle$
As	4.0	5.2	4.4
B	2.9	3.7	3.2
N	3.0	4.0	3.4
P	3.5	4.5	3.8

* The critical angle decreases with increasing energy of the ion.

After an incident ion has dissipated its kinetic energy, it typically lodges in an interstitial position of the target lattice; many target atoms are displaced by the incident ion during its penetration of the crystal and the lattice becomes supersaturated with interstitials and vacancies. Recombination of these interstitials and vacancies is enhanced under supersaturating conditions. The implanted ions move more readily than lattice atoms from their original interstitial sites to substitutional sites. This reaction is further enhanced with increasing temperature. High-temperature annealing can be considered as a trapping process for vacancies followed by recombination with interstitials.

Ion bombardment damage shifts the semiconductor Fermi level towards the center of the energy gap, i.e., towards a more intrinsic value, and is thus equivalent to a compensating doping.

When a target is continuously bombarded by an ion beam, the concentration of the implanted ions will tend to saturate rather than increase indefinitely. This saturation involves an equilibrium between bombardment, sputtering, scattering, and the lattice properties.

During the bombardment, not all ions will be retained by the target. Retention probability increases with increasing ion mass and energy. Not only may the incident ions be rejected by the target, but they may also cause the target atoms to be expelled. The ratio of the number of the ejected target atoms to that of the incident ions (sputtering yield) may vary by several orders of magnitude. Energetic incident ions can displace target atoms from one site to another within the target. The number of displaced atoms per incident ion (damage ratio) may vary by several orders of magnitude. Typical values of sputtering yield and damage ratio range from 10^{-1} to 10^2.

Through the bombardment, the incident ions may become incorporated into the surface layer of the target resulting in certain chemical effects. However, since the damage ratio is typically greater than unity and the concentration of displaced atoms in the surface layer is even higher, the damage effects usually shadow the chemical effects of the implanted ions.

Semiconductor doping can be achieved by implanting with ions which are virtually insoluble in the host material so that there is the possibility of introducing new materials to solid-state device technology. Incident ions which can conceivably be used for implantation are alkali metals and certain gases.

2. ENERGY LOSS OF PROJECTILE

An ion with an initial kinetic energy at the semiconductor surface ($E_{I_o} > 0$) upon entering the semiconductor will lose its energy to the target and eventually come to a stop. The energy loss within the target is due to

(a) interaction of the ion with the electrons (both bound and free) in the target;
(b) collisions of the ion with the nuclei of the target.

Both the electronic and the nuclear scattering mechanisms are assumed to be independent of each other. The average energy loss per unit distance in the target is

$$-dE_I/dx = C_L[S_e(E_I) + S_n(E_I)] \qquad (2.93)$$

where E_I = energy of particle at distance x from point of incidence (i.e., approximately from semiconductor surface)
C_L = average density of target (for silicon $C_L = 5 \cdot 10^{22}$ cm^{-3})
$S_e(E_I)$ = electronic stopping power as function of particle energy
$S_n(E_I)$ = nuclear stopping power as function of particle energy.

In a first-order approximation the stopping powers (dimension of energy times cross-sectional area) are:

$$S_e(E_I) = k_I E_I^{1/2} [\text{eV cm}^2] \qquad (2.94)$$

where k_I is an empirical factor ($k_I \approx 2 \cdot 10^{-16}$ eV$^{1/2}$ cm^2 for most projectiles in silicon);

$$S_n(E_I) = 2.8 \cdot 10^{-15} \frac{Z_I Z_T}{(Z_I^{2/3} + Z_T^{2/3})^{1/2}} \frac{M_I}{M_I + M_T} \quad [\text{eV cm}^2] \quad (2.95)$$

which, in this approximation (Rutherford scattering), are independent of energy. In the more exact case of Thomas-Fermi scattering, $S_n(E_I)$ is a function of energy. Z and M are atomic number and mass, subscripts T refers to target, subscript I to implanted ion.

At high energy electronic stopping is dominant and at low energy nuclear stopping is dominant; i.e., upon entering the target, the ion will first suffer electron scattering close to the semiconductor surface and, after losing much of its kinetic energy, suffer nuclear stopping. At some distance from the semiconductor surface (indicated in Figure 2.31) both scattering mechanisms will be equal:

$$S_e(E_I) = S_n(E_I).$$

At this point the energy of the ion in the target has been reduced to E_I^*. For silicon

$$E_I^* = \frac{3.83 \cdot 10^4 Z_I^2}{14^{2/3} + Z_I^{2/3}} \left(\frac{M_I}{M_I + 28} \right)^2 \quad (2.96)$$

3. ENERGY TRANSFER DURING COLLISION

The maximum energy that can be transferred from an ion whose energy at the moment of collision is E_I to a target of energy E_T is given by:

$$\gamma_E = E_T/E_I = 4 \frac{M_I M_T}{(M_I + M_T)^2} \quad (2.97)$$

where M_T is the mass of the target (electron or nucleus) with which the ion is colliding.

For electronic stopping ($M_T = m_n$):

$$M_I \gg m_n, \quad \text{i.e., } \gamma_E \ll 1;$$

for nuclear stopping ($M_T = M_{Si}$):

$$M_I \approx M_T, \quad \text{i.e., } \gamma_E \approx 1.$$

Since during a collision with a nucleus the ion thus loses most of its energy, it will come to rest shortly after the first nuclear collision.

4. RANGE OF PROJECTILE

The total distance that the ion travels within the target in coming to rest is called its range; the projection of this distance on the direction of incidence is the projected range (x_P). Because both the number of collisions and the energy

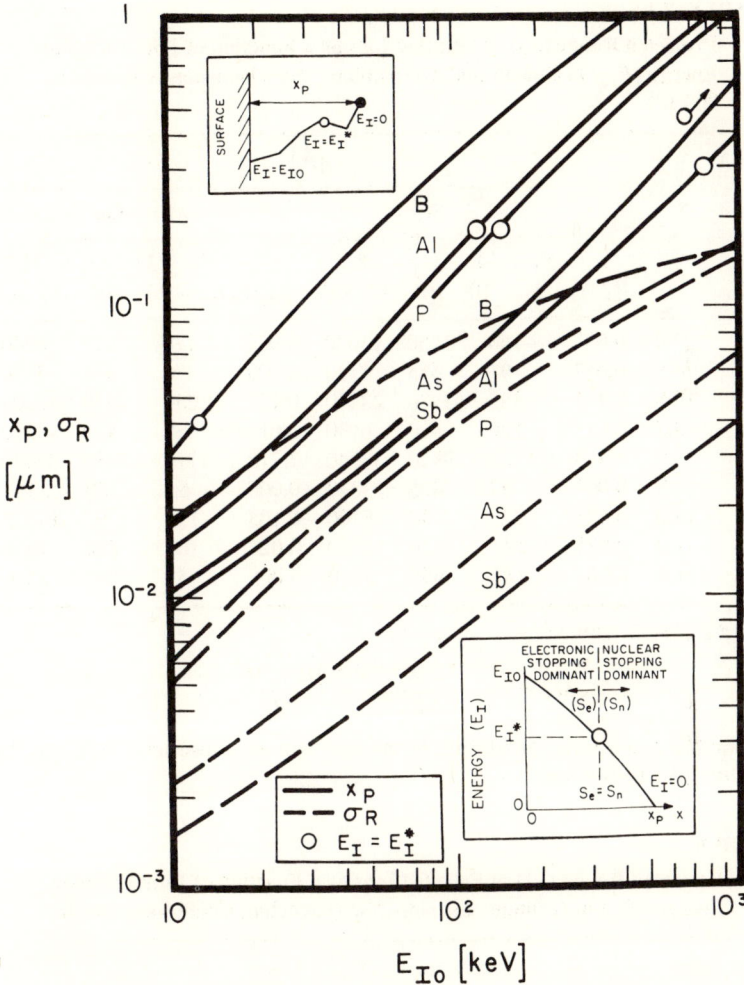

FIGURE 2.31

Projected range (x_p) and standard deviation (σ_R) vs. energy of projectile at semiconductor surface (E_{Io}). Various impurities in silicon at 300°K. The insets show the variation of energy with distance from the semiconductor surface and define the projected range. The standard deviation is defined in Figure 2.32. Figures 2.31, 2.32 and 2.33 show experimental data on silicon targets. The incident angle was 7° from the $\langle 111 \rangle$ orientation, i.e., the target appeared to be amorphous. No channeling was assumed. Beam area $A_I = 25$ cm².

TABLE 2.14

Projected Range (x_P) of Selected Ions as a Function of Initial Kinetic Energy (E_{Io}) in Silicon and Germanium. Nonchanneling Conditions, 300°K*

Ion				x_P[Å]					
				Si			Ge		
				$Z_T = 14$,	$M_T = 28.1$		$Z_T = 32$,	$M_T = 72.6$	
	Z_I	M_I	Z_*M_*	$E_{Io}=10$	10^2	10^3 keV	Z_*M_* $E_{Io}=10$	10^2	10^3 keV
Al	13	27.0	0.037	160	1450	13020	0.033 120	950	8940
As	33	74.9	0.012	95	585	5730	0.008 70	395	3650
B	5	10.8	0.151	385	3980	23230	0.177 255	2550	18300
Ga	31	69.7	0.013	100	610	6130	0.010 70	420	3910
H	1	1.0	5.413	1720	9920	38340	7.718 1110	8410	37050
In	49	114.8	0.007	90	465	3820	0.005 60	300	2380
N	7	14.0	0.093	275	2840	18700	0.105 190	1790	14030
P	15	31.0	0.031	145	1230	11760	0.027 105	820	7930
Sb	51	121.8	0.007	85	455	3680	0.005 55	295	2280

* The term Z_*M_* is defined as

$$Z_*M_* \equiv \frac{(Z_I^{2/3}+Z_T^{2/3})^{1/2}}{Z_I Z_T} \frac{M_I + M_T}{M_I}$$

Atomic number and atomic weight are given as Z and M, subscript T refers to the target, subscript I to the ion.

TABLE 2.15

Projected Range (x_P) of Selected Ions as a Function of Initial Kinetic Energy (E_{Io}) in Gallium Arsenide. Nonchanneling Conditions, 300°K

Ion			x_P[Å], GaAs		
			$Z_T = 32$,	$M_T = 72.3$	
	Z_I	M_I	$E_{Io} = 10$	10^2	10^3 keV
Cd	48	112.4	60	305	2,420
Ge	32	72.6	70	405	3,760
H	1	1.0	4,500	13,000	100,000
Se	34	79.0	65	385	3,510
Si	14	28.1	110	880	8,220
Sn	50	118.7	60	300	2,320
Te	52	127.6	55	290	2,220
Zn	30	65.4	70	430	4,040

2-5 ION IMPLANTATION

transferred per collision are random variables, all ions of a given type and of the same initial energy will not have the same range so that there will be a range distribution (defined by the standard deviation, σ_R). The range distribution determines the impurity gradient in the semiconductor; it is a function of the energy, mass and atomic number of the projectile, of the mass, atomic number, density and temperature of the target, and of the ion dose rate and the total dose. In a single-crystalline target the range distribution will also depend critically upon the target orientation with respect to the incident ion beam and the surface conditions.

In a first-order approximation x_P can be estimated under the assumption that actual range and projected range are approximately equal as follows. If electronic stopping dominates (i.e., if $E_{I_o} \gg E_I^*$):

$$x_P \approx 20 \, E_{I_o}^{1/2} \tag{2.98}$$

and if nuclear stopping dominates (i.e., if $E_{I_o} \ll E_I^*$):

$$x_P \approx 0.7 \frac{(Z_I^{2/3} + Z_T^{2/3})^{1/2}}{Z_I Z_T} \cdot \frac{M_I + M_T}{M_I} E_{I_o}. \tag{2.99}$$

TABLE 2.16

Projected Range (x_P) of Selected Ions as a Function of Initial Kinetic Energy (E_{I_o}) in Al, SiO$_2$, and Si$_3$N$_4$. Nonchanneling Conditions, 300°K

Ion	x_P[Å]								
	Al			SiO$_2$			Si$_3$N$_4$		
	$E_{I_o}=10$	10^2	10^3 keV	$E_{I_o}=10$	10^2	10^3 keV	$E_{I_o}=10$	10^2	10^3 keV
As	85	510	5080	75	475	4730	60	365	3660
B	330	3100	15910	305	3090	17140	235	2370	13240
Cd	75	410	3450	70	385	3250	55	300	2510
Ga	85	535	5430	80	495	5060	60	380	3910
Ge	85	520	5240	80	485	4880	60	375	3770
H	1220	7570	34570	1300	6920	25910	1000	5330	19990
In	75	405	3380	70	380	3190	55	295	2460
N	230	2230	13260	215	2250	14080	165	1730	10880
P	125	1080	10280	110	990	9280	85	765	7160
Sb	75	400	3250	70	375	3070	55	290	2370
Se	85	500	4910	75	465	4580	60	360	3540
Si	130	1180	10710	115	1070	9650	90	830	7440
Sn	75	405	3310	70	380	3120	55	295	2410
Te	75	400	3180	70	375	3000	55	290	2320
Zn	85	550	5630	80	505	5240	60	390	4050

These expressions will yield the projected range in Å if the ion energy E_{I_o} is given in eV. Figure 2.31 gives the projected range (x_P), i.e., the distance from the semiconductor surface ($x = 0$) at which the peak impurity concentration (C_P) is found. The standard deviation (σ_R) which is also given and which is defined as the distance from x_P at which the impurity concentration has fallen to $C_P/\sqrt{e}$ (i.e., to 60.7% of the peak concentration) describes the impurity distribution and the impurity gradient. Tables 2.14 through 2.17 show projected range and standard deviation of selected ions in the most important semiconductors and related materials.

Theoretically, the curve $C(x)$ vs. x is symmetrical around $x = x_P$. In practice, deviations are observed in crystals

(a) close to the surface ($x \approx 0$) where the impurity concentration is frequently higher than expected due to ions being unable to enter the target;
(b) far beyond the range ($x > 2x_P$) due to some ions being scattered into a channeling axis by which they achieve a greater penetration depth.

TABLE 2.17

Standard Deviation (σ_R) as a Function of Initial Kinetic Energy (E_{I_o}) of Selected Ions in Si, Ge, and GaAs. Nonchanneling Conditions, 300°K

Ion	σ_R[Å]								
	Si			Ge			GaAs		
	$E_{I_o}=10$	10^2	10^3 keV	$E_{I_o}=10$	10^2	10^3 keV	$E_{I_o}=10$	10^2	10^3 keV
Al	65	420	1610	130	425	1830			
As	25	125	820	30	150	955			
B	190	940	1810	220	1010	2370			
Cd							20	95	605
Ga	25	135	890	30	160	1020			
Ge							30	155	980
H	580	935	1020	545	1440	1730			
In	15	75	490	20	95	595			
N	130	750	1730	150	765	2130			
P	55	355	1530	90	365	1720			
Sb	15	75	465	20	90	560			
Se							25	140	910
Si							80	395	1750
Sn							20	90	570
Te							20	85	530
Zn							30	170	1060

5. IMPURITY DISTRIBUTION

The impurity distribution in a semiconductor after nonchanneling ion implantation is determined by random scattering of ions. The highest impurity concentration is not found at the semiconductor surface (as in most diffusion processes) but at a distance x_P from the surface. Around x_P the impurity concentration falls off approximately symmetrically in a Gaussian-type distribution. Because of differing masses the impurity profiles are different for different impurity ions.

The impurity distribution can be described by the empirical relationship

$$C(x) = C_P \exp\left[-(x-x_P)^2/2\sigma_R^2\right] - C_B \tag{2.100}$$

where C_P, x_P, and σ_R depend upon the properties of the projectile in addition to those of the target. C_P itself depends upon the standard deviation and is given for a beam area at the target $A_I = 25$ cm^2 by the empirical relationship

$$C_P = 10^{14} It/8\sigma_R^2 [\text{cm}^{-3}], \tag{2.101}$$

where It is the beam current density (measured in μA sec), σ_R is the standard deviation (measured in μm), so that $C(x)$ as a function of σ_R:

$$C(x) = 1.25 \cdot 10^{13}(It/\sigma_R^2) \exp\left[-(x-x_P)^2/2\sigma_R^2\right] - C_B. \tag{2.102}$$

Usually $C(x) \gg C_B$ so that C_B can be neglected.

6. PEAK IMPURITY CONCENTRATION

Figure 2.32 shows the peak concentration (C_P) for a given ion beam current density (It). For a given ion and a given ion energy C_P is proportional to It, i.e., proportional to the beam current and the implantation time. The effect of increasing It on C_P is an upward shift of the $C(x)$-curve along the x_P-line with only a slight effect on x_P and σ_R. The following empirical equation relates C_P and It:

$$C_P = 3 \cdot 10^{14} It/A_I \sigma_R^2 [\text{cm}^{-3}] \tag{2.103a}$$

where A_I is the beam area at the target in cm^2, It is the beam current density in μA sec, and σ_R is the standard deviation in μm. For $A_I = 25$ cm^2

$$C_P = 1.2 \cdot 10^{13} It/\sigma_R^2 [\text{cm}^{-3}]. \tag{2.103b}$$

The implantation dosage, Q_i, i.e., the number of implanted ions per unit area, is related to the impurity concentration, $C(x)$, i.e., the density of implanted ions per unit volume, by

$$C(x) = (1/\sqrt{2\pi})(Q_i/\sigma_R) \exp\left[-(x-x_P)^2/2\sigma_R^2\right] \tag{2.104}$$

218 IMPURITIES IN SEMICONDUCTORS

which leads to the peak concentration at $x = x_P$

$$C_P = (10^4/\sqrt{2\pi})Q_i/\sigma_R \tag{2.105}$$

where σ_R is again expressed in μm.

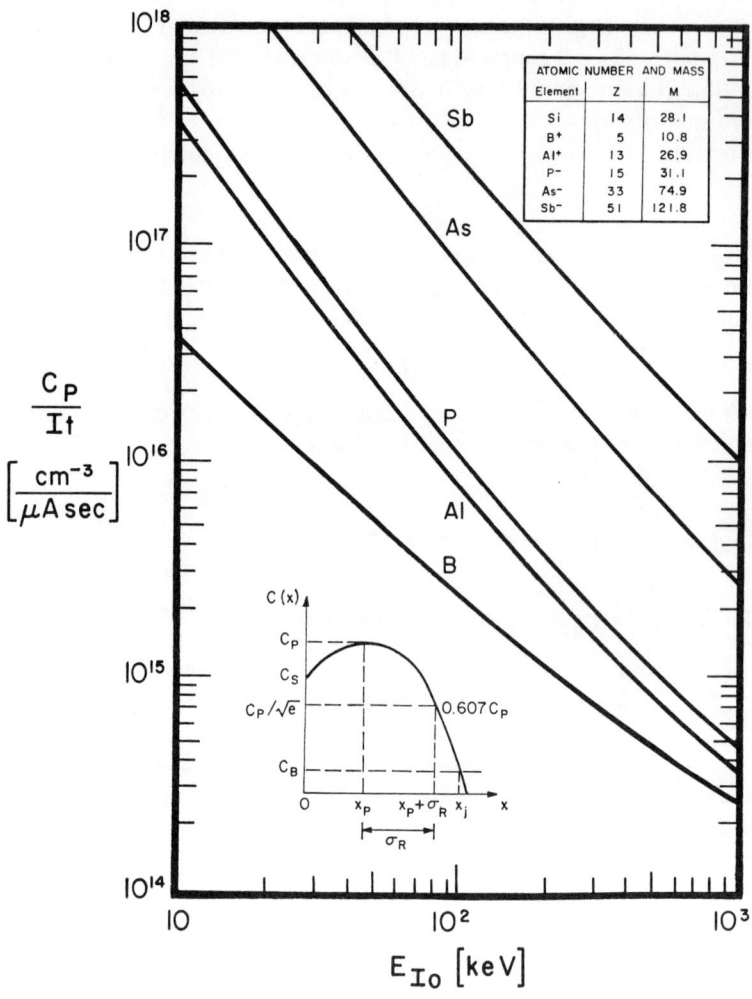

FIGURE 2.32

Ratio of peak impurity concentration (C_P) to beam current density (It) vs. energy of projectile at semiconductor surface (E_{I_o}). Various impurities in silicon at 300°K. The upper inset gives atomic number and mass of common impurities in silicon. The lower inset shows the profile of implanted impurities and defines standard deviation and relevant impurity concentrations.

7. IMPURITY CONCENTRATION GRADIENT

The impurity gradient at the junction (x_j) below the peak concentration is shown in Figure 2.33 for arbitrary target and projectile assuming a Gaussian impurity distribution. The use of this illustration requires knowledge of

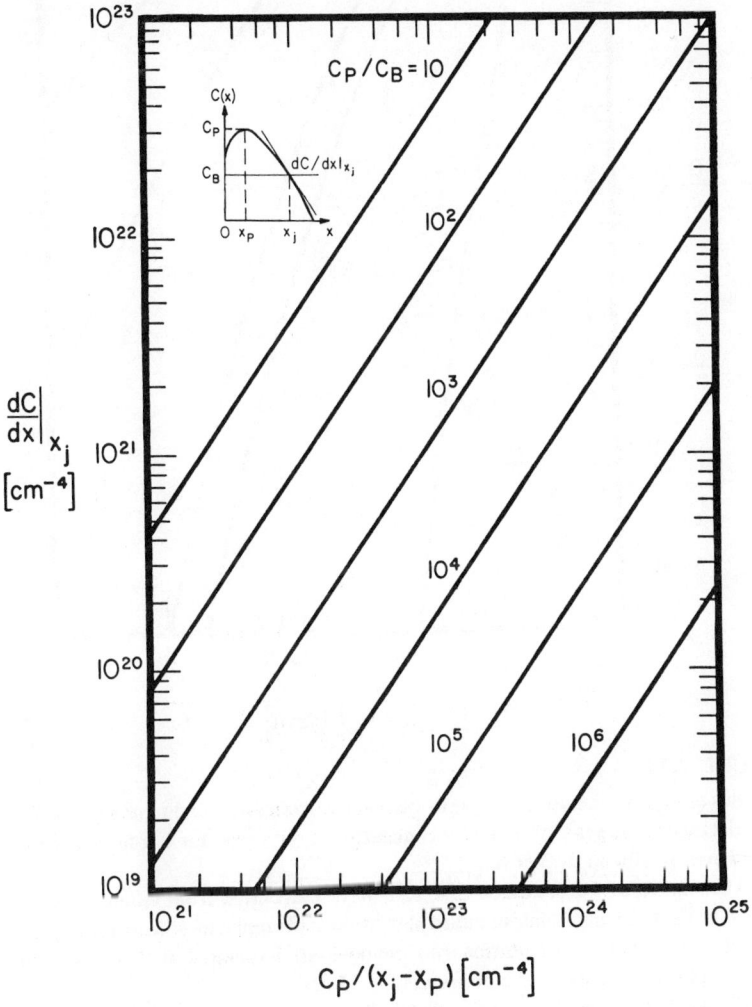

FIGURE 2.33

Impurity concentration gradient at the junction (dC/dx) vs. ratio of peak impurity concentration (C_P) to the difference between junction depth (x_j) and depth at peak impurity concentration (x_P) for various ratios C_P/C_B. Arbitrary semiconductor.

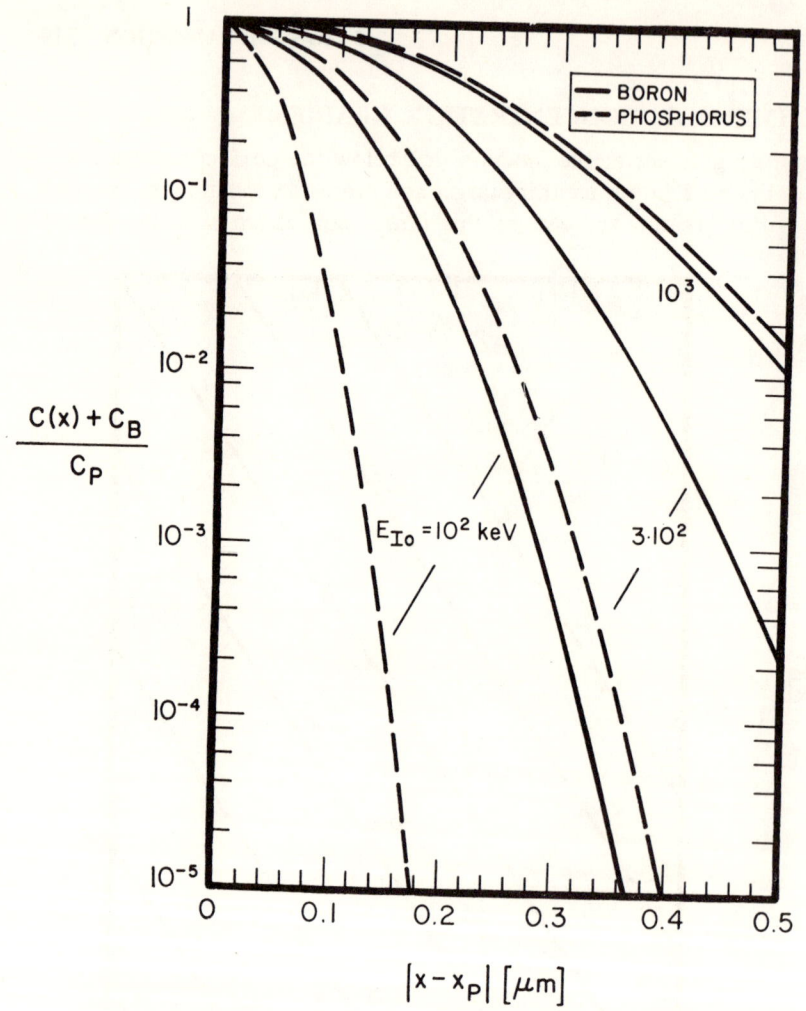

FIGURE 2.34

Impurity concentration ($C(x)$) at distance x from semiconductor surface vs. distance from peak of impurity concentration ($x - x_P$) for boron and phosphorus in silicon at 300°K.

The curves of Figures 2.34 and 2.35 were calculated from empirical data of projected range and standard deviation and apply to silicon targets 7° off the <111> orientation (no channeling) implanted with boron and phosphorus at 300°K. Beam area $A_I = 25$ cm².

Figure 2.34 shows the impurity distribution as function of distance from x_P. This presentation indicates that the lighter the projectile the sharper the impurity distribution peak, i.e., the steeper the impurity gradient. A higher initial energy results in a smaller impurity gradient for the same impurity. Since the impurity distribution is assumed to be symmetrical around x_P this illustration can be used to determine the impurity concentration below the peak as well as the impurity concentration at the semiconductor surface if x_P is known.

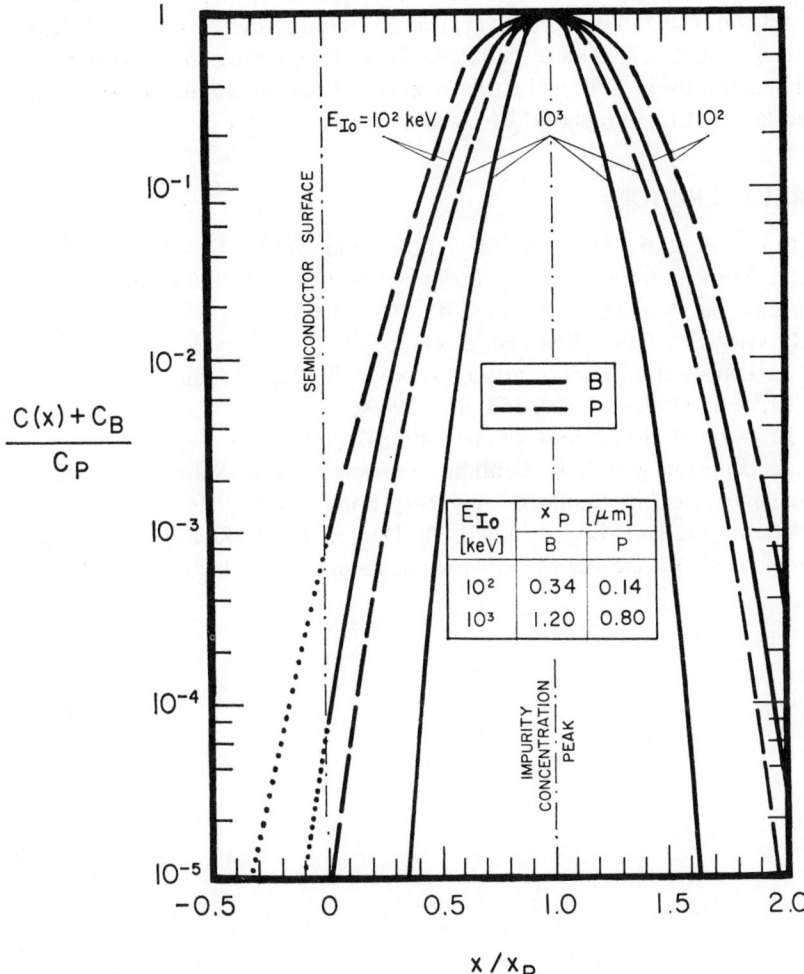

FIGURE 2.35

Impurity concentration ($C(x)$) at distance x from semiconductor surface vs. distance from surface (x) and location of impurity concentration peak (x_P). Boron and phosphorus in silicon at 300°K. This illustration shows the impurity concentration as function of normalized distance from x_P where the semiconductor surface is at $x/x_P = 0$ and the peak impurity concentration at $x/x_P = 1$.

junction depth, as given in Figure 2.34 for specific elements, and of peak concentration (C_P) and peak depth (x_P), as given in Figures 2.31 and 2.32. The impurity gradient curves were calculated by differentiating the equation for impurity distribution and evaluating it at x_j. They are similar to those of diffused layers if C_S is replaced by C_P and x_j by $(x_j - x_P)$.

8. REFERENCES

(1) J. Lindhard et al., *Mat. Fys. Medd. Dan. Vid. Sclsk.*, **33**, 1 (1963).
(2) K. E. Manchester et al., *Nucl. Instr. and Methods*, **38**, 169 (1965).
(3) J. A. Davies et al., *Can. J. Phys.*, **45**, 4053 (1968).
(4) J. F. Gibbons, *Proc. IEEE*, **56**, 295 (1968).
(5) D. E. Davies et al., *Solid-State Technology*, 33 (Oct. 1968).
(6) L. Eriksson et al., *Science*, **163**, 627 (1969).
(7) L. Eriksson et al., *J. Appl. Physics*, **40**, 842 (1969).
(8) W. S. Johnson and J. F. Gibbons, *Projected Range Statistics in Semiconductors*, dist. by Stanford University Book Store, 1969.
(9) J. W. Mayer, L. Eriksson, and J. A. Davies, *Ion Implantation in Semiconductors—Silicon and Germanium*, Academic Press, New York, 1970.

2-6 Energy Levels of Impurities

1. ACTIVATION ENERGY OF SHALLOW IMPURITIES

Shallow impurity levels (i.e., energy levels close to either conduction or valence band edge, but within the energy gap) arise from the substitution of a lattice atom by an atom from the adjacent column of the Periodic System. Thus, a donor atom has a charge $+q$ relative to the lattice ion it replaces. Similarly, shallow acceptor levels arise from substituent ions of charge $-q$ relative to the one replaced. Therefore, the model of the hydrogen atom can be used to obtain in a first-order approximation the activation energy of shallow donors and acceptors. Typical values range from 10^{-3} to 10^{-1} eV for most semiconductors. Although donor and acceptor levels have a spin degeneracy of two, a given center can capture only one electron, of either spin.

Generally, the activation energy is defined as

$$E_{aD} = E_c - E_D \quad \text{for donors,} \quad (2.106)$$

$$E_{aA} = E_A - E_v \quad \text{for acceptors.} \quad (2.107)$$

The activation energy (ionization energy, E_a) of shallow impurities (which for Group IV semiconductors are impurities of Groups III and V) is related to the effective carrier mass (m_n/m_o or m_p/m_o) and the dielectric constant of the semiconductor (ε_s) by

$$E_{aD} = E_H(m_n/m_o)/\varepsilon_s^2 = q^4 m_n/8h^2(\varepsilon_o \varepsilon_s)^2, \quad (2.108)$$

$$E_{aA} = E_H(m_p/m_o)/\varepsilon_s^2 = q^4 m_p/8h^2(\varepsilon_o \varepsilon_s)^2, \quad (2.109)$$

where E_H is the ionization energy of a hydrogen atom

$$E_H = m_o q^4/8(\varepsilon_o h)^2 - 13.59 \text{ eV}. \quad (2.110)$$

The model of the hydrogen atom can be used in this simple calculation since the activation mechanism of Groups III and V elements in Group IV semiconductors due to their lack or excess of one electron is similar to the mechanism of an electron moving around a fixed proton in a hydrogen atom, except that the mass of the electron is not the rest mass m_o but m_n and that the electron moves in a medium of dielectric constant $\varepsilon_s \varepsilon_o$. In this first approximation all elements of Groups III and V should have the same activation energy.

224 IMPURITIES IN SEMICONDUCTORS

For example, for shallow impurities (P, Sb, As; Ga, Al, B) in silicon ($m_n/m_o = 0.23$, $m_p/m_o = 0.12$, $\varepsilon_s = 11.7$):

$$E_{aD} \approx 0.023 \text{ eV}, \quad E_{aA} \approx 0.012 \text{ eV}$$

which is in fair agreement with experimentally determined activation energies.

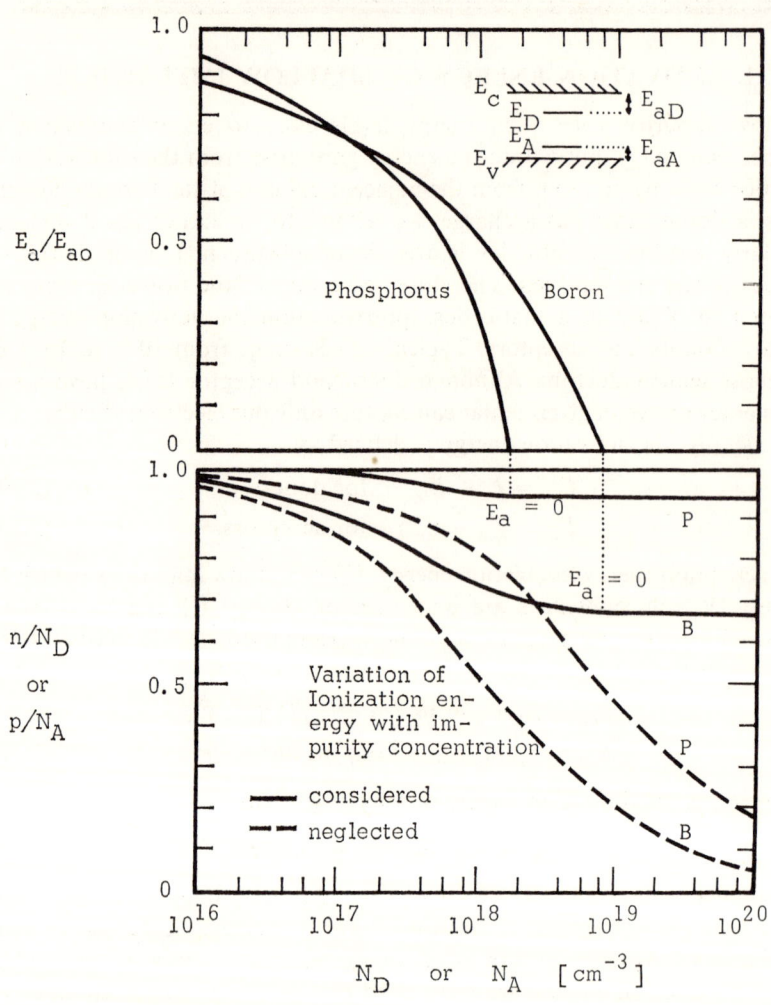

FIGURE 2.36
Decrease of activation energy (E_a) and of carrier density (n or p) from their nondegenerate values with impurity concentration (C_B; N_D or N_A) for boron (B) and phosphorus (P) in silicon. 300°K.

TABLE 2.18
Activation Energies of Impurities in Si, Ge, and GaAs

Group	Impurity	Ionization Energy [eV]					
		Si		Ge		GaAs	
		E_{aD} $E_G = 1.12$ eV	E_{aA}	E_{aD} $E_G = 0.67$ eV	E_{aA}	E_{aD} $E_G = 1.43$ eV	E_{aA}
I	Li	0.033 0.88	0.49	0.009	0.32 0.45 0.62		0.063
	Cu	0.88	0.49				0.023
	Ag	0.33	0.34		0.13 0.38 0.58		
	Au	0.77	0.58	0.62	0.16 0.47 0.63		
II	Be				0.07		
	Zn		0.31 0.55		0.03 0.09		0.26
	Cd				0.06 0.20		0.04
III	B		0.045 0.092 0.126		0.010		
	Al		0.057 0.087		0.010		
	Ga		0.065		0.011		
	In		0.16		0.011		
	Tl		0.26		0.010		
V	P	0.044		0.012			
	As	0.049		0.013			
	Sb	0.039		0.010			
	Bi	0.067					

TABLE 2.18 (*Continued*)

Group	Impurity	Si E_{aD} $E_G = 1.12$ eV	Si E_{aA}	Ge E_{aD} $E_G = 0.67$ eV	Ge E_{aA}	GaAs E_{aD} $E_G = 1.43$ eV	GaAs E_{aA}
VI	O		0.03	0.01		0.002	
	S	0.18 0.37		0.18		0.03	
	Cr				0.07 0.12		0.70
	Se	0.37		0.14 0.28		0.004	
	Te			0.11 0.30		0.004	
VII	Mn	0.53			0.16 0.30		0.095
VIII	Fe	0.55 0.72			0.34 0.40		0.36 0.59
	Co			0.58	0.25		0.54
	Ni		0.22 0.77		0.36		0.53
	Pt	0.37					

For impurities other than those of neighboring Groups of the Periodic Table the hydrogen model and above relationship are not applicable.

Semiconductors with a large energy gap (which is due to a larger amplitude of the periodic variations of the crystal potential) are normally characterized by carriers of a large effective mass (which tends to reduce carrier mobility) and a small polarizability (hence small dielectric constant). Therefore, the ionization energy of shallow impurities, as given above, will normally increase with energy gap; for this reason materials for which $E_G > 3$ eV are usually poor conductors unless the temperature is raised or if nonhydrogen-like impurities are introduced.

The activation energy varies slightly with temperature because of its dependence upon band edge energy (E_c or E_v) and impurity level (E_D or E_A) both of which are temperature-dependent. E_a varies also with impurity concentration since at high concentration a broadening of the donor or acceptor levels will take place (Figure 2.36) so that bands from shallow levels may eventually overlap the conduction or valence band, resulting in a diminishing ionization energy.

Table 2.18 lists activation energies of impurities in nondegenerate semiconductors at 300°K. The donor levels (E_{aD}) are given with respect to the conduction band energy, the acceptor levels (E_{aA}) with respect to the valence band energy.

2. ACTIVATION ENERGY OF DEEP IMPURITIES

While impurities of neighboring columns of the Periodic System have energy levels close to either conduction or valence band and consequently facilitate the generation of free carriers, impurities of other columns have energy levels deep within the energy gap and, consequently, do not substantially contribute to carrier generation. Their importance, however, lies in the fact that they act as carrier recombination centers and that they are able to reduce carrier lifetime. In silicon, the most important impurities of this type are gold and nickel. In the case of deep impurities the hydrogen model is not applicable.

3. CARRIER DENSITY AND ACTIVATION ENERGY

The density of free carriers is a function of the activation energy (E_a) of a shallow impurity:

(a) n-type semiconductor:

$$n = (1/2) N_c \exp(-E_a/kT)\{[1 + 4(N_D/N_c) \exp(E_a/kT)]^{1/2} - 1\}; \quad (2.111)$$

(b) p-type semiconductor:

$$p = (1/2) N_v \exp(-E_a/kT)\{[1 + 4(N_A/N_v) \exp(E_a/kT)]^{1/2} - 1\}. \quad (2.112)$$

At $N_D \ll (1/4) N_c \exp(-E_a/kT)$ or $N_A \ll (1/4) N_v \exp(-E_a/kT)$ the free carrier concentration is equal to N_D or N_A. At higher impurity concentration, however, the free carrier concentration is reduced and proportional to $N_D^{1/2}$ or $N_A^{1/2}$, if the dependence of the ionization energy upon impurity concentration is neglected ($dE_a/dC_B = 0$). If the variation of ionization energy with impurity concentration is taken into account ($dE_a/dC_B \neq 0$) then the reduction in free carrier concentration is considerably less, i.e., in the degenerate case $n/N_D = 0.94$ and $p/N_A = 0.66$ at all impurity concentrations. The degeneracy temperature at which the transition from nondegenerate to degenerate semiconductor occurs is given in Figure 1.15.

In a nondegenerate semiconductor the number of free carriers is assumed to be equal to the number of ionized impurity atoms. If $N_A \ll N_D$ or $N_D \ll N_A$, the donor or acceptor concentration can be replaced by C_B.

4. DECREASE OF ACTIVATION ENERGY WITH IMPURITY CONCENTRATION

At high impurity concentration the energy levels corresponding to donors and acceptors broaden to form localized bands of their own until they finally overlap with conduction or valence band, i.e., the activation energy diminishes. The value for ionization energy is then taken as that of the impurity band edge closest to conduction or valence band (Figure 2.36).

The decrease of ionization energy with increasing impurity concentration is due to a decrease in the average potential energy of electrons or holes. Since the conduction electrons will shield the ions from one another so that, on the average, small regions will be electrically neutral, there will be a resultant potential energy of attraction which will increase with increasing concentration. Therefore the ionization energy will vary with impurity concentration as

$$E_a = E_{ao} - a_* C_B^{1/3} \qquad (2.113)$$

where E_{ao} is the activation energy at very low impurity concentration and a_* is an impirically determined material constant which is related to the radius of the electrically neutral sphere around the ion (i.e., the Debye length). In the case of boron in silicon

$$E_a = E_A - E_v = 0.045 - 4.3 \cdot 10^{-8} N_A^{1/3} [\text{eV}],$$

i.e., $E_{ao} = 0.045$ eV and $a_* = 4.3 \cdot 10^{-8}$ eV cm.

A similar dependence of activation energy upon impurity concentration for near-degenerate silicon exists for phosphorus and the other shallow impurities.

5. REFERENCES

(1) G. L. Pearson and J. Bardeen, *Phys. Rev.*, **75**, 865 (1949).
(2) E. M. Conwell, *Proc. IRE*, **46**, 1281 (1958).
(3) S. M. Sze and J. C. Irvin, *Solid-State Electronics*, **11**, 599 (1968).
(4) H. F. Wolf, *Silicon Semiconductor Data*, Pergamon Press, Oxford, 1969.
(5) V. I. Fistul', *Heavily Doped Semiconductors*, Plenum Press, New York, 1969.

2-7 Growth of Epitaxial Film

1. FILM GROWTH CONDITIONS

The growth of semiconducting thin films can be accomplished by several different methods whose choice depends upon the required crystal order, impurity concentration, and crystal perfection. The most widely used methods are as follows.

(a) Vacuum evaporation:
   This method is used mainly for compound semiconductors in conjunction with condensation of the compound or its synthesis in vacuum by the evaporation of its elemental constituents. It is the simplest method available and does not require accurate control of the thermodynamic parameters of the crystallization process. The two main types are:
   (i) *Flash evaporation*:
      In this method a fine-mesh powder of the semiconductor is evaporated in vacuum at elevated temperature.
   (ii) *Three-temperature method*:
      In this method three crucibles are maintained at different temperatures—two contain one of the constituent elements of the compound semiconductor and one contains the substrate.
(b) Chemical vapor phase synthesis:
   This method includes the closed-tube or open-tube epitaxial growth of single- or poly-crystalline semiconductor films grown by chemical vapor phase transport reactions at temperatures below the melting points of the elemental semiconductor or of the constituents of compound semiconductors.
(c) Chemical liquid phase synthesis:
   This method consists of dissolving a compound semiconductor into a suitable metal solvent below the melting point of the compound.

The growth rate of an epitaxial film (film deposition rate) depends, among others, on deposition temperature, rate of gas flow, and gas composition. In terms of reaction rate constant, mass transfer coefficient, gas and solid concentrations, the growth rate of an epitaxial film

$$r_g = \frac{dx_f}{dt} = \frac{k_S h_G}{k_S + h_G} \frac{C_T}{C_L} Y = \frac{k_S h_G}{k_S + h_G} \frac{C_G}{C_L} \qquad (2.114)$$

230 IMPURITIES IN SEMICONDUCTORS

and the surface impurity concentration of the epitaxial film

$$\frac{C_S}{C_G} = \frac{1}{1 + k_S/h_G} \qquad (2.115)$$

where x_f = epitaxial film thickness

k_S = chemical surface reaction rate constant; $k_S = 10^7 \exp(-E_a/kT)$ cm/sec

h_G = gas-phase mass-transfer coefficient in terms of concentration in the gas; h_G is insensitive to temperature

h_S = gas-phase mass-transfer coefficient in terms of concentration in the solid; h_S is temperature-dependent, $h_S = h_G/H_G kT$

H_G = Henry's Law constant

C_S = film surface impurity concentration

C_T = total number of molecules per cm³ in the gas

C_G = SiCl$_4$ concentration in the bulk of the gas

C_L = number of semiconductor atoms incorporated per cm³ of film; for silicon, $C_L = 5 \cdot 10^{22}$ cm^{-3}

E_a = activation energy; for silicon, $E_a \approx 1.9$ eV

Y = mole fraction, $Y = C_G/C_T$.

Limiting cases:

(a) $h_G \ll k_S$ (mass transfer control):

$$r_g = (C_T/C_L)h_G \; Y = (C_G/C_L)h_G$$
$$C_S = 0$$

(b) $h_G \gg k_S$ (surface reaction control):

$$r_g = (C_T/C_L)k_S \; Y = (C_G/C_L)k_S$$
$$C_S = C_G.$$

At low temperature (surface reaction control) the growth rate increases exponentially with temperature:

$$r_g \sim \exp(-E_a/kT).$$

At high temperature (mass transfer control) the growth rate becomes less temperature-dependent. Epitaxial growth usually takes place at high temperature, i.e., the growth is mass-transfer-controlled.

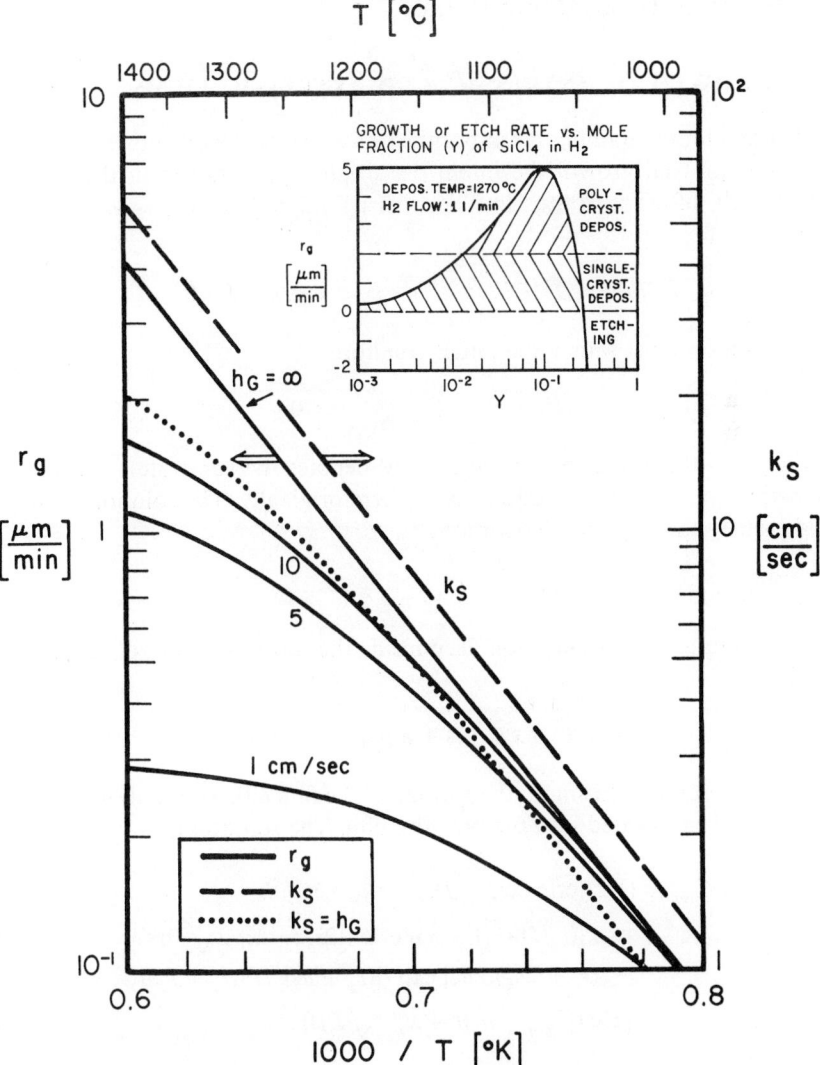

FIGURE 2.37

Growth rate (r_g) of epitaxial film and surface reaction rate constant (k_S) vs. film deposition-temperature (T) and gas-phase mass-transfer coefficient (h_G). Si–SiCl$_4$ system, mole fraction $Y = 0.005$. The dotted curve corresponds to the condition $k_S = h_G$ which separates mass transfer control (r_g below the dotted curve) and surface reaction control (r_g above the dotted curve). The inset shows the dependence of growth rate ($r_g > 0$) or etch rate ($r_g < 0$) upon mole fraction of SiCl$_4$ in H$_2$ under the conditions given. A high growth rate results in a polycrystalline film, whereas a single-crystalline film is obtained only at modest growth rates.

2. GROWTH OF INTRINSIC AND EXTRINSIC FILMS

(a) Intrinsic film (containing no impurities) on a highly doped substrate: During epitaxial growth the impurities originally contained in the substrate will diffuse into the film. Their distribution $C_1(x, t)$ is given by a solution of

$$\partial C_1/\partial t = D\partial^2 C/\partial x^2 \qquad -\infty < x < x_f, t > 0. \qquad (2.116)$$

The solution must satisfy the initial condition

$$C_1(x, 0) = C_{\text{sub}} \qquad x < 0$$

(where $x > 0$ denotes the film and $x < 0$ denotes the substrate); i.e., the substrate impurity concentration is uniform originally. The solution must also satisfy the boundary conditions

$$C_1(x, t) \to C_{\text{sub}} \qquad x \to -\infty,$$

i.e., the impurity concentration deep inside the substrate is not disturbed, and

$$-D\partial C_1/\partial x = (h_S + r_g)C_1 \qquad x = x_f,$$

i.e., the material diffusing to the surface of the film either escapes into the gas or is incorporated into the growing film. The solution is

$$C_1(x, t)/C_{\text{sub}} = (1/2)\{\text{erfc}\,(x/2\sqrt{Dt}) - (h_S + r_g)/2h_S \cdot$$
$$\cdot \exp\,[(r_g/D)(r_g t - x)]\,\text{erfc}\,[(2r_g t - x)/2\sqrt{Dt}]$$
$$+ [(r_g + 2h_S)/2h_S]\,\exp\,([(r_g + h_S)/D][(r_g + h_S)t - x])\cdot$$
$$\cdot \text{erfc}\,([2(r_g + h_S)t - x]/2\sqrt{Dt})\}. \qquad (2.117)$$

Approximations:

(i) If $r_g \gg \sqrt{Dt}/t$ (very rapid epitaxial growth), then

$$C_1(x, t)/C_{\text{sub}} \approx [\text{erfc}\,(x/2\sqrt{Dt})]/2 \qquad (2.118)$$

(ii) If $r_g \approx 0$ (insignificant epitaxial growth), then

$$C_1(x, t)/C_{\text{sub}} \approx \text{erf}\,(-x/2\sqrt{Dt})$$
$$+ \exp\,[(h_S/D)(h_S t - x)]\,\text{erfc}\,[(2h_S t - x)/2\sqrt{Dt}] \qquad (2.119)$$

(b) Extrinsic film (doped to impurity concentration C_f) on a highly doped substrate:

During epitaxial growth the impurities deposited with the film will diffuse into the substrate. The resulting impurity distribution $C_2(x, t)$ will again be described by the differential equation given above, however, with the initial condition

$$C_2(x, 0) = 0 \quad x < 0,$$

i.e., the concentration of the external impurity is originally zero everywhere within the substrate, and the boundary conditions

$$C_2(x, t) \to 0 \quad x \to -\infty$$

i.e., the concentration of the external impurity deep inside the substrate is negligible and

$$C_2(x_f, t) = C_f,$$

i.e., the concentration of the external impurity at the growing film surface is constant.

The solution is

$$C_2(x, t)/C_f = (1/2)\{1 + \text{erf}(x/2\sqrt{Dt}) \\ + \exp[(r_g/D)(r_g t - x)] \,\text{erfc}\,[(2r_g t - x)/2\sqrt{Dt}]\} \quad (2.120)$$

Approximations:

(i) If $r_g \gg \sqrt{Dt}/t$ (very rapid epitaxial growth), then

$$C_2(x, t)/C_f = [1 + \text{erf}(x/2\sqrt{Dt})]/2. \quad (2.121)$$

(ii) If $x > 2\sqrt{Dt}$ (i.e., for a thick epitaxial film), then

$$C_2(x, t)/C_f \approx 1; \quad \text{i.e.,}\ C_2(x, t) \approx C_f. \quad (2.122)$$

The net impurity distribution $C(x, t)$ due to out-diffusion of impurities from both substrate and film is

(a) If both impurities are of the same conductivity type:

$$C(x, t) = C_1(x, t) + C_2(x, t); \quad (2.123\text{a})$$

(b) if both impurities are of the opposite conductivity type:

$$C(x, t) = C_1(x, t) - C_2(x, t). \quad (2.123\text{b})$$

234 IMPURITIES IN SEMICONDUCTORS

In the latter case a *p-n* junction is formed.
Assumptions:

(a) Flux of the reaction product is neglected.
(b) Mole fraction $Y \ll 1$.
(c) Differences in the properties of the gas between heated substrate and cold glass wall are neglected.

The inset of Figure 2.37 gives the growth rate vs. mole fraction. Positive values of r_g correspond to film deposition, negative values to etching of the semiconductor surface. For deposition rates above a critical value the epitaxial film is of polycrystalline structure since the film has not sufficient time during growth to follow the crystalline pattern of the substrate.

3. GAS-PHASE MASS-TRANSFER COEFFICIENT

The gas-phase mass-transfer coefficient h_G can be estimated from

$$h_G = (3/2)(D_G/L_G)(d_G v_G L_G/\eta_G)^{1/2}(\eta_G d_G D_G)^{1/3}$$
$$\approx (3/2)(D_G/L_G)(d_G v_G L_G/\eta_G)^{1/2} \qquad (2.124)$$

where D_G = diffusivity of the active species in the gas
L_G = length of gas-semiconductor interaction region (typically 1 to 10 cm)
v_G = velocity of the free gas stream (typically 10 to 30 cm/sec)
d_G = density of carrier gas
η_G = viscosity of carrier gas.

For most gases

$$\eta_G/d_G D_G = 0.6 \text{ to } 0.8$$

independent of temperature so that

$$(\eta_G/d_G D_G)^{1/3} \approx 1.$$

The Reynolds number is

$$d_G v_G L_G/\eta_G \approx 20.$$

Values of d_G and η_G for H_2, O_2, and N_2 are given in Figure 2.38. Diffusivities of these gases range typically from 2 to 20 cm^2/sec between 1000 and 1200°C.

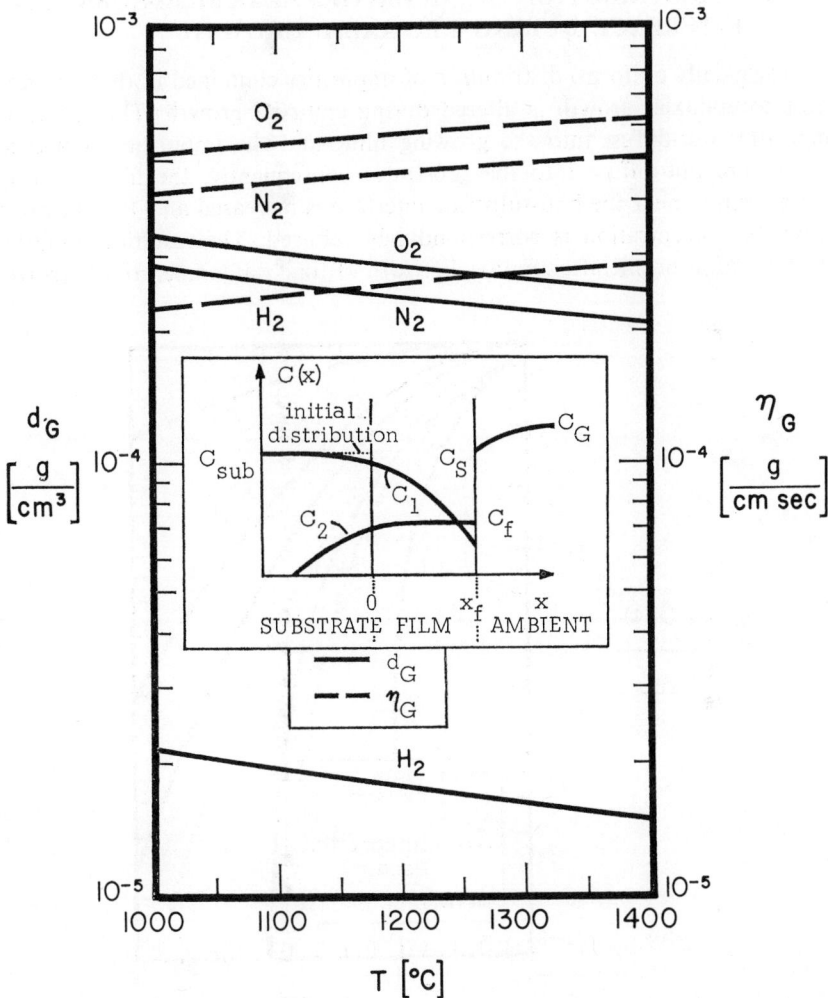

FIGURE 2.38
Density (d_G) and viscosity (η_G) of carrier gas vs. epitaxial deposition temperature (T) for commonly used gases. The inset shows the impurity distribution in semiconductor substrate, epitaxial film, and ambient gas. The impurity from the substrate is denoted C_1, the external doping impurity C_2.

4. REDISTRIBUTION OF IMPURITIES NEAR FILM-SUBSTRATE INTERFACE DURING EPITAXIAL GROWTH

The (originally uniform) distribution of impurities contained in the substrate prior to epitaxial growth is altered during epitaxial growth. The substrate impurities out-diffuse into the growing film; and the impurities contained in the film out-diffuse into the substrate; consequently, the film impurity concentration near the film-substrate interface is increased and the substrate impurity concentration is correspondingly reduced. The resulting impurity concentration profile after epitaxial growth of time t at temperature T (corre-

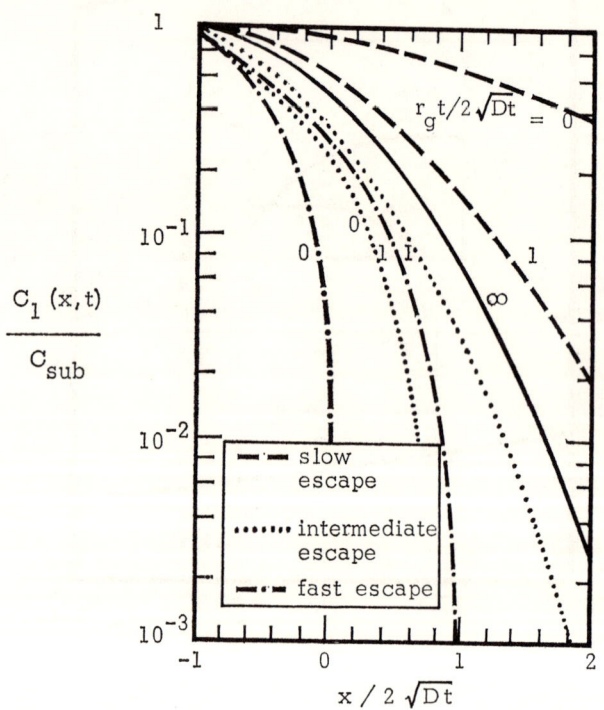

FIGURE 2.39

Impurity concentration (C_1) in an epitaxial film after epitaxial growth vs. distance from substrate–film interface (x) for slow, intermediate, and rapid escape rates ($h_s t/2\sqrt{Dt}$) and growth rates ($r_g t/2\sqrt{Dt}$). The substrate–film interface is located at $x/2\sqrt{Dt} = 0$. The solid curve corresponds to the erfc distribution, i.e., $C_1/C_{sub} = (1/2)$ erfc $(x/2\sqrt{Dt})$. The diffusion length $\sqrt{Dt}$ corresponds to the epitaxial growth temperature and the growth duration (t) for the impurity in the substrate.

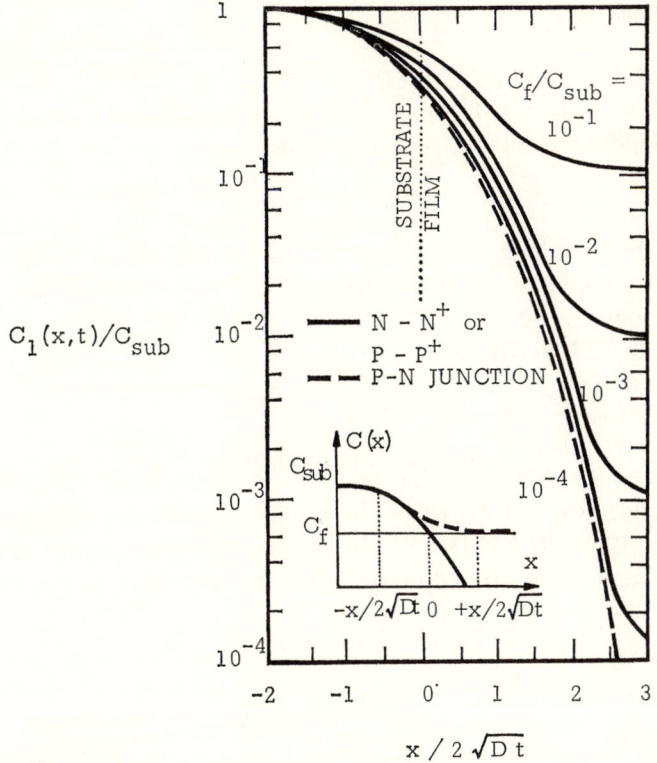

FIGURE 2.40
Impurity concentration (C_1) in the epitaxial film after epitaxial growth vs. normalized distance from the substrate–film interface (x) for various ratios of film to substrate impurity concentration (C_f/C_{sub}). Intermediate escape rate. The concentrations C_f and C_{sub} are taken at distances $x > |2\sqrt{Dt}|$ from the film–substrate interface (assumed to be at $x = 0$). The solid curves apply if substrate and film are of the same conductivity type, the dashed curve applies if they are of the opposite type.

sponding to a diffusion coefficient D of the out-diffusing species) is generally

$$\frac{C_1(x,t)}{C_{sub}} = \frac{1}{2}\left\{\operatorname{erfc}\left(\frac{x}{2\sqrt{Dt}}\right)\right.$$
$$- \frac{r_g + h_S}{h_S} \exp\left[\frac{r_g}{D}(r_g t - x)\right] \cdot \operatorname{erfc}\left[\frac{2r_g t - x}{2\sqrt{Dt}}\right]$$
$$\left. + \frac{r_g + 2h_S}{h_S} \exp\left(\frac{(r_g + h_S)[(r_g + h_S)t - x]}{D}\right) \cdot \operatorname{erfc}\left(\frac{2(r_g + h_S)t - x}{2\sqrt{Dt}}\right)\right\}.$$
(2.125)

238 IMPURITIES IN SEMICONDUCTORS

Two limiting cases:

(a) Large r_g (i.e., $r_g t/2\sqrt{Dt} \gg 1/2$):
Equation (2.125) reduces to a simple complementary error function distribution regardless of the value of h_S:

$$C_1(x,t)/C_{sub} = (1/2)\,\text{erfc}\,(x/2\sqrt{Dt}); \qquad (2.126)$$

i.e., if the growth rate is high, the epitaxial film will always be significantly thicker than the extent of the region affected by out-diffusion.

(b) Small r_g (i.e. $r_g t/2\sqrt{Dt} \approx C$):
Equation (2.125) reduces to

$$C_1(x,t)/C_{sub} = \text{erf}\,(-x/2\sqrt{Dt}) \\ + \exp\,[h_S(h_S t - x)/D]\,\text{erfc}\,[(2h_S t - x)/2\sqrt{Dt}]; \qquad (2.127)$$

i.e., if the growth rate is small, the impurity distribution in the vicinity of the substrate-film interface will be independent of the growth rate.

If $h_S t/2\sqrt{Dt} \gg 1/2$ (rapid escape), then this equation reduces further to

$$C_1(x,t)/C_{sub} = \text{erf}\,(-x/2\sqrt{Dt}) \qquad (2.128)$$

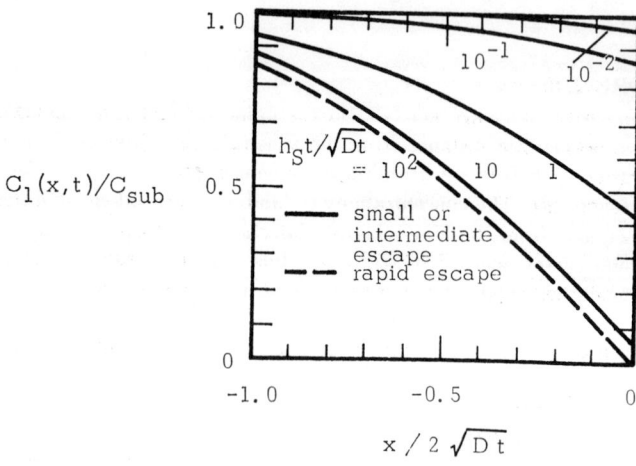

FIGURE 2.41

Impurity concentration (C_1) in the epitaxial film after epitaxial growth vs. distance from substrate–film interface (x) for various escape rates ($h_S t/\sqrt{Dt}$). The film growth rate is assumed to be very small (i.e., $r_g t/2\sqrt{Dt} \approx 0$).

Definitions:

$C_1(x, t)$ = concentration of impurities originally contained in the substrate after epitaxial growth at distance x from substrate-film interface

$C_2(x, t)$ = concentration of impurities originally contained in the epitaxial film after epitaxial growth

C_{sub} = initial uniform substrate concentration.

The term $h_S t/2\sqrt{Dt}$ is the escape rate of impurities at the film-ambient interface:

escape very slow: $h_S t/2\sqrt{Dt} = 0$

escape intermediate: $h_S t/2\sqrt{Dt} = 1/2$

escape very rapid: $h_S t/2\sqrt{Dt} = \infty$.

The transfer coefficient h_S depends upon gas flow rate, gaseous diffusion constant, gas flow profile, reactor properties and temperature; it is an empirically determined quantity. It is typically in the order of 10^{-6} cm/sec. For boron in silicon $h_S = 2 \cdot 10^{-6}$ cm/sec.

5. REFERENCES

(1) R. C. Reid and T. K. Sherwood, *The Properties of Gases and Liquids*, McGraw-Hill Book Co., New York, 1958.
(2) H. C. Theuerer, *J. El. Chem. Soc.*, **108**, 649 (1961).
(3) W. H. Shepherd, *J. El. Chem. Soc.*, **112**, 988 (1965).
(4) A. S. Grove et al., *J. Appl. Phys.*, **36**, 802 (1965).
(5) A. S. Grove, *Physics and Technology of Semiconductor Devices*, John Wiley and Sons, New York, 1967.
(6) W. R. Runyan, *First International Symposium on Silicon Materials, Science and Technology*, Abstract No. 288, New York, May 5–9, 1969.
(7) H. H. Wieder, *Intermetallic Semiconducting Films*, Pergamon Press, Oxford, 1970.
(8) V. K. Jain and S. K. Sharma, *Solid-State Electronics*, **13**, 1145 (1970).

2–8 Phase Diagrams

1. PHASE DIAGRAMS

A phase is a homogeneous portion of a system physically distinct and mechanically separable from the other portions of the system. Examples of a single phase are: any mixture of gases, any liquid or solid solution, any pure crystalline or amorphous solid. A mixture of two pure solids or solid solutions has two phases.

Phase diagrams give the melt composition in equilibrium with a solid, the maximum solid solubility and eutectic composition, and the compound formation.

The temperatures corresponding in a phase diagram to 0 and 100% give the melting points of the pure components A and B. A mixture of the components has a reduced melting point; the lowest possible melting point of the mixture is the eutectic temperature for a given pressure. Crystals of an alloy of the eutectic composition are homogeneous.

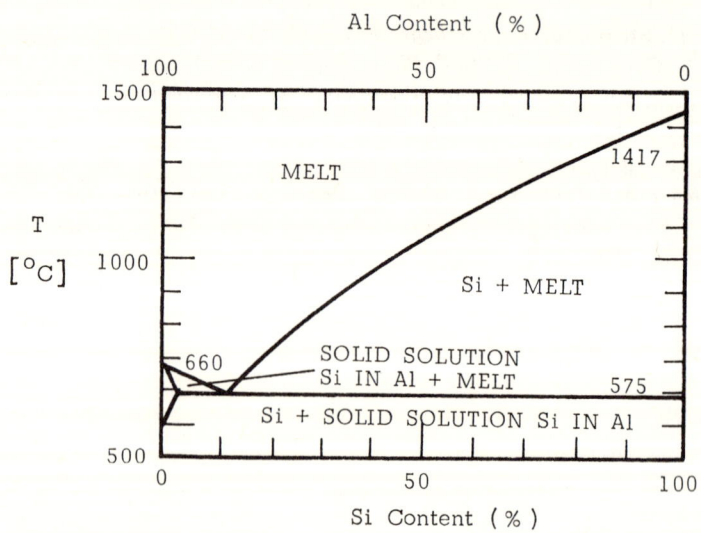

FIGURE 2.42

Phase diagram of the Si–Al system.

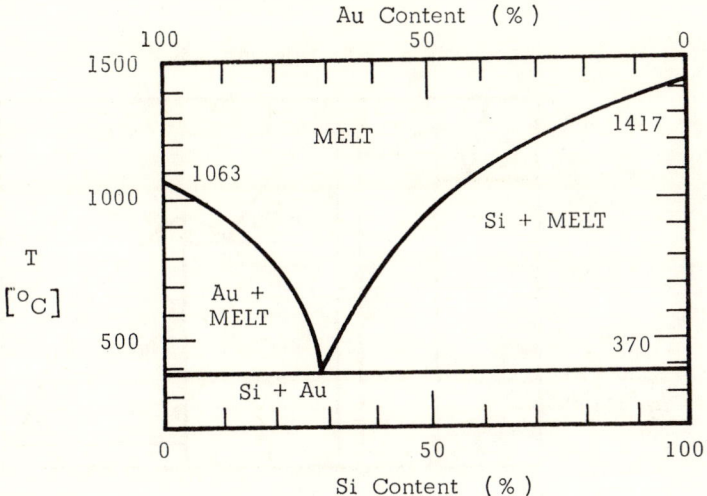

FIGURE 2.43
Phase diagram of the Si–Au system.

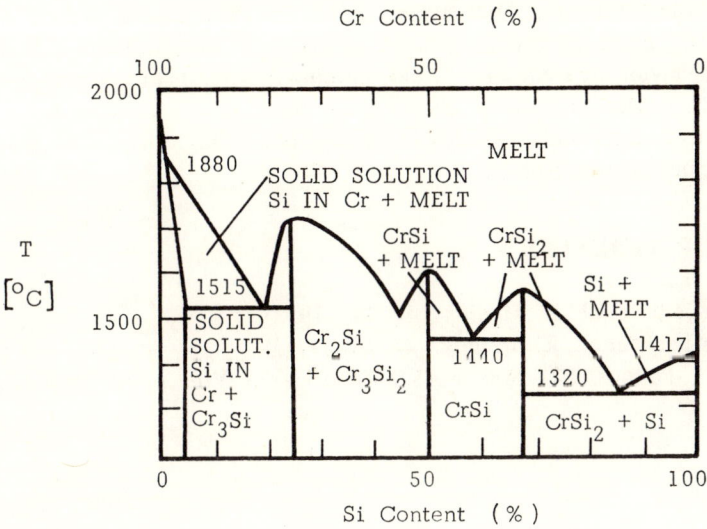

FIGURE 2.44
Phase diagram of the Si–Cr system.

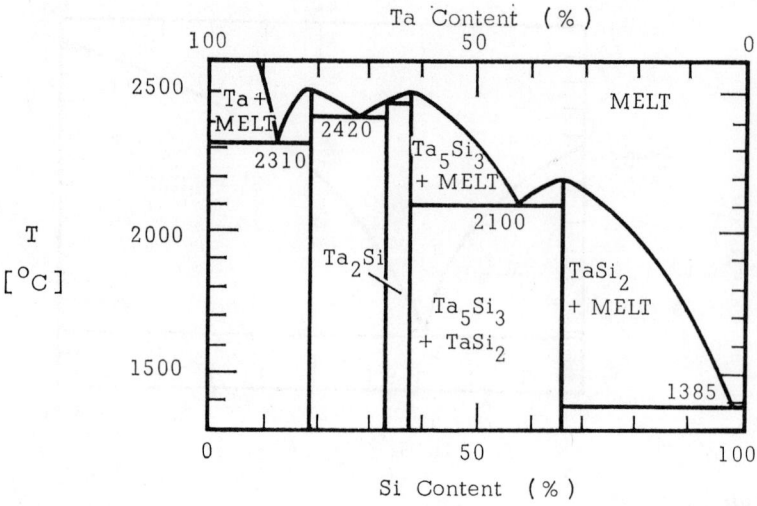

FIGURE 2.45

Phase diagram of the Si–Ta system.

Below the eutectic temperature, at all composition ratios $(A:B)$ solid A + solid B are present. Above the eutectic temperature but below the melting point of the mixture one finds either solid A + melt or solid B + melt. Above the melting point the entire mixture is in the melt. Modifications of this simplified behavior are frequently found resulting in several eutectic points which are characterized by different chemical composition of the mixture.

Figures 2.42 through 2.45 are examples of phase diagrams involving silicon and frequently used metals.

2. REFERENCES

(1) R. Kieffer et al., *Z. Metallk.*, **44**, 242 (1953).
(2) R. Kieffer et al., *Z. Metallk.*, **44**, 437 (1953),
(3) B. N. Padnos, *RTI Report ASD-TDR-63-316*, Vol. X. November, 1965.

2-9 Examples

PROBLEMS AND SOLUTIONS

1. Problem: Determine the diffusion time required to achieve saturation with gold at the surface of a 200 μm-thick *n*-type dislocation-free silicon wafer of initial resistivity $\rho_b = 1.0\,\Omega$ cm if gold is deposited on the wafer backside and is diffused in at 1100°C.

Solution: Assuming an infinitely thick wafer and an infinite source, saturation can be achieved only after an infinite diffusion time. Reasonable approximations in the above example are as follows.

(a) Assume that the gold concentration is required to be 50% of the maximum solid solubility at the given temperature; i.e.,

$$C(x)/C_{Au} = 0.5,$$

corresponding to

$$x/2\sqrt{Kt} \approx 10^{-5} \text{ and } K = 3.25 \text{ cm}^2/\text{sec};$$

hence

$$t \approx 85 \text{ hr}.$$

(b) Assume 10% saturation, i.e.,

$$C(x)/C_{Au} = 0.1,$$

corresponding to

$$x/2\sqrt{Kt} \approx 6.5 \cdot 10^{-5};$$

hence

$$t \approx 2 \text{ hr}.$$

2. Problem: Determine the implantation conditions (initial energy of projectile, E_{I_o}, and beam current density, It) if phosphorus and boron are to be implanted at a 7°angle into silicon wafers to a peak impurity concentration $C_P = 10^{18}$ cm^{-3} and a peak depth $x_P = 0.3$ μm at 25°C.

244 IMPURITIES IN SEMICONDUCTORS

Solution: First determine the required energy from Figure 2.31, then determine It from Figure 2.32.

(a) Boron:

$$E_{I_0} = 90 \text{ keV};$$
$$C_P/It = 2.6 \cdot 10^{15} \text{ cm}^{-3}/\mu\text{A sec}.$$
$$It = 385 \ \mu\text{A sec}.$$

If an ion beam current $I = 10 \ \mu\text{A}$ is used, the implantation time that is required to give the desired impurity concentration is

$$t = 38.5 \text{ sec}.$$

(b) Phosphorus:

$$E_{I_0} = 240 \text{ keV};$$
$$C_P/It = 2.6 \cdot 10^{15} \text{ cm}^{-3}/\mu\text{A sec},$$
$$It = 385 \ \mu\text{A sec}.$$

For $I = 10 \ \mu\text{A}$ the required time is also

$$t = 38.5 \text{ sec}.$$

3. **Problem:** Boron ions have been implanted into silicon samples of uniform impurity concentration $C_B = 5 \cdot 10^{15}$ cm^{-3} at a current density $It = 10^2$ μA sec and at ion energies $E_{I_0} = 60$ and 100 keV. Determine peak depth x_P, peak concentration C_P, and surface concentration C_S.

Solution: (a) $E_{I_0} = 60$ keV,

$$x_P = 0.21 \ \mu\text{m};$$
$$C_P = 4 \cdot 10^{17} \text{ cm}^{-3}.$$

Hence at the surface

$$C_S = 4 \cdot 10^{-3} \ C_P = 1.6 \cdot 10^{15} \text{ cm}^{-3}.$$

This is less than the background concentration C_B. Therefore, the impurity concentration will be constant, $C_B = 5 \cdot 10^{15}$ cm^{-3}, from $x = 0$ to $x = 0.03$ μm at which point (i.e., 0.18 μm above the peak depth) $C_P/C_B = 76.9$; from $x = 0.3$ μm to $x = 0.39$ μm the impurity concentration will rise above the background concentration; and at $x \geq 0.39$ μm the impurity concentration will again be constant and equal to C_B. The width of the buried layer will be 0.36 μm.

(b) $E_{I_0} = 100$ keV,

$$x_P = 0.34 \ \mu\text{m};$$
$$C_P = 2.4 \cdot 10^{17} \text{ cm}^{-3}.$$

Hence at the surface

$$C_S = 4.5 \cdot 10^{-5} \, C_P = 1.1 \cdot 10^{13} \text{ cm}^{-3}.$$

Since this is less than C_B, the surface impurity concentration is again that of the background. The impurity concentration begins to rise above the background concentration at $x = 0.21$ μm above the peak or 0.13 μm below the surface at which point $C_P/C_B = 47.6$; it will have decreased again to the background concentration at 0.21 μm below the peak, i.e., at $x = 0.55$ μm. The width of the implanted layer will be 0.42 μm.

4. **Problem:** Determine the surface concentration of a uniformly doped surface layer of thickness $x_j = 1.0$ μm in which a carrier mobility $\mu = 100$ cm²/V sec has been measured. The V/I measurement gave 40 Ω.

Solution: $x_j(V/I) = 40 \, \Omega \, \mu\text{m}$,
hence

$$C_S = 3 \cdot 10^{18} \text{ cm}^{-3}.$$

5. **Problem:** For a diffusion at 1100°C the surface impurity concentration has been calculated to be $C_S = 3 \cdot 10^{20}$ cm⁻³. Determine the correction factor that applies due to the built-in electric field in order to establish the true surface concentration.

Solution: At 1100°C the intrinsic carrier concentration

$$n_i = 3.5 \cdot 10^{18} \text{ cm}^{-3}.$$

Hence for $C_S/n_i = 86$:

$$C_S^*/n_i = 13,$$
$$C_S^* = 4.5 \cdot 10^{19} \text{ cm}^{-3},$$
$$C_S^*/C_S = 0.15.$$

This means that the true surface concentration is almost one order of magnitude lower than the calculated surface concentration which neglects the built-in field.

6. **Problem:** Determine the mass-transfer coefficient h_G for carrier gas H_2 in an epitaxial reactor for which $L_G = 10$ cm, $v_G = 20$ cm/sec, $T = 1200°C$, $D_G = 10$ cm²/sec.

Solution: $d_G = 1.75 \cdot 10^{-5}$ g/cm³,

$$\eta_G = 2.6 \cdot 10^{-4} \text{ g/cm sec}.$$

Hence

$$h_G = 5.5 \text{ cm/sec}$$

7. **Problem:** Determine the free carrier concentration at 25°C in a silicon sample of boron concentration $N_A = 10^{18}$ cm^{-3} and in another silicon sample of phosphorus concentration $N_D = 10^{18}$ cm^{-3}. Consider the decrease of activation energy with impurity concentration.

Solution: (a) Activation energy depends upon concentration: In this case $dE_a/dC_B \neq 0$.
Boron:

$p/N_A = 0.74$, $p = 7.4 \cdot 10^{17}$ cm^{-3};

i.e., at $N_A = 10^{18}$ cm^{-3} only 74% of all boron atoms are ionized.
Phosphorus:

$n/N_D = 0.95$, $n = 9.5 \cdot 10^{17}$ cm^{-3};

i.e., at $N_D = 10^{18}$ cm^{-3} only 95% of all phosphorus atoms are ionized.
(b) Activation energy is independent of concentration: In this case $dE_a/dC_B = 0$.
Boron:

$p/N_a = 0.51$, $p = 5.1 \cdot 10^{17}$ cm^{-3};

i.e., 51% of all boron atoms are ionized if the concentration dependence of the activation energy is neglected.
Phosphorus:

$n/N_D = 0.83$, $n = 8.3 \cdot 10^{17}$ cm^{-3};

i.e., 83% of all phosphorus atoms are ionized under the above assumption.

PROBLEMS FOR WHICH A SOLUTION IS NOT GIVEN

1. Compare the diffusion coefficients of Au, Al, B, and As in Si and Ge at 900°C. Relate them to the diffusion coefficients in SiO_2 where possible.

2. Given an ion-implanted p-type layer of peak concentration $C_P = 2.4 \cdot 10^{17}$ cm^{-3} and peak depth $x_P = 0.34$ μm which is formed in n-type silicon of uniform impurity concentration $C_B = 5 \cdot 10^{15}$ cm^{-3}. The junction depth has been found to be $x_j = 0.55$ μm. Determine the impurity gradient at the junction.

3. After a boron diffusion into silicon of uniform impurity concentration $C_B = 10^{15}$ cm^{-3} at 1100°C for 1 hr, a surface concentration $C_S = 10^{18}$ cm^{-3} has been determined. Find the impurity concentration at $x = 1.0$ μm below the silicon surface for erfc and Gaussian impurity distributions; assume Q = constant.

4. Determine the time required to establish the final surface impurity concentration C_S if $D = 10^{-12}$ cm^2/sec, $h_G = 5$ cm/sec, and $T = 1100°$C are given for boron.

5. Determine the influence of the built-in electric field on the impurity distribution at 1000°C and at $x = 0.5$ μm if $\sqrt{Dt} = 0.5$ μm and $x_j = 2.0$ μm.

6. Determine vertical and lateral junction depths at a mask opening of width $y_m = 1.0$ μm for $C_S/C_B = 10^4$ and $\sqrt{Dt} = 0.5$ μm.

7. Assuming a mole fraction $Y = 0.005$ and a mass-transfer coefficient $h_G = 5$ cm/sec, determine the epitaxial growth rate of silicon at 1150 and 1250°C.

8. An epitaxial film is grown under the following conditions:
$\sqrt{Dt} = 0.5$ μm, $r_g = 1.0$ μm/min, $t = 30$ min, $C_{sub} = 10^{15}$ cm^{-3}. Determine the impurity concentration at the substrate-film interface for intermediate escape rate.

3
ELECTRICAL BEHAVIOR OF SEMICONDUCTORS

3–1 Carrier Density
3–2 Minority Carrier Drift Velocity and Ionization Rate
3–3 Carrier Mobility and Carrier Diffusion Coefficient
3–4 Hall Coefficient and Hall Mobility
3–5 Semiconductor Resistivity
3–6 Minority Carrier Recombination and Lifetime
3–7 Examples

Foreign atoms (impurities) in a semiconductor affect its properties. Impurities may supply either free electrons or holes or may remain electrically neutral. The simple energy band structure consisting of two allowed energy bands which are separated by the forbidden band gap applies strictly to an ideal (undisturbed) intrinsic semiconductor with infinite dimensions. Any disturbance of the crystal order by crystal defects or impurity atoms introduces additional energy levels either within the energy gap or within the bands. These energy levels have a significant effect on the crystal properties only if they occur within the energy gap.

An impurity atom is incorporated in the crystal either on a lattice site which would ideally be occupied by a regular semiconductor atom (substitutional impurity) or it occupies a position between regular lattice atoms (interstitial impurity) accompanied by stresses within the crystal and crystal distortion. An interstitial impurity moves much more rapidly than a substitutional impurity. The substitutional diffusion coefficient may be greatly enhanced by the presence of dislocations. The lattice damage caused by very high impurity concentration creates an excess of lattice vacancies near the diffusion front. The crystal lattice can accommodate only a limited number of impurity atoms. This number corresponds to the maximum solid solubility at a given temperature and depends upon the diffusion mechanism and whether the impurity atoms enter the crystal interstitially or substitutionally.

Energy levels of elements which have one valence electron more or one valence electron less than the semiconductor atoms form donor or acceptor levels close to conduction or valence band (hydrogen-like impurities). For Group IV semiconductors the most important elements of this type are P, As, Sb (Group V); and B, Al, Ga, In (Group III). Most elements of other Groups (e.g., Groups II and VI) are associated with energy levels deep within the energy gap. In Group IV semiconductors impurities of Groups III and V move substitutionally, most other impurities move mainly interstitially. Gold and copper move both interstitially and substitutionally; the effective diffusion coefficient is given by a weighted combination of the two individual coefficients and varies with concentration.

The deep-lying energy levels which do not result in the generation of donors and acceptors are caused by either interstitial impurities, substitutional impurities of greater charge difference than $\pm q$, lattice defects due to radiation damage, or dislocations. The density of deep-lying levels is usually small so that they have a negligible influence on the Fermi level; they have, however, large transition probabilities to both bands, i.e., they act as generation-recombination centers.

Although the donor levels have a spin degeneracy of two, a given recombination center can capture only one electron, of either spin; a second electron would not be bound to the center. The equilibrium occupation probability, i.e., the fraction of neutral donors, is

$$f_D^o = \{1 + (1/2) \exp[(E_D - E_F)/kT]\}^{-1} \tag{3.1}$$

and the fraction of ionized (positively charged) donors

$$f_D^+ = 1 - f_D^o = \{1 + 2 \exp[-(E_D - E_F)/kT]\}^{-1}. \tag{3.2}$$

Similarly, an acceptor, when neutral, contains one electron of either spin and may become ionized (negatively charged) by the capture of another electron, which must be of the opposite spin. Hence, the fraction of ionized acceptors

$$f_A^- = 1 - f_A^o = \{1 + 2 \exp[-(E_F - E_A)/kT]\}^{-1}. \tag{3.3}$$

The energies E_D and E_A depend on the impurity concentration and occupancies of the various centers present due to interactions between centers. The centers may also have excited states whose density must be taken into account. Usually, however, these states are either within the bands or close to the band edges so that they may be ignored.

A semiconductor is called intrinsic if no impurities are present or if the impurity density is negligible at a given temperature compared to the carrier density; otherwise it is called extrinsic; if the two carrier types in an extrinsic semiconductor are present in approximately equal amounts, the semiconductor is called compensated. If in an extrinsic semiconductor the number of free carriers is less than the number of impurity atoms (i.e., if not all impurity atoms are ionized) the semiconductor is degenerate.

Semiconductors are characterized by two ranges of conductivity depending upon the availability of free carriers:

(a) Intrinsic semiconductor:

At $T = 0°K$ all of the covalent bonds (which interconnect the crystal nuclei) are ideally intact and thus all the valence electrons are tightly bound. Hence, there are no mobile carriers present and the semiconductor acts as an insulator.

At $T > 0°K$ the thermal energy is capable of breaking some of the bonds. Each broken bond results in the thermal generation of an electron-hole pair. Both electron and hole act as carriers and the intrinsic semiconductor shows some conductivity.

(b) Extrinsic semiconductor:

Atoms of Group III (B, Al, Ga, In, Tl) have only three valence electrons. In a Group IV semiconductor crystal they, therefore, miss one electron for a complete bond. The energy required to remove one electron out of a complete semiconductor bond and add it to the Group III impurity atom is much smaller than the energy required to disrupt a covalent semiconductor bond. Thus, at moderate temperature (room temperature) a Group III impurity will very probably be ionized (has added one electron) and will leave a hole.

Atoms of Group V (P, As, Sb, Bi) have an additional electron which is not required for the bond. Little energy is required to completely free this electron so that at moderate temperature (room temperature) it is likely to be a mobile or free carrier.

In addition to doping with impurity atoms, the semiconductor resistivity can be affected by light and thermal generation. In both cases electrons and holes are generated at the same rate, i.e., due to an increase in energy some valence electrons are removed from the semiconductor atoms and become free electrons leaving behind free holes (carrier generation); within the carrier diffusion length these carriers will recombine again. In equilibrium, generation and recombination rates are equal.

Much of the statistical theory of semiconductors is based on the assumption that a semiconductor is in thermal equilibrium. In this case Boltzmann or Fermi statistics apply depending upon the impurity density. If the thermal equilibrium of the semiconductor is disturbed, the system tends to return to equilibrium. In the case of conductivity, equilibrium is restored by lattice collisions. If the disturbance is manifested by a change in carrier density, equilibrium is restored by either recombination or diffusion and drift of carriers into or out of the affected region. The rate of return to equilibrium depends upon the process by which the equilibrium state is attained.

Two recombination mechanisms are of importance: recombination through a trap and band-to-band recombination. In trap recombination deep-level impurities are involved. In band-to-band recombination electron and hole from conduction and valence bands recombine directly without the aid of traps. When an electron is captured by a deep trap or in band-to-band recombination, the energy difference between initial and final electron energy is removed from the electron in the form of either a photon (radiative recombination) or a series of phonons or by energy transfer to another electron (Auger process). The recombination rate depends upon the recombination mechanism and the magnitude of the disturbance. The corresponding carrier lifetime of an electron is

$$\tau_n = \frac{\Delta n}{dn/dt} \tag{3.4}$$

where Δn is the excess number of electrons and dn/dt is the rate of disappearance of electrons by recombination. Carrier lifetime depends upon majority and minority carrier densities, density and nature of available recombination centers, and temperature. Recombination may also take place at the semiconductor surface. The surface states that hold charge and act as donors or acceptors can also act as recombination centers.

GENERAL REFERENCES

(1) J. L. Moll, *Physics of Semiconductors*, McGraw-Hill Book Co., New York, 1964.
(2) E. M. Conwell, *High Field Transport in Semiconductors*, Academic Press, New York, 1967.
(3) D. R. Frankl, *Electrical Properties of Semiconductor Surfaces*, International Series of Monographs on Semiconductors, Vol. 7, Pergamon Press, Oxford, 1967.

3-1 Carrier Density

1. CARRIER CONCENTRATION IN AN INTRINSIC SEMICONDUCTOR

In an intrinsic semiconductor (which can be considered as a special case of an extrinsic semiconductor) the density of foreign (impurity) atoms is negligible compared to the carrier densities at a given temperature; the carrier density in an intrinsic semiconductor is a property of the lattice itself and is not caused by crystal imperfections or by impurities. The number of free electrons and free holes is equal and the total carrier concentration is significantly larger than the impurity density, i.e.,

$$n_i \gg N_D, \quad n_i \gg N_A$$

at a given temperature. Only a few semiconductors are intrinsic at low temperature. At high temperature an impurity (extrinsic) semiconductor may become intrinsic if the intrinsic carrier density exceeds the impurity density, i.e., if $n_i > C_B$.

The carrier concentration in an intrinsic semiconductor is given by

$$\begin{aligned}n_i &= (np)^{1/2} \\ &= (N_c N_v)^{1/2} \exp(-E_G/2kT) \\ &= 2(2\pi kT/h^2)^{3/2} (m_n m_p/m_o^2)^{3/4} \exp(-E_G/2kT) \\ &= 4.9 \cdot 10^{15} (m_n m_p/m_o^2)^{3/4} T^{3/2} \exp(-E_G/2kT).\end{aligned} \quad (3.5)$$

Specifically for silicon:

$$\begin{aligned}n_i &= 3.73 \cdot 10^{16} T^{3/2} \exp(-0.605 q/kT) \\ &= 3.73 \cdot 10^{16} T^{3/2} \exp(-7014/T).\end{aligned} \quad (3.6)$$

At

$$T = 300°K:$$
$$n_i = 1.45 \cdot 10^{10} \text{ cm}^{-3}.$$

In an intrinsic semiconductor at $T > 0$ there is continuous thermal agitation that results in the excitation of electrons from the valence band to the conduction band and leaves an equal number of holes in the valence band, i.e.,

$$n = p = n_i. \quad (3.7)$$

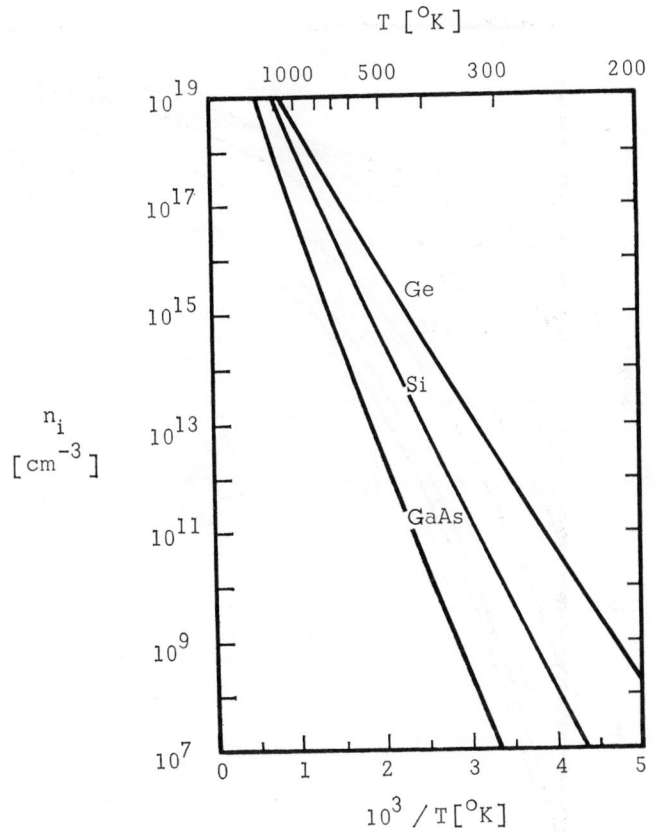

FIGURE 3.1
Intrinsic carrier concentration (n_i) as function of temperature (T) for Si, Ge, and GaAs.

This process is balanced by recombination of electrons in the conduction band with holes in the valence band. The quantity n_i represents half of the total carrier density in an intrinsic semiconductor; the total carrier density is the sum of electron and hole densities.

$$\text{Total carrier density} = n + p = 2n_i. \qquad (3.8)$$

The charge balance equation

$$n - p = N_D - N_A \qquad (3.9)$$

holds for an intrinsic as well as an extrinsic nondegenerate semiconductor; it applies, since in equilibrium the semiconductor as a whole has to remain neutral.

256 ELECTRICAL BEHAVIOR OF SEMICONDUCTORS

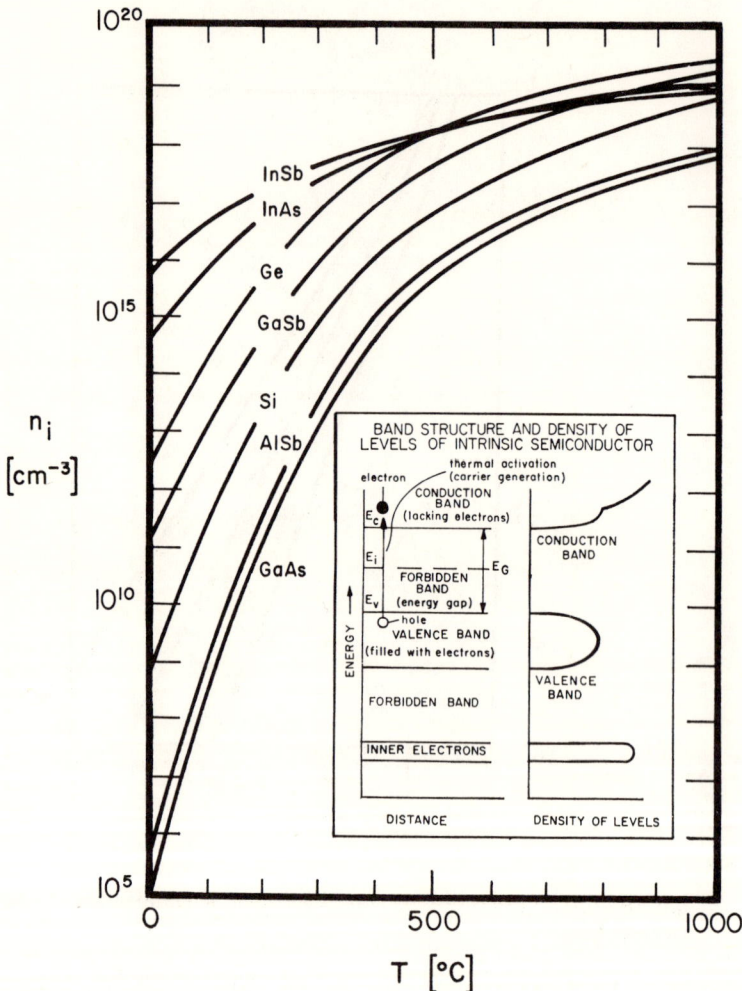

FIGURE 3.2

Carrier concentration (n_i) in various intrinsic semiconductors vs. temperature (T). The inset shows the band structure and density of levels of an intrinsic semiconductor.

The thermal generation of carriers (the transition of carriers across the band gap) is enhanced by an increase in thermal energy (kT); as a result, n_i will increase sharply with temperature. In addition, the thermal generation is a function of the width of the energy gap, which in turn is temperature-dependent. Since n_i increases very rapidly with temperature, a semiconductor which is extrinsic at low temperature may become intrinsic at high temperature so that, while the impurity concentration remains constant, at high temperature $n_i \gg C_B$ and at low temperature $n_i \ll C_B$.

In an intrinsic semiconductor the effective densities of states, N_c and N_v, are equal if the effective masses of the carriers are equal; in this case the Fermi

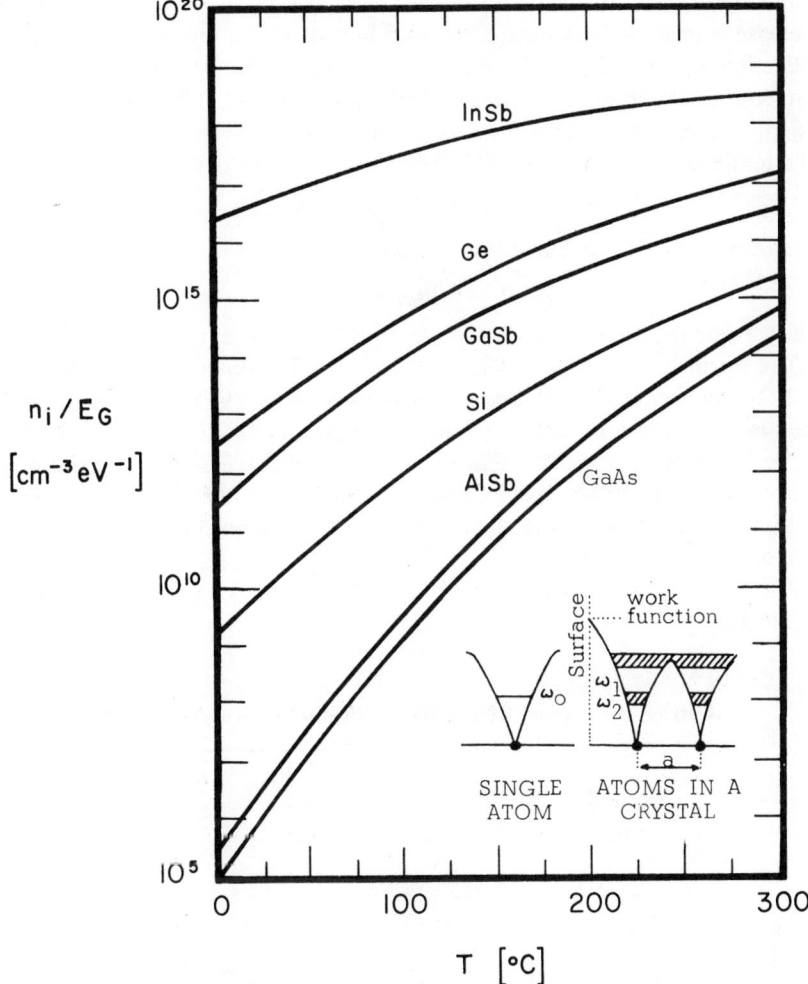

FIGURE 3.3

Ratio of carrier concentration (n_i) to energy gap (E_G) of various intrinsic semiconductors vs. temperature (T). The inset shows a model of energy bands for a single atom and atoms in a crystal.

258 ELECTRICAL BEHAVIOR OF SEMICONDUCTORS

level lies in the center of the energy gap. The energy gap (E_G) is the activation energy for the reaction

$$\text{Electron} + \text{Hole} \rightleftarrows \text{Filled valence band level}$$

in an analogy to the chemical mass-action law.

Figure 3.3 shows the ratio n_i/E_G as function of temperature for various semiconductors. This ratio increases very rapidly with temperature; the increase is steepest for semiconductors which display a low n_i/E_G ratio within the temperature range considered. At low temperature the n_i/E_G curves of all semiconductors show the same slope; they are displaced along the temperature axis by an amount characteristic of the semiconductor.

The temperature dependence of n_i near 300°K can be described by

$$d(n_i^2)/dT = (3 + E_G/kT)n_i^2/T. \tag{3.10}$$

The resistivity of an intrinsic semiconductor is

$$\rho_i = 1/[n_i q(\mu_n + \mu_p)] \tag{3.11}$$

where μ_n and μ_p are electron and hole mobilities in the intrinsic material.

The intrinsic resistivity is determined mainly by the ratio E_G/kT. If $E_G/kT \gg 1$, the concentration of ionized carriers is small. As the temperature increases, E_G/kT becomes rapidly smaller because of the term kT and because E_G decreases with increasing temperature; in this case electrons are thermally excited from the valence band to the conduction band and both the electrons in the conduction band and the holes left in the valence band will contribute to electrical conductivity.

The intrinsic resistivity of selected semiconductors at 300°K is listed in Table 3.1.

TABLE 3.1

Carrier Density and Resistivity of Intrinsic Semiconductors (300°K)

Semi-conductor	n_i [cm^{-3}]	ρ_i [Ω cm]
Si	$1.5 \cdot 10^{10}$	$2.3 \cdot 10^5$
Ge	$2.4 \cdot 10^{13}$	$5.0 \cdot 10$
AlSb	$5.0 \cdot 10^6$	$5.0 \cdot 10^6$
GaAs	$9.0 \cdot 10^7$	$7.0 \cdot 10^7$
GaP	$3.0 \cdot 10^{16}$	1.7
GaSb	$5.0 \cdot 10^{11}$	$3.5 \cdot 10^3$
InAs	$1.0 \cdot 10^{15}$	$2.8 \cdot 10^{-1}$
InP	$1.0 \cdot 10^{14}$	$1.9 \cdot 10$
InSb	$1.1 \cdot 10^{16}$	$1.1 \cdot 10^{-2}$

2. CARRIER CONCENTRATION IN AN EXTRINSIC SEMICONDUCTOR

In an extrinsic semiconductor the density of impurity atoms exceeds the intrinsic carrier density at a given temperature. If the semiconductor is nondegenerate, the carrier density is equal to the density of impurity atoms and exceeds n_i.

$$\text{Total carrier density} = n + p > 2n_i. \tag{3.12}$$

In an extrinsic nondegenerate semiconductor majority and minority carrier densities in equilibrium are as follows.

(a) *n*-Type semiconductor:
Majority carriers (electrons)

$$\begin{aligned} n &= N_c \exp\left[-(E_c - E_F)/kT\right] \\ &= n_i \exp\left[(E_F - E_i)/kT\right] \\ &= (1/2)\{(N_D - N_A) + [(N_D - N_A)^2 + 4n_i^2]^{1/2}\} \\ &= (1/2)[C_B + (C_B^2 + 4n_i^2)^{1/2}] \\ &\approx N_D - N_A \\ &\approx C_B \end{aligned} \Bigg\} \text{if } N_D - N_A \gg n_i \tag{3.13}$$

Minority carriers (holes)

$$p = n_i^2/(N_D - N_A) = n_i^2/C_B \tag{3.14}$$

(b) *p*-Type semiconductor:
Majority carriers (holes)

$$\begin{aligned} p &= N_v \exp\left[-(E_F - E_v)/kT\right] \\ &= n_i \exp\left[(E_i - E_F)/kT\right] \\ &= (1/2)\{(N_A - N_D) + [(N_A - N_D)^2 + 4n_i^2]^{1/2}\} \\ &= (1/2)[C_B + (C_B^2 + 4n_i^2)]^{1/2} \\ &\approx N_A - N_D \\ &\approx C_B \end{aligned} \Bigg\} \text{if } N_A - N_n \gg n_i \tag{3.15}$$

Minority carriers (electrons)

$$n = n_i^2/(N_A - N_D) = n_i^2/C_B \tag{3.16}$$

For an intrinsic semiconductor these equations reduce to

$$n = n_i, \quad p = n_i.$$

In an extrinsic nondegenerate semiconductor the majority carrier density (n or p) is identical to the impurity concentration and the minority carrier density is quite small. Hence, the majority carrier density is essentially independent of temperature, whereas the minority carrier density displays a very strong dependence upon temperature owing to the temperature dependence of the exponential term used in Boltzmann statistics. At higher temperature (where a considerable number of carriers are excited across the band gap) a tendency toward intrinsic behavior occurs and the carrier density may exceed the impurity density; hence at high temperature the majority carrier density will also be temperature-dependent.

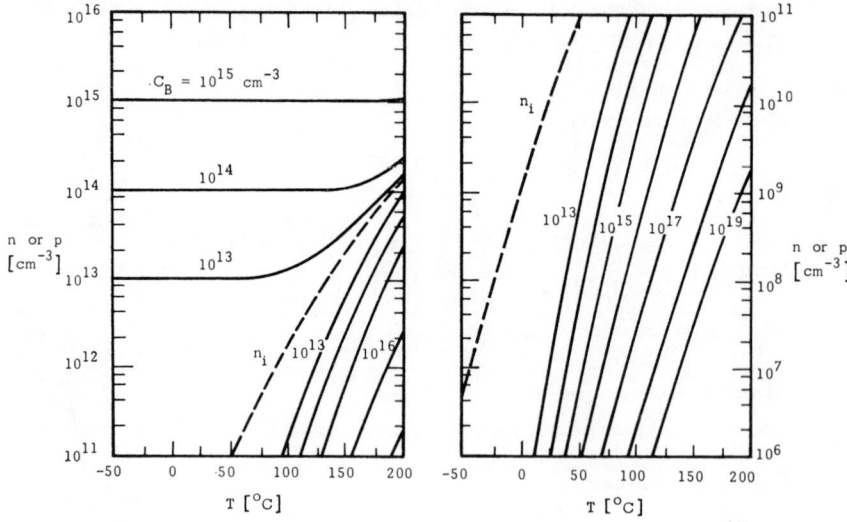

FIGURE 3.4

Majority and minority carrier densities (n and p) and intrinsic carrier density (n_i) in silicon as function of semiconductor impurity concentration (C_B). The dashed line corresponds to the intrinsic carrier density.

In Figure 3.4 the curves above the n_i curve represent majority carrier densities, curves below minority carrier densities. Figures 3.5 and 3.6 give the temperature dependence of carrier densities in nondegenerate and degenerate silicon and of N_c and N_v. The distinction between a degenerate and nondegenerate semiconductor has been made at $E_c - E_F = 2kT$ or $E_F - E_v = 2kT$, i.e., at a carrier concentration at which the Fermi level approaches the corresponding band edge within $2kT$. In a nondegenerate semiconductor the number of free carriers is equal to the number of impurity atoms present, i.e., all impurity atoms are ionized. In a degenerate semiconductor the number of free carriers is less than that of the impurity atoms; the degeneracy correction factor (ξ) describes the reduction of the number of ionized impurities

3-1 CARRIER DENSITY 261

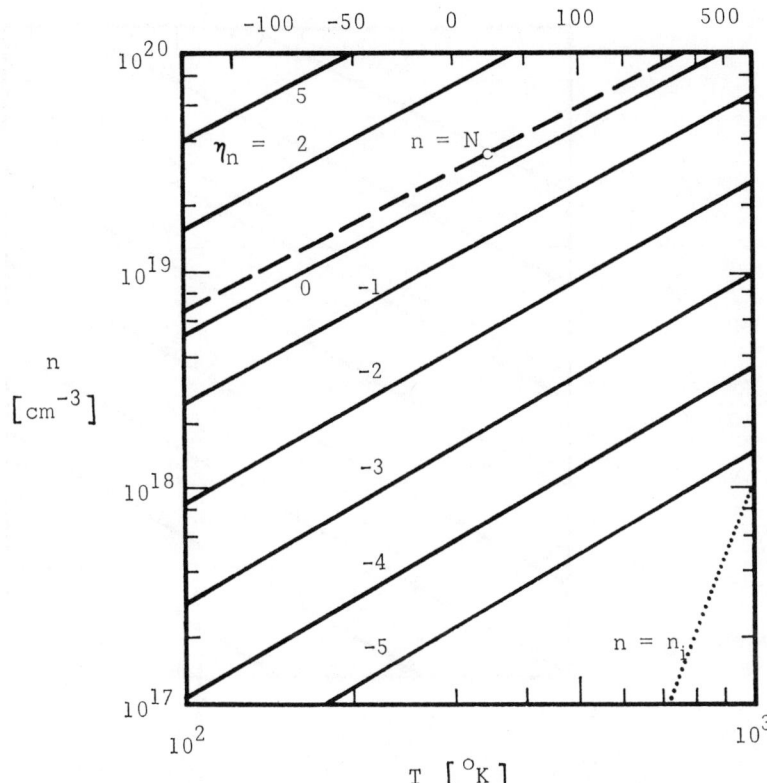

FIGURE 3.5

Electron density (n) in silicon vs. temperature (T) and reduced Fermi level (η_n).

The semiconductor is assumed to be nondegenerate if $\eta_n = (E_c - E_F)/kT < 2$ and degenerate if $\eta_n > 2$.

due to degeneracy (see 1-5). For a nondegenerate semiconductor, for which the reduced Fermi energies

$$\eta_n = (E_F - E_c)/kT < -2 \tag{3.17a}$$

or

$$\eta_p = (E_v - E_F)/kT < -2, \tag{3.17b}$$

the degeneracy correction factor ξ_n or $\xi_p = 1$. Intrinsic silicon is nondegenerate below about 1400°K. Above this temperature the intrinsic Fermi energy

262 ELECTRICAL BEHAVIOR OF SEMICONDUCTORS

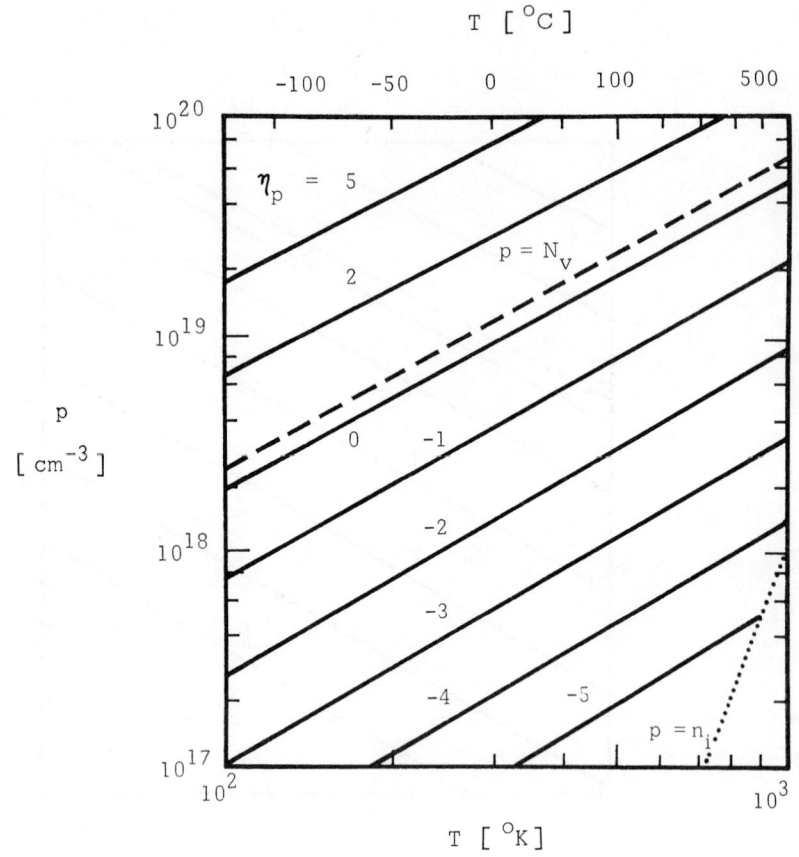

FIGURE 3.6

Hole density (p) in silicon vs. temperature (T) and reduced Fermi level (η_p). The semiconductor is assumed to be nondegenerate if $\eta_p = (E_F - E_v)/kT < 2$ and degenerate if $\eta_p > 2$.

approaches closely enough to the conduction or valence band edge that Fermi statistics must be employed and the correction factor ξ_n or ξ_p (>1) has to be used.

For a nondegenerate semiconductor the carrier densities in terms of the Fermi level

$$n = N_c \exp \eta_n \quad \text{(for } \eta_n < -2\text{)} \tag{3.18a}$$

or

$$p = N_v \exp \eta_p \quad \text{(for } \eta_p < -2\text{)}. \tag{3.18b}$$

The reduced Fermi energy describes the energy difference of the Fermi level from the appropriate band edge. If η_n or $\eta_p < 0$, then the Fermi level is within the energy gap; if η_n or $\eta_p > 0$, then it is within the energy band; and if η_n or $\eta_p = 0$, then Fermi level and band edge coincide.

Table 3.2 gives carrier densities in an arbitrary semiconductor at 300°K as function of the reduced Fermi level or of the energy difference between the Fermi level and the closest band edge.

TABLE 3.2
Carrier Densities as Function of Reduced Fermi Level (300°K)

η_n or η_p	$E_F - E_c$ or $E_v - E_F$ [eV]	n [cm^{-3}]	p [cm^{-3}]
−6	−0.156	$7.5 \cdot 10^{16}$	$3.0 \cdot 10^{16}$
−4	−0.104	$6.0 \cdot 10^{17}$	$2.5 \cdot 10^{17}$
−2	−0.052	$4.5 \cdot 10^{18}$	$1.5 \cdot 10^{18}$
0	0	$2.5 \cdot 10^{19}$	$9.0 \cdot 10^{18}$
2	0.052	$9.0 \cdot 10^{19}$	$3.8 \cdot 10^{19}$
4	0.104	$2.0 \cdot 10^{20}$	$7.0 \cdot 10^{19}$

3. NEUTRALITY CONDITION

Since in equilibrium the semiconductor as a whole has to remain neutral under all circumstances, the neutrality condition requires

$$n - p = N_D - N_A \begin{cases} = 0 & \text{for an intrinsic semiconductor,} \\ \neq 0 & \text{for an extrinsic semiconductor.} \end{cases} \quad (3.19)$$

N_D and N_A are the densities of ionized donors and acceptors which are assumed to exist simultaneously within the same semiconductor region, although one will usually be predominant; in this case, only the net impurity concentration is important, i.e., the difference between donor and acceptor concentration; on the other hand, carrier mobility is affected by impurity scattering on both types of scattering centers, i.e., it depends on the sum rather than on the difference between both types of impurity densities. From the neutrality equation for an intrinsic semiconductor ($N_D - N_A = 0$) follows $n = p$.

4. REFERENCES

(1) F. J. Morin and J. P. Maita, *Phys. Rev.*, **96**, 28 (1954).
(2) E. M. Conwell, *Proc. IRE*, **46**, 1281 (1958).
(3) W. W. Gärtner, *Semiconductor Products*, 29 (July 1960).
(4) R. N. Hall and J. H. Racette, *J. Appl. Phys.*, **35**, 379 (1964).
(5) W. M. Bullis, *Solid-State Eelctronics*, **9**, 143 (1966).
(6) H. D. Barber, *Solid-State Electronics*, **10**, 1039 (1967).

3–2 Minority Carrier Drift Velocity and Ionization Rate

1. CARRIERS IN AN ELECTRIC FIELD

Under the influence of an electric field, for example, in the depletion layer of a *p-n* junction, carriers will attain a drift velocity, v_d, which, similar to the excitation of phonons, depends upon the semiconductor temperature T and which is superimposed on a random Maxwell velocity characterized by an electron temperature T_e. As the field and thus the carrier energy increase, the rate of collisions with phonons increases and the decay of current due to collisions with impurity ions decreases. It is assumed that carrier mobility is limited by phonon collisions only.

Under the additional assumptions that the electron energy distribution is Maxwellian and that the mean free path is independent of T_e, the energy balance, i.e., the variation of carrier energy with time,

$$dE_n/dt = qEv_d n - dE_n/dt|_{phonons}$$
$$= (8/\sqrt{\pi})(m_{n,p}^{1/2} c_{ac}^2 / l_{ac})(2kT_e)^{1/2}(1 - T_e/T),$$

i.e., dE_n/dt is the difference between the rate per unit volume of acquisition of energy from the field and the net rate of loss to the phonons. At $T = T_e$ the net rate of change of energy due to collisions is zero.

The electron temperature is defined by

$$T_e - (q^2/3k)(l_{ph}E)^2/E_{ph}. \tag{3.21}$$

2. CARRIER MOBILITY AND DRIFT VELOCITY

Three regions of electric field have to be distinguished in which the mobility shows distinctly different behavior:

(a) Low electric field, i.e., $E < c_{ac}/\mu_o$:
    Carrier mobility is

$$\mu = q l_{ac}(8/9\pi m_{n,p} kT_e). \tag{3.22}$$

Carrier drift velocity is proportional to mobility:

$$v_d = \mu E$$
$$= \mu_o E(T/T_e)^{1/2}$$
$$\approx \mu_o E[1 - (3\pi/64)(\mu_o E/c_{ac})^2] \quad (3.23)$$

(b) Medium field, i.e., $E \approx (8/3)(c_{ac}/\mu_o)$:
Carrier mobility has dropped to approximately $\mu = 0.7\,\mu_o$ and drift velocity to $v_d = 0.7\,\mu_o E$.

(c) High field, i.e., $E > E_{\text{crit}} = (32/3\pi)(c_{ac}/\mu_o)$:
Carrier mobility is

$$\mu \approx \mu_o[(32/3\pi)(c_{ac}/\mu_o E)]^{1/2}. \quad (3.24)$$

At the critical field, E_{crit}, the drift velocity increases as the square root of field rather than as the field itself. At high field the dominant scattering is by emission of optical phonons. In this case, the drift velocity

$$v_d = (2\sqrt{2}/3\sqrt{\pi})q l_{ph}(m_{n,p}kT_e)^{-1/2}E$$
$$= (8/3\pi)^{1/2}(E_{ph}/m_{n,p})^{1/2} \approx 10^7 \text{ cm/sec.} \quad (3.25)$$

An electron drift velocity in the order of 10^7 cm/sec corresponds to a wavelength of approximately $\lambda = 60$ Å, i.e., the electron wavelength is larger than the lattice spacing of the semiconductor so that the electrons cannot be expected to behave like particles. The carrier drift velocities in Si, Ge, and GaAs are shown in Figures 3.7 and 3.8. The thermal variation of the electron drift velocity in silicon is given in the inset of Figure 3.10.

In the above equations the following quantities have been introduced:
c_{ac} = velocity of sound for longitudinal acoustic waves
E_{ph} = optical phonon energy ($E_{ph} \approx 0.07$ eV)
l_{ac} = mean free path for acoustical scattering, dependent on T and T_e ($l_{ac} \approx 1000$ Å)
l_{ph} = mean free path for optical phonon scattering ($l_{ph} \approx 50$ to 100 Å)
$m_{n,p}$ = effective mass of electron or hole, respectively
μ_o = carrier mobility at small field.

Due to increase of electron temperature the mobility decreases at high field as the field increases, resulting in a constant drift velocity at high field. The electron temperature is

$$T_e = \frac{(q l_{ph} E)^2}{3k E_{ph}} \approx \left(\frac{3\pi}{32}\right)^{1/2} T \frac{\mu_o E}{c_{ac}}. \quad (3.26)$$

3–2 DRIFT VELOCITY AND IONIZATION RATE

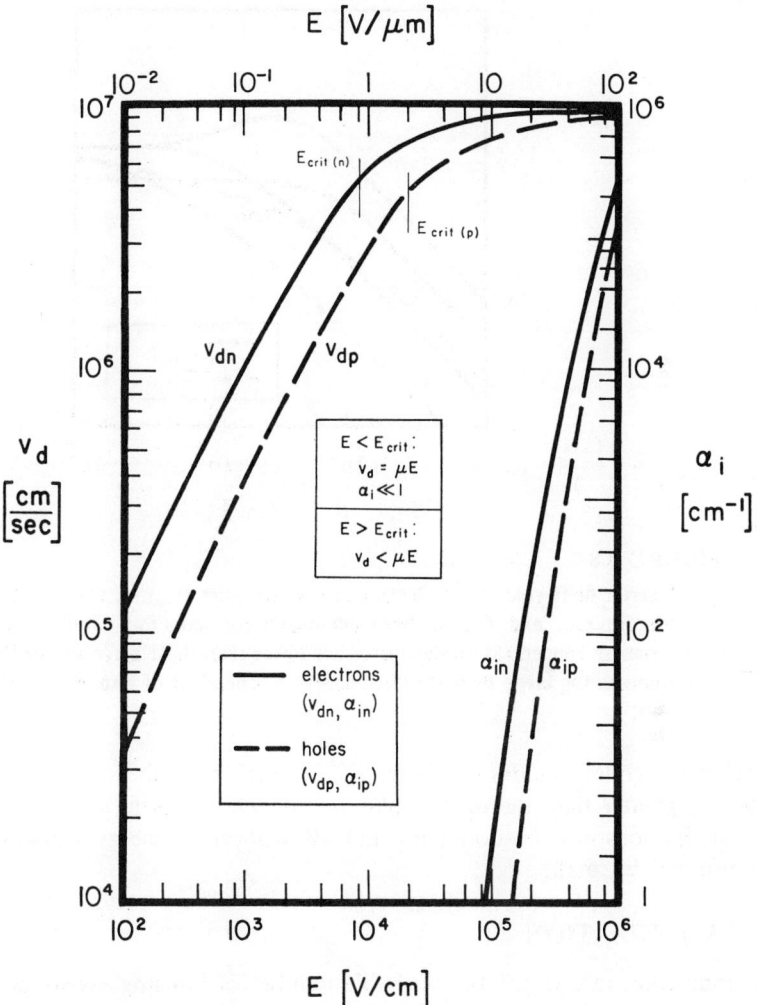

FIGURE 3.7

Limiting carrier drift velocity (v_d) and ionization rate (α_i) vs. electric field (E). Silicon, 300°K.

If the electric field is sufficiently high so that the carriers have reached their limiting drift velocity, $v_{d\max}$, an increase in the field results in an increase in the random thermal velocity or the electron temperature.

Discrepancies between this simple model which considers only phonon collisions and experimental data are due to the influence of optical and short-

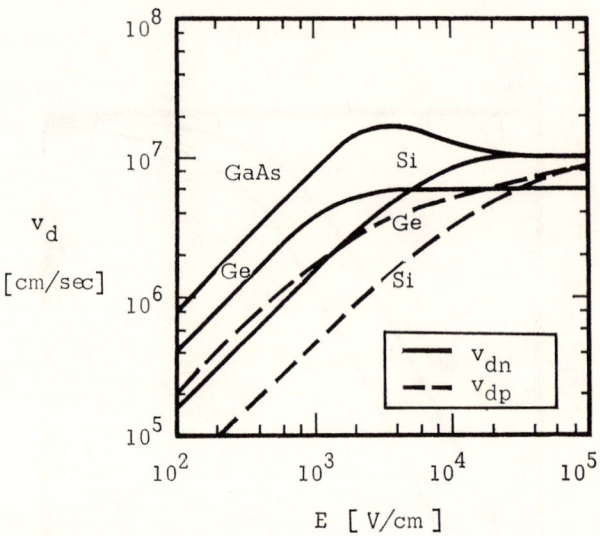

FIGURE 3.8

Carrier drift velocity of electrons (v_{dn}) and holes (v_{dp}) vs. electric field (E) for Si, Ge, and GaAs. Near-intrinsic semiconductors, 300°K. At high impurity concentration the curves are lower than those given for small field, whereas for large field they are nearly independent of impurity concentration.

wavelength acoustic modes on electron scattering. When the average electron energy is greater than the optical phonon energy, scattering by emission of optical phonons will be dominant and absorption of energy from optical phonons will be small.

3. IONIZATION

The ionization rate (α_i) is the probable number of ionizing events per unit length of carrier travel. Since the depletion layer width (x_d) of a *p-n* junction is usually significantly larger than l_{ph}, the ionization rate is a function of the applied field only:

if $kT_e \ll E_{\text{ion}}$, then $l_i = \infty$;

if $kT_e \gg E_{\text{ion}}$, then $l_i =$ finite and independent of energy,

E_{ion} is the ionization threshold energy, l_i is the mean free path for ionization, kT_e is the kinetic energy of an electron. At low electric field: $kT_e \ll E_{\text{ion}}$. In this case, any electron that ionizes must fall through a potential energy of E_{ion}; the required distance is $l_i = E_{\text{ion}}/qE$.

3-2 DRIFT VELOCITY AND IONIZATION RATE

During the length of this travel the electron has sufficient energy to emit an optical phonon. The probability of gaining sufficient energy to ionize is

$$P_{ion} \approx \exp(-E_{ion}/qEl_{ph}). \tag{3.27}$$

But even if an electron acquires the energy E_{ion} required to ionize, it may still not have an ionizing collision, but may have a succession of phonon collisions that reduce its energy below E_{ion} so that ionization cannot take place. If the electron, however, acquires sufficient energy for ionization by optical phonon collision it may emit one or more optical phonons before ionization.

During an ionizing collision the electron loses the energy E_{ion} which lies between the minimum possible energy loss E_G, the semiconductor energy gap, and the maximum energy loss kT_e, the kinetic energy of the electron at the time of ionization; i.e.,

$$E_G < E_{ion} < kT_e.$$

The ionization rate. α_i, is given by

$$\alpha_i = P_{ion} N_{pho} \tag{3.28}$$

where N_{pho} is the number of optical phonon collisions per unit length that occur at low energy.

In terms of the electric field the ionization rate can be expressed by

$$\alpha_i = \frac{qP_{ion}}{P_r + (P_i + r_l P_r)P_{ion}} E \tag{3.29}$$

where r_l = ratio of the relative probabilities of phonon emission to ionizing collision.

$$r_l = l_i/l_{ph}$$

P_i, P_r = ionization probabilities

$$qE = \alpha_i P_i + N_{ph} P_r$$

N_{ph} = number of optical phonon emissions per unit length.

The above relationship suggests a linear dependence of α_i upon E; this is true at low field, but deviations at high field resulting in a lower ionization rate are observed because the mean free path for ionization is also a function of energy and because of uncertainties in energy loss to optical phonons and in electronic behavior from band edges to the ionization energies.

The variation of the ionization rate with electric field and temperature is shown in Figures 3.9 and 3.10 for common semiconductors.

For any given collision the probability of phonon emission

$$P_{ph} = r_l/(1 + r_l) \tag{3.30}$$

and the probability of ionization

$$P_{ion} = 1/(1 + r_l) \tag{3.31}$$

The term r_l is the ratio of phonon emission probability to ionization probability and is

$$r_l = P_{ph}/P_{ion} = l_i/l_{ph} \tag{3.32}$$

where l_i is the mean free path for ionization ($l_i = E_{ion}/qE$) and l_{ph} is the mean free path for optical phonon scattering ($l_{ph} \approx 50$ to 100 Å).

Hence

$$P_{ph} = P_{ion}(l_i/l_{ph}). \tag{3.33}$$

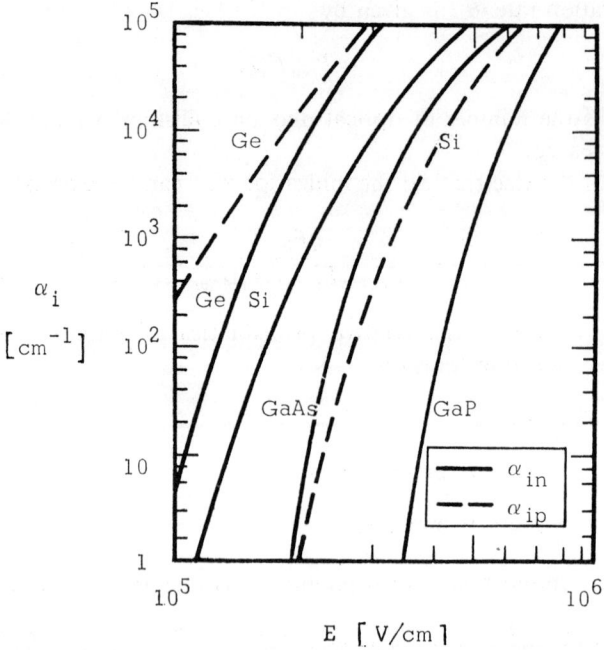

FIGURE 3.9

Ionization rate for electrons (α_{in}) and holes (α_{ip}) vs. electric field (E) for Si, Ge, GaAs, and GaP. Near-intrinsic semiconductors, 300°K. The ionization rates for electrons and holes are identical in GaAs and GaP.

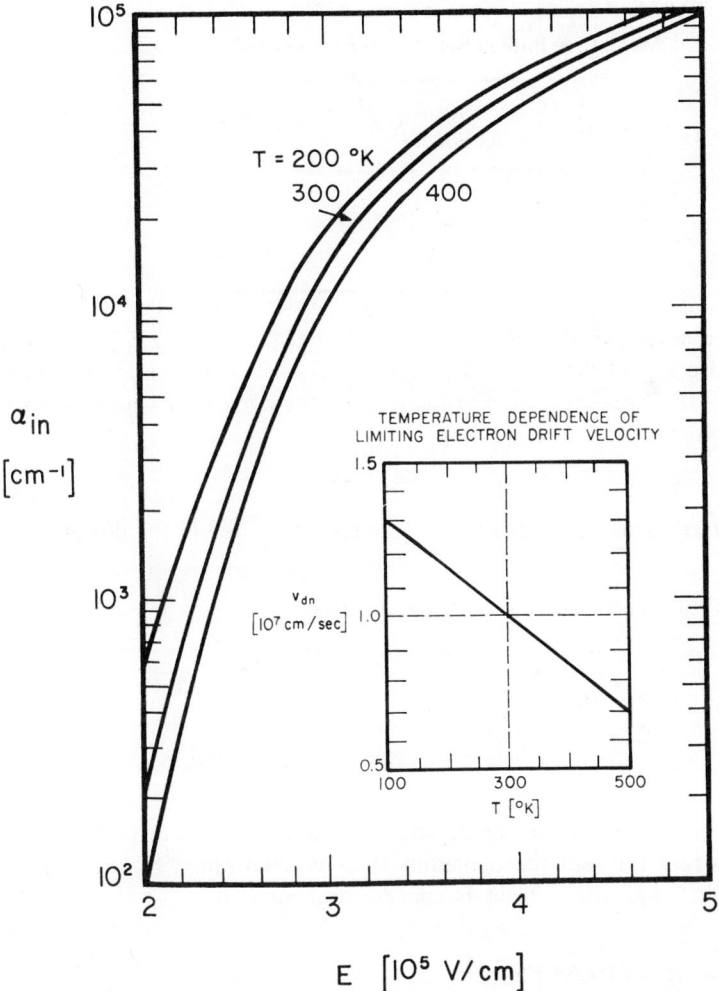

FIGURE 3.10

Variation of electron ionization rate (α_{in}) with electric field (E) and temperature (T). The inset shows the dependence of the limiting electron drift velocity (v_{dn}) with temperature (T), referred to 300°K. Near-intrinsic silicon.

Mean free paths (l_{ph}) for carrier generation at large energy associated with a momentum normal to the barrier are given in Table 3.3.

In a nondegenerate semiconductor the carrier concentration is equal to the density of impurity centers (C_B) only if the dielectric relaxation time (τ_d) is much smaller than the transit time through the drift region (τ_t). This means

TABLE 3.3

Mean Free Path in Selected Semiconductors

Semi-conductor	l_{ph} [Å]
Si	76
Ge	105
GaAs	58

that, due to transit time effects, the current through a *p-n* junction will not saturate as will the carrier drift velocity ($v_d \to v_l \approx 10^7$ cm/sec) if the electric field is increased beyond a critical value; the current at which this departure will take place is

$$j_{\text{crit}} = q v_l C_B ; \tag{3.34}$$

at current densities higher than j_{crit} charge is stored in the depletion layer.

(a) $\tau_d \ll \tau_t$:
If the electric field $E < 10^4$ V/cm, then

$$\tau_d/\tau_t = (\varepsilon_s \varepsilon_o/q) V/C_B x_d , \tag{3.35a}$$

and if $E > 10^4$ V/cm, then

$$\tau_d/\tau_t = (\varepsilon_s \varepsilon_o/q)(v_d/C_B x_d)(dE/dv_d) . \tag{3.35b}$$

(b) $\tau_d \gg \tau_t$:
In this case the carrier density will exceed the impurity density and the current through the depletion layer is determined by the space charge effect, i.e., the current is space-charge limited.

4. REFERENCES

(1) E. J. Ryder, *Phys. Rev.*, **90**, 766 (1953).
(2) K. G. McKay and K. B. McAfee, *Phys. Rev.*, **91**, 1079 (1953).
(3) K. G. McKay, *Phys. Rev.*, **94**, 877 (1954).
(4) S. L. Miller, *Phys. Rev.*, **105**, 1246 (1957).
(5) A. C. Prior, *J. Phys. Chem. Solids*, **12**, 175 (1960).
(6) J. L. Moll, *Physics of Semiconductors*, McGraw-Hill Book Co., New York, 1964.
(7) C. A. Lee, et al., *Phys. Rev.*, **134**, 761 (1964).
(8) R. A. Logan and S. M. Sze, *J. Phys. Soc. Japan Suppl.*, **21**, 434 (1966).
(9) R. A. Logan and H. G. White, *J. Appl. Phys.*, **36**, 3945 (1965).

(10) T. E. Seidel and D. L. Scharfetter, *J. Phys. Chem. Solids*, **28**, 2563 (1967).
(11) C. B. Norris and J. F. Gibbons, *IEEE Transact. Electr. Dev.*, **ED-14**, 38 (1967).
(12) C. Y. Duh and J. L. Moll, *IEEE Transact. Electr. Dev.*, **ED-14**, 46 (1967).
(13) J. G. Ruch and G. S. Kino, *Appl. Phys. Lett.*, **10**, 40 (1967).
(14) V. Rodriguez and M. A. Nicolet, *J. Appl. Phys.*, **40**, 496 (1969).
(15) P. L. Hower and V. G. K. Reddi, *IEEE Transact. Electr. Dev.*, **ED-17**, 320 (1970).
(16) R. Van Overstraeten and H. DeMan, *Solid-State Electronics*, **13**, 583 (1970).

3-3 Carrier Mobility and Carrier Diffusion Coefficient

1. CARRIER MOBILITY AND MEAN FREE PATH

At a finite temperature a semiconductor contains a number of free carriers of mass m_n or m_p (denoted as $m_{n,p}$ for both electrons and holes) which are in a state of thermal agitation and which experience frequent collisions with the crystal lattice and other obstacles in their path during which the carriers lose some of their acquired energy. If a small external electric field (E) is applied, the carriers are accelerated during the time between collisions (τ_c); this results in an average drift velocity ($\bar{v}_d$) of the carriers which is superimposed on their random thermal velocity (v_{th}) and which is

$$\bar{v}_d = \pm qE\tau_c/m_{n,p} = \pm \mu E. \tag{3.36}$$

The positive sign refers to holes, the negative sign to electrons. The term μ is defined as the mobility of the carrier and is the average drift velocity per unit field. The drift velocity is usually much smaller than the thermal velocity which is

$$v_{th} = (3\,kT/m_{n,p})^{1/2}. \tag{3.37}$$

At 300°K

$$v_{th} \approx 10^7 \text{ cm/sec}.$$

Carriers are on the average not accelerated by an electric field and their average velocity is constant and proportional to the field. However, if the electric field exceeds a certain critical value, E_{crit}, the carrier velocity is no longer proportional to E because the mobility is no longer a constant, but is reduced significantly. At a field much higher than E_{crit} the average carrier drift velocity approaches the thermal velocity (i.e., $\bar{v}_d \approx v_{th}$) and remains constant (in the order of 10^7 cm/sec) and is independent of the field.

The mean free path (l_c) between collisions is related to the time between collisions by

$$l_c = v_{th}\tau_c \tag{3.38}$$

3–3 CARRIER MOBILITY AND DIFFUSION COEFFICIENT

if it is assumed that l_c is much smaller than the sample dimensions and that it is independent of carrier velocity. The mean free time of the carriers between collisions (carrier relaxation time) is

$$\tau_c = \frac{m_{n,p}^{3/2} w_{at} a^3}{h^2} \frac{T_D}{T} \frac{(E_G/2)^{3/2}}{E_F - E_i} \left(\frac{E_G}{kT}\right)^2 \qquad (3.39)$$

where w_{at} = average atomic weight of semiconductor
 a = lattice constant
 T_D = Debye temperature
 E_i, E_F = Fermi energy in an intrinsic or extrinsic semiconductor, respectively.

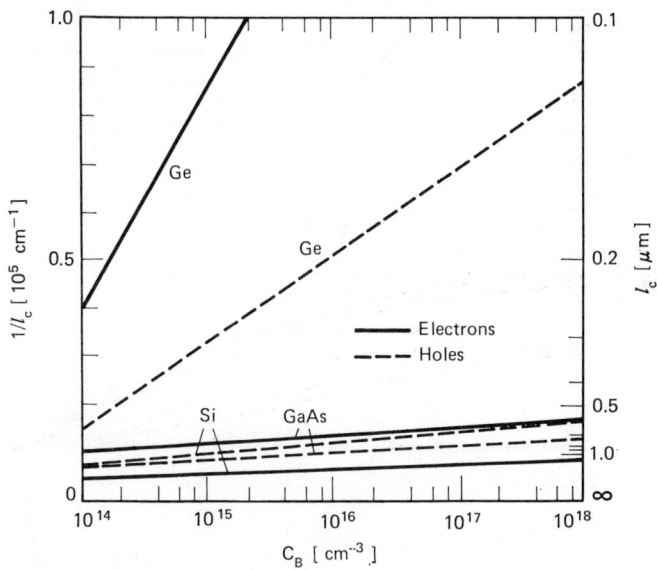

FIGURE 3.11

Carrier mean free path (l_c) as function of impurity concentration (C_B), assuming impurity scattering as the only scattering mechanism, for Si, Ge, and GaAs. 300°K. The curves are calculated from
$$l_c = (3/8)^{1/2}(m_{n,p}^{3/2} w_{at} a^3/h^2)(T_D/T) E_G^{7/2}/[(E_F - E_i)(kT)^{3/2}].$$
If the semiconductor thickness is smaller than the mean free path, carrier mobility reduction is observed; i.e., the mean free path approximately represents the lower limit of the semiconductor thickness in order to obtain bulk mobility. The mean free path approaches infinity if $C_B = n_i$.

276 ELECTRICAL BEHAVIOR OF SEMICONDUCTORS

Hence the mean free path under the assumption that impurity scattering is the dominant scattering mechanism

$$l_c = \left(\frac{3}{8}\right)^{1/2} \frac{m_{n,p}^{3/2} w_{at} a^3}{h^2} \cdot \frac{T_D}{T} \cdot \frac{E_G^{7/2}}{(E_F - E_i)(kT)^{3/2}} \tag{3.40}$$

The mean free path as a function of semiconductor impurity concentration is shown in Figure 3.11 and in Table 3.4. If the semiconductor thickness is less than the mean free path, then a significant carrier mobility reduction takes place, as shown in Figure 3.15 and discussed below.

If it is assumed that the semiconductor has a single conduction band only and that it is either cubic or isotropic, that the density of one carrier type is

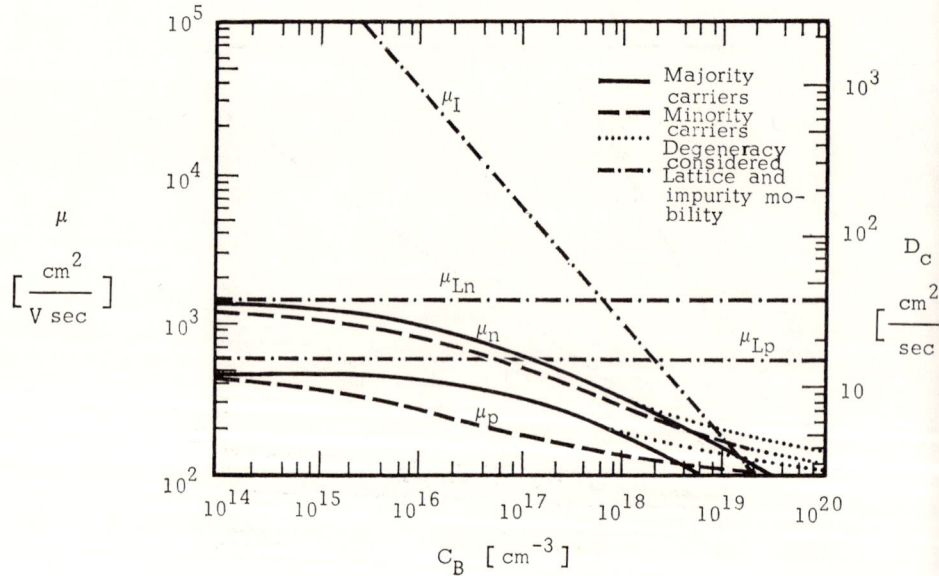

FIGURE 3.12

Majority and minority carrier mobility (μ_n, μ_p), lattice scattering mobility (μ_{Ln} and μ_{Lp}), and impurity scattering mobility (μ_I), and carrier diffusion coefficient (D_c) vs. semiconductor background impurity concentration (C_B). Silicon, 300°K. The majority carrier mobility (solid curves) applies to electrons in n-type material and to holes in p-type material; the minority carrier mobility (dashed curves) applies to electrons in p-type material and to holes in n-type material. The onset of degeneracy (dotted curves) is characterized by a reduction of carrier density compared to the density of impurity atoms present.

3-3 CARRIER MOBILITY AND DIFFUSION COEFFICIENT

TABLE 3.4

Mean Free Path of Electrons and Holes as a Function of Semiconductor Impurity Concentration

Semiconductor	l_c [μm]				
	$C_B = 10^{14}$ cm^{-3}	10^{15} cm^{-3}	10^{16} cm^{-3}	10^{17} cm^{-3}	10^{18} cm^{-3}
			n-Type		
Si	2.43	1.85	1.62	1.39	1.22
Ge	0.30	0.12	0.08	0.06	0.04
GaAs	0.99	0.85	0.74	0.66	0.59
			p-Type		
Si	1.27	1.01	0.85	0.73	0.63
Ge	0.80	0.31	0.20	0.15	0.11
GaAs	1.27	1.09	0.95	0.85	0.76

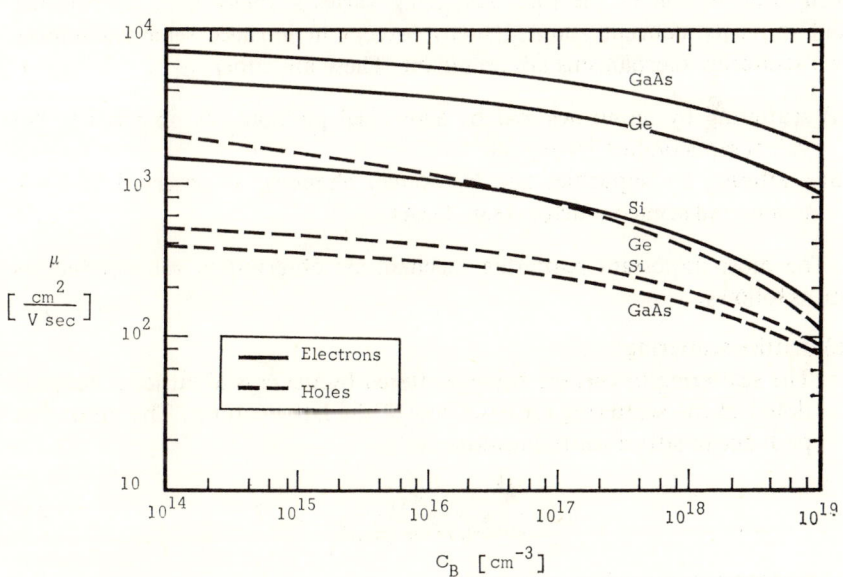

FIGURE 3.13

Majority carrier mobility (μ_n, μ_p) in Si, Ge, and GaAs at 300°K vs. semiconductor impurity concentration (C_B).

significantly higher than that of the other type, and that the semiconductor is nondegenerate, then carrier mobility and semiconductor resistivity are

$$\mu = 4ql_c/3(2\pi m_{n,p}kT)^{1/2} \tag{3.41}$$

$$\begin{aligned}\rho &= 1/q\mu n \\ &= m_{n,p}v_{th}/q^2nl_c \\ &= m_{n,p}/q^2n\tau_c \\ &= 3(2\pi m_{n,p}kT)^{1/2}/4q^2nl_c.\end{aligned} \tag{3.42}$$

2. SCATTERING MECHANISMS

In a perfectly periodic (ideal) lattice in which no vibrations take place, no scattering of charge carriers should occur, i.e., $l_c = \infty$. In practice, departures from perfect periodicity give rise to various scattering mechanisms. Basically the scattering mechanisms can be classified as those due to lattice vibrations and those due to lattice imperfections (i.e., due to the presence of ionized and nonionized impurities).

In a real crystal, several scattering mechanisms occur simultaneously; their contribution to the total scattering varies strongly with temperature and impurity concentration. Usually (except in intrinsic semiconductors), two scattering mechanisms are dominant. These are either

(a) scattering by impurities and by acoustical phonons, as observed in covalent semiconductors (e.g., Si, Ge), or
(b) scattering by impurities and by optical phonons, as observed in III–V compound semiconductors (e.g., GaAs).

The most important scattering mechanisms observed in semiconductors are as follows:

(a) Lattice scattering:
The scattering of carriers being scattered by the crystal lattice is proportional to the scattering cross-section of the lattice atoms. The mean free path due to lattice scattering alone is

$$l_L = \frac{h^4}{16\pi^3} \frac{c_{11}}{kT(m_{n,p}\varepsilon_D)^2} = A_L/T \tag{3.43}$$

and the corresponding resistivity

$$\rho_L = \frac{12\sqrt{2}\pi^{7/2}}{q^2 h^4} \frac{m_{n,p}^{5/2} \varepsilon_D (kT)^{3/2}}{c_{11} C_B}, \tag{3.44}$$

3-3 CARRIER MOBILITY AND DIFFUSION COEFFICIENT

where c_{11} = average longitudinal elastic constant of semiconductor
ε_D = displacement of the edge of the band per unit dilation of the lattice (deformation potential).

The proportionality factor A_L depends upon the mass of the vibrating atoms and on the frequency of lattice vibrations. The carrier mobility due to lattice vibrations (acoustic phonons) is theoretically

$$\mu_L = \frac{\pi^{-7/2}}{12\sqrt{2}} \frac{qh^4 c_{11}}{\varepsilon_D m_{n,p}^{5/2}(kT)^{3/2}} = q\tau_c/m_{n,p}, \quad (3.45)$$

i.e., μ_L is proportional to $m_{n,p}$ and T as

$$\mu_L \sim m_{n,p}^{-5/2} T^{-3/2}.$$

Empirically it has been found for silicon that

$$\mu_{Ln} = 2.1 \cdot 10^9 \, T^{-2.5} \text{ cm}^2/\text{V sec} \quad (3.46a)$$

$$\mu_{Lp} = 2.3 \cdot 10^9 \, T^{-2.7} \text{ cm}^2/\text{V sec.} \quad (3.46b)$$

Lattice mobility depends upon the crystal orientation of the semiconductor because of a different atom packing density for different orientations.

(b) Ionized scattering;

Two types of lattice imperfections with differing scattering characteristics may affect carrier scattering in a semiconductor—ionized and neutral scattering centers.

Charged ionized centers (impurity ions) affect mobility and resistivity as follows.

$$\mu_I = \frac{2^{7/2}(\varepsilon_s \varepsilon_o)^2 (kT)^{3/2}}{\pi^{3/2} q^3 m_{n,p}^{1/2} C_B} \frac{1}{\ln\{[1 + 9(\varepsilon_s \varepsilon_o d_I kT)^2]/q^4\}}, \quad (3.47)$$

i.e., μ_I is proportional to $m_{n,p}$ and T as

$$\mu_I \sim m_{n,p}^{-1/2} T^{3/2}.$$

The resistivity due to impurity scattering alone is

$$\rho_I = \frac{\pi^{3/2} q^2 m_{n,p}^{1/2} C_B}{2^{7/2}(\varepsilon_s \varepsilon_o)^2 (kT)^{3/2}} \ln\left(\frac{1 + 9(\varepsilon_s \varepsilon_o d_I kT)^2}{q^4}\right) \quad (3.48)$$

where d_I is the average distance between neighboring scattering centers. The mean free path, l_I, is a function of the particle velocity. It is assumed that collisions are perfectly elastic and that each scattering center has an infinite mass. The logarithmic term is relatively independent of temperature so that the temperature dependence of ρ_I can be expressed by

$$\rho_I = A_I T^{-3/2}. \quad (3.49)$$

280 ELECTRICAL BEHAVIOR OF SEMICONDUCTORS

Theoretically, the temperature dependences of ρ_I and ρ_L and of μ_I and μ_L cancel each other.

(c) Nonionized scattering:

Neutral scattering centers (of density N_N) affect mobility and resistivity as follows:

$$\mu_N = \frac{2\pi^3 q^3 m_{n,p}}{5\varepsilon_s \varepsilon_o N_N h^3}, \qquad (3.50)$$

$$\rho_N = \frac{5\varepsilon_s \varepsilon_o N_N h^3}{2\pi^3 q^4 C_B m_{n,p}}. \qquad (3.51)$$

Nonionized scattering is only of secondary importance since mobile carriers would have to approach extremely closely the core of a neutral atom in order to be scattered by it and because a localized strain field caused by the neutral atom distorts the path of the carriers in its vicinity.

(d) Optical mode (optical-phonon) scattering:

In ionic (i.e., polar) crystals, scattering depends mainly on lattice vibrations in which negative and positive ions move in opposite directions (polarization waves), i.e., on optical vibrational modes.

(e) Dislocation scattering:

In a semiconductor with a high dislocation density (N_d) the contribution by dislocation scattering mobility (μ_D) may not be negligible. It can be expressed as

$$\mu_D = \frac{32}{3\pi} \frac{kT h q}{(\varepsilon_D l_s)^2} \left(\frac{1-P}{1-2P} \right)^2 \frac{1}{m_{n,p} N_d} \qquad (3.52)$$

where ε_D = deformation potential (for Si: $\varepsilon_D = 2.68$)
l_s = crystallographic slip distance
P = Poisson ratio.

The dislocation scattering mobility depends upon the carrier concentration because the effective carrier mass is a function of concentration. The deformation potential is relatively independent of carrier concentration. The dislocation scattering mobility is proportional to temperature and inversely proportional to the dislocation density.

In addition to the scattering mechanisms considered above which are of most significant importance, some other scattering mechanisms play a more secondary role in affecting carrier mobility in a semiconductor:

(a) Intravalley scattering:

In this case an electron is scattered within an energy ellipsoid and only long-wavelength phonons are involved.

(b) Intervalley scattering:
   In this case an electron is scattered from the vicinity of one minimum to another minimum and an energetic phonon is involved.

3. TOTAL CARRIER MOBILITY

In practice several scattering mechanisms may take place simultaneously. In this case, the total mean free path of the carriers is

$$\frac{1}{l_c(v)} = \frac{1}{l_L} + \frac{1}{l_I(v)} + \frac{1}{l_N(v)} \tag{3.53}$$

where v is the average particle velocity. The total mean free path determines the overall carrier mobility and the semiconductor resistivity which become complicated functions of temperature. Scattering by lattice, ionized centers and nonionized centers are not additive because they depend in different ways on the carrier velocities.

The total carrier mobility is related to the various scattering mobilities by

$$\frac{1}{\mu} = \frac{1}{\mu_L} + \frac{1}{\mu_I} + \frac{1}{\mu_N} + \frac{1}{\mu_{opt}} + \frac{1}{\mu_D} \tag{3.54}$$

In nonpolar semiconductors (e.g., Si and Ge) lattice and impurity scattering are of primary importance so that

$$\frac{1}{\mu} \approx \frac{1}{\mu_L} + \frac{1}{\mu_I}. \tag{3.55}$$

In polar semiconductors (e.g., GaAs) optical mode scattering is also significant.

Electron and hole mobilities are slightly different in semiconductors of opposite conduction type. For small impurity concentrations (lattice-scattering range) majority and minority carrier mobilities are comparable because both types of carriers are affected by essentially the same number of charge centers. For high impurity concentrations (impurity-scattering range) the majority carriers see only about half as many charge centers as the minority carriers, hence they have a higher mobility.

Carrier mobility depends on the total number of ionized impurities, i.e., on the sum of donor and acceptor concentrations. The carrier concentration, however, depends upon the difference of donor and acceptor concentrations.

4. TEMPERATURE DEPENDENCE OF MOBILITY

(a) Intrinsic semiconductor:
   Here scattering collisions of carriers occur with lattice atoms, while carrier-carrier collisions are insignificant. At very low temperature the lattice atoms have only zero point energies, i.e., their carrier scattering

cross-sections are small. With increasing temperature the scattering cross-sections are increased due to thermal vibrations of greater amplitude; this results in an increased probability of electron-phonon collisions. The momentum distribution of the free electrons is therefore affected by two competing processes—the acceleration of electrons by an electric field and the scattering of moving electrons by phonons. The theoretical temperature variation of μ is expected to be $\mu \sim T^{-3/2}$. Experimentally it has been found to range from $\mu \sim T^{-3/2}$ to $T^{-5/2}$.

(b) Extrinsic Semiconductor:

Here carrier scattering occurs at the ionized donors and acceptors. Since they have a large scattering cross-section, their influence on

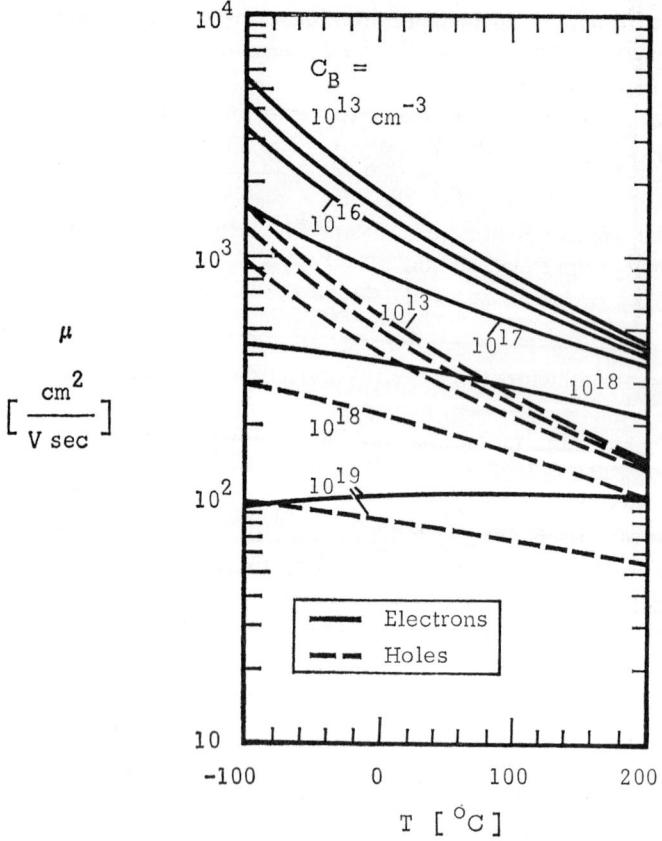

FIGURE 3.14

Minority carrier mobility (μ) vs. temperature (T) and semiconductor impurity concentration (C_B). Silicon.

mobility is significant. This is most pronounced at low temperature where the effect of the impurity atoms is noticed so that scattering mobility is affected most by impurity scattering; at high temperature the dominance of the semiconductor atoms over the impurity atoms (many orders of magnitude) results in the significant influence of lattice scattering on the carrier mobility. At low temperature, impurity scattering reduces the mean free path between collisions. This corresponds to the influence of impurity scattering mobility on the semiconductor resistivity at low temperature. An increase in impurity concentration at a given temperature results in a reduction of carrier mobility.

Table 3.5 gives the variation of electron and hole mobility with temperature for various semiconductors.

TABLE 3.5
Temperature Dependence of Mobility of Selected Semiconductors

Semi-conductor	$\mu_n{}^*$ [cm²/V sec]	$\mu_p{}^*$ [cm²/V sec]	Mobility variation**	
			w_n	w_p
Si	1,350	475	2.5	2.7
Ge	3,900	1,900	1.7	2.3
AlSb	200	420		1.8
GaAs	8,500	400	1.0	2.1
GaP	110	75	1.5	1.5
GaSb	4,000	1,400	2.0	0.9
InAs	33,000	460	1.2	2.3
InP	4,600	150	2.0	2.4
InSb	80,000	750	1.6	2.1

* At 300°K
** $\mu \sim T^{-w}$

5. IMPURITY CONCENTRATION DEPENDENCE OF MOBILITY

Generally, as the impurity concentration is increased (assuming that at room temperature all impurity atoms are ionized), the total carrier mobility decreases essentially as expected from the impurity scattering mobility. As the effective mass of the carriers increases, the total mobility decreases, thus, for a given impurity concentration, the electron mobility is larger that the hole mobility.

284 ELECTRICAL BEHAVIOR OF SEMICONDUCTORS

A general (empirical) relationship between carrier mobility and impurity concentration is

$$\mu_c \approx \frac{\mu_{io}}{1 + (C_B/n_i)^{-\alpha_1}} \tag{3.56}$$

where for silicon the parameters μ_{io} and α_1 are as follows.

	Electrons	Holes
μ_{io} [cm²/V sec]	1350	475
α_1	0.06	0.05

6. MOBILITY IN A DEGENERATE SEMICONDUCTOR

In a strongly degenerate semiconductor (in which the free carrier concentration is less than the impurity concentration) the impurity scattering mobility

$$\mu_I = \frac{3}{16\pi^2} \left(\frac{h}{q}\right)^3 \left(\frac{\varepsilon_s \varepsilon_o}{m_{n,p}}\right)^2 \frac{1}{\ln(1+f_n) - f_n/(1+f_n)} \tag{3.57}$$

where

$$f_n = \left(\frac{h}{q}\right)^2 \frac{\varepsilon_s \varepsilon_o}{m_{n,p}} \left(\frac{3n}{8\pi}\right)^{1/3} \tag{3.58}$$

The term f_n is a function of the carrier concentration; in silicon for $n = 10^{18}$ cm^{-3}: $f_n \approx 2$; for $n = 10^{21}$ cm^{-3}: $f_n \approx 20$. Since the effective electron mass is also a function of carrier concentration, its variation has to be taken into account. As a result, the impurity scattering mobility in the degenerate case is proportional to $n^{-2/3}$. Although the above equations are valid for electrons, similar expressions apply to holes.

The optical lattice scattering mobility (μ_{opt}) in a degenerate semiconductor is proportional to $n^{-1/3}$.

7. MOBILITY IN A THIN FILM

The scattering mechanisms above discussed apply to semiconductor samples which are macroscopic in all three dimensions. If, however, one or more dimensions are microscopic, then boundary scattering has to be considered in addition to other scattering mechanisms since the boundaries of the semiconductor become important scattering centers.

3-3 CARRIER MOBILITY AND DIFFUSION COEFFICIENT 285

The properties of a thin semiconductor film differ from those of a bulk semiconductor mainly because of the large surface-to-volume ratio of the film. Surface properties are, therefore, expected to play a dominant role in films, but only a secondary role in bulk semiconductor transport mechanisms.

Free carriers are scattered by the film boundary surfaces in addition to the scattering mechanisms observed in a semiconductor bulk. The carriers striking a surface can be reflected either in a specular or in a diffuse way.

(a) Specular reflection:
In this case only the component of the momentum of the carriers normal to the surface is changed. The energy of the carriers and the parallel momentum components remain constant. Consequently, the carrier mobility remains unaffected.

(b) Diffuse reflection:
In this case the velocity of the carriers is independent of their velocity prior to collision with the surface and the carriers will obtain a Maxwell–Boltzmann distribution after reflection. Consequently, carrier mobility is reduced within a distance comparable to their mean free path l_c.

The mean relaxation time of the carriers is

$$1/\overline{\tau^*} = 1/\tau_B^* + 1/\tau_S^* \qquad (3.59)$$

where τ_B^* is the relaxation time in the interior of the film assumed to be identical to the bulk relaxation time, and τ_S^* is the relaxation time of carriers colliding with the surface and losing all memory of their momenta.

The mean unilateral velocity $(\overline{v_d})$ of electrons or holes is defined by

$$\overline{v_d} = (kT/2\pi m_{n,p})^{1/2}, \qquad (3.60)$$

the unilateral mean free path by

$$l_c = \tau_B^* \overline{v_d}$$
$$= \tau_B^* (kT/2\pi m_{n,p})^{1/2}, \qquad (3.61)$$

and the surface relaxation time by

$$\tau_S^* \approx x_f/2\overline{v_d}$$
$$= \tau_B^*(x_f/2l_c), \qquad (3.62)$$

where x_f is the thickness of the semiconductor film. With these, the total mean relaxation time

$$\overline{\tau^*} = (l_c + x_f/2)(2\pi m_{n,p}/kT)^{1/2}. \qquad (3.63)$$

The mean carrier mobility in the film is

$$\overline{\mu_f} = q\overline{\tau^*}/m_{n,p} = q(l_c + x_f/2)(2\pi/m_{n,p}kT)^{1/2}, \qquad (3.64)$$

in analogy with the corresponding expression for the semiconductor bulk

$$\mu_B = q\tau_B^*/m_{n,p}. \tag{3.65}$$

Therefore

$$\overline{\mu_f}/\mu_B = 1/[1 + (2l_c/x_f)] \tag{3.66}$$

which can be further reduced if $x_f \gg l_c$:

$$\overline{\mu_f}/\mu_B \approx 1 - (2l_c/x_f). \tag{3.67}$$

If the film surface scatters carriers partly by specular and partly by diffuse scattering processes, and if $p^* = 1$ is the probability per unit time that a carrier reaching the surface will be specularly scattered and $p^* = 0$ is the probability per unit time that it will be diffuse scattered, then

$$\overline{\mu_f}/\mu_B = 1/[1 + (2l_c/x_f)(1 - p^*)]. \tag{3.68}$$

The lattice boundaries of a conductive thin film have a significant effect on its electrical and some other properties. The boundaries of the lattice can be considered as crystal imperfections; consequently, scattering of the carriers by the boundaries will affect the mobility in a similar manner. If one or two of the dimensions are small, the boundary imperfection will play an increasing role compared to other lattice defects. The total mobility in a thin film is mainly due to three electron scattering mechanisms (Matthiessen's rule):

(a) scattering by lattice imperfections due to thermal motion of impurities,
(b) scattering by lattice defects such as missing atoms (vacancies), interstitials, impurity atoms, etc.,
(c) scattering by the film boundaries.

In ferromagnetic materials a fourth type of scattering mechanism is due to the scattering by walls of magnetic domains which may be considered as a type of a boundary.

The carrier mobility in a thin film consequently consists of three components:

(a) An ideal component (μ_1):
   It depends strongly upon the amplitude of the thermal motions of the ions and depends therefore in a reversible way on temperature.
(b) A residual component (μ_2):
   It depends strongly on the density of lattice defects and is independent of temperature as long as the lattice defects are not affected by temperature variations. If lattice defects are affected by temperature changes, irreversible changes of mobility will occur.

3-3 CARRIER MOBILITY AND DIFFUSION COEFFICIENT

(c) A thickness-dependent component (μ_3):
In films of large thickness this component is negligible. The temperature coefficient of mobility is also a function of film thickness.

The total carrier mobility in a thin film is therefore

$$1/\mu_f = 1/\mu_1 + 1/\mu_2 + 1/\mu_3 \tag{3.69}$$

and the total film resistivity

$$\rho_f = \rho_1 + \rho_2 + \rho_3. \tag{3.70}$$

The carrier mobility in a thin semiconductor or metal film of thickness x_f differs from the mobility in bulk material (μ_B for $x_f = \infty$) if the film thickness is comparable to the carrier mean free path (l_c). The thin film mobility (μ_f) is given for

(a) very thin film ($x_f < l_c$):

$$\mu_f/\mu_B \approx (3/4)[\ln(l_c/x_f) + 0.423]/(l_c/x_f), \tag{3.71}$$

(b) thick film ($x_f > l_c$):

$$\mu_f/\mu_B \approx 1/[1 + (3/8)(l_c/x_f)], \tag{3.72}$$

(c) critical film thickness ($x_f \approx l_c$):

$$\mu_f/\mu_B \approx 0.67. \tag{3.73}$$

The mobility reduction in a thin film is shown in Figure 3.15. An example may illustrate this. Assuming single-crystalline Si, Ge, and GaAs films of thickness $x_f = 1.0$ μm and of impurity concentration $C_B = 10^{15}$ cm^{-3}, the electron mean free path for Si is $l_c = 1.95$ μm, for Ge, $l_c = 0.12$ μm, for GaAs, $l_c = 0.85$ μm, and the mobility ratio is

$$\mu_f/\mu_B = 0.42 \quad \text{or} \quad 42\% \text{ for Si,}$$
$$= 0.95 \quad \text{or} \quad 95\% \text{ for Ge,}$$
$$= 0.72 \quad \text{or} \quad 72\% \text{ for GaAs.}$$

This means that the bulk and film mobilities compare as follows, making the above assumptions.

Semi-conductor	μ_B [cm^2/V sec]	μ_f [cm^2/V sec]
Si	1200	504
Ge	4000	3800
GaAs	7000	5040

288 ELECTRICAL BEHAVIOR OF SEMICONDUCTORS

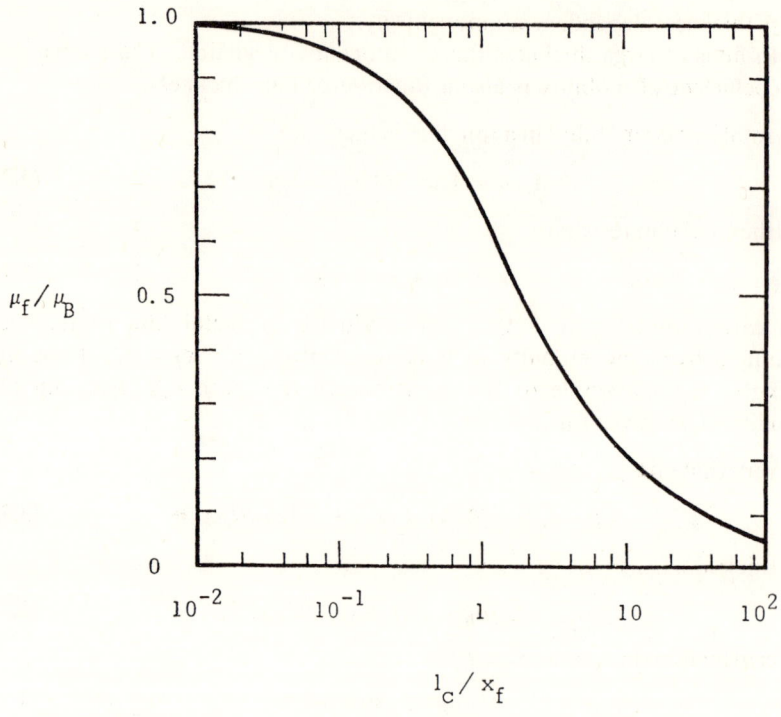

FIGURE 3.15

Variation of the ratio of carrier mobility in a thin film (μ_f) to that in a bulk semiconductor (μ_B) as function of the ratio of carrier mean free path (l_c) to film thickness (x_f) for arbitrary semiconductor. This illustration can be applied to polycrystalline bulk semiconductors since single-crystalline domains in polycrystalline material can be considered as regions of microscopic dimensions; in this case the film thickness is replaced by the mean diameter of the single-crystalline domains and μ_f corresponds to the mean carrier mobility in the polycrystalline material. Because of statistical fluctuations of the domain size the mean value of the domain diameter has to be used.

Among the three semiconductors, the mobility in Ge is least affected by film thickness or polycrystalline domain size, that in Si is most affected. Considering also the differences in mobility in bulk material, the electron mobility in a 1-μm thick single-crystalline film is approximately an order of magnitude higher in GaAs than in Si of comparable impurity concentration. It is almost as high in Ge as it is in GaAs of otherwise similar properties.

The discussion of the mobility reduction in a thin semiconductor (where one of the dimensions is microscopic, i.e., comparable to or less than the carrier mean free path) is applicable to polycrystalline semiconductors (either bulk or thin film, but nonamorphous) since here all three dimensions are microscopic. However, since all three dimensions rather than one are microscopic, the carrier mobility in a polycrystalline semiconductor is reduced beyond that in a single-crystalline film and is expected to be approximately one-third of that observed in a single-crystalline film, assuming that the film thickness and the mean domain diameter are comparable in size.

A reduction in carrier mobility in an epitaxial semiconductor film on an insulating substrate compared to the mobility in the semiconductor bulk is caused mainly by the space charge around edge dislocations, and by dislocation and neutral scattering.

8. CARRIER DIFFUSION COEFFICIENT

In thermal equilibrium the diffusion coefficient for any semiconductor

$$D_n = 2\frac{kT}{q}\mu_n \frac{F_{1/2}[(E_c - E_F)/kT]}{F_{-1/2}[(E_c - E_F)/kT]}$$

$$= 2\frac{kT}{q}\mu_n \frac{F_{1/2}(\eta_n)}{F_{-1/2}(\eta_n)} \qquad (n\text{-type}), \quad (3.74a)$$

$$D_p = 2\frac{kT}{q}\mu_p \frac{F_{1/2}[(E_F - E_v)/kT]}{F_{-1/2}[(E_F - E_v)/kT]}$$

$$= 2\frac{kT}{q}\mu_p \frac{F_{1/2}(\eta_p)}{F_{-1/2}(\eta_p)} \qquad (p\text{-type}), \quad (3.74b)$$

where $F_{1/2}$ and $F_{-1/2}$ are the Fermi-Dirac integrals (see Table 1.25).

The expressions for D_n and D_p can be reduced for nondegenerate semiconductors to the Einstein relationship

$$D_n = \mu_n kT/q \qquad (n\text{-type}), \qquad (3.75a)$$

$$D_p = \mu_p kT/q \qquad (p\text{ type}). \qquad (3.75b)$$

Generally, the carrier diffusion coefficient is given as D_c; i.e., $D_c = D_n$ for electrons, $D_c = D_p$ for holes. Figure 3.12 gives the diffusion coefficient at 300°K as based on the Einstein relationship. At other temperatures D_c is shown in Figure 3.16.

If a large number of minority carriers are injected into a semiconductor (i.e., if $n \approx p$ in the injection region) the density of majority carriers in this injection region will also be affected, and the space charge region of minority carriers will be compensated by a space charge region of majority carriers.

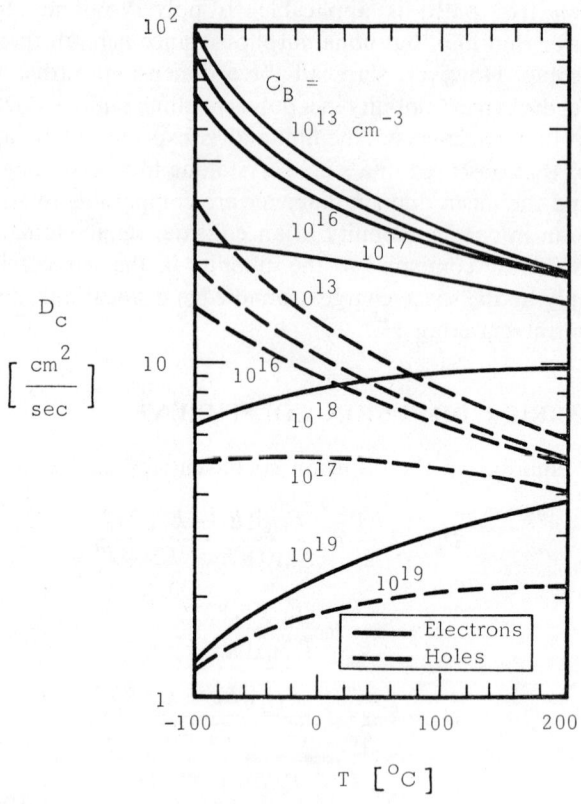

FIGURE 3.16

Minority carrier diffusion coefficient (D_c) vs. temperature (T) and semiconductor impurity concentration (C_B). Silicon.

In this case, the minority carrier diffusion coefficient has to be replaced by an ambipolar diffusion coefficient

$$D_{\text{amb}} = 2 \frac{D_n D_p}{D_n + D_p}. \tag{3.76}$$

In intrinsic silicon at 300°K

$$D_{\text{amb}} = 18 \text{ cm}^2/\text{sec}.$$

9. CARRIER DIFFUSION CURRENT

In a semiconductor of uniform carrier distribution (i.e., $dn(x)dx = 0$) carriers move only under the influence of an external electric field. If, however, there is a carrier concentration gradient (i.e., $dn(x)dx \neq 0$; e.g., due to nonuniform

3-3 CARRIER MOBILITY AND DIFFUSION COEFFICIENT

impurity distribution, irradiation with light, injection of excess carriers through contacts, nonuniform temperature profile, etc.) carriers will diffuse; in this case, the carrier flux will contain a term which is dependent upon the concentration gradient while the proportionality constant is the carrier diffusion coefficient, D_n or D_p (in analogy to but to be distinguished from the diffusion of ions under the influence of the impurity charge). This carrier diffusion will result in a diffusion current which may flow even if no external electric field is present in the semiconductor.

The total current density in a semiconductor may be composed of contributions by

(a) carrier diffusion:

$$j_D = q D_c \operatorname{grad} n, \tag{3.77}$$

(b) electric field:

$$j_E = q\mu n E, \tag{3.78}$$

(c) magnetic field:

$$j_H = R_H H[(j_D + j_E) \times k_H]/\rho. \tag{3.79}$$

With these contributions the total current density

$$j_T = j_D + j_E + j_H. \tag{3.80}$$

In the above equations electron and hole densities are represented by n.

The individual contributions are as follows. Only majority carriers are considered.

(a) Influence of carrier diffusion:
   Electron and hole diffusion current densities are

$$j_{Dn} = q D_n \operatorname{grad} n, \tag{3.81a}$$

$$j_{Dp} = -q D_p \operatorname{grad} p. \tag{3.81b}$$

(b) Influence of electric field:
   Electron and hole current densities under the influence of an electric field are

$$j_{En} = q\mu_n n E, \tag{3.82a}$$

$$j_{Ep} = q\mu_p p E. \tag{3.82b}$$

292 ELECTRICAL BEHAVIOR OF SEMICONDUCTORS

(c) Influence of magnetic field:
If in addition to an external electric field a magnetic field of magnitude H is applied, the then occurring Hall voltage adds an additional term to the total carrier current density, i.e., a transverse current component which for electrons or holes

$$j_{Hn} = R_{Hn}|H|[(j_{Dn} + j_{En}) \times k_H]/\rho_n, \quad (3.83a)$$

$$j_{Hp} = R_{Hp}|H|[(j_{Dp} + j_{Ep}) \times k_H]/\rho_p, \quad (3.83b)$$

where R_{Hn}, R_{Hp} = Hall coefficients; $R_{Hn} \approx -3\pi/8qn$, $R_{Hp} \approx 3\pi/8qp$
ρ_n, ρ_p = resistivity in n-type, p-type semiconductor, respectively
k_H = unit vector in direction of magnetic field.

10. DEPENDENCE OF DRIFT VELOCITY AND MOBILITY ON ELECTRIC FIELD

Carrier mobility (defined as the variation of drift velocity with electric field) and in turn semiconductor conductivity or resistivity depend upon electric field (E) and impurity concentration (C_B). Mobility and resistivity data usually are those taken at small field. At high field, however, the carrier velocity is less than proportional to the electric field which results in a reduced carrier mobility.

(a) Minority carrier drift mobility:
The minority carrier drift velocity as a function of the electric field is generally

$$\frac{v_d}{v_{d\,\max}} = \frac{E/E_{\text{crit}}}{[1 + (E/E_{\text{crit}})^{k_1}]^{1/k_1}}. \quad (3.84)$$

Specifically,
(i) *for electrons* ($k_1 = 2$):

$$\frac{v_{dn}}{v_{d\,\max(n)}} = \frac{E/E_{\text{crit}(n)}}{[1 + (E/E_{\text{crit}(n)})^2]^{1/2}}, \quad (3.85a)$$

(ii) *for holes* ($k_1 = 1$):

$$\frac{v_{dp}}{v_{d\,\max(p)}} = \frac{E/E_{\text{crit}(p)}}{1 + E/E_{\text{crit}(p)}}. \quad (3.85b)$$

3-3 CARRIER MOBILITY AND DIFFUSION COEFFICIENT

(b) Mobility:
  (i) *Low impurity concentration* $(C_B \approx n_i)$:
    Carrier mobility as a function of electric field is

$$\mu_i(E) = dv_d(E)/dE = v_{d\ max}/E_{crit}$$
$$= \mu_{io}/[1 + (E/E_{crit})^{k_1}]^{1+1/k_1} \qquad (3.86)$$

where the corresponding values for μ_{io}, E_{crit}, and k_1 are used for electrons and holes.
At small electric field $(E \ll E_{crit})$:

$$\mu_i(E) = \mu_{io} \qquad (3.87)$$

  (ii) *High impurity concentration* $(C_B \gg n_i)$:
    Carrier mobility as a function of electric field and impurity concentration is

$$\mu(E, C_B) = dv_d(E)/dE$$
$$= \mu^*/[1 + (E/E_{crit})^{k_1}]^{1+1/k_1}$$
$$= \mu^*/[1 + (E\mu^*/v_{d\ max})^{k_1}]^{1+1/k_1}. \qquad (3.88)$$

The critical electric field (at which carrier multiplication is observed) is a function of impurity concentration:

$$E_{crit}(C_B) = v_{d\ max}/\mu^*. \qquad (3.89)$$

The modified mobility (through which the concentration dependence comes about) is defined as

$$\mu^* = \frac{\mu_{io} - \mu_{\infty o}}{1 + (C_B/C_B^*)^{k_2}} + \mu_{\infty o} \qquad (3.90)$$

C_B^* corresponds to the point of inflection of the $\mu - C_B$ curve (see inset of Figure 3.17) at which the slope of the curve is described by k_2. Subscript o refers to small electric field, subscript ∞ to high impurity concentration, subscript i to near-intrinsic conditions.
At small electric field, i.e., at $E/v_{d\ max} \ll (\mu^*)^{-k_1}$:

$$\mu(C_B) = \mu^*. \qquad (3.91)$$

Figures 3.17 and 3.18 illustrate the influence of the electric field on carrier mobility and semiconductor resistivity. They indicate that a significant influence of the electric field is found only at $E > E_{crit}/10$. The influence of the field is less pronounced at high impurity concentration. Table 3.6 lists the parameters used above for silicon at 300°K.

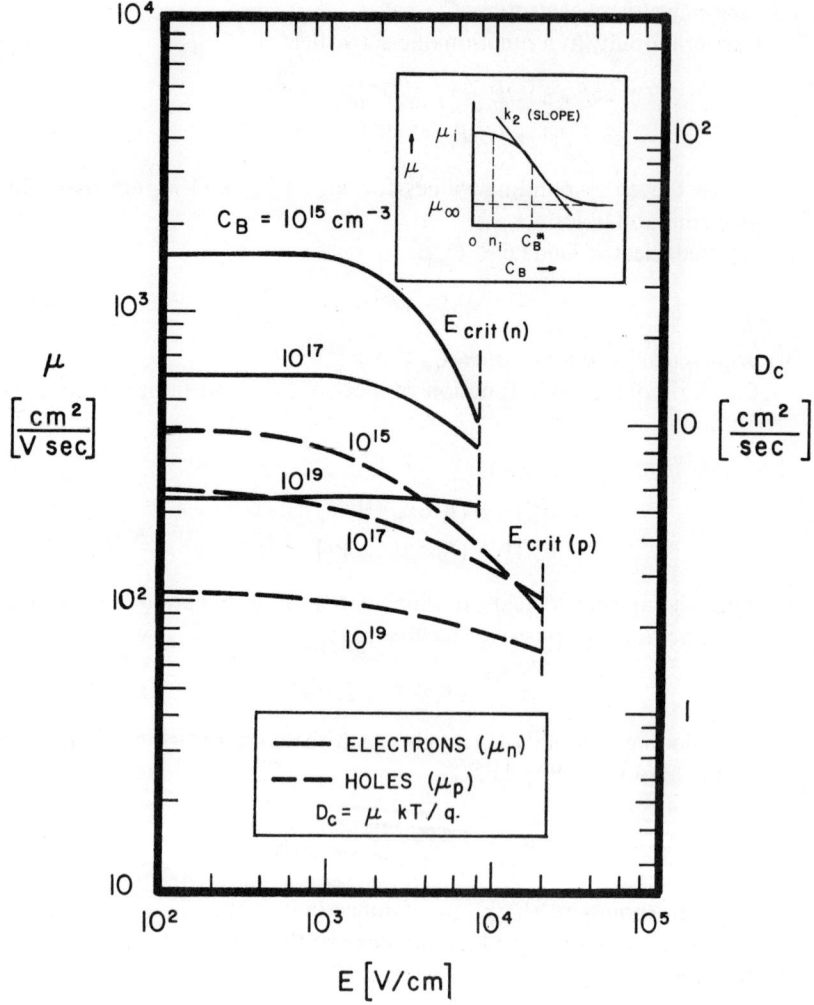

FIGURE 3.17

Minority carrier mobility (μ) and diffusion coefficient (D_c) vs. electric field (E) and semiconductor impurity concentration (C_B). Silicon, 300°K.

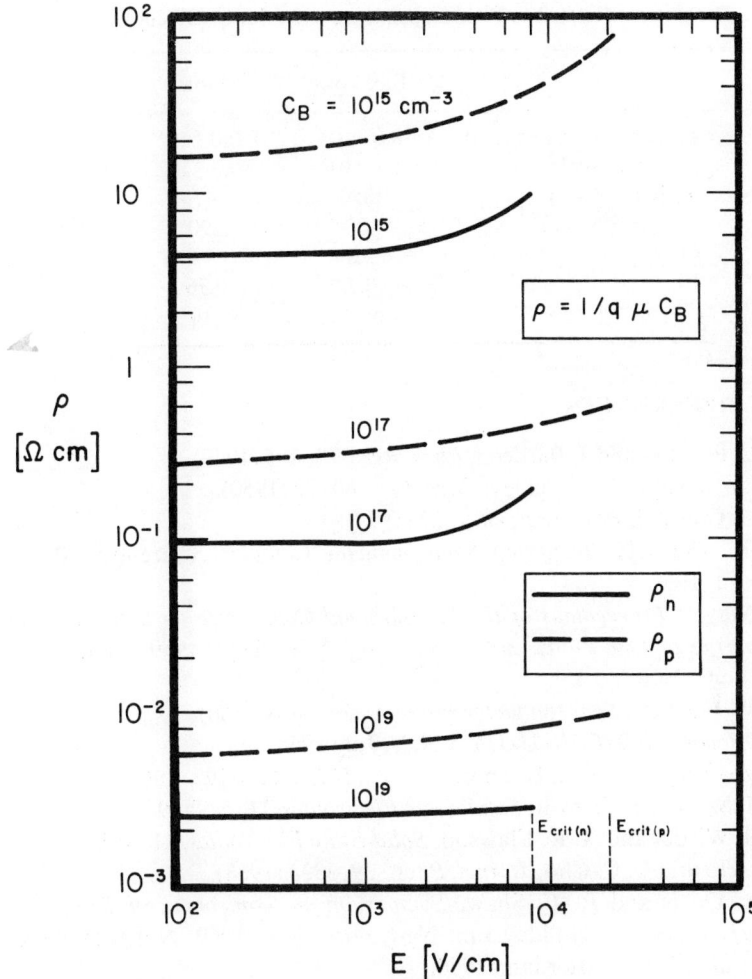

FIGURE 3.18

Resistivity (ρ) of silicon vs. electric field (E) and background impurity concentration (C_B). 300°K.

TABLE 3.6
Parameters Related to Field Dependence of Mobility (Silicon)

	Electrons	Holes
E_{crit}, at low C_B [V/cm]	$8.0 \cdot 10^3$	$2.0 \cdot 10^4$
$v_{d\,max}$ [cm/sec]	$1.1 \cdot 10^7$	$9.5 \cdot 10^6$
μ_{io} [cm²/V sec]	1350	475
$\mu_{\infty o}$ [cm²/V sec]	130	90
k_1	2.0	1.0
k_2	0.72	0.76
C_B^* [cm⁻³]	$5 \cdot 10^{17}$	$3 \cdot 10^{16}$

11. REFERENCES

(1) G. L. Pearson and J. Bardeen, *Phys. Rev.*, **75**, 865 (1949).
(2) J. Bardeen and W. Shockley, *Phys. Rev.*, **80**, 72 (1950).
(3) E. M. Conwell, *Proc. IRE*, **40**, 1327 (1952).
(4) H. K. Henisch, *Rectifying Semiconductor Contacts*, Clarendon Press, Oxford, 1957.
(5) H. Mayer, *Proceedings of the International Conference on Structure and Properties of Thin Films*, Bolton Landing, N.Y., Sept. 1959, John Wiley and Sons, New York.
(6) W. W. Gärtner, *Semiconductor Products*, 29, (July 1960).
(7) *RTI Report ASD-TDR-63-316*, Vol. V, July 1964.
(8) D. M. Caughey and R. E. Thomas, *Proc. IEEE*, **55**, 2192 (1967).
(9) S. M. Sze and J. C. Irvin, *Solid-State Electronics*, **11**, 599 (1968).
(10) H. H. Wieder and A. R. Clawson, *Solid-State Electronics*, **11**, 887 (1968).
(11) B. L. Ho and C. C. Cho, *J. Appl. Phys.*, **39**, 3333 (1968).
(12) E. A. Davis and R. F. Shaw, *Symposium on Semiconductor Effects in Amorphous Solids*, Holiday Inn, New York, May 1969, North-Holland Publishing Co., Amsterdam, 1970.
(13) H. H. Wieder, *Intermetallic Semiconducting Films*, Pergamon Press, Oxford, 1970.

3-4 Hall Coefficient and Hall Mobility

1. HALL EFFECT

The Hall effect is the occurrence of a (transverse) current component perpendicular to control current I_z and magnetic field H_x due to the Lorentz force. Charge carriers are deflected to the side by the magnetic field; the Hall field is required to ensure that the transverse current is zero.

If $H_x = 0$, the two points y_1 and y_2 (see Figure 3.19) have the same potential, i.e., the voltage difference between the two points is zero:

$$V_H = 0. \tag{3.92}$$

If $H_x \neq 0$ there will be a voltage difference between points y_1 and y_2 which is proportional to the magnetic field:

$$V_H = R_H(I_z H_x/x_w), \tag{3.93}$$

where I_z = current in direction z,

x_w = semiconductor thickness,

R_H = proportionality factor (Hall coefficient).

The Hall coefficient is inversely proportional to the density of free carriers; therefore, the above relationship can be used to determine the number of free carriers in a semiconductor; if the carrier density is known, it can be used to determine the magnitude of the magnetic field.

To obtain a large Hall voltage, a material with a small carrier density is required. For metals, carrier density is usually equal to the atom density of the crystal and is therefore fixed. For semiconductors, carrier density is several orders of magnitude less than the atom density; therefore semiconductors may display a very large Hall voltage.

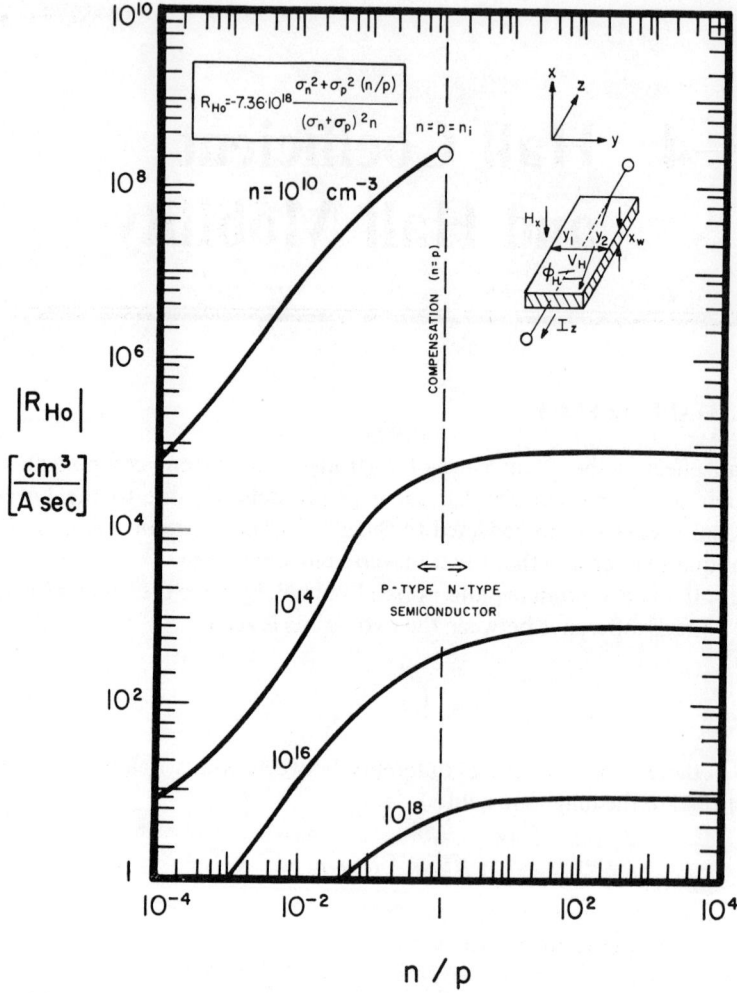

FIGURE 3.19

Hall coefficient (R_H) vs. ratio of carrier concentrations (n/p) in n-type or p-type silicon and electron concentration (n). 300°K. R_{Ho} is the Hall coefficient at small magnetic field. The inset illustrates terms used in defining Hall voltage (V_H) and Hall angle (ϕ_H); the Hall voltage is measured between points y_1 and y_2.

The curves of Figures 3.19, 3.20, and 3.21 apply to nondegenerate n-type and p-type silicon at small magnetic field at 300°K.

Figure 3.19 shows R_{Ho} as function of the concentration ratio of electrons and holes for various electron concentrations. In an uncompensated nondegenerate semiconductor the majority carrier density at 300°K is equal to the impurity concentration. For $n/p \gg 1$ the impurity concentration C_B corresponds to n; for $n/p \ll 1$ it corresponds to n divided by the appropriate ratio n/p. The center of the illustration applies to a compensated semiconductor where electron and hole densities are comparable ($n/p \approx 1$). The highest Hall coefficient is achieved for an intrinsic semiconductor. In silicon, n-type material gives a higher Hall coefficient than p-type material.

2. HALL COEFFICIENT

The relation between electric field (E) and current can be written in component form as

$$I_i = \sigma_{ij} E_j \qquad (3.94)$$

where i, j = coordinates
σ_{ij} = tensor components of conductivity.

Therefore the Hall field

$$E_y = R_H I_z H_x$$
$$= -E_z(\sigma_{yz}/\sigma_{yy}) \qquad (3.95)$$

where $I_y = \sigma_{yz} E_z + \sigma_{yy} E_y$
$I_z = \sigma_{zz} E_z + \sigma_{zy} E_y$.

The symmetry properties apply for isotropic and cubic crystals, i.e.,

$$\sigma_{yz} = -\sigma_{zy}, \quad \sigma_{yy} = \sigma_{zz},$$

so that

$$R_H = (1/H_x)/(\sigma_{yz}^2 + \sigma_{zz}^2). \qquad (3.96)$$

In polyvalent crystals there are usually two contributions to the Hall coefficient—one from electrons and one from holes—with opposite signs. These contributions are σ_n and σ_p in terms of conductivity and ω_n and ω_p in terms of cyclotron frequency. In a nondegenerate semiconductor

$$\sigma_n = q^2 n \tau_n^*/m_n \qquad \sigma_p = q^2 p \tau_p^*/m_p \qquad (3.97a)$$
$$\omega_n = 3\pi q H_x/8m_n \qquad \omega_p = -3\pi q H_x/8m_p. \qquad (3.97b)$$

The relaxation time is defined as a time constant associated with the rate of return to equilibrium of the carrier distribution when the external magnetic field is turned off. It is proportional to the lattice constant a.

$$\tau_i^* = aE^{3/2}. \qquad (3.98)$$

The tensor components of conductivity as a function of magnetic field are therefore:

$$\sigma_{yz} = \frac{\omega_n \tau_n^* \sigma_n}{1 + (\omega_n \tau_n^*)^2} - \frac{\omega_p \tau_p^* \sigma_p}{1 + (\omega_p \tau_p^*)^2} \qquad (3.99a)$$

$$\sigma_{zz} = \frac{\sigma_n}{1 + (\omega_n \tau_n^*)^2} + \frac{\sigma_p}{1 + (\omega_p \tau_p^*)^2} \qquad (3.99b)$$

300 ELECTRICAL BEHAVIOR OF SEMICONDUCTORS

from which follows the general expression of the Hall coefficient for arbitrary field

$$R_H = \frac{3\pi}{8q} \frac{\sigma_n^2/n - \sigma_p^2/p + (\omega_n \omega_p)(\tau_n^* \tau_p^*)(\sigma_n \sigma_p)(n-p)/np}{(\sigma_n + \sigma_p)^2 + (\omega_n \omega_p)(\tau_n^* \tau_p^*)(\sigma_n \sigma_p)(n-p)^2/np}. \quad (3.100)$$

The dependence of the Hall coefficient upon the magnetic field comes about by the field dependence of the cyclotron frequencies; the electric field enters through the relaxation times.

In terms of Hall voltage, magnetic field, control current, and semiconductor thickness, the Hall coefficient is generally

$$R_H = V_H x_w / H_x I_z. \quad (3.101)$$

3. VARIATION OF HALL COEFFICIENT WITH IMPURITY CONCENTRATION

Three impurity ranges have to be distinguished:

(a) Intrinsic semiconductor ($n = p = n_i$):

$$R_{Hi} = \frac{3\pi}{8qn_i} \frac{\mu_n/\mu_p - 1}{\mu_n/\mu_p + 1} \quad (3.102)$$

$$\mu_{Hi} = (8/3\pi)(R_{Hi}/\rho) \quad (3.103)$$

For silicon:

$$R_{Hi} = 2.3 \cdot 10^8 \text{ cm}^3/\text{A sec}, \mu_{Hi} = 960 \text{ cm}^2/\text{V sec}.$$

(b) Extrinsic nondegenerate semiconductor:

$$R_{Hn} = -(3\pi/8qn) = -7.4 \cdot 10^{18}/n \quad (3.104a)$$

$$R_{Hp} = (3\pi/8qp) = 7.4 \cdot 10^{18}/p \quad (3.104b)$$

$$\mu_{Hn} = -0.85 R_{Hn}/\rho = 6.3 \cdot 10^{18}/n\rho \quad (3.105a)$$

$$\mu_{Hp} = 0.85 R_{Hp}/\rho = 6.3 \cdot 10^{18}/p\rho \quad (3.105b)$$

(c) Extrinsic degenerate semiconductor:

$$R_{Hn} = -1/qn = -6.3 \cdot 10^{18}/n \quad (3.106a)$$

$$R_{Hp} = 1/qp = 6.3 \cdot 10^{18}/p \quad (3.106b)$$

$$\mu_{Hn} = -R_{Hn}/\rho = 6.3 \cdot 10^{18}/n\rho \quad (3.107a)$$

$$\mu_{Hp} = R_{Hp}/\rho = 6.3 \cdot 10^{18}/p\rho \quad (3.107b)$$

4. VARIATION OF HALL COEFFICIENT WITH MAGNETIC FIELD

The two limiting cases are:

(a) Small magnetic field (i.e., $H_x \approx 0$, $-\sigma_{yz} \ll \sigma_{zz}$):

$$R_{Ho} = -(1/H_x)(\sigma_{yz}/\sigma_{zz}^2)$$
$$= (R_{Hn}\sigma_n^2 + R_{Hp}\sigma_p^2)/(\sigma_n + \sigma_p)^2$$
$$= (3\pi/8q)(\sigma_n^2 + \sigma_p^2)(n/p)/(\sigma_n + \sigma_p)^2 n. \qquad (3.108)$$

The sign of R_{Ho} depends upon the relative magnitudes of $R_{Hn}\sigma_n$ and $R_{Hp}\sigma_p$. In some materials $R_{Hn}\sigma_n \approx R_{Hp}\sigma_p$ so that $R_{Ho} \approx 0$ and $V_H \approx 0$. The low-field Hall effect can be described by the Hall angle ϕ_H which measures the change in orientation of the electric field in the plane normal to H_x:

$$\tan \phi_H = E_y/E_z = -\sigma_{yz}/\sigma_{zz}. \qquad (3.109)$$

(b) Large magnetic field (i.e., $H_x \gg 0$, $-\sigma_{yz} \gg \sigma_{zz}$):

$$R_{H\infty} = -1/H_x \sigma_{yz}$$
$$= (8q/3\pi)/(n-p). \qquad (3.110)$$

The sign of $R_{H\infty}$ depends upon the excess carrier density; i.e., in an n-type semiconductor it is negative, in a p-type semiconductor positive. Above equation is not valid in a semiconductor where $n \approx p$.

5. HALL MOBILITY

The Hall mobility (μ_H) in a nondegenerate semiconductor is defined;

$$\mu_H = (8/3\pi)(|R_H|/\rho). \qquad (3.111)$$

It is to be distinguished from carrier mobility and is given in Figure 3.20 for majority carriers.

In the degenerate case incomplete ionization has to be taken into account. The relative decrease in carrier concentration with respect to impurity concentration as a result of degeneracy is reflected in a slight relative mobility increase (dotted curves of Figure 3.20).

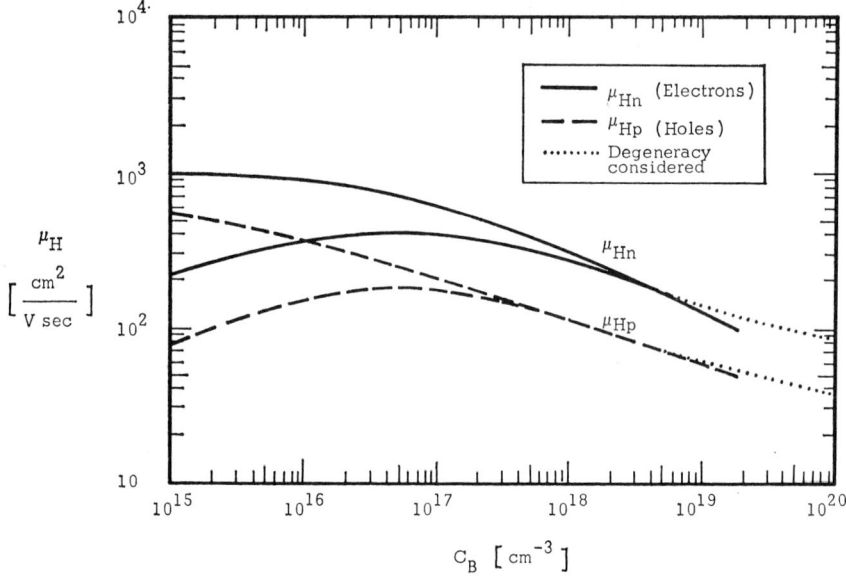

FIGURE 3.20

Majority carrier Hall mobility (μ_H) vs. semiconductor impurity concentration (C_B). Silicon, 300°K. The upper curves refer to bulk material, the lower curves to thin film material; the thin film curves were measured in 2-μm thick silicon films on an Al_2O_3 substrate and deviate from bulk mobilities at carrier concentrations less than about 10^{18} cm^{-3}.

6. HALL MOBILITY IN THIN FILMS

Experimental data indicate that carrier mobilities in thin film semiconductors are substantially less than those in bulk material. This is partially due to the influence of the semiconductor surface where mobility is reduced below that in the bulk. It is also due to dislocation scattering, neutral scattering, and the space charge around edge dislocations. For silicon, a maximum Hall mobility is observed if the impurity concentration of the thin film is approximately 10^{16} to 10^{17} cm^{-3}, i.e., the Hall mobility in thin-film silicon decreases toward lower and higher doping levels whereas that in bulk silicon increases at lower impurity concentration.

7. TEMPERATURE DEPENDENCE OF HALL MOBILITY

The low-field temperature dependence of Hall mobilities of selected semiconductors is given in Table 3.7. The mobilities given in this Table apply to temperatures close to room temperature. Electron and hole mobilities at 300°K are the numbers in front of the brackets.

3–4 HALL COEFFICIENT AND HALL MOBILITY 303

TABLE 3.7

Hall Mobilities in Selected Semiconductors

Semi-conductor	μ_{Hn} [cm²/V sec]	μ_{Hp} [cm²/V sec]	μ_{Hn}/μ_{Hp} at $T=300°$K	μ_{Hn}/μ_{Hp} at T [°K]
Si	$1300\,(300/T)^{2.0}$	$500\,(300/T)^{2.7}$	2.6	$0.01 \cdot T^{0.7}$
Ge	$4500\,(300/T)^{1.6}$	$3500\,(300/T)^{2.3}$	1.3	$0.08 \cdot T^{0.5}$
GaAs	$8500\,(300/T)^{1.0}$	$420\,(300/T)^{2.1}$	20.2	$0.05 \cdot T^{1.1}$
GaP	$110\,(300/T)^{1.5}$	$75\,(300/T)^{1.5}$	1.5	1.5
GaSb	$4000\,(300/T)^{2.0}$	$1400\,(300/T)^{0.9}$	2.9	$1101.0 \cdot T^{-1.1}$
InAs	$33000\,(300/T)^{1.2}$	$460\,(300/T)^{2.3}$	71.7	$0.19 \cdot T^{1.1}$
InP	$4600\,(300/T)^{2.0}$	$150\,(300/T)^{2.4}$	30.7	$3.95 \cdot T^{0.4}$
InSb	$78000\,(300/T)^{1.6}$	$750\,(300/T)^{2.1}$	104.0	$5.90 \cdot T^{0.5}$

Although GaP has both low electron and low hole mobility, the ratio of both mobilities is independent of temperature. Of the semiconductors considered, only GaSb has a mobility ratio which decreases with increasing temperature. The mobility ratio of GaAs and InAs is almost proportional to temperature. The highest mobility ratio at 300°K is found in InSb, the lowest in Ge. Variations in R_H and ρ tend to compensate each other over a larger temperature range so that μ_H is relatively independent of temperature within the nondegenerate range.

8. SENSITIVITY OF HALL GENERATOR

The figure of merit for the sensitivity of a Hall generator contains Hall voltage and Hall input power P_H.

$$V_H/(P_H^{1/2} H_x) = (R_H \mu_H/x_w)^{1/2}. \tag{3.112}$$

This indicates that a high figure of merit requires the product of Hall coefficient and Hall mobility to be high and the semiconductor thickness to be small. Since both Hall coefficient and Hall mobility increase with decreasing impurity concentration, a lightly doped semiconductor gives the highest Hall sensitivity.

The material figure of merit for a no-load Hall generator is defined (for a heat sink as given in Figure 3.21) by

$$M_H = \left(\frac{2A_{np}^2}{q}\right)^{1/3} \frac{(T_H - T_S)^{1/3}}{T_H}$$

$$\cdot \frac{C_B[(\mu_n/\mu_p)^2 + 1] + (C_B^2 + 4n_i^2)^{1/2}[(\mu_n/\mu_p)^2 - 1]}{[C_B(\mu_n/\mu_p - 1) + (C_B^2 + n_i^2)^{1/2}(\mu_n/\mu_p + 1)]^{4/3}} \tag{3.113}$$

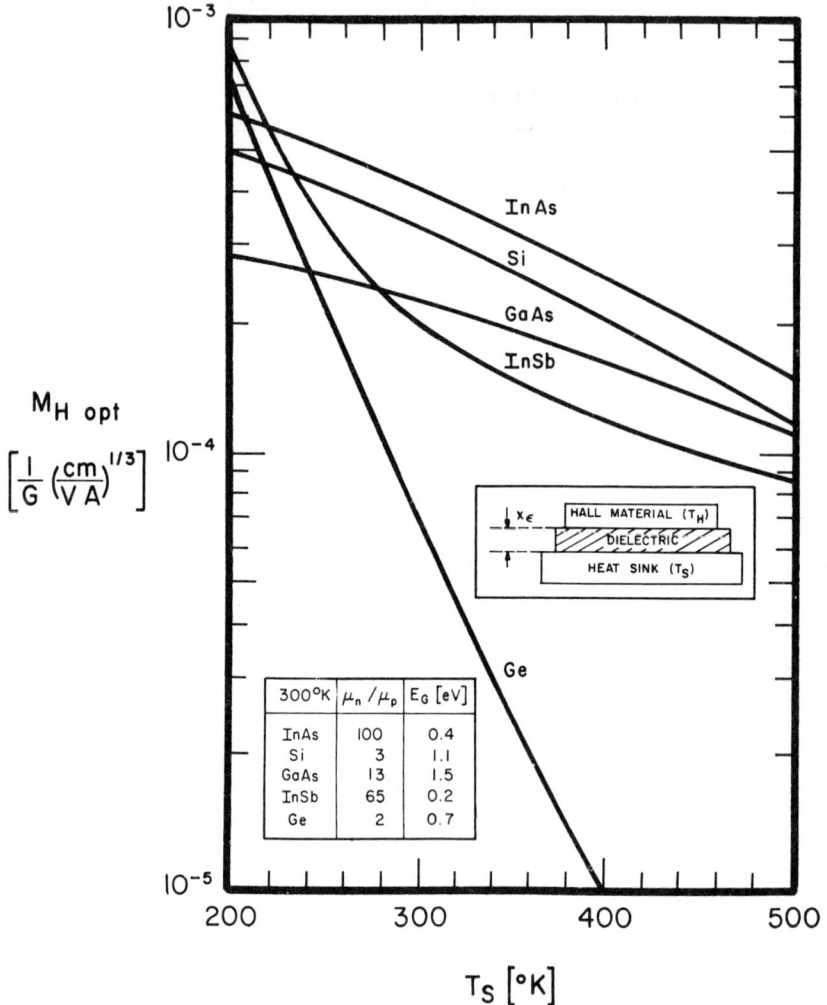

FIGURE 3.21

Material figure of merit ($M_{H\text{opt}}$) of Hall generator vs. heat sink temperature (T_S) for various semiconductors. The Hall generator is mounted as shown in the upper inset.

This illustration gives the optimum figure of merit vs. temperature for various nondegenerate n-type semiconductors under the assumption $\mu_n \gg \mu_p$. It indicates that the semiconductors considered display only minor differences of figure of merit at 300°K. The semiconductors with low impurity concentration and large energy gap have a figure of merit comparable to semiconductors whose energy gap is small and whose mobility is high. For the latter group a smaller terminal voltage is required for the same geometrical factor.

3-4 HALL COEFFICIENT AND HALL MOBILITY

where the heat sink is separated from the semiconductor by a dielectric of thickness x_ε, which, in turn, is given by

$$x_\varepsilon = V_z/E_\varepsilon,$$

and where the other symbols have the following meaning:

A_{np} = proportionality factor of electron or hole drift mobility; it is independent of magnetic field; $\mu_{n,p} = A_{np} T^{-3/2}$
T_H, T_S = temperature of Hall generator and of heat sink, respectively
E_ε = dielectric strength of dielectric separating Hall generator and heat sink
V_z = voltage difference between the two control current terminals of the Hall generator.

The optimum temperature ($T_{H\,\text{opt}}$), i.e., the temperature at which the highest Hall voltage is obtained, depends upon the energy gap of the semiconductor. At the optimum temperature the optimum material figure of merit

$$M_{H\,\text{opt}} = N_{np} N_T \frac{(\mu_n/\mu_p)^2(N_H + 1) - (N_H - 1)}{[(\mu_n/\mu_p)(N_H + 1) + (N_H - 1)]^{4/3}} \frac{1}{C_B^{1/3}} \quad (3.114)$$

where

$$N_{np} = (2A_{np}^2/q)^{1/3}$$
$$N_T = (T_{H\,\text{opt}} - T_S)^{1/3}/T_{\text{opt}}$$
$$N_H = [1 + (2n_i/C_B)^2]^{1/2}.$$

Two limiting cases:

(a) $\mu_n/\mu_p = 1$:

$$M_{H\,\text{opt}} = N_{np} N_T C_B/(C_B^2 + 4n_i^2)^{2/3} \quad (3.115a)$$

(b) $\mu_n/\mu_p \gg 1$:

$$M_{H\,\text{opt}} = N_{np} N_T [C_B + (C_B^2 + 4n_i^2)^{1/2}]^{-1/3} \quad (3.115b)$$

An ideal (n-type) Hall generator should have the following properties:

(a) High electron mobility, not restricted by lattice vibrations, with carrier velocity high but below the limiting drift velocity.
(b) High mobility ratio since it eliminates the reduction of Hall voltage due to excess minority carriers.
(c) Low impurity concentration since it is equivalent to high carrier mobility and results in a large ratio of electron velocity per unit of input power.
(d) Large energy gap of the semiconductor because of a small intrinsic carrier density.

9. REFERENCES

(1) G. L. Pearson and J. Bardeen, *Phys. Rev.*, **75**, 865 (1949).
(2) F. J. Morin and J. P. Maita, *Phys. Rev.*, **96**, 28 (1954).
(3) H. K. Henisch, *Rectifying Semi-Conductor Contacts*, Clarendon Press, Oxford, 1957.
(4) D. M. Krembs, *Tech. Rep. No. 1702-2*, Stanford Electronic Labs., October 1959.
(5) Special Issue of *Solid-State Electronics* devoted to the Hall effect and its applications, **9**, 338–583 (Oct. 1966).
(6) W. J. Patrick, *Solid-State Electronics*, **9**, 203 (1966).
(7) B. Donovan, *Elementary Theory of Metals*, Pergamon Press, Oxford, 1967.
(8) H. Weiss, *IEEE Spectrum*, **5**, 75 (Jan. 1968).
(9) V. I. Fistul', *Heavily Doped Semiconductors*, Plenum Press, New York, 1969.
(10) R. C. Bracken, *Electro-Technology*, **83**, 43 (May 1969).
(11) H. F. Mataré, *First International Symposium on Silicon Materials, Science and Technology*, Abstract No. 281, New York, May 5–9, 1969.

3-5 Semiconductor Resistivity

1. RESISTIVITY

The electrical resistivity (inverse of conductivity) of a semiconductor is due to the finite charge carrier mobility which results from various scattering mechanisms. Of these, lattice scattering and impurity scattering are most important. Only lattice scattering is dependent upon crystallographic orientation. In the absence of scattering events, carriers would accelerate under the influence of an electric field approaching the speed of light.

In a nondegenerate semiconductor (containing donors and acceptors which are all ionized) carrier mobilities and impurity concentrations are generally related to resistivity by

$$\rho = \frac{1}{q(\mu_n N_D + \mu_p N_A)} = \frac{1}{q(\mu_n n + \mu_p p)}. \qquad (3.116)$$

In the case of the dominance of one impurity type this reduces to

$$\rho_n = 1/q\mu_n N_D \quad (n\text{-type semiconductor}) \qquad (3.117a)$$

or

$$\rho_p = 1/q\mu_p N_A \quad (p\text{-type semiconductor}) \qquad (3.117b)$$

where μ_n and μ_p refer to majority carrier mobilities. The semiconductor resistivity is a function of temperature (Figure 3.23), of impurity concentration (Figure 3.22), and of electric field (Figure 3.18). Most data showing the variation of resistivity with temperature and impurty concentration refer to a small electric field.

The conductivity or resistivity of a semiconductor may be locally affected by external influences. This controlled change in the conductivity or resistivity in response to an external stimulus is called conductivity or resistivity modulation. Several types of stimuli can be used to modulate locally the resistivity of a semiconductor.

(a) Temperature:
A change in temperature of the semiconductor usually results in a logarithmic change in the conductivity characteristic of an activation

energy of the semiconductor. The temperature dependence of the conductivity is impurity-sensitive especially at low temperatures. Important applications are in low-temperature thermometry.

(b) Carrier injection:
Charge carriers can be injected into a semiconductor at a metal-semiconductor contact. The resistivity of the semiconductor is changed for a short time in the vicinity of the injection. The conductivity modulation by injection is important in transistors.

(c) Irradiation:
Photoconductivity is a modulated conductivity caused by irradiation of the semiconductor. Important applications are photoconductors and phototransistors.

(d) Magnetic field:
The conductivity of a surface region can be changed by the variation of a magnetic field perpendicular to the direction of current flow in the semiconductor (Suhl effect). In a semiconductor with a transverse electric and magnetic field, minority carriers will be concentrated on the same surface as the majority carriers by bending in the magnetic field. The concentration of minority carriers on the surface is increased by an increased magnetic field, but as the surface concentration increases the carrier recombination rate will increase. The conductivity of a surface region should increase with increasing magnetic field due to this magnetic concentration. When the increase in recombination rate begins to exceed the minority density increase, the conductivity of the surface region will begin to decrease with further increase in the magnetic field.

2. IMPURITY CONCENTRATION DEPENDENCE OF RESISTIVITY

Since the majority carrier densities in an extrinsic nondegenerate semiconductor are

$$n = (1/2)\{N_D - N_A + [(N_D - N_A)^2 + 4n_i^2]^{1/2}\}$$
(n-type semiconductor) (3.118a)

or

$$p = (1/2)\{N_A - N_D + [(N_A - N_D)^2 + 4n_i^2]^{1/2}\}$$
(p-type semiconductor), (3.118b)

the semiconductor resistivity as a function of impurity concentration can be expressed as

$$1/\rho = (q/2)\{(\mu_n + \mu_p)[(N_D - N_A)^2 + 4n_i^2]^{1/2} + (\mu_n - \mu_p)(|N_D - N_A|)\}$$
$$= (q/2)[(\mu_n + \mu_p)(C_B^2 + 4n_i^2)^{1/2} + (\mu_n - \mu_p)C_B] \quad (3.119)$$

if $|N_D - N_A| = C_B$.

The following simplifications can be made:

(a) Intrinsic semiconductor ($C_B \leq n_i$):

$$1/\rho = q(\mu_n + \mu_p)n_i. \quad (3.120)$$

(b) Extrinsic nondegenerate semiconductor ($C_B > n_i$):

$$1/\rho_e = q(\mu_n^2 - \mu_p^2)C_B/2$$
$$\approx q\mu_n C_B \quad (n\text{-type}), \quad (3.121)$$

or

$$\approx q\mu_p C_B \quad (p\text{-type}).$$

(c) Degenerate semiconductor ($C_B > N_c$ or $> N_v$):

$$1/\rho_e = q\mu_n n \quad (n\text{-type}), \quad (3.122a)$$

or

$$1/\rho_e = q\mu_p p \quad (p\text{-type}). \quad (3.122b)$$

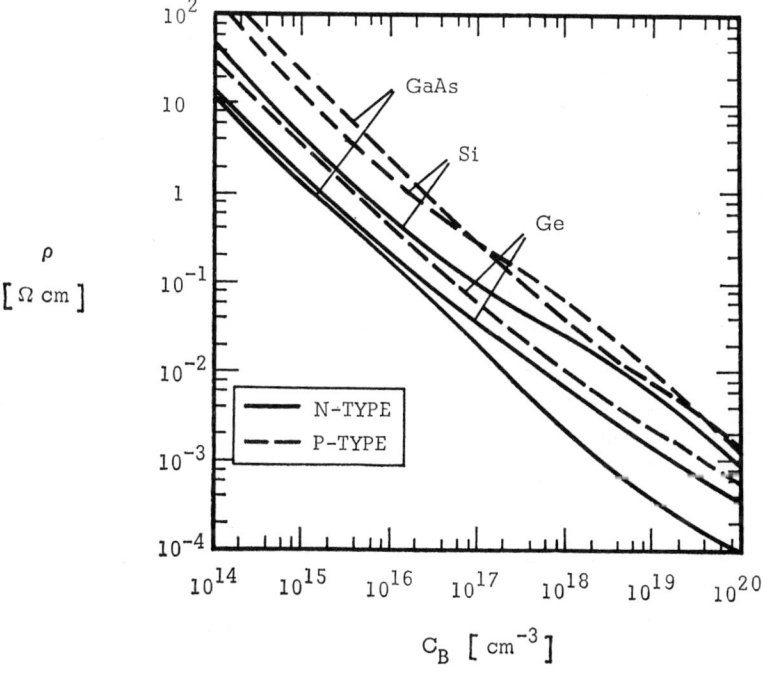

FIGURE 3.22

Resistivity (ρ) vs. semiconductor impurity concentration (C_B) for Si, Ge, and GaAs. 300°K.

TABLE 3.8

Resistivity as a Function of Impurity Concentration (300°K)

C_B [cm^{-3}]	$\rho[\Omega$ cm]					
	Si		Ge		GaAs	
	N	P	N	P	N	P
n_i	$2 \cdot 10^5$		50		$7 \cdot 10^7$	
10^{14}	40	180	15	30	12	160
10^{15}	4.5	12	1.5	2.5	0.9	22
10^{16}	0.6	1.8	0.2	0.4	0.2	2.3
10^{17}	0.1	0.3	$3.5 \cdot 10^{-2}$	$5.7 \cdot 10^{-2}$	$9.0 \cdot 10^{-3}$	0.3
10^{18}	$2.5 \cdot 10^{-2}$	$6.2 \cdot 10^{-2}$	$7.0 \cdot 10^{-3}$	$1.1 \cdot 10^{-2}$	$2.1 \cdot 10^{-3}$	$3.5 \cdot 10^{-2}$
10^{19}	$6.0 \cdot 10^{-3}$	$1.2 \cdot 10^{-2}$	$1.5 \cdot 10^{-3}$	$2.4 \cdot 10^{-3}$	$2.9 \cdot 10^{-4}$	$8.0 \cdot 10^{-3}$

At a given temperature

$$\rho_i > \rho_e.$$

The resistivity as a function of impurity concentration is shown for Si, Ge, and GaAs in Figure 3.22 and in Table 3.8.

3. TEMPERATURE DEPENDENCE OF RESISTIVITY

Carrier concentration and mobility are temperature-sensitive and affect the thermal behavior of resistivity. The carrier concentration is also a function of the Fermi level which is temperature-dependent.

Four distinct temperature ranges are observed in a given semiconductor in which resistivity is affected in different ways by temperature.

(a) Very low temperature:

The decrease in resistivity with temperature is partially due to a decreasing amount of condensation of donors or acceptors, i.e., fewer carriers fall back into their low-energy levels (freeze-out) and, thus, more carriers are free. Carrier density increases exponentially with temperature and conduction by one carrier dominates. In most cases, the energy associated with the exponent is the ionization energy of a shallow impurity rather than half the energy gap. In n-type silicon this freeze-out occurs below room temperature, in p-type silicon and in silicon containing deep-lying impurities (e.g., gold) it may occur above room temperature.

(b) Low temperature:

As the temperature increases, impurity scattering mobility increases which results in a decrease in resistivity. At the same time, the lattice scattering

mobility decreases so that the overall temperature effect on resistivity is relatively small, i.e., the temperature coefficient of resistivity is at its minimum.
(c) Intermediate temperature (extrinsic range):
In this temperature range the number of carriers is identical to the number of impurity atoms and is constant over this entire temperature range. However, the reduction of lattice scattering mobility results in an increase in resistivity. In this region most of the temperature variation of the Fermi level takes place, resulting in a high temperature coefficient of resistivity. In silicon of low impurity concentration the mobility is limited by lattice scattering; at higher impurity concentration ionized impurity scattering also reduces the mobility and increases the resistivity. In a degenerate semiconductor freeze-out will not occur and the impurities remain ionized down to very low temperature.
(c) High temperature (intrinsic range):
In this temperature range the number of electrons thermally excited across the energy gap is greater than the number of ionized donors or acceptors, i.e.,

$$n = n_i \gg N_D \quad \text{or} \quad p = n_i \gg N_A.$$

This results in a low resistivity of the semiconductor. The number of carriers increases approximately exponentially with temperature.

For a nondegenerate semiconductor the temperature dependence of resistivity in the intermediate temperature range can be described by

$$1/\rho = q(n\mu_n + p\mu_p)$$
$$= q\{\mu_n N_c \exp[-(E_c - E_F)/kT] + \mu_p N_v \exp[-(E_F - E_v)/kT]\}$$
$$= q\mu_n N_c \exp[-(E_c - E_F)/kT]\left\{1 + \frac{\mu_p}{\mu_n}\frac{N_v}{N_c}\exp[(E_c + E_v - 2E_F)/kT]\right\}.$$

The ratios μ_p/μ_n and N_v/N_c and the product $\mu_n N_c$ are essentially independent of temperature, so that the temperature dependence of ρ is mainly determined by the Fermi level and by the influence of T itself. Consequently, the resistivity of a semiconductor depends inverse-exponentially upon Fermi energy and temperature. Through this dependence, measurements of resistivity as a function of temperature permit the calculation of the semiconductor energy gap.

In silicon a minimum variation of resistivity with temperature is observed if the impurity concentration is about $C_B = 10^{19}$ cm^{-3}. The variation of resistivity with temperature for silicon is shown in Figure 3.23.

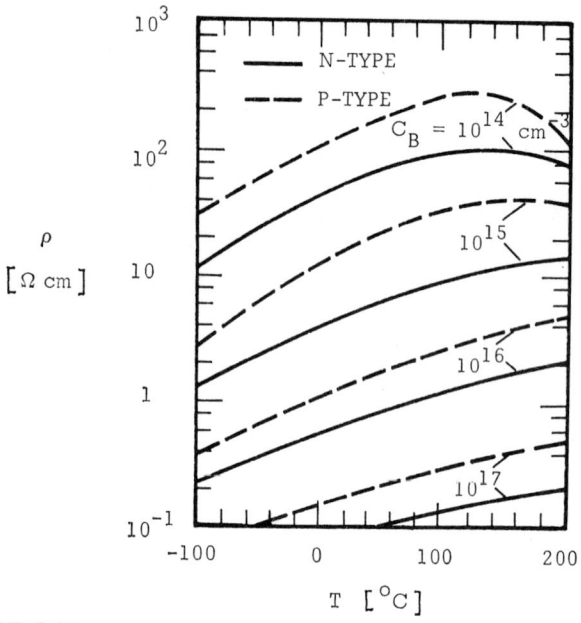

FIGURE 3.23

Resistivity (ρ) vs. temperature (T) and semiconductor impurity concentration (C_B). Silicon.

4. AVERAGE RESISTIVITY AND SHEET RESISTIVITY OF SURFACE AND SUBSURFACE LAYERS

A surface layer is defined as a diffused, ion implanted, or deposited semiconductor region of thickness x_j or x_f adjacent to the semiconductor surface which is doped with impurities of the opposite conductivity type as the semiconductor bulk or substrate, i.e., at the interface of semiconductor bulk and surface layer a *p-n* junction is formed. It is not assumed here that the surface layer is formed by surface inversion. In a diffused layer the impurity concentration is assumed to be highest at the surface; in a deposited layer the impurity concentration is assumed to be uniform throughout the layer. A subsurface layer bordered by two junctions at depth x_{j1} and x_{j2} is assumed to be imbedded by regions of opposite conductivity type.

(a) Diffused subsurface layer (Figures 3.24, 3.25):

The average resistivity ($\bar{\rho}$) of a subsurface layer of thickness ($x_{j2} - x_{j1}$) and of average carrier mobility $\bar{\mu}$ is given by

$$\bar{\rho} = (x_{j2} - x_{j1}) / \int_{x_{j1}}^{x_{j2}} q\bar{\mu}C(x)\,dx. \qquad (3.123)$$

Two limiting cases are distinguished:

(i) $x_{j1}/x_{j2} = 0$:
In this case $x_{j1} = 0$, i.e., the entire diffused region between semiconductor and *p-n* junction is considered.

(ii) $x_{j1}/x_{j2} = 1$:
In this case $x_{j1} = x_{j2}$, i.e., the layer is infinitely thin and its resistivity approaches infinity.

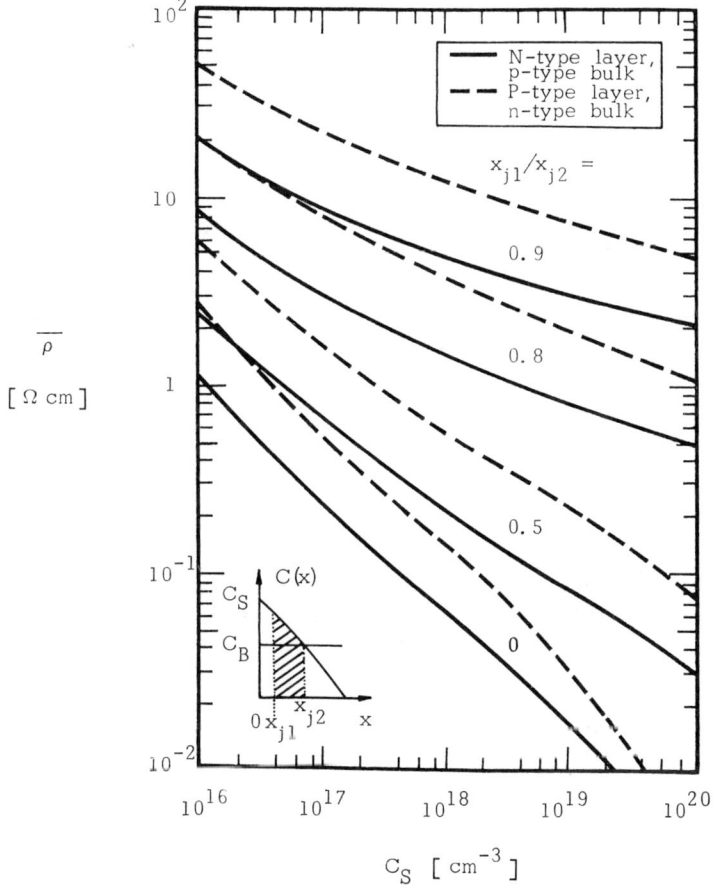

FIGURE 3.24
Average resistivity ($\bar{\rho}$) of a subsurface layer vs. surface impurity concentration (C_S) and layer thickness (x_{j1}/x_{j2}). Silicon, 300°K; Gaussian impurity distribution, $C_B = 10^{15}$ cm^{-3}.

314 ELECTRICAL BEHAVIOR OF SEMICONDUCTORS

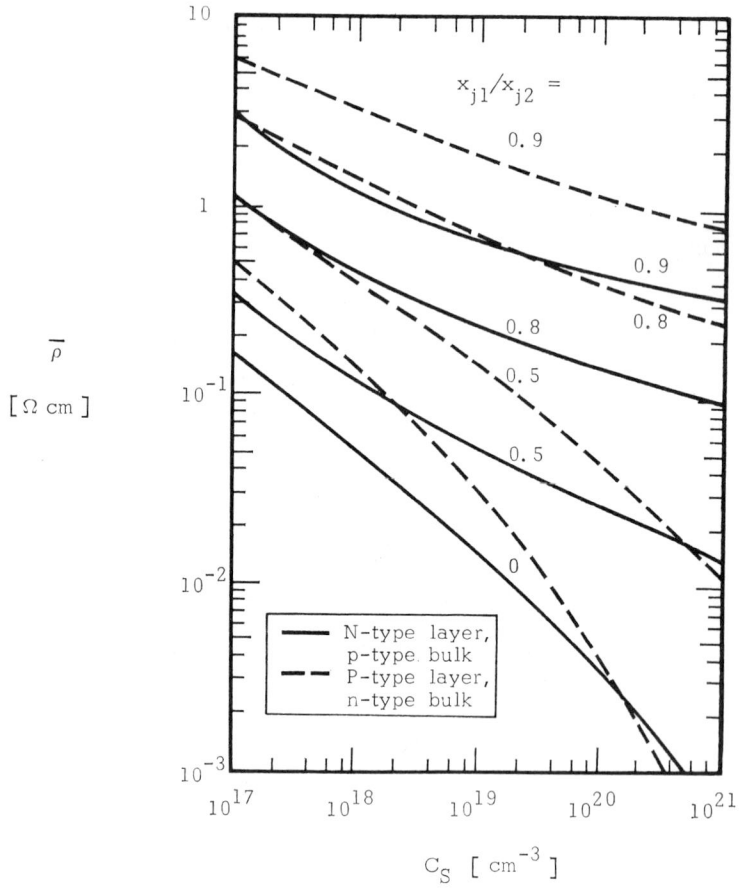

FIGURE 3.25

Average resistivity ($\bar{\rho}$) of a subsurface layer vs. surface impurity concentration (C_S) and layer thickness (x_{j1}/x_{j2}). Silicon, 300°K; Gaussian impurity distribution, $C_B = 10^{16}$ cm^{-3}.

The sheet resistivity (ρ_s, dimension Ω/square) is related to the average resistivity ($\bar{\rho}$, dimension Ωcm) by

$$\rho_s = \bar{\rho}/(x_{j2} - x_{j1}). \tag{3.124}$$

(b) Diffused surface layer (Figure 3.26):
This corresponds to the first limiting case (i.e., $x_{j1} = 0$, $x_{j2} = x_j$).

$$\bar{\rho} = x_j \bigg/ \int_0^{x_j} q\bar{\mu}C(x)\,dx, \tag{3.125}$$

$$\rho_s = \bar{\rho}/x_j. \tag{3.126}$$

(c) Uniform layer (Figure 3.27):
The average resistivity is generally

$$\bar{\rho} = x_f / \int_0^{x_f} q\bar{\mu} C(x)\, dx \qquad (3.127)$$

which for a box-type profile can be reduced to

$$\bar{\rho} = 1/q\mu(C_S - C_B) \qquad (3.128)$$

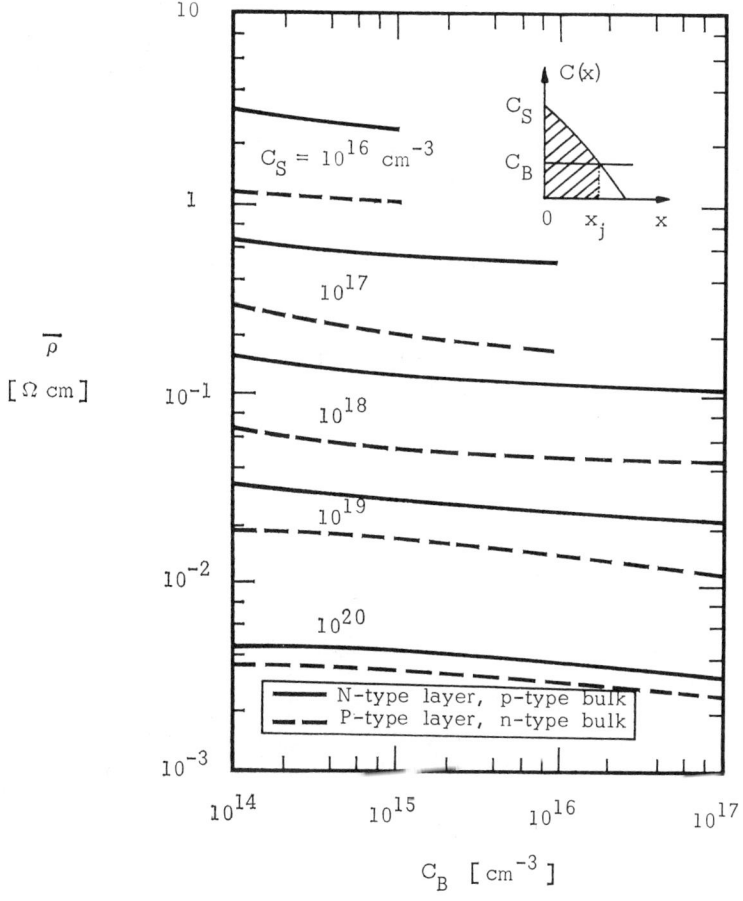

FIGURE 3.26

Average resistivity ($\bar{\rho}$) of a diffused surface layer vs. background impurity concentration (C_B) of semiconductor and surface impurity concentration (C_S) of layer. Silicon, Gaussian impurity distribution; 300°K.

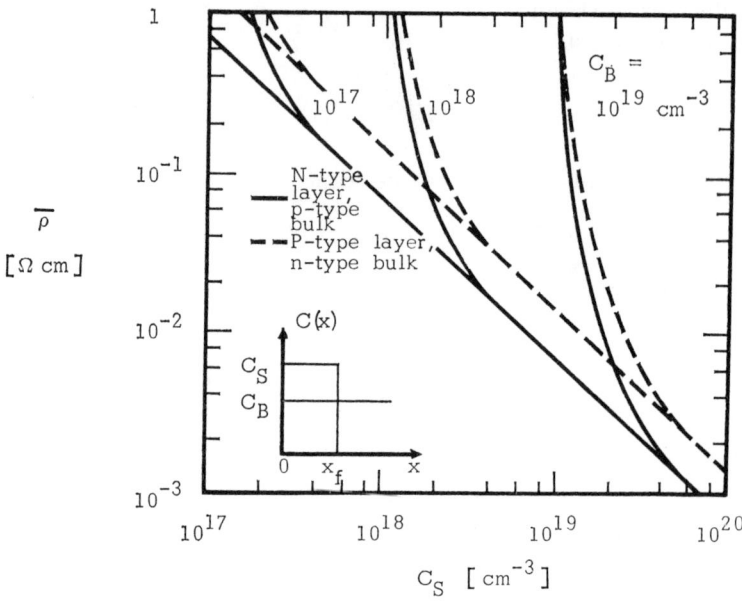

FIGURE 3.27

Average resistivity ($\bar{\rho}$) vs. surface impurity concentration (C_S) and background impurity concentration (C_B) for box-type impurity profile (see inset). Silicon, 300°K.

which is independent of x_f. Rearrangement leads to

$$C_S = C_B + 1/q\mu\bar{\rho}. \tag{3.129}$$

The curves of Figure 3.27 were calculated from this equation assuming $\mu_n = 85$ cm²/V sec and $\mu_p = 45$ cm²/V sec (silicon). The sheet resistivity is

$$\rho_s = \bar{\rho}/x_f. \tag{3.130}$$

Assumptions:

(a) The diffusion coefficient is independent of impurity concentration.
(b) Precipitation and compensation of impurities near the surface does not affect the distribution.
(c) Mobility decrease due to the proximity of the surface is neglected.

5. REFERENCES

(1) G. W. Ludwig and R. L. Watters, *Phys. Rev.*, **101**, 1699 (1956).
(2) D. B. Cuttriss, *BSTJ*, **40**, 509 (1961).
(3) J. C. Irvin, *BSTJ*, **41**, 387 (1962).

(4) R. P. Donovan, *RTI Report ASD-TDR-63-316*, Vol. I, June 1963.
(5) *RTI Report ASD-TDR-63-316*, Vol. V, July 1964.
(6) A. S. Grove, *Physics and Technology of Semiconductor Devices*, John Wiley and Sons, New York, 1967.
(7) S. M. Sze and J. C. Irvin, *Solid-State Electronics*, **11**, 599 (1968).
(8) W. M. Bullis et al., *Solid-State Electronics*, **11**, 639 (1968).

3-6 Minority Carrier Recombination and Lifetime

1. CARRIER RECOMBINATION

In a semiconductor in thermal equilibrium there is a continuous generation and recombination of electron-hole pairs with identical generation and recombination rates, resulting in an equilibrium concentration of carriers. In this case $np = n_i^2$. If this equilibrium is disturbed by an external influence, e.g., by steady illumination, generation and recombination rates are affected and new rates are established as soon as a steady-state condition has been reached; this new state does not represent equilibrium conditions but is a nonequilibrium steady state. In this case $np > n_i^2$. During the transition from one state to the other a transitional process affects generation and recombination rates. After removal of the external stimulus, the semiconductor tends to return to equilibrium conditions.

There are two basic types of recombination mechanisms.

(a) Band-to-band recombination:
This process is characterized by the transition of an electron directly from conduction to valence band. The transition is possible by either of two mechanisms:
  (i) *Auger recombination*:
  This process is characterized by an energy transfer to another free electron or hole. It is the inverse to impact ionization.
  (ii) *Radiative recombination*:
  This process is characterized by the emission of a photon. It is the inverse to direct optical transitions, which are important for many compound semiconductors with direct energy gap.
(b) Intermediate-level recombination:
This process is characterized by the transition of an electron from conduction to valence band in steps. The transition takes place either by means of a single level or of several levels.
  (i) *Single-level recombination*:
  It consists of four steps: electron capture, electron emission, hole capture, hole emission.

(ii) *Multiple-level recombination*:
The basic mechanism is similar to that of single-level recombination; differences are observed mainly at high injection level. The total average carrier lifetime is an average of the lifetimes of positively charged, negatively charged, and neutral trapping centers.

The most effective intermediate-level recombination centers are those close to the center of the energy gap, i.e., those for which $E_t - E_i \approx 0$. Table 3.9 lists some deep impurity levels in Si, Ge, and GaAs which may act as intermediate-level recombination centers.

Carrier recombination, which occurs when electrons and holes encounter each other in the crystal lattice, takes place at a rate which is directly proportional to their individual densities at any given time. During recombination the momentum (h/λ) is conserved. Direct recombination in an indirect energy gap semiconductor (e.g., silicon) is highly improbable since the annealing of electron-hole pairs during recombination can occur only when the momenta of the carriers were exactly equal and of opposite sign before the encounter, and since the carriers must be in the immediate vicinity of each other. Direct recombination is, however, possible in a semiconductor which has a narrow

TABLE 3.9

Deep-Level Recombination Centers in Si, Ge, and GaAs (300°K)

Impurity	$E_t - E_i$ [eV]		
	Si	Ge	GaAs
Ag	+0.23	+0.04	
	−0.22		
Au	+0.02	+0.13	
	−0.21	−0.18	
Co		−0.09	−0.18
Cr			−0.04
Fe	+0.01	+0.06	−0.13
	−0.16	0	−0.36
Li	−0.07	+0.18	
		+0.11	
		−0.02	
Mn	+0.03	−0.04	
		−0.18	
Ni	+0.21	+0.02	−0.19
O			−0.09
Pt	+0.19		
Se		+0.06	
Te		+0.04	

direct energy gap (e.g., InSb). In an indirect gap semiconductor, phonons are involved in the recombination process since the momentum associated with a phonon is about 10^4 times larger than that associated with a photon. The momentum of a wave quantum is h/λ; i.e., for phonons $\lambda \approx 1$ Å, for photons $\lambda \approx 10^4$ Å.

In an indirect semiconductor, recombination is possible only in the presence of one or more energy levels within the forbidden energy gap which are only partially occupied by electrons. These energy levels may result from impurities in the crystal lattice or from crystal imperfections. Recombination occurs when a free carrier of the appropriate charge is captured by these levels and recombines with a carrier of the opposite charge, which is subsequently captured. The excess momentum is transferred to the lattice as phonon vibrations. A recombination center is effective only if it is capable of existing in two different charge states and if it has equal access to both conduction and valence band. Only deep-lying energy levels, i.e., those close to the center of the energy gap, meet both requirements. Recombination removes equal amounts of electrons and holes from the total carrier density.

Crystal imperfections (e.g., lattice defects) disrupt the periodicity of the lattice and introduce energy levels within the energy gap (recombination centers) similar to those introduced by impurity atoms. They act as stepping stones in the transition of electrons and holes between the conduction and valence bands; since the transition probability depends upon the energy difference between these stepping stones, the presence of crystal imperfections enhances the probability of transitions and therefore affects the carrier lifetime.

Imperfections are introduced during crystal growth or subsequent processing or by impurities left behind after purification. High energy radiation by electrons, protons, neutrons, or gamma rays may also introduce energy levels within the energy gap due to the lattice damage they produce. For example, in Si electron irradiation introduces a donor level $E_{aD} = 0.36$ eV and an acceptor level $E_{aA} = 0.40$ eV, neutron irradiation an acceptor level $E_{aA} = 0.56$ eV, deuteron irradiation an interstitial level $E_t - E_c = 0.25$ eV. The presence of impurity atoms which have energy levels close to the center of the energy gap reduces the carrier lifetime significantly below that of crystal imperfections alone. In principle, however, there is no significant difference between recombination centers introduced by crystal imperfections or crystal damage and those introduced by deep-lying impurities in their effect on lifetime.

The recombination and generation process through intermediate levels proceeds in four phases:

(a) capture of an electron from the conduction band by the recombination center;

(b) emission of an electron from the recombination center into the conduction band;
(c) capture of a hole from the valence band by a recombination center;
(d) emission of a hole from a recombination center into the valence band.

The probability of carrier recombination at a recombination center depends upon

(a) total density of recombination centers in the semiconductor;
(b) density of recombination centers that are normally filled under equilibrium conditions;
(c) capture probability of the recombination centers;
(d) free carrier density, which, for a nondegenerate semiconductor, is equal to the density of acceptors or donors.

The minority carrier lifetime is defined as the time at which $1 - 1/e = 63.4\%$ of all carriers have recombined. In an n-type semiconductor lifetime is that of holes; in a p-type semiconductor, that of electrons. The lifetime (mean free time) of minority carriers is the reciprocal of the collision probability per unit time.

If a group of n_{to} electrons is considered at time $t = 0$, then the number of electrons which have survived without a collision until time t is

$$n_t = n_{to} \exp(-t/\tau) \tag{3.131}$$

and the rate at which collisions are removing electrons from the group of surviving carriers is

$$dn/dt = (n_t/\tau) = (n_{to}/\tau) \exp(-t/\tau). \tag{3.132}$$

In this discussion it is assumed that classical statistics are employed; voltage drops in the quasi-neutral regions, generation-recombination in the depletion region and voltage breakdown are neglected; and low current densities (small injection rate) are encountered.

Carriers usually recombine within the semiconductor bulk and at the semiconductor surface. Both of these recombination processes may be significant; their relative magnitude is a function of bulk and surface characteristics and sample dimensions. The total minority carrier lifetime is given by

$$1/\tau = 1/\tau_B + 1/\tau_S \tag{3.133}$$

where τ_B and τ_S refer to bulk and surface recombination times, respectively.

2. BULK RECOMBINATION

In steady state, carrier generation (G) and recombination (U) rates are equal if an external activation process (e.g., illumination)—assumed to inject carriers at a density which is small compared to the free carrier density—has reached equilibrium. In the absence of current and diffusion effects, the recombination rate at steady state is given for holes in an *n*-type semiconductor by

$$U_p = \frac{dp}{dt} = \frac{p_n - p_t}{\tau_p} + G_p \tag{3.134a}$$

and for electrons in a *p*-type semiconductor by

$$U_n = \frac{dn}{dt} = \frac{n_p - n_t}{\tau_n} + G_n. \tag{3.134b}$$

After removal of the external activation $G_p = 0$ or $G_n = 0$ and

$$p_t - p_n = (p_{to} - p_n) \exp(t/\tau_p) \tag{3.135a}$$

$$n_t - n_p = (n_{to} - n_p) \exp(t/\tau_n). \tag{3.135b}$$

This means that carrier decay is exponential. The above equations are valid only if $p_{to} \approx p_n$ or $n_{to} \approx n_p$, i.e., if the carrier densities do not deviate appreciably from equilibrium. Minority carrier lifetime is structure-sensitive and depends upon the degree of order within the semiconductor crystal. Direct transitions from conduction to valence band, however, are usually insignificant (they are in the order of one second) and lifetime is usually controlled by indirect transitions in stages through crystal imperfections.

In thermal equilibrium

$$U_p = G_p \quad \text{or} \quad U_n = G_n$$

and consequently

$$p_n = p_t \quad \text{or} \quad n_p = n_t.$$

Three types of crystal imperfections serve as recombination centers:

(a) interstitial or substitutional impurity atoms if they have deep-lying energy levels;
(b) dislocations as a result of mechanical stress or of processing influences;
(c) crystal disorder other than dislocations as a result of high-temperature treatment.

The steady-state net recombination rate (U) of electrons and holes in a semiconductor in which there is only one type of recombination center (a single recombination level) can be expressed in terms of minority carrier lifetime in highly extrinsic material (τ_{no} and τ_{po}) as

$$U = \frac{n_p p_n - n_i^2}{\tau_{no}(p_n + p_1) + \tau_{po}(n_p + n_1)}$$

$$= \sigma_{np} v_{th} N_t \frac{n_p p_n - n_i^2}{n_p + p_n + 2n_i \cosh[(E_t - E_i)/kT]} \tag{3.136}$$

3-6 RECOMBINATION AND LIFETIME

and the carrier generation rate

$$G = \frac{n_i}{\tau_{no} \exp\left[(q/kT)(E_t - E_i)\right] + \tau_{po} \exp\left[(q/kT)(E_i - E_t)\right]} \quad (3.137)$$

where $E_t - E_i$ = energy difference between the intrinsic Fermi level and a generation-recombination center. If $E_t - E_i > 0$, $\tau_{po}/\tau_{no} \gg 1$ and the center is a donor. If $E_t - E_i < 0$, $\tau_{po}/\tau_{no} \ll 1$ and the center is an acceptor. U approaches a maximum when $E_t \approx E_i$.

In a semiconductor where there is more than one recombination level, τ_{no} and τ_{po} are not simply the minority carrier lifetimes in highly extrinsic material.

For $n_p p_n > n_i^2$ = (i.e., $U > 0$), carriers recombine in the depletion region (forward biased junction).

For $n_p p_n < n_i^2$ = (i.e., $U < 0$), carriers are generated in the depletion region (reverse biased junction).

Lifetime is affected by injection of carriers. If the injected carrier density is Δc, then

$$\tau = \tau_{po} \frac{n_p + n_1 + \Delta c}{n_p + p_n + \Delta c} + \tau_{no} \frac{p_n + p_1 + \Delta c}{n_p + p_n + \Delta c}. \quad (3.138)$$

Usually small and large injection levels are distinguished.

(a) Small injection level:
   In this case there are only small deviations from equilibrium and

$$\Delta c \ll n_p + p_n.$$

Consequently the small-injection lifetime

$$\tau = \tau_o = \tau_{po} \frac{n_p + n_1}{n_p + p_n} + \tau_{no} \frac{p_n + p_1}{n_p + p_n} = \frac{(n_p + n_1)/\sigma_p + (p_n + p_1)/\sigma_n}{N_t v_{th}(n_p + p_n)} \quad (3.139)$$

The low-level injection lifetime in bulk silicon is shown in Figure 3.28.

(b) Large injection level:
   In this case deviations from equilibrium are considerable and

$$\Delta c \gg n_p + p_n.$$

Consequently the large-injection lifetime

$$\tau = \tau_\infty = \tau_{po} + \tau_{no} = \frac{1}{N_t v_{th}(\sigma_n + \sigma_p)}. \quad (3.140)$$

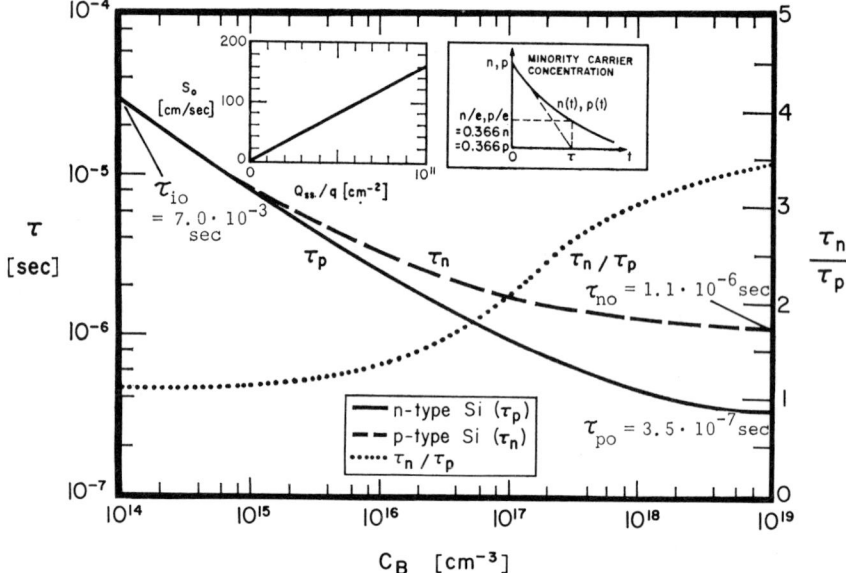

FIGURE 3.28

Minority carrier lifetime (τ) vs. semiconductor impurity concentration (C_B); silicon, 300°K. The left inset shows surface recombination velocity (S_o) vs. surface state density (Q_{ss}/q) for a capture cross section $\sigma_c \approx 2 \cdot 10^{-16}$ cm². The right inset shows the decrease of minority carrier density with time and defines the carrier lifetime.

Generally, the lifetime can be expressed in terms of the injection ratio γ_i:

$$\tau = \frac{\tau_o + \gamma_i \tau_\infty}{1 + \gamma_i} \tag{3.141}$$

where

$$\gamma_i = \Delta c/(n_p + p_n). \tag{3.142}$$

For high injection level $\gamma_i > 1$, for low injection level $\gamma_i < 1$. The excess lifetimes at high impurity levels (i.e., the lifetime of holes injected into a highly doped n-type semiconductor, τ_{po}, or the lifetime of electrons injected into a highly doped p-type semiconductor, τ_{no}) are given in the case of only one recombination level by:

$$\tau_{no} = 1/N_t \alpha_{no} = 1/N_t v_{th} \sigma_n \tag{3.143a}$$

$$\tau_{po} = 1/N_t \alpha_{po} = 1/N_t v_{th} \sigma_p. \tag{3.143b}$$

The capture probabilities α_{no} and α_{po} are related to the capture cross sections σ_n and σ_p by

$$\alpha_{no} = v_{th} \sigma_n \quad \text{and} \quad \alpha_{po} = v_{th} \sigma_p, \tag{3.144}$$

3-6 RECOMBINATION AND LIFETIME

where v_{th} is the thermal carrier velocity.

$$v_{th} = (3kT/m_{n,p})^{1/2}. \tag{3.145}$$

Numerical values for silicon at 300°K are:
For both electrons and holes:

$$v_{th} \approx 10^7 \text{ cm/sec}.$$

For gold in n-type silicon:

$\sigma_n = 5 \cdot 10^{-16} \text{ cm}^2,$ $\quad \sigma_p = 1.0 \cdot 10^{-15} \text{ cm}^2;$
$\alpha_{no} = 5 \cdot 10^{-9} \text{ cm}^3/\text{sec},$ $\quad \alpha_{po} = 1.0 \cdot 10^{-8} \text{ cm}^3/\text{sec};$

and in p-type silicon:

$\sigma_n = 3.5 \cdot 10^{-15} \text{ cm}^3$ $\quad \sigma_p = 7.0 \cdot 10^{-15} \text{ cm}^2;$
$\alpha_{no} = 3.5 \cdot 10^{-8} \text{ cm}^3/\text{sec},$ $\quad \alpha_{po} = 7.0 \cdot 10^{-8} \text{ cm}^3/\text{sec}.$

The presence of deep-lying impurity atoms (e.g., gold in silicon) increases the density of recombination centers and thus reduces the minority carrier lifetime as follows.

(a) Minority carrier (hole) lifetime in an n-type semiconductor:

$$\tau_p = \frac{1 + (p_n/n_n)(\beta_p/\beta_n) + (n_p/p_p)(\alpha_n/\alpha_p)}{C_{Au}(\beta_p + \alpha_n n_p/p_p)}. \tag{3.146a}$$

(b) Minority carrier (electron) lifetime in a p-type semiconductor:

$$\tau_n = \frac{1 + (p_n/n_n)\beta_p/\beta_n + (n_p/p_p)(\alpha_n/\alpha_p)}{C_{Au}(\alpha_n + \beta_n p_n/n_n)}. \tag{3.146b}$$

The ratio of minority carriers to majority carriers is p_n/n_n in an n-type semiconductor and n_p/p_p in a p-type semiconductor. The reduction of minority carrier lifetime with recombination center density is shown in Figures 3.29 and 3.30.

A qualitative discussion of the dependence of lifetime upon impurity concentration is as follows. At high donor concentration (n-type semiconductor) the Fermi level lies close to the conduction band edge (E_c) and the recombination centers are almost completely filled; hence the lifetime is limited entirely by the capture of holes by the many filled recombination centers and is given by τ_{po}. As the donor concentration decreases, the Fermi level moves toward the center of the energy gap and the recombination centers are becoming less filled with electrons while the hole density increases. This increases the lifetime of holes. At the intrinsic level the Fermi level is at the center of the energy gap and the lifetime shows a maximum. At increasing acceptor concentration (p-type semiconductor) the Fermi level is moving

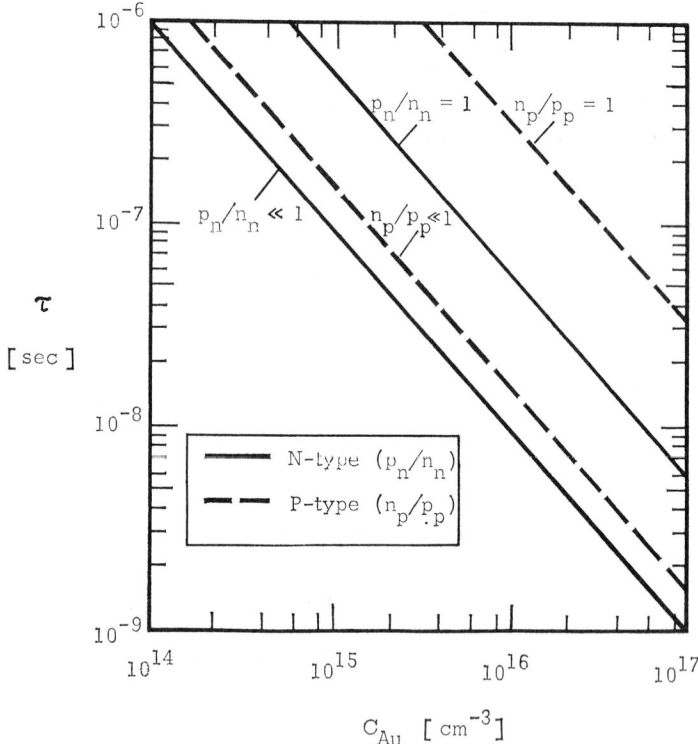

FIGURE 3.29

Minority carrier lifetime (τ) vs. gold concentration in saturation (C_{Au}) and ratio of minority to majority carrier concentration (p_n/n_n for n-type silicon, n_p/p_p for p-type silicon). 300°K. Under most conditions the lower set of the curves apply, i.e., if the ratio of minority carrier density to majority carrier density is $< 10^{-3}$.

toward the valence band edge (E_v) and fewer recombination centers become filled with electrons while the hole density increases; this reduces the lifetime of electrons. At high acceptor concentration the recombination centers are almost completely empty; hence the lifetime is limited entirely by the capture of electrons by the unfilled centers and is given by τ_{no}.

Since Fermi level and carrier densities are temperature-dependent, lifetime is also a function of temperature. Temperature increase changes an impurity semiconductor toward a more intrinsic behavior which results in exponential lifetime increase with temperature. As the intrinsic point is reached, lifetime begins to decrease again due to an increase in n_i with temperature and a resulting increase in the recombination rate.

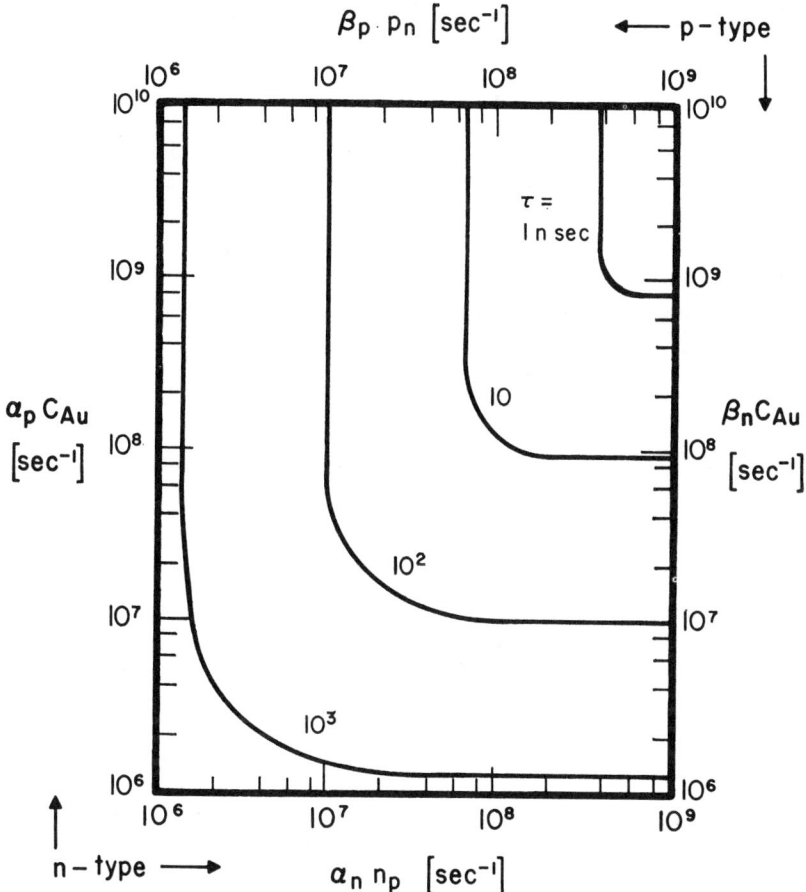

FIGURE 3.30

Minority carrier lifetime (τ) vs. carrier capture probabilities (α_n, α_p, β_n, β_p), minority carrier densities (n_p, p_n), and gold concentration (C_{Au}). Silicon, 300°K. The curves represent contours of constant lifetime. The capture probabilities are defined in the text. For n-type silicon use the left-hand and the lower scales, for p-type silicon use the right-hand and the upper scales.

Deviations from the model used arise in heavily doped (degenerate) semiconductors where there is no constant relationship between mobility and carrier diffusion coefficient and where changes in the band structure and impurity levels are observed. In this case, the impurity ionization energy becomes negligible and the impurity recombination levels coincide with the conduction or valence band energy.

Recombination and generation of carriers is accompanied by a current whose density is given by:

$$j_U = -q \int_{-x_n}^{-x_p} U \, dx \qquad (3.147)$$

where x_n, x_p are the fractional depletion layer widths on n- or p-side of the junction, respectively; (total depletion layer width $x_d = x_n + x_p$).
The recombination current displays a different behavior for reverse and forward bias.

(a) Reverse bias:
If it is assumed that $qV/kT \ll 1$; $p_n \ll p_1$ and $n_p \ll n_1$ in the depletion layer, then

$$j_U = \frac{qn_i^2 x_d}{\tau_{no} p_1 + \tau_{po} n_1}. \qquad (3.148)$$

If there is only one recombination level (at the intrinsic Fermi level, $E_t - E_i = 0$), then

$$p_1 = n_1 = n_i \qquad (3.149)$$

and τ_{no} and τ_{po} are equal to the lifetimes for the respective quasi-neutral regions.

(b) Forward bias:
If it is assumed that $qV/kT \gg 1$; $p_n \gg p_1$ and $n_p \gg n_1$ in the depletion layer, then

$$j_U = -\frac{qn_i^2 x_d}{\tau_{no} p_1 + \tau_{po} n_1} \exp(qV/2kT). \qquad (3.150)$$

If there is only one recombination level (at the intrinsic Fermi level, $E_t - E_i = 0$), then

$$p_n = n_p = n_i \exp(qV/kT). \qquad (3.151)$$

Band-to-band Auger recombination is the inverse process for band-to-band impact generation of free carriers and consists of two processes involving either electron-electron or hole-hole collisions. Auger recombination is important only in very narrow band gap semiconductors. In these semiconductors the complex recombination mechanism accounts for the strong temperature dependence of lifetime at high temperature.

Assuming a nondegenerate semiconductor with parabolic bands, the thermal equilibrium generation rate for band-to-band electron Auger transitions is

$$G_n = \frac{8(2\pi)^{5/2} q^4}{h^3} \frac{m_n/m_o}{(1 + m_n/m_o)^{1/2}(1 + 2m_n/m_o)} \frac{n_p}{(\varepsilon_s \varepsilon_o)^2} \cdot \left(\frac{kT}{E_G}\right)^{3/2}$$

$$(F_{1n} F_{2n})^2 \exp\left(-\frac{1 + 2m_n/m_o}{1 + m_n/m_o} \frac{E_G}{kT}\right). \qquad (3.152)$$

F_1 and F_2 are overlap integrals of the periodic part of Bloch wave functions; n_p is the equilibrium electron concentration.

If the electron transition is not from band to band but from band to recombination level (of energy E_t), then, in thermal equilibrium, the generation rate for electrons

$$G_n = 8\pi q^4 \frac{m_n/m_o}{(m_p/m_o)^{3/2}} \frac{n_p N_t}{(\varepsilon_s \varepsilon_o)^2} \frac{1}{(E_c - E_t)^{3/2}} (F_{1n} F_{2n})^2 \cdot \exp\left(-\frac{E_c - E_t}{kT}\right).$$
(3.153)

Similar expressions hold for holes.

3. SURFACE RECOMBINATION

At a semiconductor surface there are surface energy states with energies within the energy gap of the semiconductor; these energy states can be divided into layer states (owing to the characteristics of the oxide layer; they arise from absorbed ions, molecules, or imperfections) and interface states (owing to characteristics of the semiconductor-oxide interface; they arise from initial semiconductor surface treatment before oxide formation and are similar in behavior to the bulk recombination centers). The density of interface states is usually much smaller than that of the layer states, but the interface states have capture probabilities several orders of magnitude higher than the layer states. Therefore, the surface recombination is attributed mainly to recombination at interface centers. Surface recombination velocity (S)—defined as the number of carriers recombining per second per unit surface area divided by the excess concentration over the equilibrium value at the surface—is a function of density and energy of interface states, capture probabilities, carrier concentrations and degree of occupancy of the interface states.

In the absence of a space charge region (denoted by the subscript o) surface recombination velocity (under the assumption that surface recombination is independent of majority carriers, i.e., if $n_S \ll p_p$ and $p_S \ll n_n$) is given for an n-type semiconductor by

$$S_{op} = j_{Sp}/q(p_S - p_n) \quad (3.154a)$$

and for a p-type semiconductor by

$$S_{on} = j_{Sn}/q(n_S - n_p). \quad (3.154b)$$

In equilibrium, surface recombination rate and surface generation rate are equal, i.e., $j_{Sn} = 0$ and $j_{Sp} = 0$.

The surface recombination velocity in the absence of a space charge region can also be expressed as

$$S_o = \sigma_c v_{th} N_{tS} \quad (3.155)$$

where σ_c is the carrier capture cross-section.

The associated surface lifetime, τ_S, is related to the sample dimensions; for a rectangular sample of dimensions a_1 and a_2 and for small S_o the surface lifetime in an n-type semiconductor

$$\tau_{Sp} = A'/(\sigma_p v N_{thtS}) = A'/S_{op} \tag{3.156a}$$

and in a p-type semiconductor

$$\tau_{Sn} = A'/(\sigma_n v_{th} N_{tS}) = A'/S_{on} \tag{3.156b}$$

where $A' = a_1 a_2/(a_1 + a_2)$.

For large sample dimensions: $\tau_t \approx \tau_B$; for a sample with a large surface-to-volume ratio and a long bulk lifetime: $\tau_t \approx \tau_S$. The surface recombination rate is structure-sensitive and depends upon the chemical and mechanical surface treatment. Example: a silicon transistor base of dimensions $a_1 = a_2 = 10\mu m$ which has been surface-treated with a 1% NaOH/99% H_2O etchant may have a surface recombination velocity $S_o \approx 250$ cm/sec. This results in $\tau_S = 2 \cdot 10^{-6}$ sec and is comparable to observed bulk lifetimes.

In the presence of a surface space charge region the surface recombination velocity is related to carrier and impurity concentration by

$$S_p = S_{op} N_D/(n_S + p_S + 2n_p) \quad (n\text{-type}) \tag{3.157a}$$

$$S_n = S_{on} N_A/(n_S + p_S + 2p_n) \quad (p\text{-type}) \tag{3.157b}$$

and in a near-intrinsic material (where $n_S + p_S$ is at a minimum) by

$$S_p = S_{p\max} = S_{op} N_D/4n_i \text{ or } S_n = S_{n\max} = S_{on} N_A/4n_i. \tag{3.158}$$

In thermally oxidized silicon $1 \leq S_o \leq 100$ cm/sec.
A relationship between S_o and surface state density (Q_{ss}/q) is shown in the inset of Figure 3.28 for a capture cross-section $\sigma_c \approx 2 \cdot 10^{-16}$ cm^2.

4. RECOMBINATION RADIATION

In a direct-gap semiconductor, recombination may result in the emission of photons. The total energy emitted by photons of a direct-gap semiconductor of energy gap E_G as a result of radiative recombination per unit volume is

$$E_r = E_G U, \tag{3.159}$$

where the recombination rate in equilibrium is

$$U = P_p(v)P_\alpha(v)$$
$$= 2\pi\alpha n^{*2}(E_G/hc)^2/[\exp(E_G/kT) - 1]$$
$$= (8\pi/h^3)(n^*/c)^2\alpha(kT)^3 \exp(-E_G/kT) \cdot [0.8(E_G/kT) - (E_G/kT + \sqrt{2})^2]$$
$$\tag{3.160}$$

and the energy gap in terms of radiation frequency v

$$E_G = hv, \tag{3.161}$$

where $P_p(v)$ = photon density distribution
$P_\alpha(v)$ = probability of absorption of a photon per unit time
n^* = refractive index
α = optical absorption coefficient
c = velocity of light.

5. DEFINITIONS

n_p, p_n = equilibrium minority carrier concentration
n_{to}, p_{to} = minority carrier concentration at time zero
n_t, p_t = minority carrier concentration at time t
n_1, p_1 = density of electrons in conduction band or holes in valence band when Fermi level coincides with recombination energy level

$$n_1 = N_c \exp\left[(E_c - E_t)/kT\right] = N_c \exp\left(E_{aD}/kT\right)$$
$$p_1 = N_v \exp\left[(E_t - E_v)/kT\right] = N_v \exp\left(E_{aA}/kT\right)$$

n_S, p_S = minority carrier density in the immediate vicinity of the semiconductor surface
n_i = intrinsic carrier concentration
τ_{no}, τ_{po} = limiting lifetime of excess minority carriers injected into a heavily doped semiconductor; τ_{no} and τ_{po} are a function of the density of recombination centers and capture cross-sections
σ_n, σ_p = capture cross-section of a recombination center for electron or hole; it is a measure of how close the carrier has come to the recombination center to be captured
N_t, N_{tS} = density of recombination centers within the semiconductor bulk, at the semiconductor surface, respectively
C_{Au} = density of gold atoms in saturation
G_n, G_p = electron and hole generation rate; it is a function of the intensity of the external activation
j_{Sn}, j_{Sp} = total drift and diffusion current density of minority carriers into the surface.

Carrier capture probabilities:

α_n = probability of electron capture by a neutral Au atom; $\alpha_n = 1.65 \cdot 10^{-9}$ cm^3/sec
α_p = probability of hole capture by a negative Au atom; $\alpha_p = 1.15 \cdot 10^{-7}$ cm^3/sec
β_n = probability of electron capture by a positive Au atom; $\beta_n = 6.3 \cdot 10^{-8}$ cm^3/sec

β_p = probability of a hole capture by a neutral Au atom; $\beta_p = 2.4 \cdot 10^{-8}$ cm^3/sec

α_{no}, α_{po} = capture probabilities of the recombination centers for holes and electrons; α_{no} and α_{po} are different because of the different masses of electrons and holes

6. REFERENCES

(1) W. Shockley and W. T. Read, *Phys. Rev.*, **87**, 835 (1952).
(2) B. Ross and J. R. Madigan, *Phys. Rev.*, **108**, 1428 (1957).
(3) C. T. Sah et al., *Proc. IRE*, **45**, 1228 (1957).
(4) G. Bemski, *Proc. IRE*, **46**, 990 (1958).
(5) A. B. Phillips, Transistor Engineering, McGraw Hill Book Co. New York, 1962.
(6) K. L. Ashley and A. G. Milnes, *J. Appl. Phys.*, **35**, 369 (1964).
(7) J. M. Fairfield and B. V. Gokhale, *Solid-State Electronics*, **8**, 685 (1965).
(8) R. M. Burger, *RTI Report ASD- TDR-63-316*, Vol. VIII, March 1966.
(9) S. C. Choo, *Solid-State Electronics*, **11**, 1069 (1968).
(10) A. B. Grebene, *J. Appl. Phys.*, **39**, 4866 (1968).
(11) S. C. Choo and A. C. Sanderson, *Solid-State Electronics*, **13**, 609 (1970).
(12) M. C. Collet, *J. El. Chem. Soc.*, **117**, 259 (1970).

3-7 Examples

PROBLEMS AND SOLUTIONS

1. **Problem:** Compare the carrier concentrations in intrinsic Si and GaAs at 1100°C.

   Solution: Si: $n_i = 1.2 \cdot 10^{19}$ cm^{-3},
   GaAs: $n_i = 1.6 \cdot 10^{18}$ cm^{-3}.
   This means that the intrinsic carrier concentration in Si is 7.5 times higher than that in GaAs.

2. **Problem:** Compare the ratio of majority to minority carrier densities in n-type silicon of impurity concentrations $C_B = 10^{15}$ and 10^{18} cm^{-3}. Assume room temperature.

   Solution: (a) $C_B = 10^{15}$ cm^{-3}:
   $n = 1 \cdot 10^{15}$ cm^{-3}, $p = 2 \cdot 10^5$ cm^{-3}; i.e., $n/p = 5 \cdot 10^9$.
   (b) $C_B = 10^{18}$ cm^{-3}:
   $n = 1 \cdot 10^{18}$ cm^{-3}, $p = 2 \cdot 10^3$ cm^{-3}; i.e., $n/p = 5 \cdot 10^{14}$.
   This means that the carrier density ratio increases by five orders of magnitude if the impurity concentration is increased by three orders of magnitude.

3. **Problem:** In an n-type silicon Hall generator of thickness $x_w = 2.0$ μm and of donor concentration $C_B = 10^{15}$ cm^{-3} a control current $I_z = 1$ mA is flowing. It is embedded in p-type silicon of impurity concentration 10^{14} cm^{-3}. Determine the Hall voltage for a magnetic field $H_x = 10^2$ Gauss.
   Solution: The carrier density ratio is $n/p = 10$; it corresponds to a Hall coefficient $R_{H_0} = -7 \cdot 10^3$ cm^3/A sec. From this the Hall voltage (using the conversion 1 Gauss $= 10^{-8}$ V sec/cm^2)

   $$V_H - 3.5 \cdot 10^{-2} \text{ V}.$$

PROBLEMS FOR WHICH A SOLUTION IS NOT GIVEN

1. Compare the bulk lifetimes of electrons and holes in silicon samples of opposite conductivity type of resistivities 0.1, 1, and 10 Ω cm at 25°C.

2. Into n-type and p-type silicon samples of impurity concentration $C_B = 10^{15}$ cm^{-3} gold is diffused to saturation at 1050°C. Determine the expected minority carrier lifetimes at 25°C.

3. Determine the reduction of electron and hole mobility compared to the low-field values if $E = 5 \cdot 10^3$ V/cm for Si, Ge, and GaAs.

4. Determine the reduction in electron and hole mobility compared to the values in bulk material for thin films of intrinsic Si, Ge, and GaAs if the film thickness is 1 and 3 μm.

4
PROPERTIES OF SILICON DIOXIDE

4–1 Atomic Structure and Energy Diagram
4–2 Thermal Oxidation
4–3 Diffusion in SiO_2
4–4 Other Characteristics
4–5 Examples

Silicon dioxide has gained importance in the fabrication of semiconductor devices due to its masking and surface passivating capabilities, its role in MOS structures, and its characteristics as dielectric in the formation of capacitors and multi-level interconnections. More recently, some other dielectrics, e.g., Al_2O_3, Si_3N_4, have taken over some of the functions of SiO_2, whereas the masking capabilities of certain metals are sometimes utilized in low-temperature operations, e.g., in ion implantation.

The amorphous form of glassy SiO_2 (silica glass) is one of the twenty-two phases of silica. It is usually assumed that the characteristics of bulk amorphous silica are similar to those of SiO_2 thermally grown at a silicon surface. Amorphous SiO_2 forms a random three-dimensional network, consisting of silicon-oxygen tetrahedra which are joined only at the corners and share no faces or edges. Each Si atom forms the center of the tetrahedron, the vertices of which are defined by the four associated O atoms which themselves are shared between the Si atoms. Short range order (regions of crystallinity 10 to 100 Å in size) may exist. The SiO_2 network has the following dimensional properties.

Average Si–O distance: 1.62 Å;
Average O–O distance: 2.65 Å;
Average Si–Si distance: 3.00 Å;
Si–O–Si angle: $143° \pm 17°$.

SiO_2 tetrahedra are joined to one another by bridging oxygen ions. In fused silica some of the vertices have nonbridging oxygen ions which belong to only one tetrahedron. The degree of cohesion between the tetrahedra and the network as a whole is a function of the ratio of bridging to nonbridging oxygen ions.

In silica glass a silicon atom moves by the rupture of four Si—O bonds whereas an oxygen atom moves by the rupture of only two Si—O bonds; thus oxygen atoms can move more freely in SiO_2 than silicon atoms. This enhanced probability of oxygen movement leads to bridging and nonbridging oxygen ion vacancies; from binding energy considerations the formation of nonbridging vacancies is more probable. Oxygen ion vacancies form positively charged defects in the SiO_2 structure which may influence the characteristics of an underlying semiconductor surface. The movement of impurity atoms in SiO_2 follows diffusion laws although the network is not a crystalline structure.

Non-uniformities and voids exist in the loose disordered network. Consequently, the density of silica glass is less than that of crystalline quartz. Silica glass is thermodynamically unstable below 1710°C; below 1000°C devitrification is negligible, so that it is quite stable at room temperature. Imperfections in SiO_2 are usually oxygen vacancies, oxygen excesses, or other foreign ions at interstitial positions; or metal ions at either substitutional or interstitial positions.

An amorphous SiO_2 layer, as, for example, obtained by the thermal oxidation of silicon, has many properties similar to those of crystalline quartz. Among others, the short-range atomic order, the angles between bonds, and the ionic and covalent binding forces in SiO_2 and fused quartz closely resemble those formed in quartz; effects of ionizing radiation and the annealing characteristics of damage centers of oxide and quartz are almost identical. The important difference is the absence of long-range structural order in amorphous SiO_2. On the other hand, thin SiO_2 films have many properties different from those of bulk SiO_2.

The properties of SiO_2 depend upon impurity content if the impurities are ionized. Otherwise, they merely occupy holes in the network. Impurities in SiO_2 are usually those used as impurities in semiconductor technology, e.g., phosphorus, boron, aluminum, arsenic, antimony. Most glasses contain also lead, potassium, and sodium. Pure SiO_2 is an insulator. Its electric conductivity ($\approx 10^{-16}$–$10^{-20} \Omega^{-1} cm^{-1}$) is due to ion migration and can be altered by the addition of impurities (semiconducting and conducting glasses). The electric conductivity of SiO_2 depends also upon temperature and electric field; the thermal conductivity is about one order of magnitude less than that of crystalline silicon.

Ionized impurities in SiO_2 are either network formers or network modifiers, some impurities (e.g., aluminum) can act as both. Network formers (e.g., boron, phosphorus, aluminum) have small atomic radii, enter the network substitutionally (substituting for Si atoms) in forming the network, or can form glasses by themselves. Network modifiers (e.g., sodium, potassium, lead, calcium, barium, aluminum) enter the network interstitially; when introduced as oxide they ionize, donate an oxygen atom to the network while the atom occupies an interstitial position—resulting in a weakening of the network and a lower melting point and other variations in properties. In addition, the existence of H_2O and OH in SiO_2 weakens the network structure and affects most oxide properties, most significantly infrared absorption, dielectric constant, and dissipation factor.

The preferred method of SiO_2 formation on silicon is the thermal oxidation. The oxidation proceeds by the inward motion of the oxidizing species. The resultant oxide thickness is usually proportional to the square root of oxidation time. On other semiconductors SiO_2 is formed by low-temperature oxide deposition.

During the initial growth phase of thermal SiO_2 the molecular oxygen, upon entering the oxide, dissociates to form a negative superoxide ion (O_2^-) and a hole. The hole, which has a significantly higher mobility than the oxygen ion runs ahead of it, resulting in the formation of a space charge region near the gas-oxide interface; the associated electric field, in turn, will enhance the diffusion of the oxygen ions in the oxide. The thickness of this space charge region is of the order of the Debye length and depends inversely upon the square root of the concentration of the oxidizing species in the layer. This Debye length is about 200 Å for dry oxidation and about 5 Å for wet oxidation; the difference is due to the higher (by three orders of magnitude) solid solubility of water in silicon compared to that of oxygen in silicon. As a result, a significantly increased oxidation rate is observed for the first 200 Å of dry oxidation while for wet oxidation this is not found. Even the surface of freshly cleaned silicon is covered at all times by a thin (10 to 15 Å) layer of SiO_2 due to oxidation in air.

The inward motion of the $Si-SiO_2$ interface during thermal oxidation leads to an impurity redistribution in the semiconductor near the semiconductor-oxide interface. It depends upon the segregation coefficient of the impurity at the interface, the ratio of impurity diffusion coefficients in semiconductor and oxide, and the ratio of oxidation rate constant to impurity diffusion coefficient in the semiconductor.

GENERAL REFERENCES

(1) R. M. Burger and R. P. Donovan, *Fundamentals of Silicon Integrated Device Technology*, Vol. 1, Prentice-Hall, Englewood Cliffs, N.J., 1968.
(2) S. K. Ghandhi, *The Theory and Practice of Microelectronics*, John Wiley and Sons, New York, 1968.
(3) C. W. Gwyn, *J. Appl. Phys.*, **40**, 4886 (1969).

4-1 Atomic Structure and Energy Diagram

1. NONCRYSTALLINE SOLIDS

Nonperiodic arrays of atoms are either noncrystalline or semicrystalline in nature. In an amorphous solid the atoms are arranged in a completely random fashion; this can result in a solid under conditions where there is insufficient mobility for the atoms to arrange themselves in a crystalline way.

Glasses are common examples of amorphous materials. In the glassy state the molecular arrangement of the liquid is frozen in because the very high viscosity of the material makes the growth of regular crystals very difficult as the liquid is solidified. The most common example is SiO_2 which consists of three-dimensional networks of SiO_4 tetrahedra; the random arrangement of the chains facilitates entry of other ions into the large openings of the solid. The irregular network structure leads to variations in the interatomic spacings and consequently to varying bonding strengths within the solid. Hence all the bonds do not break at the same temperature so that there is no sharp melting point; the gradual breaking of bonds leads to an extensive temperature range in which the glass softens and the viscosity gradually decreases. Under certain conditions a glass may begin to crystallize ("devitrification").

Organic polymers have a semicrystalline structure. Within a given polymer there are many structural variations from molecule to molecule; parameters which vary are chain length and branching structure. Polymers are molecular solids, with Van der Waals cohesive forces between the polymer molecules. Because of the great length of the polymer molecules they also intertwine in a way which provides extra cohesive strength for the solid.

2. STRUCTURE OF SiO_2

Amorphous SiO_2 forms a random three-dimensional network which consists of silicon-oxygen tetrahedra; each Si atom forms the center of the tetrahedron. Even in amorphous SiO_2 many of the tetrahedra cluster in the six-sided ring pattern characteristic of crystalline quartz. Since the loose disordered networks of amorphous SiO_2 allow non-uniformities to exist, the density of amorphous SiO_2 (2.20 g/cm^3) is less than that of crystalline quartz (2.65 g/cm^3) and the amorphous form has no well-defined melting point. In

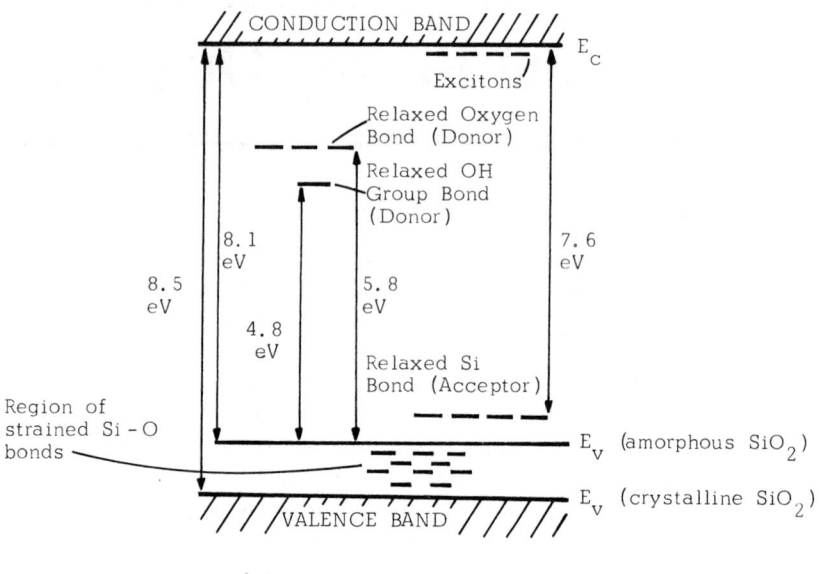
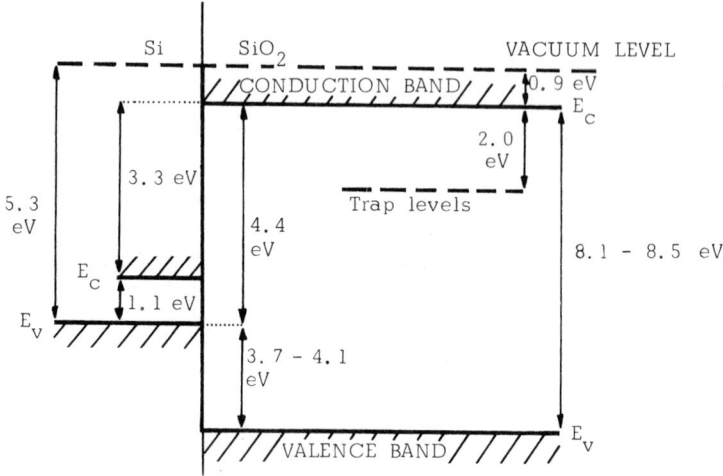

FIGURE 4.1

Energy band diagram of amorphous and crystalline SiO$_2$ (upper part) and of the Si–SiO$_2$ system (lower). The SiO$_2$ diagram shows the various energy levels of trapping centers with respect to the band edges. The Si–SiO$_2$ diagram was obtained from photoemission measurements.

crystalline SiO_2 all oxygen atoms are shared by two Si atoms and are called bridging oxygen atoms; in amorphous SiO_2 some of the oxygen atoms are associated with only one Si atom and are called nonbridging.

3. ENERGY BANDS IN SiO_2

Amorphous and crystalline SiO_2 can be characterized by an energy band structure, similar to crystalline semiconductors. Figure 4.1 shows the energy band diagram of grown silicon dioxide and of crystalline quartz, both of chemical composition SiO_2. It indicates the position of the strained Si–O bonds relative to a sharp energy band edge for crystalline quartz and the energy levels of the broken bond hole and electron trapping centers. A narrowing of the band gap as the crystalline structure becomes amorphous is observed, i.e., the width of the energy gap decreases as the long-range order in the crystal decreases. Also, both the conduction and the valence bands broaden and therefore cause a corresponding decrease of the energy gap.

4. REFERENCES

(1) R. Williams, *Phys. Rev.*, **140**, A 569 (1965).
(2) R. M. Burger and R. P. Donovan, *Fundamentals of Silicon Integrated Device Technology*, Vol. I, Prentice-Hall, Englewood Cliffs, N.J. 1967.
(3) C. W. Gwyn, *J. Appl. Phys.*, **40**, 4886 (1969).

4-2 Thermal Oxidation

1. METHODS OF OXIDE FORMATION

An SiO_2 layer can be formed on a silicon surface by various methods, for example, by thermal oxidation at elevated temperature, reactive sputtering, anodic deposition, or the pyrolytic decomposition of oxysilane compounds. The most important methods are:

(a) Thermal oxidation:
  This is the most frequently used and most convenient method of oxide formation. Thermal oxidation in an oxygen ambient (dry oxidation) and in a water ambient (wet oxidation) are distinguished. The chemical reactions taking place in these cases are

  $$Si_{(solid)} + O_2 \longrightarrow SiO_{2(solid)}$$

  for dry oxidation or

  $$Si_{(solid)} + 2H_2O \longrightarrow SiO_{2(solid)} + 2H_2$$

  for wet oxidation. During thermal oxidation part of the silicon surface is used up by the oxide; i.e., if the final oxide thickness is x_o, silicon of thickness $0.4x_o$ is converted to oxide. Since the resulting silicon surface is never exposed to the ambient this is of considerable importance in obtaining extremely clean Si–SiO_2 interfaces. Thermal oxidation is usually taking place in an oxidation furnace (i.e., in a quartz tube) at temperatures above 1000°C in an oxygen-containing environment.

(b) Anodic oxidation:
  This is an electrolytic method at low temperature resulting in the migration of ions under the influence of a voltage. In a constant-voltage anodic oxidation, oxide thickness increases negative-exponentially, oxidation rate and efficiency decrease with time and thickness. Although the advantage of anodic oxidation over thermal oxidation is the lower temperature required in the former case preventing unwanted impurity profile variations in the silicon, the disadvantages of limited maximum oxide thickness ($x_o < 0.3$ μm) and a higher density of interface states make anodic oxidation less attractive than thermal oxidation.

(c) Vapor-phase oxide deposition:
  SiO_2 can be formed by evaporation in vacuum, by sputtering, and by various vapor-phase chemical reactions. The simplest method of vapor-

phase deposition is thermal decomposition of a silicon-organic compound at elevated temperature. Oxidation of silane (SiH_4) or reaction of silicon-halogen compounds (e.g., $SiCl_4$) with water vapor at elevated temperature also results in SiO_2 formation. SiO_2 films obtained by vapor-phase reaction are characterized by a high deposition rate, but the oxide is usually porous and a high density of interface states is observed unless a special post-deposition heat treatment is applied. The structure of vapor-deposited oxide is such that there is very little short-range order so that the composition is given as SiO_x, where $x = 1$ to 2.

For semiconductors other than silicon the conventional methods of oxide formation are by anodic deposition and by vapor-phase deposition.

2. THERMAL OXIDATION

Thermal oxidation is the principal method of oxide formation on a silicon surface. The thermal oxidation of silicon can be divided into three temperature ranges:

(a) $T \approx 25 - 500°C$:
A very thin oxide (<100 Å) forms at a logarithmic rate.
(b) $T \approx 700 - 900°C$:
Oxidation takes place at a linear rate.
(c) $T > 1000°C$:
Oxidation takes place at a parabolic rate.

At elevated temperature the activation energy associated with the oxidation reaction and the diffusion is supplied by thermal energy. If the oxidizing gas molecules are in an excited state or are ionized, then less thermal energy is needed.

Various impurities, mostly sodium and water, participate in the thermal oxidation of silicon. Some of the impurities contribute to the disorder at the $Si–SiO_2$ interface region; they can also contribute to the ion migration and interface instability. Part of this disorder is a result of the oxidation process itself because some bonds remain unsaturated or defects (e.g., trivalent silicon, oxygen vacancies, etc.) are present in the oxide close to the interface; the disorder has an influence on the density of interface states. It is less for lower final oxidation rates and higher temperatures. These interface states (which can be eliminated by high-temperature annealing) can be divided into the following groups:

(a) Donor states close to or above the conduction band:
These states are permanently ionized and represent a fixed positive charge.

(b) Donor or acceptor states within the energy gap:
Impurities in the oxide may contribute under certain conditions to these states.
(c) Hole traps close to the valence band:
These states can trap holes from the valence band causing slow-trapping instability. Sodium in the oxide does not contribute to these states.

Silicon dioxide acts as a protective layer against contamination of the silicon surface (i.e., of the Si–SiO$_2$ interface) since the molar volume of the amorphous SiO$_2$ is about 2.2 times larger than that of Si. Therefore, the oxide not only masks against certain impurities, but also (unless cracks and other defects develop) retards further oxidation which is taking place at the Si–SiO$_2$ interface (and not at the oxide surface). The oxidation mechanism is determined by how fast the reacting species is supplied to the interface and by the interface reaction. These, in turn, depend upon the nature and concentration gradient of the diffusing species, electric field in the oxide (which is due to adsorption of ions, space charges, and applied voltage), imperfections in the oxide, etc. The interaction among these factors determines the structure of oxide and interface.

There are three methods of thermal oxide growth.

(a) Dry oxidation:
Dry oxidation is the formation of an SiO$_2$ layer at the semiconductor surface due to the controlled flow of oxygen over the surface. For dry oxidation at $T > 1100°C$ the oxide thickness is proportional to $t_o^{1/2}$ as follows:

$$x_o = 4.6 t_o^{1/2} \exp(-E_a/kT), \qquad (4.1)$$

where the activation energy $E_a = 0.66$ eV; x_o is given in μm and t_o in min. The term $[4.6 \exp(-E_a/kT)]$ corresponds to the parabolic oxidation rate constant.

(b) Wet oxidation:
Wet oxidation is the formation of an SiO$_2$ layer if oxygen is flowing through water at elevated temperature before reaching the semiconductor surface, thus carrying water molecules to the surface. The water temperature is usually slightly below the boiling point to prevent undue depletion. For wet oxidation the oxide thickness is proportional to $t_o^{1/2}$ and to T_B^2.

(c) Steam oxidation:
Steam oxidation is accomplished by the use of water steam flowing directly over the semiconductor surface.

In practice, the growth of an oxide layer is accomplished by either dry or wet oxidation or by their sequential combination. The use of steam oxidation is less favored because of poor oxide quality owing to the etching action of the excess water. The oxide density is a function of oxidation process and oxidation temperature. The thickness of an oxide layer formed by thermal oxidation is approximately 2.3 times the thickness of the consumed silicon.

Table 4.1 compares oxide densities and dielectric strengths for various oxidation conditions.

TABLE 4.1
Characteristics of Thermal SiO_2

Oxidation method	Density [g/cm³]		Dielectric strength [V/μm]	
	1000°C	1200°C	1000°C	1200°C
Dry O_2	2.27	2.15	550	515
Wet O_2	2.18	2.21	525	535
Steam	2.08	2.05	500	490

3. CHEMISTRY AND KINETICS OF THERMAL OXIDATION

The oxidizing species (i.e., either oxygen or water molecules) moves through the growing oxide layer until reaching the receding silicon surface; consequently, the oxidation process proceeds at a decreasing rate as the oxide thickness increases.

(a) Dry oxidation:
   The oxidizing species are oxygen ions. Permeability of oxygen in SiO_2 is assumed to be insignificant. One molecule of oxygen is used to form one molecule of SiO_2:

$$Si + O_2 \longrightarrow SiO_2.$$

(b) Wet oxidation:
   The oxidizing species are water molecules. Two molecules of water are used to form one molecule of SiO_2:

$$Si + 2H_2O \longrightarrow SiO_2 + 2H_2.$$

The hydrogen molecule moves rapidly through the oxide layer and enters the ambient. Wet oxidation proceeds in three stages as follows:
   (i) Water vapor reacts with the bridging oxygen ions and forms non-bridging OH groups; this results in a weakening of the SiO_2 network.
   (ii) At the Si–SiO_2 interface, the OH groups react with the silicon lattice and form SiO_2 polyhedra and hydrogen.

(iii) Hydrogen leaves the SiO_2 and reacts further with bridging oxygen ions in the SiO_2 network to form OH groups, resulting in a further weakening of the SiO_2 structure.

The maximum solid solubility of oxygen and water molecules in SiO_2 at 1000°C is:

$C_{B\,max} = 5.2 \cdot 10^{16}$ cm^{-3} for oxygen

$C_{B\,max} = 3.0 \cdot 10^{19}$ cm^{-3} for water (at 1 atm).

The kinetics of thermal oxidation of a semiconductor is given by the diffusion of the oxidizing species through the growing oxide layer. Transport of the species occurs by diffusion and drift. The flux density of the oxidant arriving at the oxide-semiconductor interface is

$$F_O = D_O(\partial C_O/\partial x) \approx D_O(C'_O - C''_O)/x_o, \qquad (4.2)$$

where D_O is the effective diffusion coefficient of the oxidizing species in the oxide, C'_O and C''_O are the oxidant concentrations at the two surfaces of the oxide, and $C_O(x)$ is the oxidant concentration within the oxide at distance x from the outer oxide surface. At the semiconductor surface the species enters into a chemical reaction with it which proceeds at a rate proportional to the concentration of the species and which results in the formation of the semiconductor oxide.

4. GENERAL RELATIONSHIP FOR THERMAL OXIDATION

The thermal oxidation of silicon can generally be described by the equation

$$x_o^2 + Ax_o - B(t_o + t^*) = 0 \qquad (4.3)$$

with the solution

$$x_o = \frac{A}{2}\left[\left(\frac{t_o + t^*}{A^2/4B} + 1\right)^{1/2} - 1\right], \qquad (4.4)$$

where the terms A and B are functions of oxidation conditions and material properties. This equation has been plotted in Figure 4.2 (dashed curve) as

$$\frac{x_o}{A/2} \quad \text{vs.} \quad \frac{t_o + t^*}{A^2/4B}$$

where x_o is the resulting oxide thickness and t_o is the oxidation time. The oxidation rate constants A and B are given in Figure 4.3. For very short and very long oxidation times, above equation can be reduced as follows:

(a) short oxidation time, i.e., $t_o \ll A^2/4B$:

$$\frac{x_o}{A/2} \approx \frac{1}{2}\frac{t_o + t^*}{A^2/4B} \qquad (4.5a)$$

or
$$x_o \approx (B/A)(t_o + t^*) \tag{4.5b}$$

and further, if $t_o \gg t^*$,
$$x_o \approx (B/A)t_o; \tag{4.5c}$$

(b) long oxidation time, i.e., $t_o \gg A^2/4B$:
$$\frac{x_o}{A/2} \approx \left(\frac{t_o + t^*}{A^2/4B}\right)^{1/2} \tag{4.6a}$$

or
$$x_o \approx [B(t_o + t^*)]^{1/2} \tag{4.6b}$$

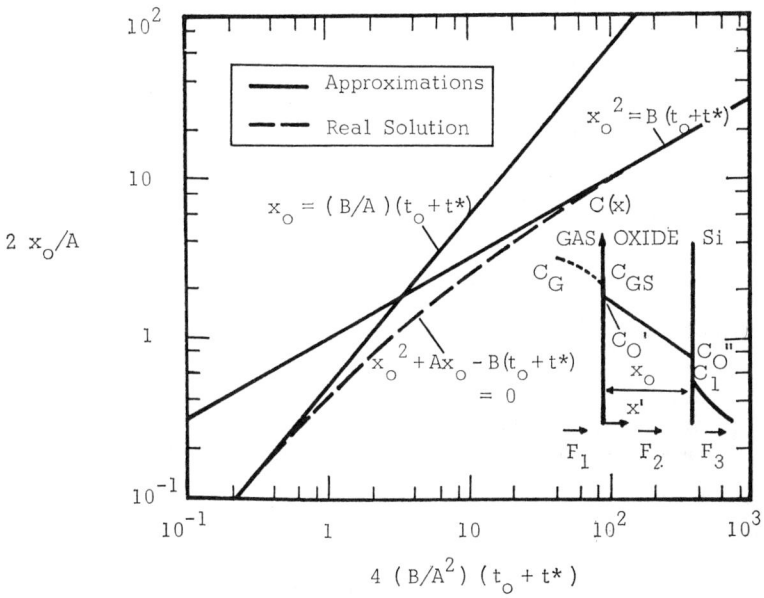

FIGURE 4.2

General relationship for the thermal oxidation of silicon. Oxide thickness (x_o) vs. oxidation rate constants (A and B). and oxidation time (t_o). The quantity t^* represents an initial oxidation time (see text). The inset defines impurity concentrations in semiconductor, oxide, and ambient. n-type and p-type Si of $\langle 111 \rangle$ orientation. The flux of the oxidant in gas, oxide, and semiconductor, respectively, is, assuming the direction of gas flow normal to the plane of the paper,

$$F_1 = h_G(C_G - C_{GS}); F_2 = D_o(C_o' - C_o'')/x_o; F_3 = k_S C_o''.$$

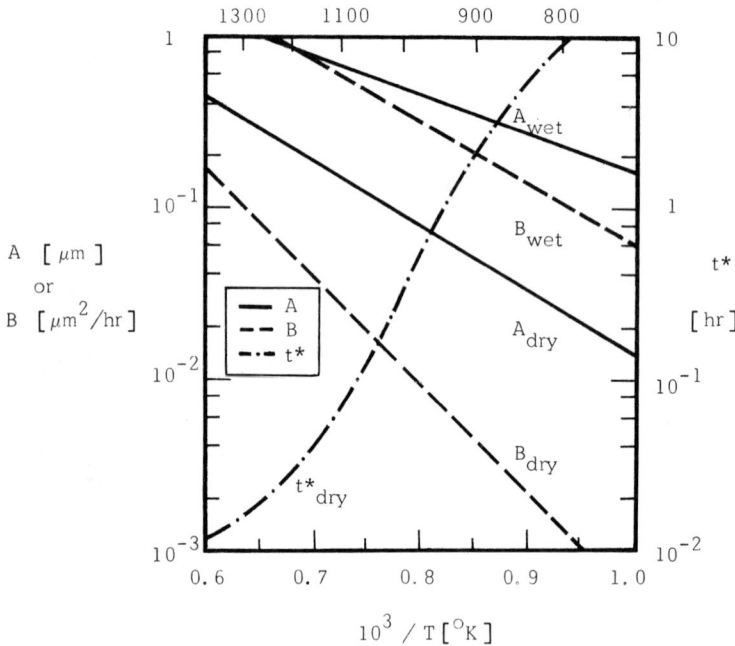

FIGURE 4.3

Linear (A) and parabolic (B) oxidation rate constants and initial oxidation time (t^*) vs. oxidation temperature (T). The parabolic rate constant for wet oxidation has been corrected for 760 Torr. The initial oxidation time for wet oxidation $t^*_{wet} = 0$ at all temperatures.

and further, if $t_o \gg t^*$,

$$x_o \approx (Bt_o)^{1/2}. \tag{4.6c}$$

The dashed curve of Figure 4.2 is a composite of the two limiting cases.

5. RATE OF OXIDE GROWTH

The flux of oxidant reaching the semiconductor-oxide interface is

$$F_O = N_0(dx_o/dt) = \frac{k_S C^*}{1 + k_S x_o/D_O + k_S/h_S} \tag{4.7}$$

where N_O is the number of oxidant molecules incorporated into a unit volume of oxide.

$$N_O = 2.2 \cdot 10^{22} \text{ cm}^{-3} \quad \text{for dry oxidation,}$$
$$N_O = 4.4 \cdot 10^{22} \text{ cm}^{-3} \quad \text{for wet oxidation.}$$

The solution of the above equation yields

$$x_o^2 + 2D_O(1/k_S + 1/h_S)x_o - 2(t_o + t^*)D_O C^*/N_O = 0, \tag{4.8}$$

which corresponds to

$$x_o^2 + Ax_o - B(t_o + t^*) = 0. \tag{4.3}$$

Furthermore,

$$\frac{B}{A} = \frac{k_S h_S}{k_S + h_S} \frac{C^*}{N_O}. \tag{4.9}$$

The gas-phase mass-transfer coefficients h_S and h_G are related by

$$h_G/h_S = H_G kT. \tag{4.10}$$

6. OXIDANT CONCENTRATION WITHIN OXIDE

Under steady-state conditions ($F_1 = F_2 = F_3 = F_O$; see inset of Figure 4.2) the concentrations of the oxidant at the two outside interfaces of the oxide

$$C_O' = \frac{(1 + k_S x_o/D_O)C^*}{1 + k_S x_o/D_O + k_S/h_S} \tag{4.11}$$

$$C_O'' = \frac{C^*}{1 + k_S x_o/D_O + k_S/h_S} \tag{4.12}$$

and

$$\frac{C_O'}{C_O''} = 1 + \frac{k_S x_o}{D_O}; \tag{4.13}$$

where D_O is the diffusivity of the oxidant in the oxide.
Limiting cases:

(a) $D_O \approx 0$:
In this case
$$C_O' \to C_G \quad \text{and} \quad C_O'' \to 0.$$
This is the diffusion-controlled case.

(b) $D_O \gg 0$:
In this case
$$C_O' = C_O'' = C^*/(1 + k_S/h_S).$$
This is the reaction-controlled case.

In the above equations the following terms have been used:
k_S = chemical surface reaction rate constant for oxidation
h_G, h_S = gas-phase mass-transfer coefficient in terms of concentration in the gas or solid, respectively
H_G = Henry's Law constant; $H_G \approx 10^{25}$ atm/cm^3.

The impurity concentrations are defined as follows:
C_B = impurity concentration in the semiconductor bulk
C_2 = final impurity concentration at semiconductor surface
$\overline{C_O}$ = average impurity concentration in oxide
C'_O, C''_O = impurity concentration at outside or inside oxide surface, respectively; $C'_O \approx 0$ if the impurity has the tendency to escape from solid into ambient
C^* = equilibrium concentration of oxidant in oxide;

$$C^* \approx \frac{C'_O + C''_O}{2} = \frac{C'_O}{2}\left(1 + \frac{1}{1 + k_S x_o / D_O}\right) = H_G p_G. \qquad (4.14)$$

7. OXIDATION RATE CONSTANTS AND TIME CONSTANTS

(a) Rate constants:

Because of their presence in the limiting cases for very short and very long oxidation times which correspond to linear and parabolic oxidation rates, respectively, the terms A and B are referred to as linear or parabolic oxidation rate constants. Values of A and B are given in Figure 4.3. The rate constants can be described in terms of an effective diffusion coefficient (D_O) which incorporates effects of ionic space charges in enhancing the transport rate:

$$A = 2D_O \left(\frac{1}{k_S} + \frac{1}{h_S}\right) \qquad (4.15)$$

$$B = 2D_O \frac{C^*}{N_O} \qquad (4.16)$$

Usually

$$D_O = 2D_o \qquad (4.17)$$

where D_o is the diffusion coefficient of oxygen in SiO$_2$. Because of the temperature dependence of the effective diffusion coefficient, the rate constants display an exponential temperature dependence which can be described by

$$A = A^o \exp(-E_{aA}^o / kT) \qquad (4.18a)$$

$$B = B^o \exp(-E_{aB}^o / kT) \qquad (4.18b)$$

where A^o and B^o are process- and slightly temperature-dependent coefficients and E_{aA}^o and E_{aB}^o are activation energies which are given in Table 4.2 for silicon oxidation conditions at 1 atm pressure.

TABLE 4.2
Rate Constants and Activation Energies

Ambient	A^o [μm]	B^o [μm^2/hr]	E_{aA}^o [eV]	E_{aB}^o [eV]
O_2	$8.8 \cdot 10^{-5}$	760	0.76	1.24
H_2O	$7.2 \cdot 10^{-9}$	20	1.90	0.70
$O_2 + Na$	$5.5 \cdot 10^{-2}$	220	0.07	1.32

(b) Time constants:
The time constant t^* is defined as

$$t^* = (x_i^{*2} + Ax_i^*)/B \tag{4.19}$$

and corresponds to a shift in the time coordinate which corrects for the presence of an initial oxide layer, x_i^*. The quantity x_i^* can be regarded as the oxide thickness grown before the approximations given become valid.
For wet oxidation: $x_i^* \approx 0$.
For dry oxidation: $x_i^* \approx 200$ Å.
Hence, $t_{wet}^* = 0$.

8. DEPENDENCE OF RATE CONSTANTS ON IMPURITY CONCENTRATION, CRYSTAL ORIENTATION, AND PARTIAL PRESSURE

(a) Impurity concentration:
At semiconductor impurity concentrations below $C_B = 10^{19}$ cm^{-3} the oxidation rate constants as given in Figure 4.3 apply. At higher impurity concentrations an enhanced rate constant B is observed. Figure 4.4 gives the appropriate multiplication factor (f_R). It is defined as

$$B_{true} = f_B B, \tag{4.20}$$

i.e., values of B as taken from other illustrations have to be multiplied by f_B in order to take near-degenerate conditions into account. Rate constant A is relatively insensitive to impurity concentration variations. The average impurity concentration in oxide is

$$\overline{C_O} = C_O' + \left(\frac{C_2(x_*, t)}{m} - C_O'\right)\left(1 + \frac{i \operatorname{erfc}[(B/4D_o)^{1/2}] - \pi^{1/2}}{(B/4D_o)^{1/2}}\right) \tag{4.21}$$

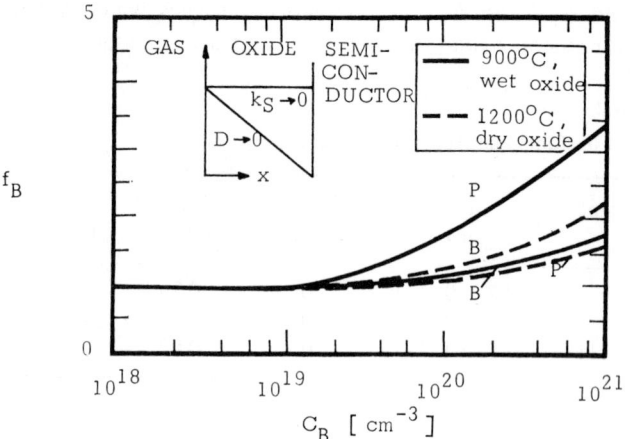
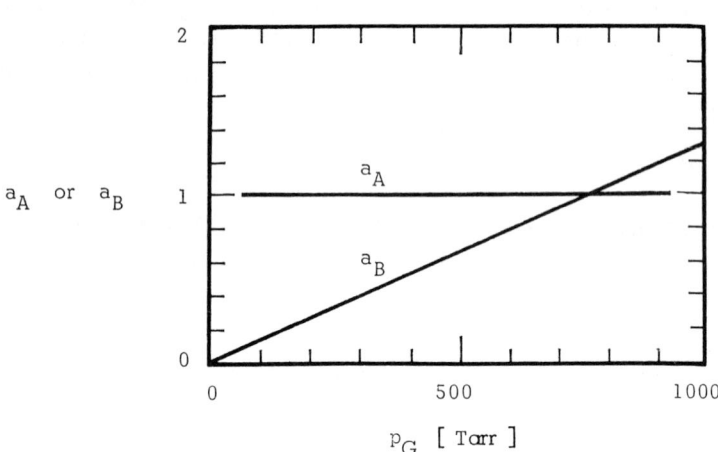

FIGURE 4.4

Correction factors (f_B; a_A, a_B) by which the oxidation rate constants (A and B) have to be multiplied vs. semiconductor impurity concentration (C_B) and partial pressure of the oxidant (p_G). n-type or p-type silicon, $\langle 111 \rangle$ orientation. The inset shows the distribution of the oxidant for reaction-controlled ($k_S \to 0$) and diffusion-controlled ($D \to 0$) oxidations.

The term i erfc indicates the integral of the complementary error function. Extreme cases:
(i) $B/4D_o = \infty$: $\overline{C_o} = C_2(x_*, t)/m$.
(ii) $B/4D_o = 0$: $\overline{C_o} = C'_o$.

For 1200°C dry oxidation, $B/4D_o$ is a maximum and for 900°C wet oxidation, $B/4D_o$ is a minimum. These conditions approach the extreme cases within the temperature range of Figure 4.4.

(b) Crystal orientation:

The crystallographic orientation of the silicon substrate affects the dependence of oxidation rate upon surface-state density; this influence can be attributed to a greater tendency to order in the interface region of $\langle 100 \rangle$ silicon than of $\langle 111 \rangle$ silicon. The orientation dependence of the surface-state density is the same as that of the linear oxidation rate (in the range where oxide growth is surface-controlled); i.e., surface-state density and oxidation rate increase in the order $\langle 100 \rangle$, $\langle 110 \rangle$, $\langle 111 \rangle$ approximately in the ratio of 1 : 2 : 3.

The oxidation reaction takes place at the Si–SiO$_2$ interface and is between a silicon bond and a water molecule which migrates from an interstitial site in the SiO$_2$. The linear oxidation rate increases in the same order as the available bond density and in reverse order as the surface energy of the crystallographic planes which are given in Table 4.3.

TABLE 4.3

Bond Density and Surface Energy for Different Silicon Orientations

Orientation	Bond density [10^{14} cm^{-2}]	Surface energy [10^{14} eV/cm^2]
$\langle 100 \rangle$	6.77	13.30
$\langle 110 \rangle$	9.38	9.44
$\langle 111 \rangle$	11.76	7.64

(c) Partial pressure:
  (i) *Linear rate constant (A)*:
     The linear rate constant is independent of partial pressure.
  (ii) *Parabolic rate constant (B)*:
     The parabolic rate constant is proportional to the equilibrium concentration of the oxidant in the oxide (C^*) which is (Henry's Law)

$$C^* = H_G p_G, \qquad (4.22)$$

i.e., B is dependent upon partial pressure assuming no association or dissociation of the oxidant at the outer surface of the oxide.

9. OXIDATION COEFFICIENT

The oxidation coefficient O_s (which has the same dimension as the diffusion coefficient) is given in Table 4.4 for wet oxidation of silicon (see inset of Figure 4.6). It is defined as

$$O_s = d(x_o^2)/dt_o \qquad (4.23)$$

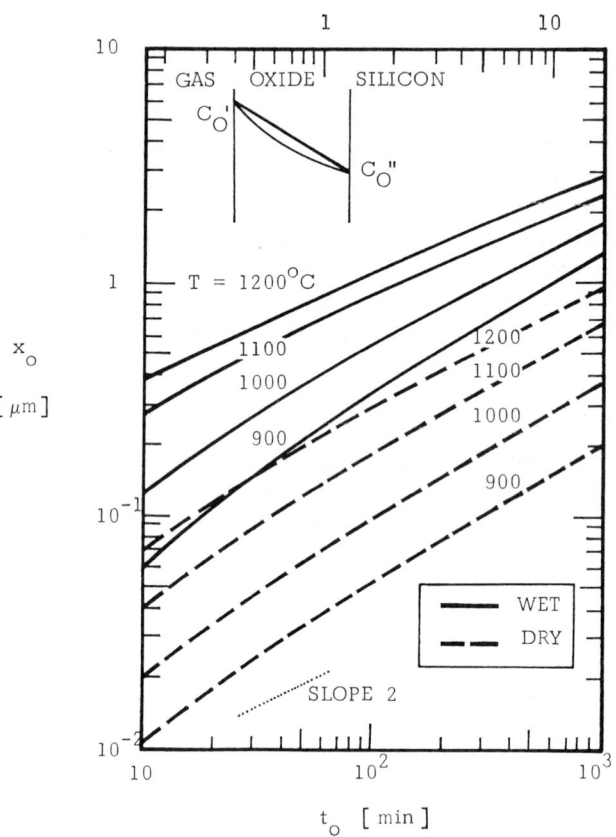

FIGURE 4.5

Oxide thickness (x_o) as function of oxidation time (t_o) and oxidation temperature (T) for the thermal oxidation (dry and wet, i.e., 95°C H_2O) of silicon of ⟨111⟩ orientation. The inset shows the distribution of the oxidizing species during oxidation.

and gives the variation of the oxidation rate with oxidation time. The oxide thickness as a function of oxidation conditions is shown in Figures 4.5 and 4.6.

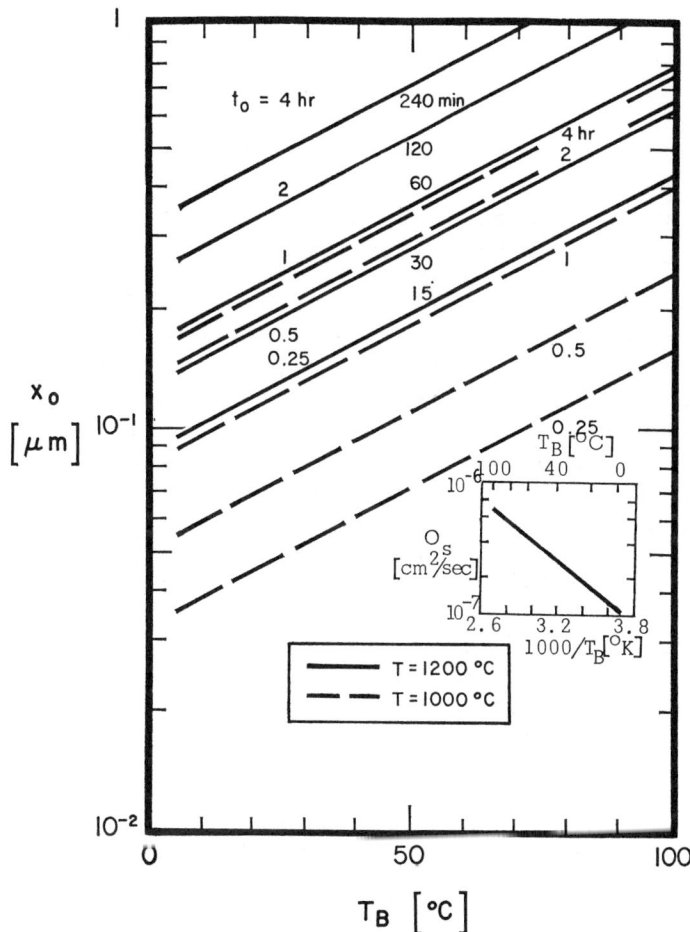

FIGURE 4.6

Oxide thickness (x_o) vs. water bath temperature (T_B) and oxidation time (t_0) for wet oxidation of silicon at 1000 and 1200°C. The inset shows the variation of the oxidation coefficient (O_s) with water bath temperature.

TABLE 4.4
Oxidation Coefficient

T_B [°C]	$d(x_o^2)/dt_o$ [cm²/sec]
dry O_2	$1.5 \cdot 10^{-7}$
25	$1.9 \cdot 10^{-7}$
85	$5.6 \cdot 10^{-7}$
95	$6.5 \cdot 10^{-7}$
97	$6.6 \cdot 10^{-7}$

10. REDISTRIBUTION OF IMPURITIES DURING SEMICONDUCTOR OXIDATION

Due to the inward motion of the semiconductor–oxide interface during thermal oxidation, a redistribution of impurities originally contained in the semiconductor region being oxidized takes place. The impurity profile in semiconductor and oxide in the proximity of their interface is a function of the oxidation rate constant, the diffusion coefficient of the impurity in the semiconductor, and the segregation coefficient. Depending upon whether impurity atoms can escape from the oxide surface into the ambient or not, two cases of redistribution are distinguished.

(a) No escape of impurities from oxide at oxide-ambient interface:

$$\frac{C_2(x_*, t)}{C_B} = \frac{2b^3 B/4D_s}{b - 1/m + (2/m)b^2 B/4D_s}$$

$$= \frac{2b^3 r_3^2}{b - 1/m + (2/m)b^2 r_3^2}$$

$$= \frac{1 + r_1(C_0'/C_B)}{1 + \pi^{1/2}(1/m - b)r_3 \exp(b^2 r_3^2)\, \text{erfc}\,(br_3) + r_1/m} \qquad (4.24)$$

where $r_1 = \dfrac{4 \exp\,[(b^2 r_2 - 1)r_3^2/r_2]\,\text{erfc}\,(br_3)}{\text{erf}\,(r_4)}$

$r_2 = D_o/D_s$

$r_3 = (B/4D_s)^{1/2}$

$r_4 = (B/4D_o)^{1/2}$.

4-2 THERMAL OXIDATION 357

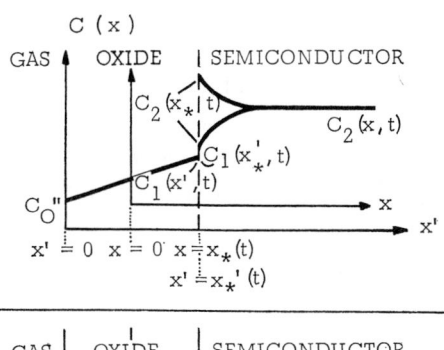

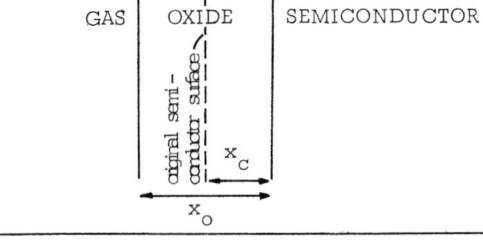

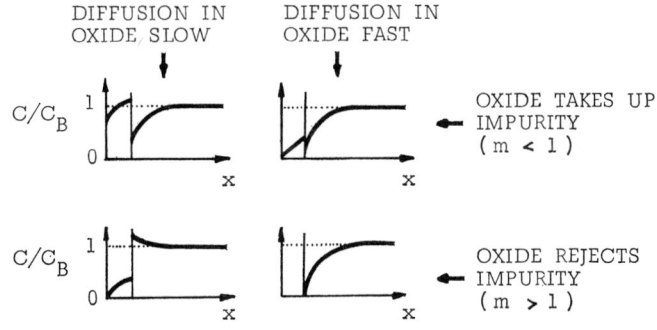

FIGURE 4.7
Definition of terms used in the redistribution of impurities near the Si-SiO$_2$ interface.

Since in most practical cases for the Si–SiO$_2$ system

$$D_o \ll D_s, \quad r_1 = 0, \quad r_2 = 0; \quad b \approx 0.44,$$

the above equation can be reduced to

$$\frac{C_2(x_*, t)}{C_B} = \frac{1}{1 + \pi^{1/2}(1/m - 0.44)r_3 \exp(0.19r_3^2) \operatorname{erfc}(0.44r_3)} \quad (4.25)$$

(b) Finite escape of impurities from oxide at oxide–ambient interface:
In this case the impurity concentration in the oxide at the oxide–ambient interface (C'_0) is constant, i.e., the impurity concentration in the oxide is in equilibrium with the solute in the ambient.

$$\frac{C_2(x_*, t)}{C_B} = \frac{\exp(-b^2 B/4D_s)}{\exp(-b^2 B/4D_s) - \pi^{1/2}(b - 1/m)(B/4D_s)^{1/2} \operatorname{erfc}[(b^2 B/4D_s)^{1/2}]}$$

$$= \frac{\exp(-b^2 r_3^2)}{\exp(-b^2 r_3^2) - \pi^{1/2}(b - 1/m)r_3 \operatorname{erfc}(br_3)} \quad (4.26)$$

The impurity redistribution in the semiconductor is limited to a distance

$$(x - x_*)/2\sqrt{D_s t} < 2,$$

i.e., to a region less than $4\sqrt{D_s t}$ away from the semiconductor–oxide interface.

The velocity at which the semiconductor–oxide interface moves into the semiconductor while it is being oxidized is

$$v_o = (1 - b)\, dx_o/dt \approx 0.6\, dx_o/dt \quad (4.27)$$

Assumptions:

(a) Initial impurity distribution is uniform.
(b) Impurity concentration at oxide–ambient interface (C'_0) is constant. If the impurity has a tendency to escape into the ambient, $C'_0 \approx 0$.
(c) Semiconductor sample is thick enough so that the redistribution does not affect the bulk.
(d) The impurity concentrations at the two sides of the semiconductor–oxide interface are a function of the segregation coefficient.

Explanation of other terms: $C_1(x', t)$ is the impurity concentration in the oxide layer, $C_2(x, t)$ is that in the semiconductor. The semiconductor–oxide interface is at $x = x_*(t)$ and $x' = x'_*(t)$, where x designates the coordinate system within the semiconductor (with the original semiconductor surface at $x = 0$) and x' that within the oxide (with the oxide surface at $x' = 0$). The impurity concentration at the resulting silicon surface after oxidation (C_S) is equivalent to the term $C_2(x_*, t)$. The average impurity concentration in the oxide layer after impurity redistribution is denoted as $\overline{C_0}$.

D_o, D_s = impurity diffusion coefficient in oxide and semiconductor, respectively

B = parabolic oxidation rate constant

b = thickness of that portion of the semiconductor which is being oxidized to total thickness of the oxide produced; $b = x_c/x_o$; specifically, $b = 0.44$ for dry oxidation of silicon, $b = 0.41$ for wet oxidation of silicon.

The general expression of relating semiconductor surface concentration after oxidation to original semiconductor impurity concentration can be simplified for boron and phosphorus in silicon as follows.

$$C_2(x_*, t)/C_B = 1 + B_m B \exp(0.05 B^2/D_s) \operatorname{erfc}(0.22 B/\sqrt{D_s})/\sqrt{D_s}. \quad (4.28)$$

For boron ($m = 0.3$): $B_m = 2.57$.
For phosphorus ($m = 10$): $B_m = 0.29$.

Figure 4.8 is based on these equations. The curves of this illustration may be used to determine the impurity concentration ratio at the silicon surface after and before oxidation without knowledge of $B/4D_s$ by reading the impurity ratio at the appropriate oxidation conditions for boron or phosphorus.

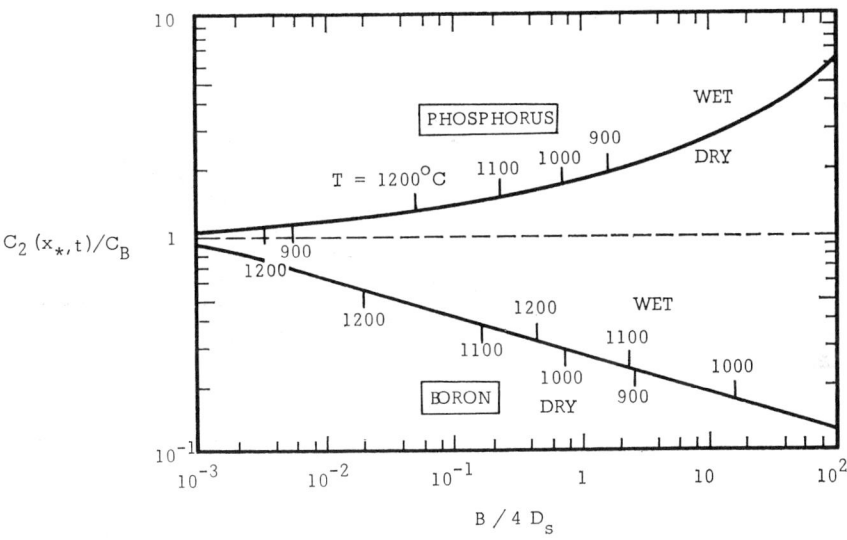

FIGURE 4.8
Ratio of surface concentration at Si-SiO$_2$ interface (C_2) to semiconductor bulk concentration (C_B) as function of parabolic rate constant (B), impurity diffusion coefficient in the semiconductor (D_s), and oxidation temperature (T). No escape of impurities from oxide surface into the ambient. Phosphorus and boron in silicon of impurity concentration $C_B = 10^{16}$ cm^{-3}.

11. SEGREGATION COEFFICIENT

The segregation coefficient (m) is defined as the ratio of equilibrium concentration of the impurity in the semiconductor to that in the oxide at the semiconductor–oxide interface:

$$m = C_2(x_*, t)/C_1(x'_*, t) \quad (4.29)$$

(a) $m < 1/b$:

The oxide contains more impurity atoms than the original semiconductor and the impurity concentration in the immediate vicinity of the interface will be depleted.

(b) $m > 1/b$:

The oxide contains less impurity atoms than the original semiconductor and impurity atoms will be piled up in the semiconductor near the interface.

The segregation coefficient is temperature-dependent. The impurity distribution after oxidation as a function of the segregation coefficient is shown in Figure 4.9. The range of m for common impurities in silicon is given in Table 4.5 which also indicates if impurity pile-up or depletion at the oxide-semiconductor interface takes place during thermal oxidation.

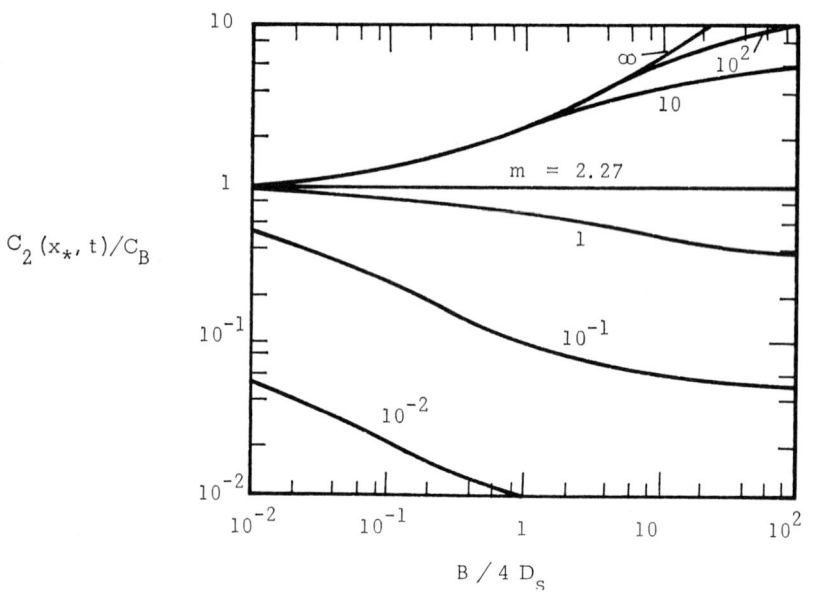

FIGURE 4.9

Ratio of impurity concentration at the semiconductor surface after thermal oxidation (C_2) to semiconductor bulk concentration (C_B) as function of parabolic rate constant (B), impurity diffusion coefficient in the semiconductor (D_s), and segregation coefficient (m). Si-SiO$_2$ system. Curves were calculated for dry and wet oxidation of n-type and p-type silicon between 700 and 1300°C. No escape of impurities from the oxide surface into the ambient is assumed.

TABLE 4.5

Segregation Coefficient of Impurities in Silicon

Impurity	m	Impurity redistribution
Al	$<10^{-3}$	depletion
As	10 to 10^3	pile-up
B	10^{-2} (to 10^3)	depletion
Ga	$>10^3$	pile-up
In	$>10^3$	pile-up
P	10 to 10^3	pile-up
Sb	10 to 10^3	pile-up

12. SURFACE EFFECTS

Thermal oxidation of a silicon surface results in the following surface effects:

(a) a vacancy concentration far above the thermal equilibrium concentration will be created at the surface;
(b) most of the unsaturated bonds at the surface will be tied up resulting in a reduction of the surface state density;
(c) quenching the interface will cause super-saturation and aggregations which under compression will vanish by the creation of dislocations and interaction with them;
(d) after stress annihilation, additional vacancies will be created.

13. HENRY'S LAW

Generally, Henry's Law states that the solubility of a gas in a liquid or solid is directly proportional to the pressure of the gas above liquid or solid at a given temperature. Specifically for the thermal oxidation of a semiconductor, in equilibrium the concentration of a species within the semiconductor is proportional to the partial pressure of that species in the surrounding ambient; i.e., the concentration at the outer surface of oxide (C_O') is proportional to the partial pressure of the oxidant at the oxide–ambient interface (p_S)

$$C_O' = H_G p_S \tag{4.30}$$

where the proportionality factor H_G is Henry's Law constant.

14. REFERENCES

(1) J. R. Ligenza, *J. Phys. Chem.*, **65**, 2011 (1961).
(2) J. R. Ligenza, *J. El. Chem. Soc.*, **109**, 73 (1962).

(3) B. E. Deal, *J. El. Chem. Soc.*, **110**, 527 (1963).
(4) A. S. Grove et al., *J. Appl. Phys.*, **35**, 2695 (1964).
(5) H. Y. Ku, *J. Appl. Phys.*, **35**, 3391 (1964).
(6) B. E. Deal et al., *J. El. Chem. Soc.*, **112**, 308 (1965).
(7) B. E. Deal and M. Sklar, *J. El. Chem. Soc.*, **112**, 430 (1965).
(8) B. E. Deal and A. S. Grove, *J. Appl. Phys.*, **36**, 3770 (1965).
(9) R. P. Donovan, *RTI Report ASD-TDR-63-316*, Vol. VII, June 1965.
(10) W. A. Pliskin, *IBM J. Res. Devel.*, **10**, 198 (1966).
(11) A. S. Grove, *Physics and Technology of Semiconductor Devices*, John Wiley and Sons, New York, 1967.
(12) A. G. Revesz and K. H. Zaininger, *RCA Review*, **29**, 22 (1968).
(13) S. K. Ghandhi, *The Theory and Practice of Microelectronics*, John Wiley and Sons, New York, 1968.
(14) C. H. Lane, *IEEE Transact. Electr. Dev.*, **ED-15**, 998 (1968).
(15) K. E. Bean and P. S. Gleim, *Proc. IEEE*, **57**, 1469 (1969).

4-3 Diffusion in SiO$_2$

1. DIFFUSION MECHANISMS

The diffusion mechanism of impurities in SiO$_2$, an amorphous material, is similar to that in semiconductors, although they are usually crystalline. In analogy to diffusion in a crystal (see 2-1), the diffusion coefficient of an impurity in silicon dioxide can be described by a diffusion coefficient of the form

$$D_o = D_\infty \exp(-E_a^*/kT) \qquad (4.31)$$

where D_∞ is the apparent diffusion coefficient at infinite temperature ($1/T = 0$) and E_a^* is the activation energy of the diffusing impurity. Empirical values of D_∞ and E_a^* for selected impurities are given in Table 4.6.

TABLE 4.6
Diffusion Coefficient and Activation Energy of Impurities in SiO$_2$

Impurity	D_∞ [cm^2/sec]	E_a^* [eV]
H$_2$	$9.5 \cdot 10^{-4}$	0.69
O$_2$	$1.5 \cdot 10^{-2}$	3.09
H$_2$O	$1.0 \cdot 10^{-6}$	0.79
Au	$1.5 \cdot 10^{-7}$	2.14
B	$3.0 \cdot 10^{-6}$	3.50
Na	5.0	1.50
P	$1.0 \cdot 10^{-8}$	1.75

As in silicon, a distinction between interstitial and substitutional impurities is made where these terms refer to a network rather than to a crystalline structure. In addition, the presence of water vapor alters the network.

(a) Interstitial impurities:
These are network modifiers and are the oxides of large metal ions. Na, K, Pb, Ba and sometimes Al fall into this category. They enter the SiO$_2$ network between the polyhedra, give up their oxygen to it, and produce two nonbridging oxygen ions in place of the original bridging ions. This weakens the network and makes it more porous to other diffusing impurity atoms.

364 PROPERTIES OF SILICON DIOXIDE

(b) Substitutional impurities:
These are network formers and replace the silicon in a silicon dioxide polyhedron. Boron and phosphorus and sometimes Al fall into this category; elements of Group III usually reduce the nonbridging ion concentration, elements of Group V increase the number of nonbridging ions.

(c) Water vapor:
The presence of water vapor results in a weakening of the SiO_2 network, making it more porous to diffusing impurity atoms. Water vapor combines with bridging oxygen ions and forms pairs of stable nonbridging hydroxyl groups.

Under the assumption that SiO_2 possesses at least some crystalline structure (it is usually microcrystalline), a Debye temperature T_D and a frequency of lattice vibrations v_o can be defined for the SiO_2 network so that, similar to the diffusion in a semiconductor, the apparent diffusion coefficient D_∞ can be theoretically determined. As in the case of a given semiconductor, the apparent diffusion coefficient should be the same for all impurities and for different diffusion mechanisms whose influences should show only in their activation energies. In practice, considerable deviations have been observed, partially because of measurement errors and partially because the theoretical model used has only limited validity. Theoretical values are as follows:

$$T_D = 470°K$$
$$v_o = T_D(k/h) = 9.8 \cdot 10^{12} \text{ Hz}$$
$$D_\infty = (2/3)v_o a^2 = 1.7 \cdot 10^{-3} \text{ cm}^2/\text{sec}$$

where a is taken as the average distance between silicon and oxygen ions ($a = 1.62$ Å).

Activation energies of interstitial impurities are typically in the order of

$$E_{aI}^* \approx 1.0 \text{ eV}$$

and of substitutional impurities in the order of

$$E_{aS}^* \approx 3.0 \text{ eV}.$$

Theoretically expected diffusion coefficients at 1000°C are therefore

$$D_I = 4.1 \cdot 10^{-7} \text{cm}^2/\text{sec},$$
$$D_S = 2.3 \cdot 10^{-9} \text{ cm}^2/\text{sec};$$

these diffusion coefficients are higher than those experimentally observed.

2. COMPARISON OF DIFFUSION IN Si AND SiO_2

Diffusion coefficients of several impurities in SiO_2 are shown in Figure 4.10. Figure 4.11 relates the diffusion coefficients of selected impurities in Si and SiO_2; the background impurity concentration is assumed to be small. Values for D_s and D_o were taken from Figures 2.1 and 4.10 respectively. Most impurities display an increasing ratio D_s/D_o with increasing temperature, i.e., impurity diffusion in Si will become more significant than diffusion in SiO_2 as the temperature is increased. A notable exception is gold which shows a decreasing ratio of the diffusion coefficients. Gold diffuses very rapidly along a silicon surface or along an Si–SiO_2 interface. The diffusion coefficient of gold in pure SiO_2, however, is relatively small, whereas it is very high in silicon bulk.

The small diffusion coefficient of some elements in SiO_2 is utilized to prevent their penetration of silicon dioxide. Note the high diffusion coefficient of H_2, Na, and H_2O even at low temperatures.

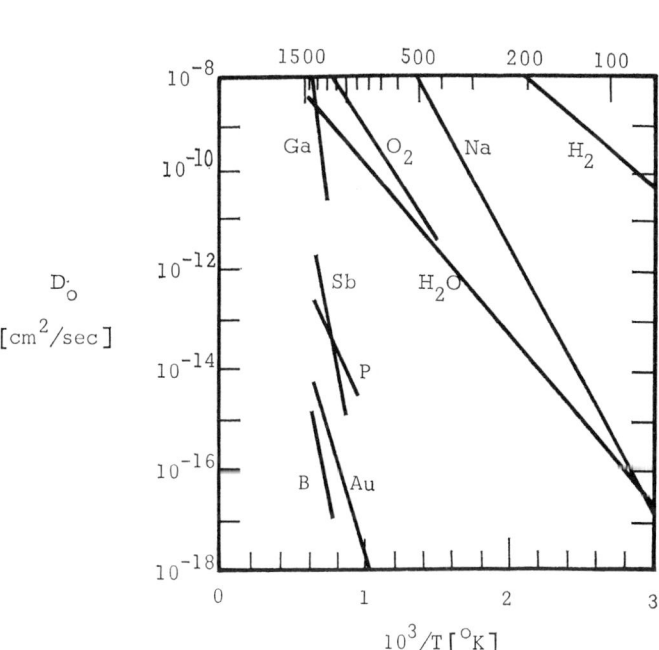

FIGURE 4.10

Diffusion coefficient (D_o) of impurities in thermally grown SiO_2 as function of diffusion temperature (T).

366 PROPERTIES OF SILICON DIOXIDE

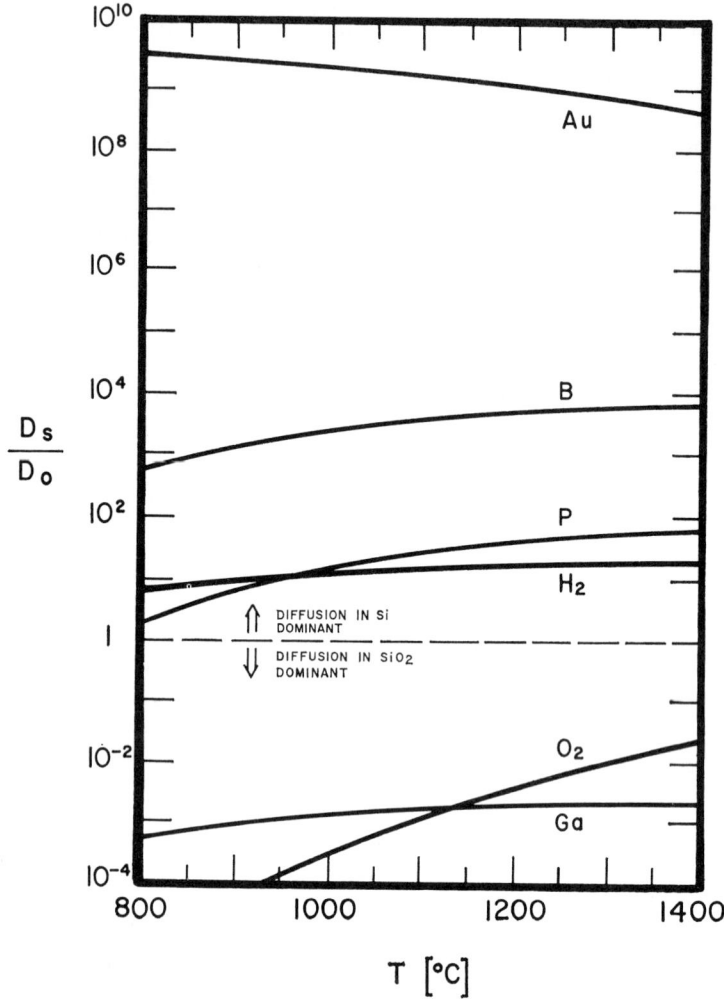

FIGURE 4.11

Ratio of diffusion coefficients of impurities in silicon (D_s) and in SiO_2 (D_o) vs. diffusion temperature (T).

3. MASKING CAPABILITIES OF A PASSIVATING LAYER

If an impurity source is at the surface of a passivating oxide layer, diffusion through the layer will take place according to diffusion laws. The diffusion characteristics can be described by the diffusion coefficient of the impurity in the layer. Depending upon diffusion conditions, diffusion coefficient, and oxide thickness, it is possible to prevent the majority of the impurity atoms

4-3 DIFFUSION IN SiO₂

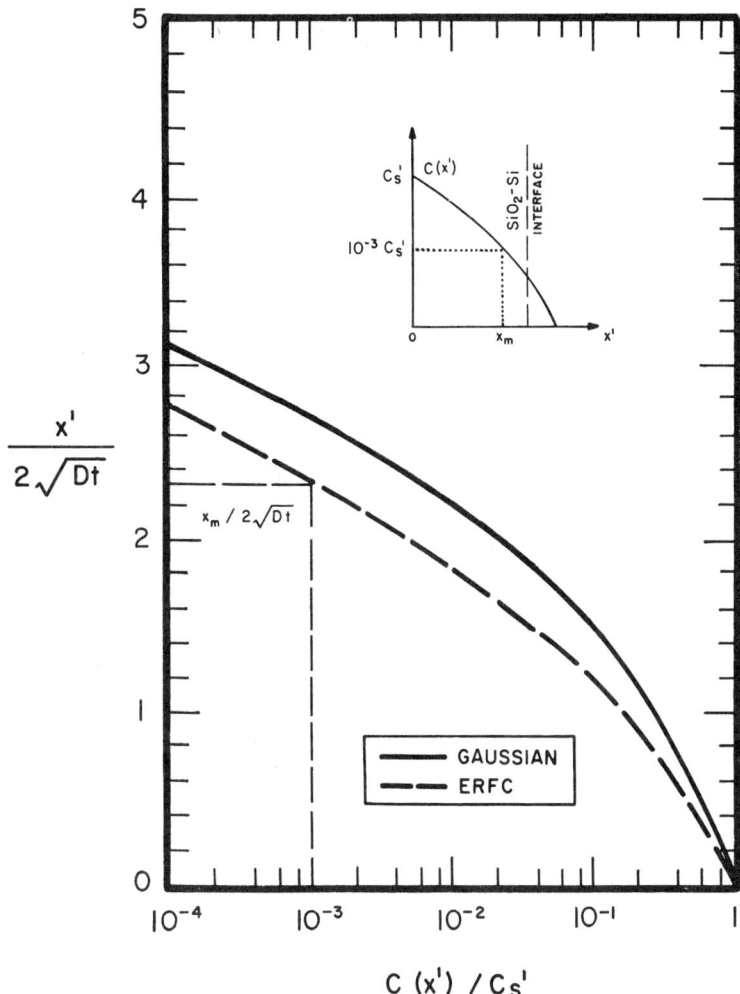

FIGURE 4.12

Theoretical impurity concentration profile in a masking layer. The minimum thickness of the layer (x_m) required to mask against impurity diffusion is the distance from the layer surface (impurity source) at which the impurity concentration has fallen to 10^{-3} of its surface value (C'_s); see inset.

from reaching the semiconductor–oxide interface. In this case the oxide acts as a mask against the impurities.

The impurity profile within the oxide may be determined by either a Gaussian or an erfc-type distribution. Curves of Figure 4.12 have been calculated from:

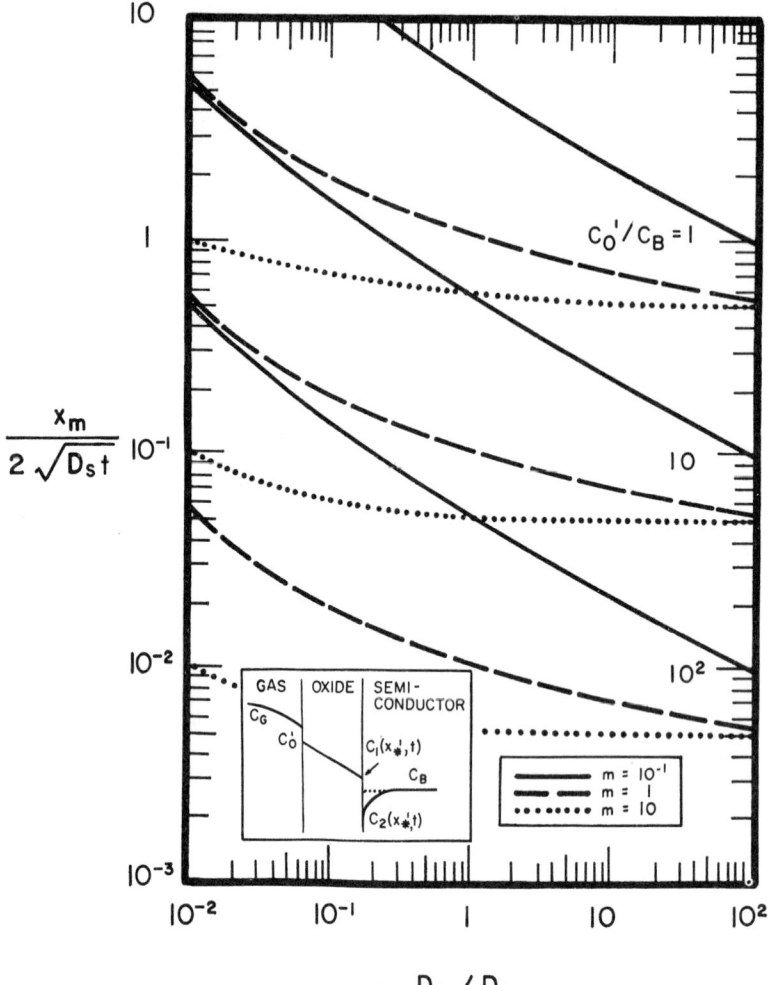

FIGURE 4.13

Normalized oxide thickness (x_m) required to mask against the penetration by impurities vs. ratio of impurity diffusion coefficients in semiconductor and in oxide (D_s/D_o), ratio of impurity concentrations at oxide surface (C_o') and semiconductor bulk (C_B), and impurity segregation coefficient (m). The impurity concentrations C_1 and C_2 refer to the oxide-semiconductor interface and give the concentration in oxide and semiconductor, respectively.

(a) Gaussian distribution:

$$x'/2\sqrt{D_o t} = [\ln C'_S/C(x')]^{1/2}$$
$$\approx \sqrt{2[(C'_S/C(x') - 1)/(C'_S/C(x') + 1)]^{1/2}}; \qquad (4.32)$$

(b) erfc distribution:

$$x'/2\sqrt{D_o t} = (\sqrt{\pi}/2)[1 - C(x')/C'_S] \qquad \text{for } x'/2\sqrt{D_o t} \ll 1 \qquad (4.33a)$$

$$x'/2\sqrt{D_o t} = (1/\sqrt{\pi})[C'_S/C(x')] \exp(-x'^2/4D_o t) \qquad \text{for } x'/2\sqrt{D_o t} \gg 1 \qquad (4.33b)$$

where x' = distance from surface normal to surface
C'_S = impurity concentration at oxide surface
$C(x')$ = impurity concentration at distance x'.

In most cases the masking ability of a passivating layer against impurities can be described by the complementary error function; it allows a theoretical determination of the minimum oxide thickness at which $C(x')$ has decreased to 10^{-3} of its value at the oxide surface, i.e., then

$$C(x') = 10^{-3} C'_S,$$
$$x_m = 4.6\sqrt{D_o t}.$$

The minimum mask thickness is related to appropriate diffusion coefficients, impurity concentrations, and segregation coefficient for arbitrary semiconductor and oxide by

$$\frac{x_m}{2\sqrt{D_s t}} = \frac{1}{2}\left(1 + \frac{1}{m}\frac{D_o}{D_s}\right)\frac{C_B}{C'_O}. \qquad (4.34)$$

This relationship indicates that the required mask thickness increases with increasing ratio D_o/D_s, decreasing segregation coefficient and decreasing oxide surface impurity concentration. Figure 4.13 is based on the above equation.

4. MASKING CAPABILITY OF SiO_2

The masking capability of SiO_2 is related to the diffusion coefficient of the impurity in the oxide (see Figure 4.10). Normally it is sufficient to assume that the diffusion of impurities in SiO_2 follows a complementary error function distribution if the masking capabilities of SiO_2 are used during an impurity predeposition at which the semiconductor and oxide surfaces are kept at a constant impurity concentration. During a subsequent drive-in diffusion at which the impurity dosage at the oxide surface is kept constant, a Gaussian function describes the resulting impurity distribution within the oxide, in analogy to diffusion laws in a semiconductor. Modifications of these

370 PROPERTIES OF SILICON DIOXIDE

laws apply at the semiconductor-oxide interface due to the influence of the segregation coefficient upon the impurity distribution. The curves of Figure 4.14 give empirical data for the minimum thickness of SiO_2 which is required to mask against boron and phosphorus diffusion.

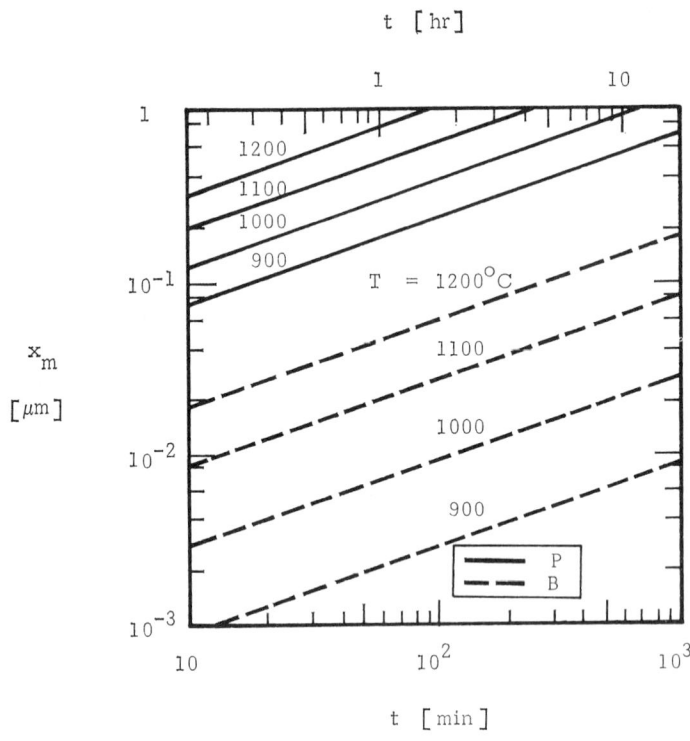

FIGURE 4.14

Miminum thickness (x_m) of SiO_2 required to mask against boron (B) and phosphorus (P) as function of diffusion time (t) and diffusion temperature (T). Ratio of surface impurity concentration at oxide surface to semiconductor impurity concentration $C_S/C_B = 10^3$.

5. REFERENCES

(1) C. J. Frosch and L. Derick, *J. El. Chem. Soc.*, **104**, 547 (1957).
(2) C. T. Sah et al., *J. Phys. Chem. Solids*, **11**, 288 (1959).
(3) S. Horiuchi and J. Yamaguchi, *Jap. J. Appl. Phys.*, **1**, 314 (1962).
(4) D. R. Collins et al., *J. Appl. Phys. Letters*, **8**, 323 (1966).
(5) A. S. Grove, *Physics and Technology of Semiconductor Devices*, John Wiley and Sons, New York, 1967.
(6) H. F. Wolf, *Silicon Semiconductor Data*, Pergamon Press, Oxford, 1969.

4–4 Other Characteristics

1. ETCHING OF SiO_2

The etching of SiO_2 is usually accomplished by the use of HF and follows the reaction

$$SiO_2 + 6HF \longrightarrow H_2SiF_6 + 2H_2O.$$

Frequently the etchant is buffered with NH_4F to avoid depletion of the fluoride ions. Figure 4.15 shows the etch rate of silicon dioxide as a function of etchant composition and temperature.

The etch rate is relatively independent of surface polish of the original silicon. However, a dense oxide (e.g., grown in dry O_2) presents a smaller surface area than a loose oxide (e.g., grown in wet O_2) and therefore has a slightly smaller etch rate. Mechanical stress usually enhances the etch rate. Lateral etching (undercutting) is facilitated by stress at the Si–SiO_2 interface generated during high-temperature thermal oxidation and subsequent cooling to room temperature; this stress is due to differences in thermal expansion coefficients of Si and SiO_2. The presence of impurities in the oxide may increase the etch rate substantially.

2. COMPRESSIBILITY

Compressibility (or volume elasticity) is a material constant which describes the volume change (du_v) with pressure. It is defined as

$$\kappa_v^* = -(1/u_v)(du_v/dp). \tag{4.35}$$

The dimension of κ_v^* is the inverse of pressure.

Isothermal compressibility (T = constant):

$$\kappa_{vT}^* = -(1/u_v)(du_v/dp)|_T \tag{4.36a}$$

Adiabatic compressibility (entropy S = constant):

$$\kappa_{vS}^* = -(1/u_v)(du_v/dp)|_S = \kappa_{vT}^*(T\gamma'^2/d_s c_p) \tag{4.36b}$$

In the above equations the terms used have the following meaning:

u_v = volume
d_s = density of insulator
c_p = specific heat at constant pressure
γ' = coefficient of volume thermal expansion.

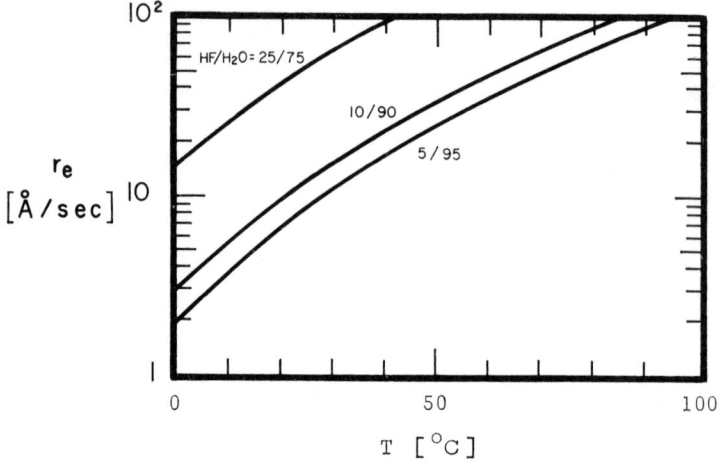

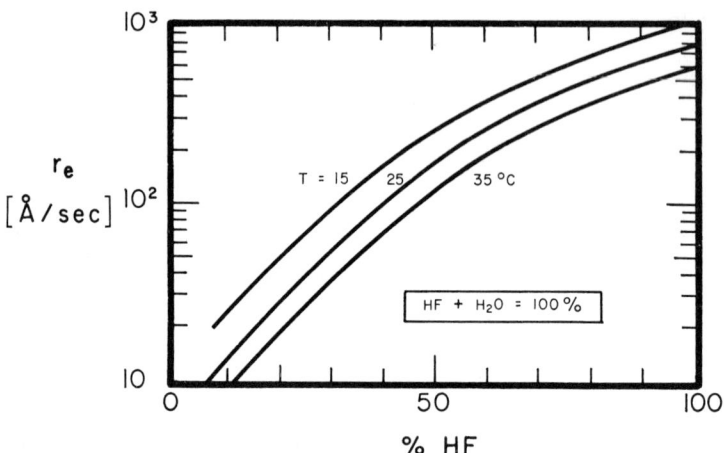

FIGURE 4.15

Etch rate of thermally grown SiO_2 as function of etchant temperature (T) and etchant concentration (by volume percent).

The compressibility of SiO_2 decreases with increasing temperature and decreasing pressure. For example, at $p = 10^8$ g/cm² for SiO_2 : $\kappa_v^* = 2.8 \cdot 10^{-9}$ cm²/g.

3. VISCOSITY

Solids subjected to an applied force show either one of two types of reaction. Either they deform until a state of equilibrium is reached when no further change of shape takes place, or there is no permanent resistance to change of shape and continuous deformation takes place as long as the force is applied. Most materials (including all semiconductors) belong to the first class, whereas most glasses (including SiO_2) belong to the second class; they can be considered as a special mode of a fluid. When flow takes place it is opposed by internal friction arising from the cohesion of the molecules; this internal friction is the viscosity of the material. The velocity gradient dv/dx is proportional to shearing stress per area F^*/A; the proportionality factor is the viscosity η^*:

$$F^*/A = \eta^*(dv/dx). \qquad (4.37)$$

The temperature dependence of η^* can be described by

$$\eta^* = \eta_o \exp(2.34/kT - 5.88) \qquad (4.38)$$

for SiO_2. The term η_o refers to the viscosity at room temperature.

The differential viscosity (η^*) is measured in poise, whereas the kinetic viscosity, i.e., the differential viscosity divided by the density (d_s), (η^*/d_s) is measured in stokes.

Useful conversions are:

$$1 \text{ poise} = 1 \text{ dyne sec cm}^{-2} = 1 \text{ g cm}^{-1} \text{ sec}^{-1}$$
$$1 \text{ stokes} = 1 \text{ cm}^2 \text{ sec}^{-1} = 1 \text{ poise } (\text{cm}^3/\text{g})$$

For comparison, the viscosity of some materials at 300°K is given in Table 4.7. The viscosity of SiO_2 decreases exponentially with temperature:

$$\text{at } 1000°C \ \eta^* = 6 \cdot 10^{15} \text{ poise},$$
$$\text{at } 1200°C \ \eta^* = 5 \cdot 10^{12} \text{ poise},$$
$$\text{at } 1400°C \ \eta^* = 2 \cdot 10^{10} \text{ poise}.$$

TABLE 4.7
Viscosity of Selected Materials

Material	Viscosity [poise]
Air	$1.7 \cdot 10^{-4}$
H_2O	10^{-2}
Glycerine	8.3
Syrup	10^3
Glass at melting point	10^3
at working temperature	10^7
at annealing temperature	10^{13}

4. OTHER PROPERTIES OF SiO_2

Electric volume conductivity of SiO_2 and glasses is due to finite ion mobility. It is a function of the impurity content. Higher ion mobility at elevated temperature results in an increase in volume conductivity, i.e., a decrease in electric resistivity, due to a weakening of the SiO_2 network. Volume resistivity of SiO_2 decreases exponentially with temperature.

The dielectric constant of pure SiO_2 is, in a first order, independent of temperature; SiO_2-P_2O_5 glass, however, shows an increase above about 100°C.

Table 4.8 lists some other properties of SiO_2 and SiO_2-P_2O_5 glass. In Tables 4.9 and 4.10 some properties of insulating materials commonly used in semiconductor technology are compared. The data of Table 4.10 are for thin

TABLE 4.8
Characteristics of SiO_2 and SiO_2-P_2O_5 Glass

Parameter	SiO_2	SiO_2-P_2O_5
Composition [mole % P_2O_5]	0	14.5–16.5
Density [g/cm^3]	2.24–2.27	2.35–2.40
Dielectric strength [V/μm]	810–870	1200–1450
Etch rate* [Å/sec]	2	500

* Etchant: 15 HF, 10 HNO_3, 300 H_2O (by volume).

TABLE 4.9
Comparison of Properties of Various Insulators

Parameter	SiO_2	Si_3N_4	BeO	Al_2O_3
Crystal structure	Random network of SiO_4 tetrahedra	Hexagonal	Hexagonal	Rhombohedral
Lattice constant [Å]		7.75	2.69	4.76
Density [g/cm^3]	2.65	3.44	2.90	2.40
Dielectric constant	3.9	7.5	6.4	5.0
Electric resistivity [Ω cm]	$>10^{16}$	10^7	$>10^{16}$	$>10^{16}$
Thermal conductivity [W/cm°K]	0.140	0.185	2.300	0.170
Specific heat [10^7 cm^2/sec^2°K]	1.00	0.17	0.26	0.18
Thermal diffusivity [cm^2/sec]	0.006	0.010	0.199	0.013
Linear thermal expansion coefficient [10^{-5}°K^{-1}]	0.05	0.28	0.60	0.70
Melting point [°C]	≈ 1700	≈ 1900	≈ 2500	≈ 2000

TABLE 4.10
Properties of Thin Film Materials

Material	E_G [eV]	n^* (0.546 μm)	ϵ	δ (1kHz)	Breakdown [V/μm]	Surface charge (at $V=0$) [cm^{-2}]	Q_{ss}/q [cm^{-2}]
Si_3N_4	4.7	2.0	5.5-9.4	$\approx 10^{-2}$	100	$1 \cdot 10^{12}$	$\approx 10^{11}$
TiO_2	>5	2.0	20-50	$\approx 10^{-2}$	50	$1.5 \cdot 10^{12}$	$9 \cdot 10^{11}$
Ta_2O_5	4.2	2.3	20	$\approx 10^{-2}$	10	$4 \cdot 10^{12}$	$5 \cdot 10^{11}$
Nb_2O_5	3.5	2.2	11	$\approx 10^{-2}$	100	$<5 \cdot 10^{11}$	$2 \cdot 10^{11}$
Al_2O_3	>5	1.7	7.6	$\approx 10^{-3}\text{-}10^{-2}$	500	$10^{11}\text{-}10^{12}$	$2 \cdot 10^{10}$

films of thicknesses between 10^2 and 10^4 Å and are partially different from those of bulk materials. The following properties are given:

E_G = energy gap
n^* = refractive index
ε = dielectric constant
δ = loss tangent (extinction coefficient)
Q_{ss}/q = surface state density.

5. REFERENCES

(1) H. J. McSkimin, *J. Appl. Phys.*, **24**, 988 (1953).
(2) R. Bruckner, *Glastechn. Ber.*, **37**, 413 (1964).
(3) E. H. Fontana and W. A. Plummer, *Physics and Chemistry of Glasses*, **7**, 139 (1966).
(4) C. C. Mai and J. C. Looney, *SCP and SST*, **9**, 19 (Jan. 1966).
(5) E. H. Snow and B. E. Deal, *J. El. Chem. Soc.*, **113**, 263 (1966).
(6) N. B. Hannay, *Solid-State Chemistry*, Prentice-Hall, Englewood Cliffs, N.J., 1967.
(7) H. F. Wolf and K. F. Greenough, *Microelectronics and Reliability*, **6**, 285 (1967).
(8) K. H. Zaininger and C. C. Wang, *Proc. IEEE*, **57**. 1564 (1969).

4-5 Examples

PROBLEMS AND SOLUTIONS

1. **Problem:** Determine the final oxide thickness after a two-step thermal oxidation of silicon.
   Step 1: $T_1 = 1000°C$, wet, $t_{o1} = 60$ min;
   Step 2: $T_2 = 1200°C$, dry, $t_{o2} = 30$ min.

   Solution: (a) Step 1:
   $$x_{o1} = 0.45 \ \mu m;$$

   (b) Step 2:
   Starting with $x_{o1} = 0.45 \ \mu m$, find the corresponding time on the 1200°C curve; this is $t_o = 5$ hr. The final oxide therefore corresponds to $t_o = 5.5$ hr.
   $$x_{o2} = 0.47 \ \mu m.$$

2. **Problem:** Determine the true parabolic rate constant if silicon of bulk concentration $C_B = 10^{20}$ cm^{-3} (phosphorus-doped) is oxidized in a wet atmosphere at 900°C.

   Solution: $f_B = 1.64$; $B = 0.25 \ \mu m^2/hr$;
   hence the corrected oxidation rate constant
   $$B_{\text{true}} = 0.41 \ \mu m^2/hr.$$

3. **Problem:** Determine the final oxide thicknesses at silicon surfaces (⟨111⟩ orientation) if the partial pressure of the oxidant is either 1.0 atm or 0.5 atm during wet thermal oxidations of short duration. Oxidation conditions; $T = 1100°C$, $t_o = 15$ min.

   Solution: $A = 0.7 \ \mu m$, $B = 0.5 \ \mu m^2/hr$.
   Since for short oxidation time $x_o = (B/A)t_o$,
   (a) at $p_G = 1.0$ atm
   $\quad x_o = 0.18 \ \mu m$,
   (b) at $p_G = 0.5$ atm
   $\quad x_o = 0.09 \ \mu m$;
   i.e., reduction in partial pressure of the ambient gas results in a corresponding reduction in final oxide thickness if the oxidation is of short duration.

PROBLEMS FOR WHICH A SOLUTION IS NOT GIVEN

1. Determine the decrease in the resulting oxide thickness if silicon wafers are oxidized for 1 hr and if the water bath temperature is reduced from 97°C to 25°C, assuming a wet oxidation at 1200°C.

2. Compare the diffusion coefficients of boron and gallium in Si and SiO_2 at 1100°C.

3. Determine the SiO_2 thickness required to mask against boron and gallium at 1100°C.

4. Discuss the redistribution of impurities at the surface of a silicon wafer for which the segregation coefficient $m = 10^{-1}$, 1, and 10. Assume ratios of the oxidation rate constant to the diffusion coefficient $B/4D_s = 10^{-1}$ and 10.

5
SEMICONDUCTOR SURFACES

5–1 The Ideal MIS System
5–2 Semiconductor Surface Depletion and Inversion
5–3 MIS Capacitance
5–4 Metal–Semiconductor Contacts
5–5 Examples

The Si–SiO$_2$ interface has gained considerable importance in semiconductor device fabrication and research. More recently, other semiconductor–dielectric interfaces have enhanced the interest in interface properties. With the improvement in the technology required for other semiconductors attention is being drawn to other semiconductor–dielectric interfaces.

Much of the original interest in SiO$_2$ related to its masking capabilities against a number of impurities. Although this is still of great significance, the importance of the Si–SiO$_2$ system, or more generally, of the semiconductor–dielectric system, is to a large degree associated with the properties of the semiconductor–dielectric interface and their influence on the electrical characteristics of the silicon surface, since active device structures (MIS and bipolar devices) are affected by the properties of the semiconductor surface. Ionic surface contamination, alkali ions within the oxide, fixed charges within the oxide, interface states, and radiation-induced charges affect the properties of a semiconductor surface. Surface conductance, capacitance, surface recombination velocity, and others are related to the surface potential.

Almost all properties of a semiconductor depend to some extent on the nature, arrangement, and surroundings of its outermost atomic layer. The semiconductor surface is defined as a three-dimensional region of discontinuity between the semiconductor and an insulating layer or an ambient gas. The discontinuity extends over a region that is narrow compared to most semiconductor dimensions and depends upon the property under consideration; e.g., for a semiconductor–gas interface the region where the type of atoms and the crystal structure change is much narrower than the region in the semiconductor where the carrier density differs from that of the bulk; for semiconductor–oxide and semiconductor–gas interfaces the discontinuities extend over atomic dimensions.

The periodicity of the crystal potential is disturbed or lost at the crystal surface; furthermore, the surface atoms are usually displaced from their ideal lattice positions, resulting in a surface structure that deviates electrically from the bulk structure. These disturbances lead to localized electronic states at the surface referred to as surface states or interface states.

At the surface of a semiconductor crystal, normally, a net positive charge is found because its valence bonds end abruptly there, with electrons missing. These charges add to the potential near the surface with an applied reverse bias and cause that part of the depletion layer near the surface to act like a low-resistivity region.

The positive charges normally observed in the dielectric may result in a depletion or inversion of an underlying *p*-type surface or in an accumulation of an underlying *n*-type surface. Associated surface charges or states can be classified as follows:

(a) fixed surface state or interface charges (of density Q_{ss}) close to the semiconductor–dielectric interface;
(b) mobile charges within the dielectric resulting from processing contamination;
(c) surface generation-recombination centers or fast surface states resulting from processing contamination;
(d) traps (usually positively charged) within the dielectric which can be ionized by radiation (e.g., by x-ray, electron or other ionizing radiation).

The electrical operation of a metal–insulator–semiconductor (MIS) structure (which consists of a semiconductor substrate covered by a dielectric layer, covered in turn by a metal plate) depends upon the charge relationship at the two interfaces. There are fast and slow surface states, i.e., localized electron energy levels, available for occupancy by electrons, which can exchange charges rapidly or slowly with the semiconductor space charge region. The slow states are at the outer surface of the insulator, the fast states are at the semiconductor–insulator interface.

Interface states, i.e., stationary electronic states, located at the semiconductor–insulator interface, are analogous to fast surface states. The energy levels of these states are either within the energy gap or within conduction or valence bands of the semiconductor.

The space charge of the dielectric is trapped in its bulk and does not interact with the semiconductor surface. At room temperature this charge exists in a mobile form (due to metal ions) and an immobile form (due to traps that are part of the intrinsic defect structure of the insulator).

Charges at the semiconductor–insulator interface will cause a displacement of the capacitance–voltage curve of an MIS capacitor along the voltage axis; this displacement is of magnitude Q_{ss}/C_o. It can be used to determine the magnitude of Q_{ss}. If the displacement is parallel, Q_{ss} is both spatially uniform under the metal electrode and independent of the semiconductor surface potential; if the displacement is not parallel, either Q_{ss} is not uniform under the electrode or there exists a nonuniform contamination or a high density of fast surface states.

Electrical conduction in SiO_2 is mainly ionic (primarily due to Na or other alkali ions); the formation of the associated space charge will induce an excess charge in the silicon, in addition to that induced by the applied voltage and will result in a shift of the capacitance–voltage characteristic (C–V curve); it depends mainly on the initial concentration and distribution of mobile ions in the oxide. Ion migration and polarization in oxide and glass films cause a shift of the C–V curve due to

(a) mobile ions initially at the metal–oxide interface;

(b) mobile ions initially uniformly distributed;
(c) orientation of uniformly distributed dipoles.

The metal–semiconductor interface finds application in simple ohmic contacts to the semiconductor if the junction is nonrectifying, and application in Schottky barrier diodes and other microwave devices if the junction is rectifying. While in a semiconductor a forbidden energy gap exists between conduction and valence bands, there is no such energy gap in a metal. In a metal the allowed energy levels are filled up to the Fermi level at $T = 0$; at $T > 0$ the state occupancy drops from 1.0 to 0 in an energy range of a few kT on either side of the Fermi level.

If a metal and an n-type semiconductor are brought into intimate contact, electrons will flow from the side of the higher level (metal) to the other side (semiconductor) since both Fermi energies must be equal. This is accomplished by a modification of the surface barriers of both sides; the electrons flowing from the metal form a negative charge layer on the metal surface which repels other electrons in the semiconductor until it is neutralized by a matching positive charge layer consisting of ionized donor atoms inside the semiconductor surface. This results in an upward bending of the band edges of the semiconductor at its surface. Such a bend keeps the normal electron density in an n-type semiconductor conduction band away from the surface. A similar situation exists for a p-type semiconductor-metal contact.

The difference in Fermi levels before contact is the contact potential difference; this difference in connection with the semiconductor energy gap is theoretically the height of the barrier between the two materials. In practice there is little correlation between measured and theoretical barrier heights; this is usually due to a permanent dipole layer of trapped charge at the semiconductor surface which affects the barrier height.

Barrier layer rectification is the rectification which appears at the contact between dissimilar materials such as a metal-semiconductor contact or at a p-n junction. It results from a readjustment of the energy levels on each side of the discontinuity, the readjustment being a consequence of the condition that in the absence of an external voltage, the Fermi level is constant throughout materials in electrical contact.

There are two possible mechanisms by which conduction across the barrier can take place—electrons near the bottom of the conduction band penetrate the barrier by quantum-mechanical tunneling, or electrons are excited thermally or otherwise and pass over the barrier. The probability that tunneling is responsible for an appreciable share of the observed conduction is much smaller than that of thermal activation; this is because tunneling predicts that the resistance of the contact is extremely sensitive to variations in the barrier width (e.g., if the width were doubled, the resistance to tunneling would rise by a factor of several million), and it predicts rectification in the opposite sense to that which is observed; specifically, it predicts that the direction of easy flow for electrons in the case under consideration would be from the metal to the semiconductor, whereas the reverse is observed experimentally.

Similar to semiconductor *p-n* junctions, the carrier flow across a metal–semiconductor interface can be affected by the application of a voltage which decreases or increases the Fermi level depending upon its polarity. Also similar to *p-n* junctions, a metal-semiconductor contact (Schottky barrier) can be utilized as an optical detector (photo absorption) where photons are absorbed by band-to-band transitions and the resulting electron-hole pairs are collected. Photoemission from metal into semiconductor is another important characteristic of a Schottky barrier. In this case photons are transmitted through the semiconductor and are absorbed in the metal adjacent to the barrier. A fraction of the excited electrons will surmount the barrier, flow into the semiconductor, and be collected. The spectral response is bound by two limiting photon energies—a lower threshold given by the barrier height ($h\nu_{\min} = q\phi_{MS}$) and an upper threshold determined by transmission through the semiconductor ($h\nu_{\max} = E_G$).

Main characteristic features of a Schottky barrier diode are:

(a) It has nearly ideal characteristics in all respects, including capacitive, transport, and optical properties.
(b) Minority carrier injection and charge storage effects can essentially be eliminated.
(c) For a given semiconductor the barrier height can be varied over a wide range by proper choice of the metal.

GENERAL REFERENCES

(1) J. T. Wallmark, *RCA Rev.*, **24**, 641 (1963).
(2) A. Many, Y. Goldstein, N. B. Grover, *Semiconductor Surfaces*, John Wiley and Sons, New York, 1965.
(3) L. J. Sevin, *Field Effect Transistors*, McGraw-Hill Book Co., New York, 1965.
(4) L. P. Hunter, *Introduction to Semiconductor Phenomena and Devices*, Addison-Wesley, Reading, Mass., 1966.
(5) M. M. Atalla, *Metal-Semiconductor Schottky Barriers, Devices, and Applications*, Mikroelektronik 2, Munich, October 24–26, 1966.
(6) A. S. Grove, *Physics and Technology of Semiconductor Devices*, John Wiley and Sons, New York, 1967.

5–1   The Ideal MIS System

1. OPERATING MODES OF AN MIS SYSTEM

In an ideal MIS structure (consisting of metal, oxide or insulator, semiconductor) the only charges that can exist under any bias condition are:

(a) those in the semiconductor;
(b) those on the metal surface at the metal–oxide interface.

Both of these charges are equal in magnitude but opposite in sign. In an ideal MIS structure there is no carrier transport through the oxide under any steady-state bias condition. The energy bands of an ideal MIS system will be affected near the insulator–semiconductor interface by an applied electric field (voltage) between metal contact (gate) and the underlying semiconductor. The width of the region that is affected by the gate voltage (i.e., the region in which band bending occurs) is identical with the width of the space charge layer. The terms used in describing the MIS system are defined in Figure 5.1.

Depending upon the magnitude and polarity of the gate voltage (V_G), four basic cases of surface conditions are distinguished (Figure 5.2). The discussion applies to an n-type semiconductor, but is valid for a p-type semiconductor as well under observation of the appropriate polarities.

(a) No applied bias ($V_G = 0$):
The Fermi levels in metal and semiconductor are equal and no band bending is observed (flat-band condition). There is no energy difference between the metal work function, ϕ_M, and the semiconductor work function, ϕ_S, i.e., $\phi_{MS} = 0$.

(b) Positive bias ($V_G > 0$):
The semiconductor energy bands in the bulk are shifted upward (by an amount of qV_G), and the energy bands near the interface are bent downward (no bending of the Fermi level occurs although it is shifted upward by qV_G). Electrons from the semiconductor bulk are attracted to the surface and make the surface more n-type; i.e., accumulation of carriers takes place. The capacitance of the system is that of the oxide (C_o).

(c) Small negative bias ($V_G < 0$):
The energy bands in the bulk are shifted downward (by an amount qV_G) and bent upward near the interface (again no bending of the Fermi level but a downward shift by qV_G). Electrons are repelled from the surface making the surface less n-type; i.e., carrier depletion of the surface takes

5-1 THE IDEAL MIS SYSTEM

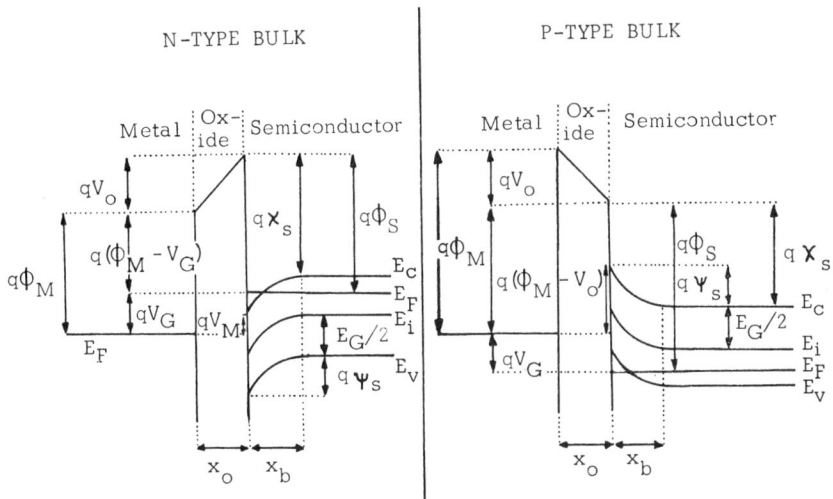

FIGURE 5.1
Definitions of terms used in the discussion of ideal metal-insulator-semiconductor and metal-semiconductor structures, shown at forward bias. The following terms are defined.

V_G = applied voltage at the metal contact,
V_o = voltage drop in the insulator,
V_M = barrier height seen by a carrier in the metal (at the Fermi level),
χ_s = electron affinity of the semiconductor,
ϕ_M = metal work function,
ϕ_S = semiconductor work function,
ψ_s = surface potential (or barrier height) in the semiconductor.

place. The resultant depletion layer has a fixed charge density equal to the semiconductor impurity density (N_D or C_B) but no free carriers and hence is an effective insulator whose capacitance (which is in series with capacitance C_o) reduces the total MIS capacitance.

(d) Large negative bias ($V_G \ll 0$):
The depletion layer is widened further until eventually holes are attracted to the semiconductor surface which will form an inversion layer, i.e., the surface will become *p*-type. Further increase in negative bias does not affect the width of the inversion layer which has its maximum width at the onset of inversion, provided the bias is altered slowly. If the bias is altered rapidly, the surface layer first becomes very wide (deep depletion) and the capacitance approaches zero; as holes are generated and the

386 SEMICONDUCTOR SURFACES

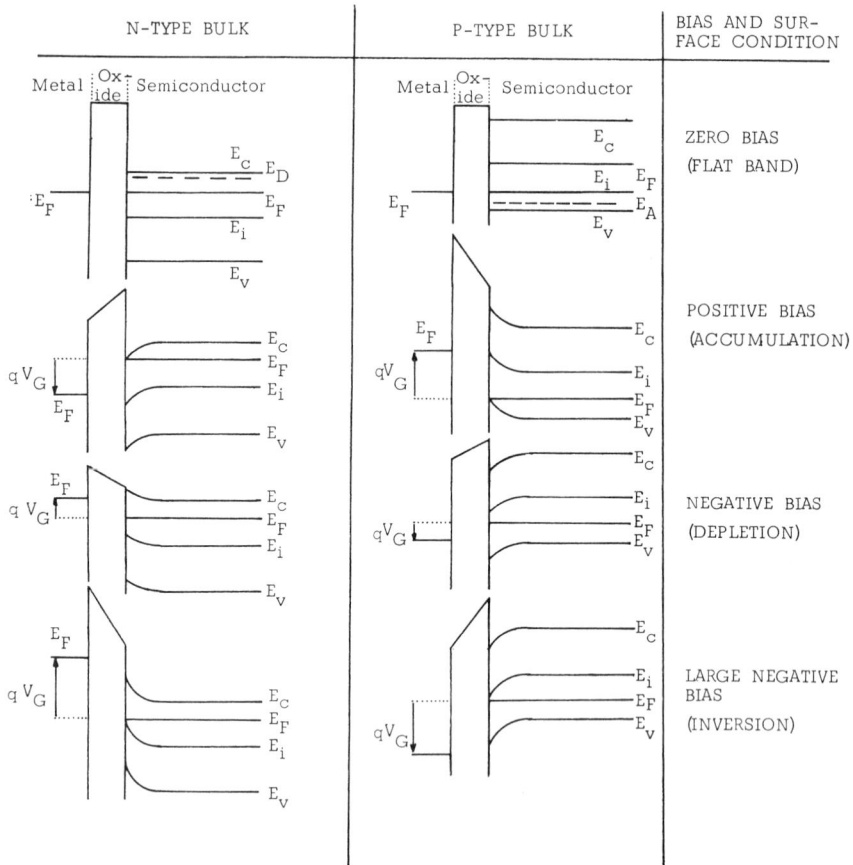

FIGURE 5.2

Energy bands of *n*-type and *p*-type bulk MIS structures under various bias conditions representing the four surface conditions: flat band, accumulation, depletion, and inversion.

surface comes to equilibrium with the bulk, the inversion layer width returns to its maximum equilibrium value and a surface inversion layer forms.

These four cases of surface conditions are summarized schematically in Figure 5.3, showing the difference between the low-frequency and the high-frequency model. The maximum space charge width is a function of the semiconductor impurity concentration; the corresponding capacitance minimum can be used to determine the impurity concentration near the sur-

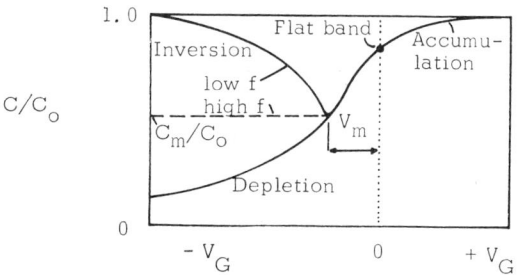

FIGURE 5.3
General capacitance–voltage relationship for an n-type semiconductor bulk.

face. At higher temperature the inversion layer tends more to stay in equilibrium with the bulk and the low-frequency $C-V$ curve persists at slightly higher frequency while the capacitance minimum rises to higher C/C_o values.

At lower temperature it becomes more difficult to sweep the bias slowly enough (i.e., a trend toward lower transition frequency is observed), and the capacitance minimum moves to lower C/C_o values. At very low temperature (40–60°K) the shallow impurities are almost completely deionized and the Fermi level lies between the impurity energy level and the majority carrier band edge, resulting in a secondary capacitance minimum around zero bias. At even lower temperature the semiconductor becomes isolating and $C-V$ curves become meaningless.

Assumptions made in presenting these ideal MIS conditions are:

(a) semiconductor bulk properties extend up to the surface;
(b) no excess charge collected at the interface;
(c) no extra surface states;
(d) no fixed oxide or interface charges;
(e) no fast interface states;
(f) oxide is a perfect insulator.

Considerable deviations from ideal conditions are usually observed in practical MIS structures.

2. REFERENCES

(1) A. S. Grove et al., *Solid-State Electronics*, **8**, 145 (1965).
(2) B. E. Deal et al, *J. El. Chem. Soc.*, **114**, 266 (1967).
(3) E. H. Nicollian and A. Goetzberger, *BSTJ*, **46**, 1055 (1967).
(4) B. E. Deal et al, *SCP and Solid State Technology*, **9**, 25 (Nov. 1966).
(5) P. V. Gray, *Proc. IEEE*, **57**, 1543 (1969).
(6) S. M. Sze, *Physics of Semiconductor Devices*, John Wiley and Sons, New York, 1959.

5–2 Semiconductor Surface Depletion and Inversion

1. SEMICONDUCTOR SURFACES

The surface of a semiconductor possesses properties which differ from those of the semiconductor bulk. This is due to the termination of the crystal lattice at the surface and the consequently unfilled bonds and due to the influence of the ambient. Two types of semiconductor surfaces are distinguished.

(a) Atomically clean surface:
   An atomically clean semiconductor surface possesses surface dangling bonds or unfilled orbitals, where two electrons can occupy a free orbital. Surface atoms may become negatively charged and thus will provide acceptor surface states by trapping bulk electrons and forming a p-type surface if there is a separate potential trough at the surface or if the energy bands arising from separate atomic levels overlap. These surface states occur within the energy gap and their density is approximately equal to the density of surface atoms. Such an atomically clean surface will be strongly p-type and sensitive to surface effects, e.g., chemisorption of some foreign species.

(b) Practical surface:
   Chemisorption of foreign atoms will reduce an atomically clean semiconductor surface to a practical surface, resulting in the following changes in surface states:
   (i) removal of some of the original acceptor states;
   (ii) introduction of new (neutral or positive) acceptor and donor states (these are fast states because of their fast relaxation time);
   (iii) introduction of outer states located beyond the interface, within or on the surface of a grown film (these are slow states because of their long relaxation time).

The potential distribution in the proximity of a practical surface (which is in equilibrium with the bulk) is a function of density, distribution and type of all surface states. A practical surface may be either n-type or p-type, depending on the existing surface states and independent of the existing bulk doping; e.g., the surface in the vicinity of a *p-n* junction with predominantly donor-type surface states will become less *p*-type (and possibly inverted) in the *p*-region and more *n*-type in the *n*-region as a result of the influence of the surface states.

2. SURFACE OPERATING CONDITIONS

Depending upon the magnitude of the electric field (E_o) at the semiconductor surface, four regions of surface space charge are distinguished (for an *n*-type semiconductor; for a *p*-type semiconductor the appropriate polarity changes have to be made).

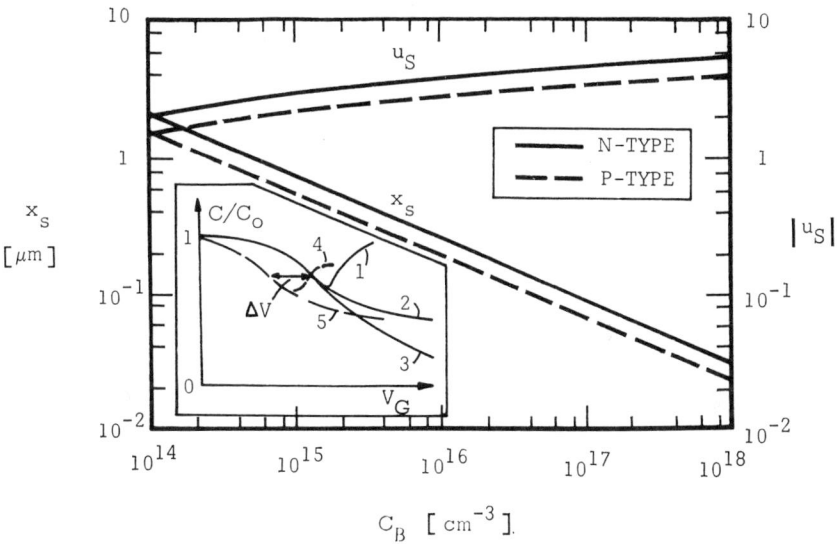

FIGURE 5.4

Width of the surface inversion region (x_s) and surface potential (u_s) as function of semiconductor bulk impurity concentration (C_B). Silicon, 300°K. The inset shows schematically the MIS capacitance as function of gate voltage (V_G) under two conditions:

(a) No surface states, no surface dipoles, no work function difference between metal and semiconductor (ideal curves; solid lines).
  1—High frequency (minority carriers follow both DC and AC signal)
  2—Low frequency (minority carriers follow DC but not AC signal)
  3—Schottky depletion layer capacitance (minority carriers cannot accumulate at the surface)

(b) Surface states are present (actual curves; dashed lines).
  4—High frequency (corresponds to curve 1)
  5—Low frequency (corresponds to curve 2)

ΔV is a measure of the density of ionized surface states.

(a) $E_o > 0$:
The majority carrier density is higher in the space charge region than in the semiconductor bulk (accumulation).

(b) $E_o = 0$:
The majority carrier density is uniform throughout the semiconductor (flat band condition).

(c) $E_o < 0$:
The majority carriers are repelled from the surface and the positive charge consists of uncompensated ionized impurity atoms (depletion).

(d) $E_o \ll 0$:
The minority carrier density at the surface equals or exceeds the majority carrier density in the bulk; thus a surface layer of opposite conductivity type is formed (inversion).

3. SURFACE STATES

Localized electronic states within the surface region can be divided into fast and slow surface states, depending upon the speed with which they interact with the semiconductor space charge region.

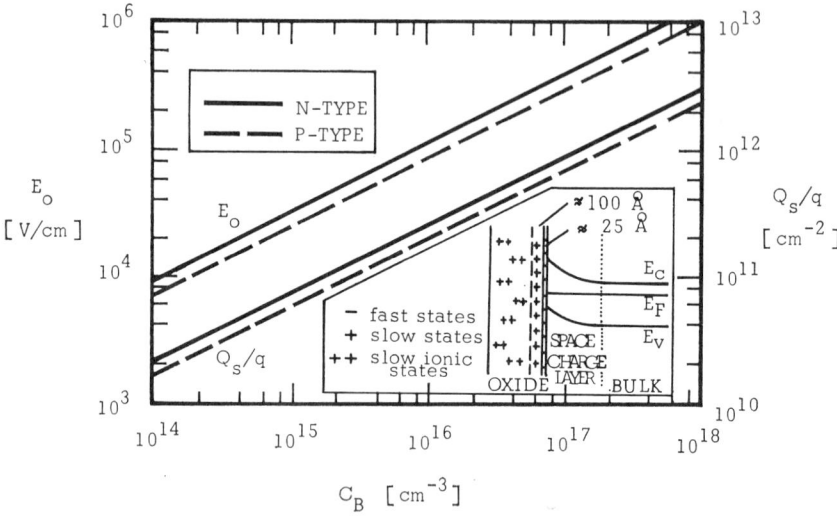

FIGURE 5.5

Electric field (E_o) in the oxide and semiconductor space charge density (Q_s/q) as function of semiconductor bulk impurity concentration (C_B). Si-SiO$_2$ system, 300°K. The inset defines the slow and fast states near the semiconductor-oxide interface of an n-type semiconductor.

5-2 SURFACE DEPLETION AND INVERSION

(a) *Slow surface states (oxide states)*:
They are found at the outer surface of the oxide layer (i.e., at the oxide–gas interface). The occupation by carriers of these states can usually be affected by the ambient, indicating that they participate or originate in adsorption processes; slow surface states can largely be eliminated or reduced by thermal oxidation.
The charge associated with these states is trapped within the oxide film and cannot move to the semiconductor surface. There are two types of oxide charges:
  (i) *Immobile charges*:
    These charges are held in traps that are part of the intrinsic defect stucture of the oxide.
  (ii) *Mobile charges*:
    These charges are due to ions that are capable of migrating through the oxide, especially at high electric field and elevated temperature.
(b) *Fast surface states (interface states)*:
They are found at the semiconductor–insulator interface. The disturbance of the periodicity of the semiconductor crystal potential at the surface results in a high density of states within the energy gap of the semiconductor in the vicinity of its surface. Their density is of the order of one fast surface state for every surface atom, resulting in a density of about 10^{15} cm^{-2}. Fast surface states result in a deviation in the shape of the capacitance–voltage characteristics of MIS structures from the ideal curve shapes or in a displacement of the curves.
Fast surface states are associated with a surface recombination velocity

$$S_o = \sigma_c v_{th} N_{tS} \tag{5.1}$$

where σ_c is the capture cross-section, v_{th} is the thermal carrier velocity, and N_{tS} is the density of recombination centers at the semiconductor surface.
Interface states are stationary electronic states. Because the insulator has a wide energy gap, the energy levels of interface states can lie either within or outside the energy gap of the semiconductor. This will determine whether or not they change their charge state when an electric field is applied between metal and semiconductor. If these levels are outside the semiconductor energy gap, then the charge in them is usually referred to as oxide or insulator charge because it does not change its magnitude when a field is applied between metal and semiconductor.

The carrier distribution in the semiconductor space charge region takes into account the occupation statistics of the surface states, Maxwell–Boltzmann statistics, and Poisson's equation.

Surface state charges have the following properties:

(a) they are located within 200 Å from the semiconductor–insulator interface within the semiconductor;
(b) they are fixed and cannot be charged or discharged over a wide variation of bending of the semiconductor energy bands;
(c) their density (Q_{ss}) is not significantly affected by insulator thickness (x_o) or by type or concentration of impurities in the semiconductor; they are, however, strongly affected by oxidation and annealing conditions and crystal orientation.

4. OXIDE OR INSULATOR CHARGES

Charges within the insulator (oxide) are due to either mobile or fixed charges.

(a) Mobile charges:
They are traps within the insulator that electronically interact with semiconductor or metal contacts. They are similar to surface states but with a three-dimensional distribution.
(b) Fixed or built-in charges:
They do not participate in electron transfer processes across the interfaces and may be associated with either ionic defects in the insulator or states located at the semiconductor–insulator interface.
  (i) *Ionic defects*:
  In this case an electronic transfer takes place between defect and semiconductor or contact metal during fabrication and not afterwards.
  (ii) *Interface states*:
  They are energetically outside the semiconductor energy gap but affect the semiconductor surface potential. They can also modify the distribution of semiconductor interface states or discharge at one of the interfaces.

5. SEMICONDUCTOR CHARGES

The total charge induced within the semiconductor surface (Q_s/q) is composed of the charge due to minority carriers (Q_n/q or Q_p/q) and the charge due to impurity ions (Q_B/q).

$$Q_s/q = Q_p/q + Q_B/q \quad (n\text{-type semiconductor}) \quad (5.2\text{a})$$

$$Q_s/q = Q_n/q + Q_B/q \quad (p\text{-type semiconductor}). \quad (5.2\text{b})$$

5-2 SURFACE DEPLETION AND INVERSION

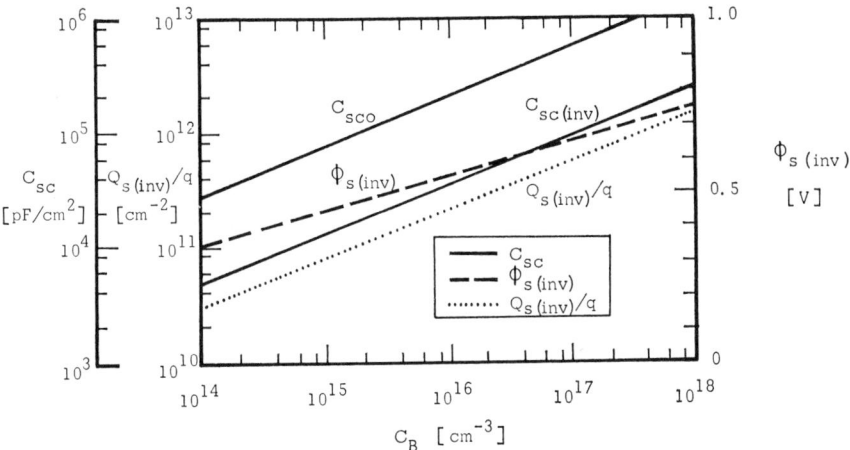

FIGURE 5.6

Space charge capacity (C_{sc}), semiconductor space charge density (Q_s/q), and surface potential (ϕ_s) as function of semiconductor bulk impurity concentration (C_B) at the onset of strong surface inversion (subscript inv) and at flat band condition ($\phi_s = 0$, subscript 0). Silicon, 300°K. For inversion $C_{sc(inv)} = C_m$.

Values of the charge due to minority carriers are given in Figure 5.8. The charge due to impurities is the product of impurity concentration (C_B) and depletion layer width (x_d).

$$Q_B/q = \pm C_B x_d, \tag{5.3}$$

where C_B represents donor or acceptor concentrations. Signs of the quantities involved are given in Table 5.1.

In the case of surface depletion the minority carrier charge is small compared to the impurity charge so that

$$Q_n/q = 0 \quad \text{or} \quad Q_p/q = 0$$

and

$$Q_s/q = \pm C_B x_d.$$

This condition is called the depletion approximation. It fails when Q_n/q or Q_p/q is not negligible compared to the fixed charge of the impurity ions within the depletion region, i.e., at the onset of surface inversion; in this case the minority carrier density equals the average impurity concentration in the surface layer, C_B or C_d. Once an inversion layer is formed, the width of the

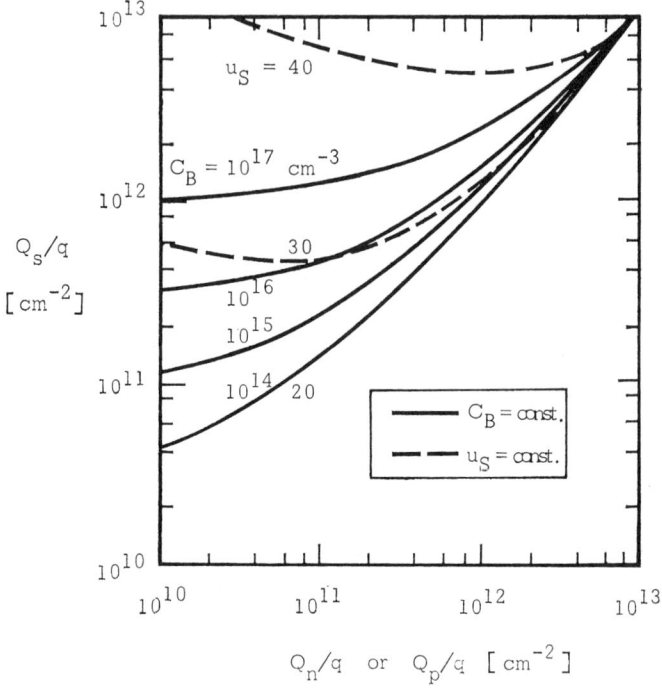

FIGURE 5.7

Semiconductor space charge density (Q_s/q) as function of minority carrier space charge density (Q_n/q or Q_p/q) in the inversion region for various semiconductors impurity concentrations (C_B) and surface potentials (u_S).

depletion layer reaches a maximum which it does not exceed even if the surface potential is further increased:

$$x_d = x_{d\,\max}.$$

At or after the onset of strong surface inversion

$$\begin{aligned}Q_s/q &= \pm C_B x_{d\,\max} \\ &= \pm 2(\varepsilon_s \varepsilon_o C_B |\phi_F|/q)^{1/2} \\ &= \pm 2 n_i L_D \,[\exp(u_F - u_S) - \exp u_F \\ &\quad + \exp(u_S - u_F - V_a^*) - \exp(-u_F - V_a^*) \\ &\quad + 2 u_S \sinh u_F]^{1/2}. \end{aligned} \quad (5.4)$$

The total voltage required to produce surface inversion is

$$\begin{aligned} V_T &= \phi_{s\,(\text{inv})} + V_o \\ &= V_a + V_o + 2\phi_F. \end{aligned} \quad (5.5)$$

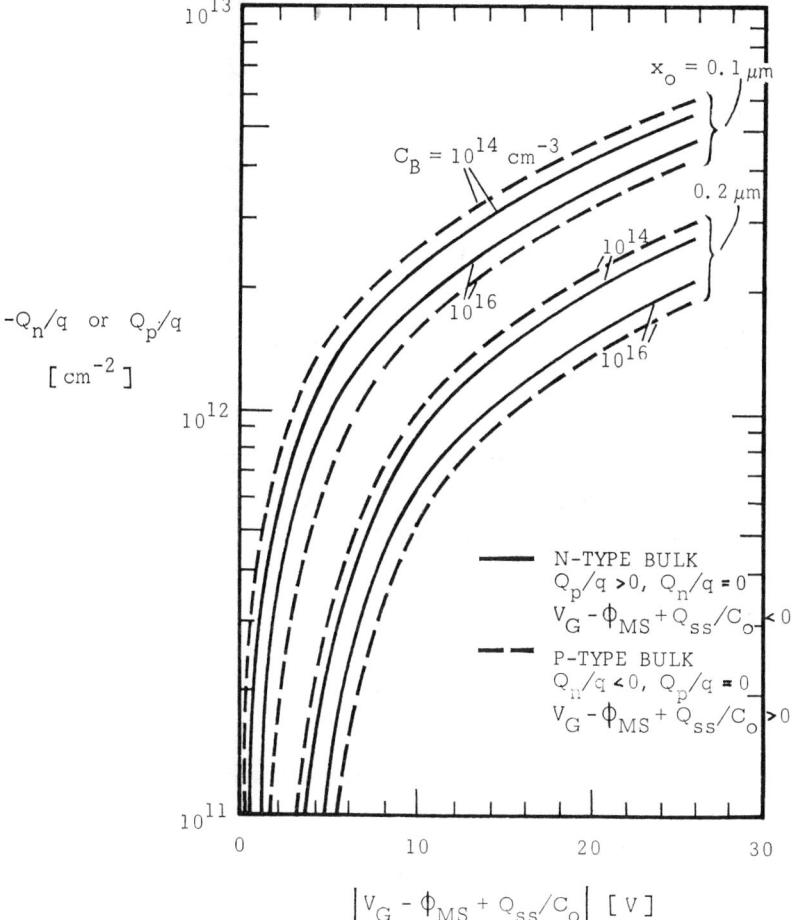

FIGURE 5.8

Charge carrier density (Q_n/q or Q_p/q) within the surface inversion layer as function of gate voltage (V_G) and semiconductor bulk impurity concentration (C_B). Si-SiO$_2$ system, 300°K. The polarities of voltage and densities are given in the inset.

TABLE 5.1

Signs of Potentials and Charges

	n-type semiconductor	p-type semiconductor
ϕ_F	<0	>0
$d\phi_F/dT$	>0	<0
V_R	<0	>0
V_F	>0	<0
dV_T/dT	>0	<0
Q_B	>0	<0

	Accumulation	Depletion	Inversion	Accumulation	Depletion	Inversion
ϕ_s	>0	<0	<0	<0	>0	>0
Q_s	<0	>0	>0	>0	<0	<0
Q_n						<0
Q_p			>0			

After inversion of the semiconductor surface has taken place the charge in the surface depletion region will remain constant.

The following terms have been used:

u_S, ϕ_s = surface potentials;

$$u_S = q\phi_s/kT$$
$$= q^2 C_B x_d^2 / 2\varepsilon_s \varepsilon_o kT \tag{5.6}$$
$$\phi_s = q C_B x_d^2 / 2\varepsilon_s \varepsilon_o$$
$$= |V_a| + |\phi_F| \quad \text{(depletion approximation)} \tag{5.7}$$

u_F, ϕ_F = Fermi potentials;

$$u_F = q\phi_F/kT$$
$$= (E_i - E_F)/kT \tag{5.8}$$
$$\phi_F = (E_i - E_F)/q \tag{5.9}$$

L_D = Debye length

V_a, V_a^* = applied voltage at the metal contact, substrate grounded;

$$V_a^* = qV_a/kT \tag{5.10}$$

V_o = voltage drop in the insulator at the onset of surface inversion;

$$V_o = E_o x_o \tag{5.11}$$

in the absence of oxide charges.

6. DEPLETION LAYER WIDTH

The width of the depletion layer (x_d) in the depletion approximation (see Figure 5.9) is

$$x_d = (2\varepsilon_s \varepsilon_o \phi_s / q C_B)^{1/2}$$
$$= [2\varepsilon_s \varepsilon_o (|V_a| + |\phi_F|)/qC_B]^{1/2}. \quad (5.12)$$

The surface potential is

(a) for strong inversion:

$$\phi_s = \phi_{s\,(inv)} = 2\phi_F, \quad (5.13)$$

$$\phi_F = (E_i - E_F)/q. \quad (5.14)$$

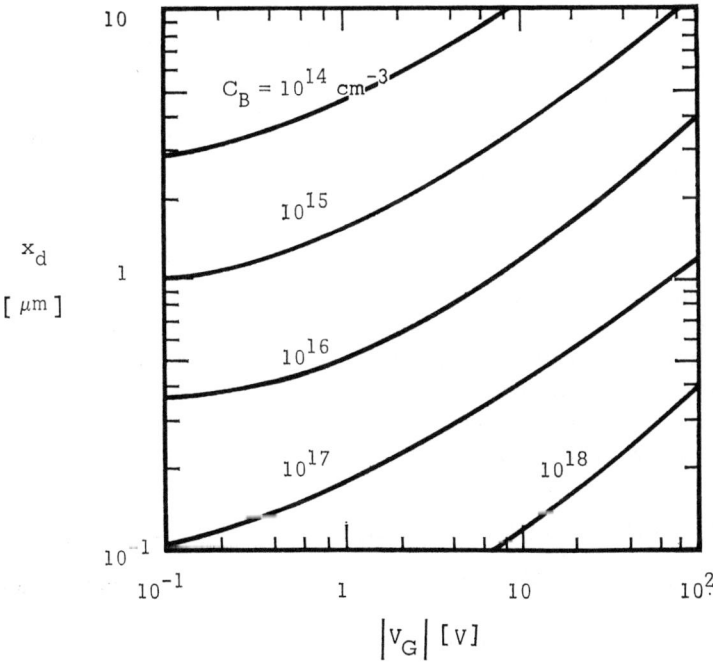

FIGURE 5.9

Width of depletion layer (x_d) at semiconductor surface as function of gate voltage (V_G) and semiconductor bulk impurity concentration (C_B). Si–SiO$_2$ system, 300°K.

(b) for non-inversion:

$$\phi_s = qC_B x_d^2 / \varepsilon_s \varepsilon_o. \qquad (5.15)$$

At the onset of strong inversion

$$\begin{aligned}x_d = x_{d\,\text{max}} &= (2\varepsilon_s \varepsilon_o \phi_{s(\text{inv})} / qC_B)^{1/2} \\ &= 2(\varepsilon_s \varepsilon_o \phi_F / qC_B)^{1/2}, \qquad (5.16)\end{aligned}$$

i.e., the maximum width of the surface depletion layer is identical to the width of the depletion layer of a one-sided step junction at $V = 0$, where $\phi_{s(\text{inv})} = 2\phi_F$ is taking the place of the built-in voltage $V_D = \phi_b$; the physical difference between the two lies in the fact that in a step junction the depletion layer is a result of the presence of impurity ions, while in a surface depletion layer the type inversion is due to the electric field generated by the field plate; thus, surface inversion results in a field-induced junction.

7. EFFECTIVE IMPURITY CONCENTRATION IN DEPLETION LAYER

The impurity concentration in the surface layer of a semiconductor (C_d) may differ from that in the semiconductor bulk. The reason for this is the redistribution of impurities close to the semiconductor surface during thermal oxidation. The region of impurity redistribution extends to approximately $2\sqrt{D_s t_o}$ away from the semiconductor-oxide interface; usually

$$2\sqrt{D_s t_o} > x_{d\,\text{max}}.$$

D_s is the diffusion coefficient of the impurity in the semiconductor and t_o is the oxidation time. The surface depletion layer is usually fully within the region of impurity distribution. Since the width of the depletion layer, on the other hand, is a function of the impurity concentration, in the above expressions for x_d the bulk impurity concentration C_B has to be replaced by the effective surface impurity concentration C_d, which is an average concentration over the entire depletion region.

8. TEMPERATURE DEPENDENCE OF TURN-ON VOLTAGE

The turn-on (threshold) voltage is defined as the gate voltage of an MIS system at the onset of strong inversion.

For an n-type semiconductor

$$V_T - \phi_{MS} + Q_{ss}/C_o = 2\phi_F < 0, \qquad (5.17a)$$

for a p-type semiconductor

$$V_T - \phi_{MS} + Q_{ss}/C_o = 2\phi_F > 0. \qquad (5.17b)$$

5–2 SURFACE DEPLETION AND INVERSION

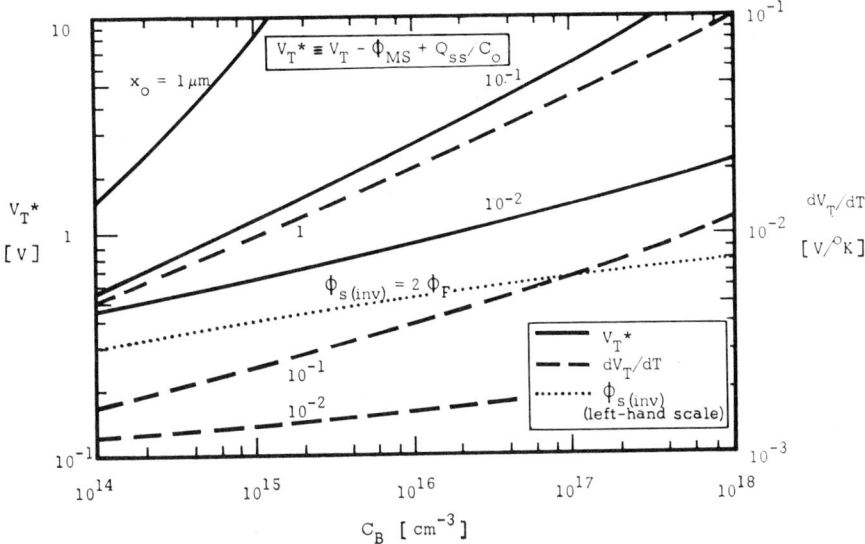

FIGURE 5.10

Turn-on voltage (V_T) of a surface channel and temperature coefficient (dV_T/dT) vs. semiconductor bulk impurity concentration (C_B) and oxide thickness (x_o). Si–SiO$_2$ system, n-type and p-type silicon; 300°K. Curves were calculated; measurements between -50 and $+125$°K indicate that the temperature coefficients given are accurate over this range.

The Fermi potential of the semiconductor

$$\phi_F = \pm(kT/q)\ln(C_B/n_i), \tag{5.18}$$

and its variation with temperature

$$d\phi_F/dT \approx (1/T)[(E_G/2q) - \phi_F]. \tag{5.19}$$

Hence, the turn-on voltage varies with temperature as follows, assuming $dQ_{ss}/dT = 0$.

$$dV_T/dT = (d\phi_F/dT)[2 - (1/C_o)(Q_B/2\phi_F)]$$
$$\approx \pm[(E_G/2q) - |\phi_F|][2 - (1/C_o)(Q_B/2\phi_F)]/T. \tag{5.20}$$

Increase in oxide thickness or semiconductor impurity concentration results in an enhanced temperature coefficient of V_T. Increase in temperature results in an increase of V_T for an n-type semiconductor and a decrease of V_T for a p-type semiconductor.

9. SEMICONDUCTOR SURFACE STATE DENSITY AS A FUNCTION OF CRYSTAL ORIENTATION

The surface state density of an Si–SiO$_2$ system is highest for the $\langle 111 \rangle$ orientation of the silicon substrate, since the (111) surface has the largest number of available bonds per cm^2, whereas the (100) surface has the smallest number, as is apparent from Table 5.2.

TABLE 5.2
Atomic Properties of the Silicon Surface

Orientation	Plane area of unit cell [cm^2]	Atoms in plane area of unit cell	Available bonds in plane area	Atoms per cm^2	Available bonds per cm^2
$\langle 100 \rangle$	a^2	2	2	$6.8 \cdot 10^{14}$	$6.8 \cdot 10^{14}$
$\langle 110 \rangle$	$\sqrt{2}\,a^2$	4	4	$9.6 \cdot 10^{14}$	$9.6 \cdot 10^{14}$
$\langle 111 \rangle$	$(\sqrt{3}/2)a^2$	2	3	$7.9 \cdot 10^{14}$	$11.8 \cdot 10^{14}$

Bonds parallel to the silicon surface react with O$_2$ or H$_2$O molecules during thermal oxidation most readily, while those at an angle to the surface react less easily because of a steric hindrance due to the position of the silicon atoms in the neighborhood of the bond. The activation energy for thermal oxidation decreases from (100) to (110) to (111) surface planes; i.e., since for the (100) surface the activation energy is largest and the number of available bonds is smallest, the oxidation rate for the (100) plane is smallest. Since the origin of the fixed surface charge is due to the excess ionic silicon in the oxide, the oxidation rate is proportional to the number of excess silicon ions, i.e., the (111) surface has the highest fixed charge density and in turn, the highest surface state density.

The ratio of semiconductor surface state densities (Q_{ss}) for different orientations is

$$Q_{ss\langle 111 \rangle} : Q_{ss\langle 110 \rangle} : Q_{ss\langle 100 \rangle} = 3 : 1.5 : 1.$$

These ratios are the same as the ratios of surface vacancies for the appropriate orientations.

10. SURFACE MOBILITY

The carrier mobility close to the semiconductor surface differs from the mobility within the semiconductor bulk mainly for these reasons:

(a) Except for flat-band conditions (i.e., for $\phi_s = 0$), carrier densities in the surface layer are different from those in the semiconductor bulk. The carrier densities in the space charge region are determined by the surface potential.

(b) Thermal oxidation of the semiconductor surface frequently leads to a redistribution of impurities near the semiconductor–oxide interface and a corresponding difference in surface and bulk impurity concentrations.

(c) For very large positive or negative values of surface potential (i.e., for $\phi_s \gg 0$ or $\phi_s \ll 0$) a potential well exits for one or the other carrier type which may increase surface scattering and consequently reduce the surface mobility.

(d) In addition to bulk scattering processes which affect semiconductor bulk mobility, the semiconductor surface may produce additional scattering sites which reduce the carrier mobility at the surface below that in the bulk. Increase in impurity concentration results in an increase in Coulomb scattering and a corresponding reduction in surface mobility.

The total carrier mobility in a semiconductor surface layer is composed of a bulk-dependent and a surface-dependent term (see Figure 5.11):

$$1/\mu_{\text{eff}} = 1/\mu_B + 1/\mu_S. \tag{5.21}$$

Near the semiconductor surface (i.e., $x \approx 0$),

$$\mu_{\text{eff}} = \mu_B \mu_S/(\mu_B + \mu_S) \leq \mu_B/2 \tag{5.22}$$

and deep inside the semiconductor (i.e., $x \gg 0$)

$$\mu_{\text{eff}} = \mu_B. \tag{5.23}$$

Generally the ratio of surface mobility (μ_S) to bulk mobility (μ_B)

$$\mu_S/\mu_B = 1 - \exp(\alpha_S{}^2)\,\text{erfc}(\alpha_S), \tag{5.24}$$

where $\alpha_S = (1/E_S)(1/\mu_B)(2kT/m_n)^{1/2}$
E_S = surface field
m_n = electron mass.

Up to about $E_S = Q_s/\varepsilon_s = 10^5$ V/cm, corresponding to $Q_s/q \sim 10^{12}$ cm^{-2} for silicon, the surface mobility is essentially constant; at higher fields the surface mobility is reduced.

Other expressions for the ratio of surface mobility to bulk mobility are,

(a) in terms of the mean free path of the carriers:

$$\mu_S/\mu_B = 1/(1 + l_B/x_{cS})$$
$$\approx 1 - l_B/x_{cS} \quad \text{if} \quad l_B \ll x_{cS}; \tag{5.25}$$

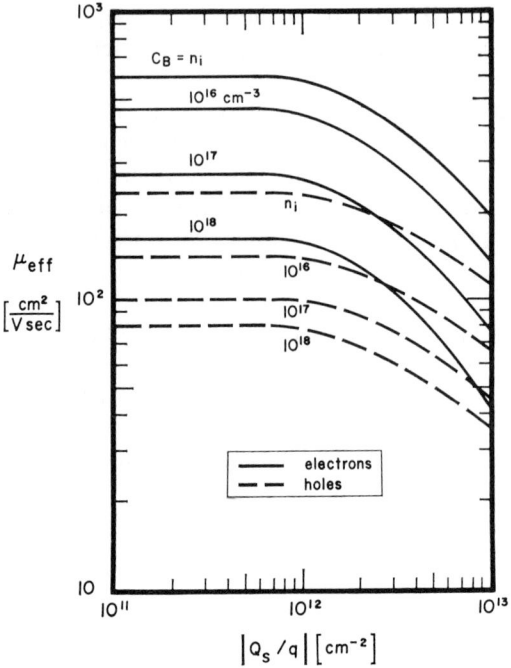

FIGURE 5.11

Effective minority carrier mobility (μ_{eff}) in inversion or surface layer vs. total charge induced in semiconductor (Q_s/q) and semiconductor impurity concentration (C_B). Silicon, 300°K.

(b) in terms of the effective bulk charge voltage:

$$\mu_S/\mu_B = 1 - V_\mu/4\phi_F$$
$$= 1 - (1/2C_o)(q\varepsilon_s \varepsilon_o C_B/\phi_F)^{1/2} ; \quad (5.26)$$

(c) in terms of surface velocity:

$$\mu_S/\mu_B = [1 + \mu_B(m_n n_i/\pi\varepsilon_s \varepsilon_o)^{1/2} F(u_S, u_b)(1 + v_S)^{1/2}/v_S]^{-1}. \quad (5.27)$$

In the above expressions the following terms have been used:
l_B = mean free path of the carriers in the bulk
x_{cS} = average distance of the carriers from the surface
V_μ = effective bulk charge voltage; $V_\mu = 2(q\varepsilon_s \varepsilon_o C_B \phi_F)^{1/2}/C_o$
v_S = carrier velocity at the semiconductor surface
$F(u_S, u_b)$ = dimensionless electric field as function of surface and bulk potentials.

5-2 SURFACE DEPLETION AND INVERSION

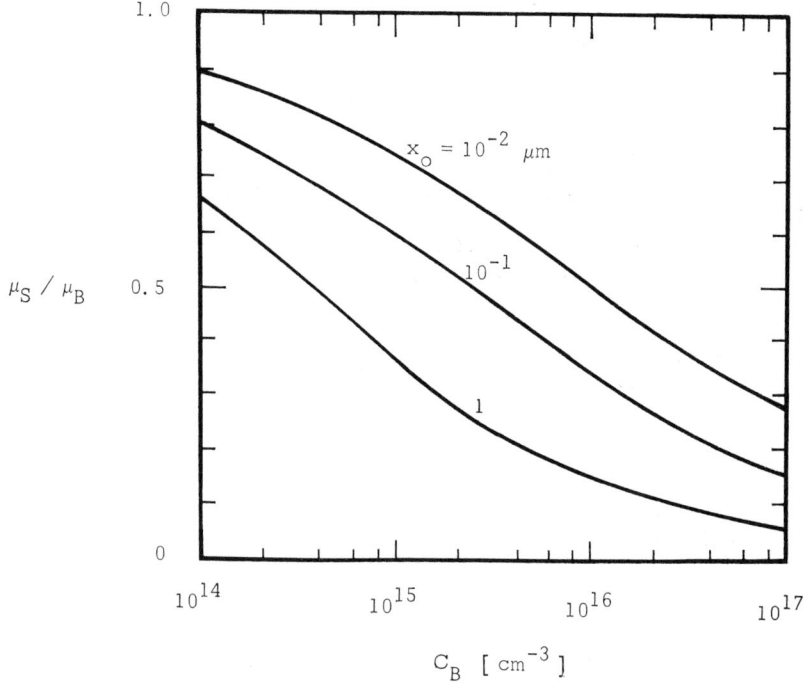

FIGURE 5.12

Ratio of minority carrier mobilities in semiconductor surface layer (μ_S) and semiconductor bulk (μ_B) as function of semiconductor bulk impurity concentration (C_B) and oxide thickness (x_o). Si-SiO$_2$ system, 300°K.

The carrier mobility in a surface inversion layer of width x_s is

$$\mu_S = \int_0^{x_s} \mu_n(x) n(x)\, dx / (Q_n/q) \quad \text{(n-layer)} \tag{5.28a}$$

or

$$\mu_S = \int_0^{x_s} \mu_p(x) p(x)\, dx / (Q_p/q) \quad \text{(p-layer).} \tag{5.28b}$$

The width of the surface layer, i.e., the width of a transition layer which would provide the desired amount of net charge if it were entirely depleted of carriers, is

$$\begin{aligned} x_s &= [2\varepsilon_s \varepsilon_o (|V| + |\phi_F| - 3kT/q)/qC_B]^{1/2} \\ &= [2q\varepsilon_s \varepsilon_o \phi_s - 3kT)/q^2 C_B]^{1/2}. \end{aligned} \tag{5.29}$$

The effective surface mobility is, as is the bulk mobility, proportional to temperature by the 3/2 power law.

$$\mu_{\text{eff}} \sim T^{-3/2}.$$

11. ELECTRIC FIELD AS A FUNCTION OF CARRIER CONCENTRATION IN THE SPACE CHARGE REGION

A relationship between electric field, potentials, and the carrier concentrations in the semiconductor space charge region is given by the dimensionless electric field F.

(a) Electric field:

Generally, the electric field is defined as

$$E = dV/dx \tag{5.30}$$

(b) Normalized (dimensionless) electric field:

The dimensional electric field (E) can be normalized with respect to thermal voltage (kT/q) and Debye length (L_D) as follows:

$$F(u_d, u_b) = E \frac{L_D}{kT/q}$$

$$= \left(\frac{\exp(u_d) - u_d - 1 + \exp(-2u_b)[\exp(-u_d) + u_d - 1]}{2[1 + \exp(-2u_b)]} \right)^{1/2}. \tag{5.31}$$

The above equation for $F(u_d, u_b)$ can be simplified under the following conditions.

(a) High impurity concentration (i.e., $n \gg n_i$ and $\exp(-2u_b) \ll 1$):

$$F(u_d, u_b) = \sqrt{2}[\exp(u_d) - u_d - 1 + \exp(-2u_b - u_d)]^{1/2}$$
$$= \sqrt{2}[n_d/n - \ln(n_d/n) - 1 + p_d/n]^{1/2} \tag{5.32}$$

(b) Heavy accumulation (i.e., $n_d \gg n$):

$$F(u_d, u_b) = \sqrt{2} \exp(u_d/2)$$
$$= \sqrt{2} (n_d/n)^{1/2} \tag{5.33}$$

(c) Strong inversion:

$$F(u_d, u_b) \approx \sqrt{2}[-(u_d + 1)]^{1/2} \tag{5.34}$$

(d) Large electric field (i.e., $F \gg 1$):

$$F(u_d, u_b) \approx \sqrt{2} \exp[-(u_d + 2u_b)/2]. \tag{5.35}$$

5-2 SURFACE DEPLETION AND INVERSION

Assumptions:

(a) All impurity atoms are ionized.
(b) Immobile charges are only due to donors and acceptors.
(c) There are no recombination centers in the space charge region.

The surface conditions in terms of carrier potentials are

(a) accumulation:
$$u_d > 0;$$

(b) depletion:
$$0 > u_d > -2u_b;$$

(c) inversion:
$$u_d < -2u_b.$$

The potentials are defined as follows:
  u = dimensionless carrier potential. It is a measure of the separation of extrinsic and intrinsic Fermi levels;

$$u = \ln (n_d/n_i) \qquad (5.36)$$

u_d, u_b = dimensionless potentials in depletion layer or semiconductor bulk, respectively;

$$u_d = \ln (n_d/n) = u - u_b \qquad (5.37)$$
$$u_b = \ln (n/n_i) \qquad (5.38)$$

if $n_d = 0$, then $u_b = u$; in an n-type semiconductor $u_b > 0$; in a p-type semiconductor $u_b < 0$.

n_d, n = carrier concentrations inside and outside the space charge region, respectively.

12. REFERENCES

(1) M. M. Atalla et al., *BSTJ*, **38**, 749 (1959).
(2) C. T. Sah, *Proc. IRE*, **49**, 1623 (1961).
(3) J. S. Blakemore, *Semiconductor Statistics*, Pergamon Press, Oxford, (1962).
(4) C. Goldberg, *Solid-State Electronics*, **7**, 593 (1964),
(5) A. S. Grove et al., *Solid-State Electronics*, **8**, 145 (1965).
(6) R. Hall and J. P. White, *Solid-State Electronics*, **8**, 211 (1965).
(7) A. S. Grove and D. J. Fitzgerald, *IEEE Transact. Electr. Dev.*, **ED-12**, 619 (1965).

(8) B. E. Deal et al., *J. El. Chem. Soc.*, **112**, 308 (1965).
(9) O. Leistiko et al., *IEEE Transact. Electr. Dev.*, **ED-12**, 248 (1965).
(10) C. T. Sah and H. C. Pao, *IEEE Transact. Electr. Dev.*, **ED-13**, 393 (1966).
(11) L. Vadasz and A. S. Grove, *IEEE Transact. Electr. Dev.*, **ED-13**, 863 (1966).
(12) A. S. Grove and D. J. Fitzgerald, *Solid-State Electronics*, **9**, 783 (1966).
(13) A. S. Grove, *Physics and Technology of Semiconductor Devices*, John Wiley and Sons, New York, 1967.
(14) B. E. Deal et al., *J. El. Chem. Soc.*, **114**, 266 (1967).
(15) A. G. Revesz and K. H. Zaininger, *RCA Review*, **29**, 22 (1968).
(16) S. M. Sze, *Physics of Semiconductor Devices*, John Wiley and Sons, New York, 1969.
(17) M. B. Das, *Solid-State Electronics*, **12**, 305 (1969).
(18) K. H. Zaininger and F. P. Heiman, *Solid State Technology*, **13**, 49 (May 1970).

5–3 MIS Capacitance

1. TOTAL CAPACITANCE OF AN MIS STRUCTURE

The total capacitance of an MIS structure consisting of semiconductor–insulator–metal layers is

$$1/C = dV_G/dQ_G + d\phi_s/dQ_G$$

$$= \frac{1}{(1/q)(\partial Q_s/\partial \phi_s)}$$

$$= \frac{1}{C_o} + \frac{1}{C_s}$$

$$= \frac{1}{C_o} + \frac{1}{C_{sc} + C_{ss}} \quad (5.39)$$

or, if referred to C_o,

$$\frac{C}{C_o} = \frac{1}{q} \frac{\partial Q_s/\partial \phi_s}{C_o}$$

$$= \frac{C_s}{C_s + C_o}$$

$$= \frac{C_{sc} + C_{ss}}{C_{sc} + C_{ss} + C_o} \quad (5.40)$$

The total (measured) capacitance (C) consists of the geometric capacitance (C_o), which would be measured if the semiconductor were replaced by a metal (see Figure 5.13), in series with the parallel combination of space charge and surface state capacities.

Since the total surface charge is

$$Q_s = Q_{sc} + Q_{ss}, \quad (5.41)$$

the total semiconductor surface capacitance

$$C_s = C_{sc} + C_{ss}. \quad (5.42)$$

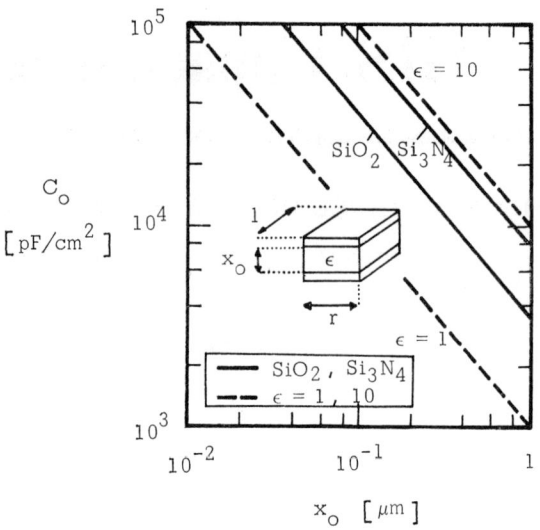

FIGURE 5.13

Capacitance per unit area (C_o) of a sheet capacitor of dielectric thickness x_o and dielectric constant ϵ.

C_s is frequency-dependent because of the finite rate of carrier generation and recombination. It is a function of the availability of carriers at the semiconductor surface.

An increase in gate voltage (i.e., the voltage at the metal contact) will produce the following changes at the semiconductor surface (see Figure 5.14). The application of a gate voltage (V_G) results in the formation of a depletion region whose width (x_d) adds effectively to the thickness of the dielectric (x_o). Hence, the capacitance will decrease with increasing voltage until an inversion region is formed in which most of the additional charge will appear. Consequently, the depletion region will reach a maximum ($x_{d\,max}$) and the capacitance simultaneously a minimum (C_m). If the measurement frequency is low compared to the generation rate of minority carriers, the carriers in the inversion region will be able to follow the variation in voltage amplitude and the capacitance will rise again to the value of C_o. If the carriers cannot follow the signal variation, the capacitance will remain approximately C_m. The maximum width of the depletion layer is independent of frequency.

The following conditions exist for the total MIS capacitance.

(a) No surface depletion or accumulation:

If for an n-type semiconductor the gate voltage is positive or for a p-type semiconductor it is negative, no depletion region exists and the semiconductor merely acts as a resistor in series with the oxide capacitance. In this case

$$C/C_o = 1. \tag{5.43}$$

(b) Surface depletion:
In this case the capacitance is voltage-dependent.

$$C/C_o = [1 + 2(\varepsilon_{SiO_2}^2 \varepsilon_o/\varepsilon_s)(V_G - \phi_{MS} + Q_{ss}/C_o)/qC_B x_o^2]^{-1/2} \tag{5.44}$$

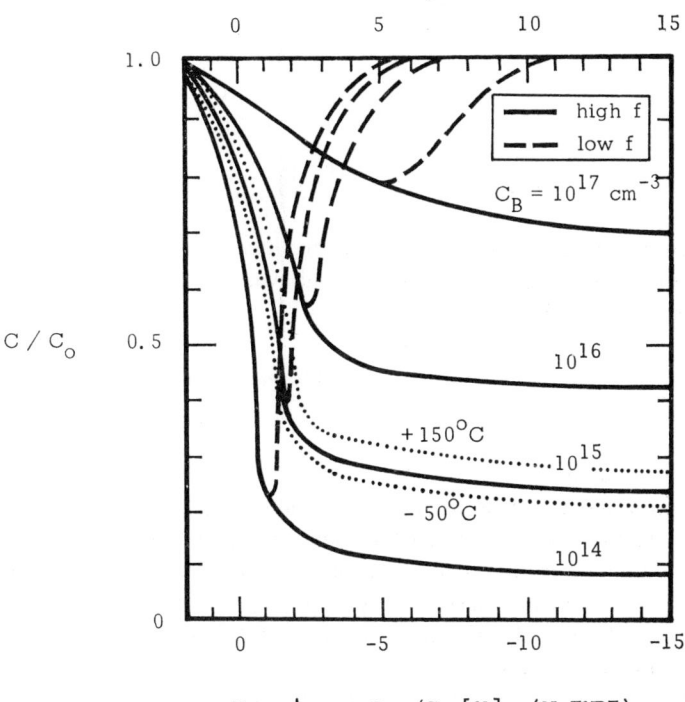

FIGURE 5.14

Normalized theoretical MIS capacitance (C/C_o) as function of gate voltage (V_G) and semiconductor bulk impurity concentration (C_B). Si-SiO$_2$ system, 300°K; low-frequency and high-frequency case. The upper scale refers to p-type silicon, the lower scale to n-type silicon. The oxide thickness is $x_o = 0.1$ μm. The variation of C/C_o with temperature is indicated for $C_B = 10^{15}$ cm^{-3} (dotted curves; high-frequency case).

(c) Inversion:

At the onset of strong inversion (i.e., when the energy bands are sufficiently bent and $\phi_s \geq \phi_F$) the depletion layer width has reached a maximum ($x_{d\,max}$) and the capacitance has reached a minimum (C_m) at voltage $V_G = V_T = V_m$. For an n-type semiconductor $V_m < 0$, for a p-type semiconductor $V_m > 0$. In the inversion region minority carriers will accumulate near the surface.

$$C/C_o = [1 + 2(\varepsilon_{SiO_2}/\varepsilon_s)(x_{d\,max}/x_o) + 2\phi_F]^{-1/2}. \qquad (5.45)$$

If

$$qC_B x_{d\,max}/2\varepsilon_{SiO_2}\varepsilon_o \gg \phi_F$$

and

$$2(\varepsilon_{SiO_2}/\varepsilon_s)(x_{d\,max}/x_o) \leq 1,$$

then

$$C/C_o = [1 + 2(\varepsilon_{SiO_2}/\varepsilon_s)(x_{d\,max}/x_o)]^{-1/2}. \qquad (5.46)$$

Further increase in voltage will not produce an additional charge within the inversion layer. The gate voltage which produces strong surface inversion (i.e., the turn-on voltage) is

$$V_G = V_T = V_m = \pm Q_B/C_o + \phi_{s(inv)}$$
$$= qC_B x_{d\,max} x_o/\varepsilon_{SiO_2}\varepsilon_o + 2\phi_F. \qquad (5.47)$$

In the above equations

$\varepsilon_s, \varepsilon_{SiO_2}$ = dielectric constants of semiconductor and insulator

C_B = semiconductor impurity concentration in the bulk; under conditions where impurity redistribution close to the surface has taken place during thermal oxidation, C_B has to be replaced by the average or effective impurity concentration C_a.

The flat-band capacitance of an MIS system (Figure 5.15) is defined as the capacitance at zero surface potential. In this case

$$\phi_s = 0 \qquad (5.48)$$
$$V_a = -\phi_F = -(E_i - E_F)/q \qquad (5.49)$$
$$C_{FB} = (\varepsilon_{SiO_2}\varepsilon_s)/x_o + (1/\sqrt{2})(\varepsilon_{SiO_2}/\varepsilon_s)L_D. \qquad (5.50)$$

For semiconductors other than silicon and insulators other than SiO_2 the illustrations have to be modified by the use of the proper dielectric constants.

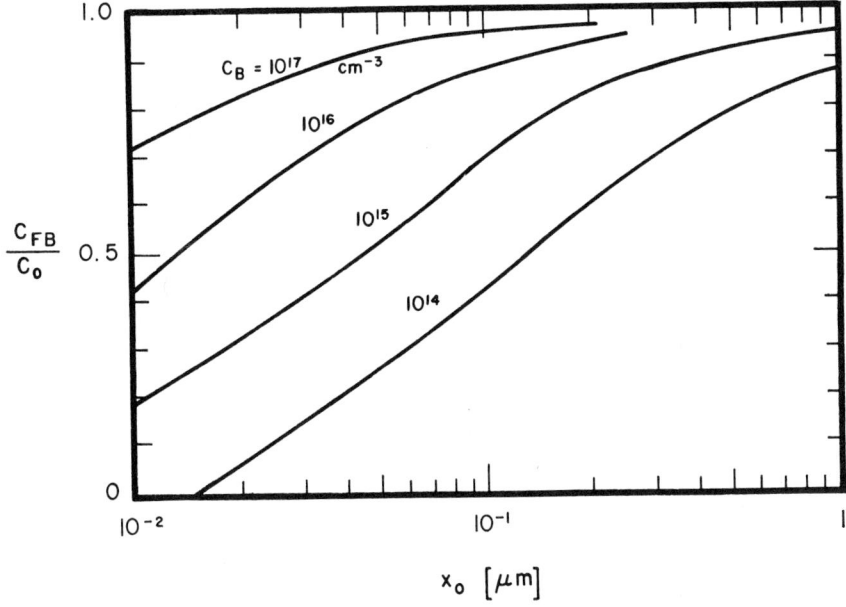

FIGURE 5.15

Ratio of flat-band capacitance (C_{FB}) to oxide capacitance (C_o) vs. oxide thickness (x_o) and semiconductor impurity concentration (C_B). Si-SiO$_2$ system, 300°K.

Note also that the Debye length L_D is a function of the dielectric constant. For dielectrics other than SiO$_2$ replace x_o by $x_\varepsilon \varepsilon_{SiO_2}/\varepsilon_{dielectric}$, where x_ε is the equivalent thickness of the dielectric.

The slope of the C-V curve is, for a given capacitance ratio C/C_o, the difference between the voltage at which for the slope $C/C_o = 0$ and that at which $C/C_o = 1$. This voltage difference is denoted as V_{01}. The slope of the C-V curve for a given C/C_o (see Figure 5.16) is

$$d(C/C_o)/dV = 1/V_{01}$$
$$= (C/C_o)^3 (q/2kT)(\varepsilon_{SiO_2}/\varepsilon_s)^2 (L_D/x_o)^2. \tag{5.51}$$

The slope is inversely proportional to $C_B x_o^2$ since L_D^2 is inversely proportional to the impurity concentration (C_B). L_D is the semiconductor Debye length. In Figure 5.16 the slope of C-V curves is presented for the point $C/C_o = 0.5$ for silicon.

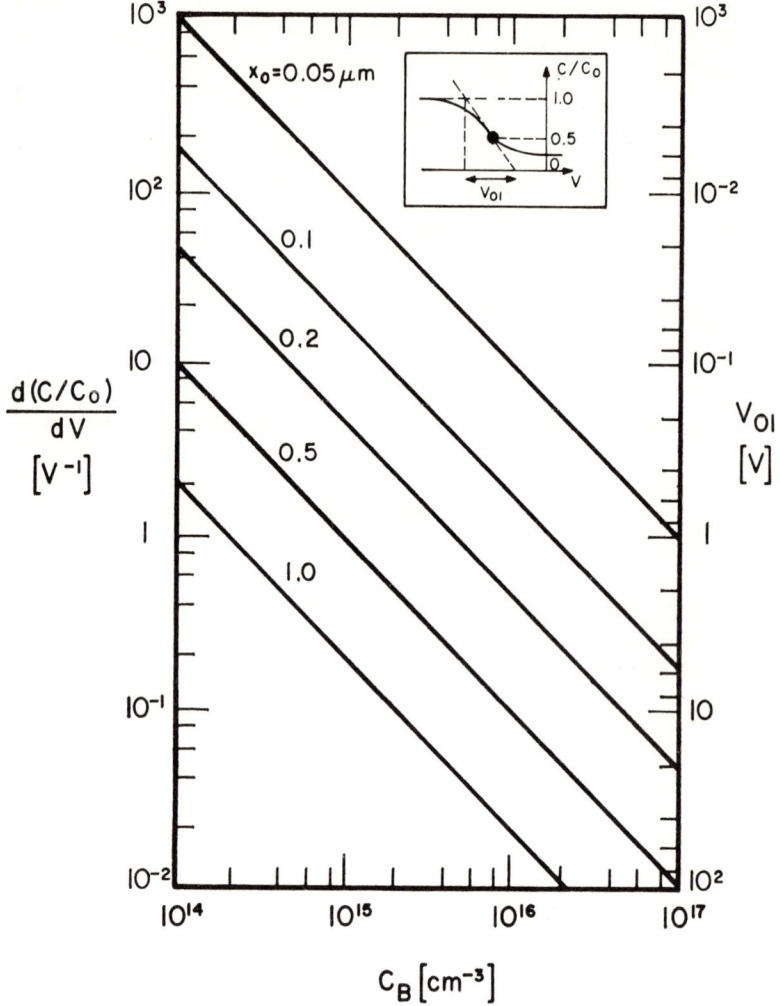

FIGURE 5.16

Slope of C-V curves at $C/C_o = 0.5$ vs. semiconductor impurity concentration (C_B) and oxide thickness (x_o). Si-SiO$_2$ system.

2. MAXIMUM AND MINIMUM CAPACITANCE

The capacitance of an MIS structure varies with applied voltage from a maximum $C/C_o = 1$ to a minimum $C/C_o = C_m/C_o < 1$. The appropriate expressions for C_o and C_m are as follows (Figure 5.17).

(a) Maximum capacitance per unit area (C_o):

$$C_o = \varepsilon_{SiO_2} \varepsilon_o / x_o \qquad (5.52)$$

(b) Minimum capacitance per unit area (C_m):

$$C_m = \varepsilon_o/(x_o/\varepsilon_{SiO_2} + x_{d\,max}/\varepsilon_s)$$
$$\approx \varepsilon_s \varepsilon_o / 5L_D$$
$$= (\varepsilon_s \varepsilon_o/5)(2kT\varepsilon_s \varepsilon_o/q^2 C_B)^{1/2} \qquad (5.53)$$

(c) Capacitance ratio:

$$C_m/C_o = 1/[1 + (\varepsilon_{SiO_2}/\varepsilon_s)(x_{d\,max}/x_o)]. \qquad (5.54)$$

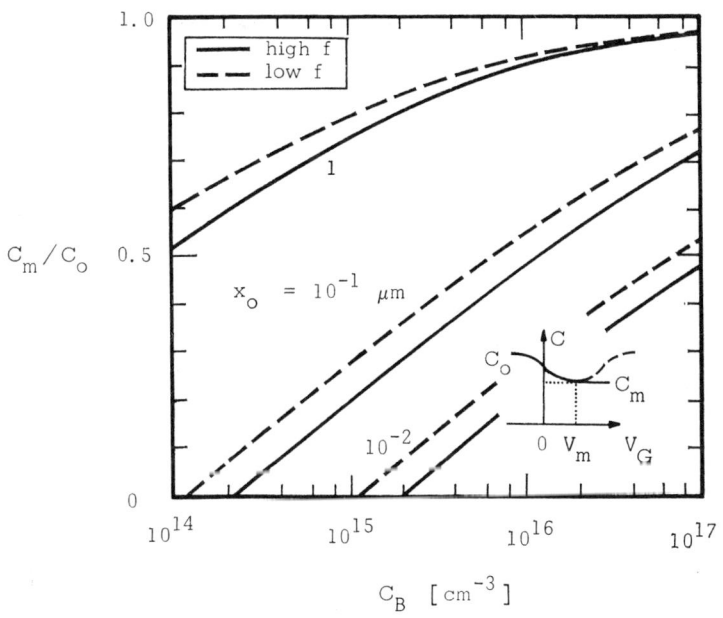

FIGURE 5.17

Ratio of minimum to maximum MIS capacitance (C_m/C_o) as function of semiconductor bulk impurity concentration (C_B) and oxide thickness (x_o). Si-SiO$_2$ system, 300°K. Low-frequency and high-frequency case.

The voltage at which the capacitance minimum occurs is denoted as V_m and is shown in Figure 5.18.

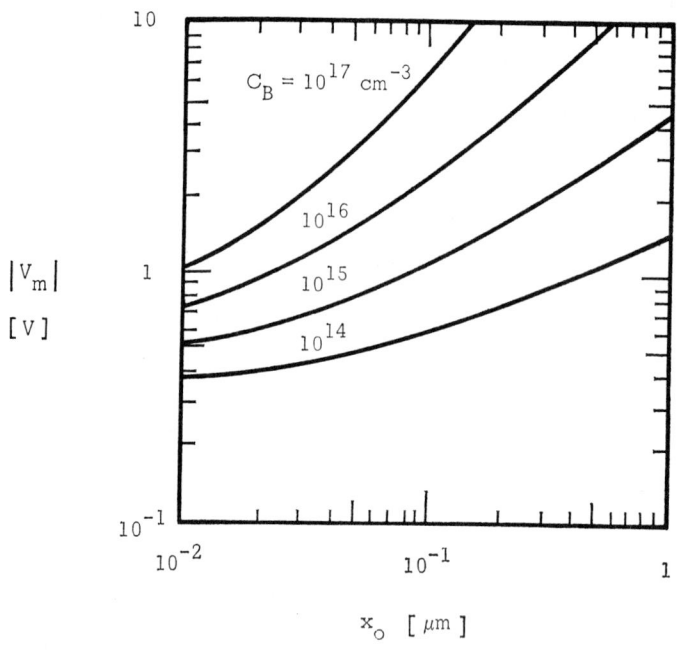

FIGURE 5.18

Gate voltage (V_m) at which the capacitance C/C_o of an MIS system is at its minimum as function of oxide thickness (x_o) and semiconductor bulk impurity concentration (C_B). Si-SiO$_2$ system, low-frequency case, 300°K.

3. OXIDE CAPACITANCE

The oxide or insulator capacitance per unit area is

$$C_o = \varepsilon_{SiO_2} \varepsilon_o / x_o \tag{5.55}$$

where ε_{SiO_2} = dielectric constant of SiO$_2$ or insulator; ε_{SiO_2} = 3.9 for SiO$_2$
ε_o = permittivity of free space; ε_o = 8.86 · 10^{-14} F/cm
$\varepsilon_{SiO_2} \varepsilon_o$ = 0.346 · 10^{-12} F/cm = 0.346 pF/cm for SiO$_2$.

The oxide capacitance is affected by the addition of impurities to the oxide owing to their influence on the dielectric constant. At very high frequencies the dielectric constant is also slightly wavelength-dependent; in addition, at certain small wavelengths, absorption maxima occur which result in significant increases in the complex dielectric constant. For small capacitor dimensions, fringe fields may not be negligible and may affect the capacitance.

A sheet capacitor has finite values of inductance and impedance which are dimension-dependent. Appropriate expressions for the electric properties of a sheet capacitor are as follows.

(a) Capacitance:
$$C_o = \varepsilon_{SiO_2}/3.6\pi x_o \quad [\text{pF/cm}^2] \tag{5.56}$$

(b) Inductance:
$$L_o = 4\pi x_o/r^2 \quad [\text{nH/cm}^2] \tag{5.57}$$

(c) Impedance:
$$Z_o = 120\pi(x_o/r^2)\varepsilon_{SiO_2}^{-1/2} \quad [\Omega/\text{cm}^2] \tag{5.58}$$

if $x_o \ll r$

$$Z_o = 120\pi(l/r)\varepsilon_{SiO_2}^{1/2} \ln(4x_o/r) \quad [\Omega/\text{cm}^2] \tag{5.59}$$

if $x_o \gg r$.

In these expressions the width of the capacitor is given as r and the length as l.

4. SURFACE CAPACITANCE

(a) Space charge capacitance (C_{sc}):

In the presence of surface states at the semiconductor-insulator interface the space charge region can be classified into the four cases: accumulation, flat band, depletion, and inversion. If the surface state density (Q_{ss}/q) is so high that the charges trapped there shield the interior of the semiconductor, not all four of these cases can be realized, i.e., the semiconductor surface behaves like a metal.

The charge in the semiconductor surface layer (Q_{sc}/q) is an unambiguous function of semiconductor surface potential (ϕ_s or u_s) and is associated with a space charge capacitance

$$C_{sc} = (1/q)(\partial Q_{sc}/\partial \phi_s). \tag{5.60}$$

Figure 5.6 shows the space charge capacitance for two extreme conditions:

(i) *Flat band*:
In this case $\phi_s = 0$, and the space charge capacitance displays a maximum, $C_{sc} = C_{sco}$.

(ii) *Inversion*:
In this case $\phi_s = \phi_{s(inv)}$, and space charge and space charge capacitance display a minimum, $C_{sc} = C_m$. The associated total space charge is $Q_{s(inv)}/q$:
for an *n*-type semiconductor: $\phi_{s(inv)} < 0$, $Q_{s(inv)} > 0$;
for a *p*-type semiconductor: $\phi_{s(inv)} > 0$, $Q_{s(inv)} < 0$.

(b) Surface state capacitance (C_{ss}):
In the presence of surface states, charges are distributed among these states and the semiconductor space charge region. The extent of the ionization of the surface states depends upon their energy level relative to the Fermi level. The associated surface state capacitance is

$$C_{ss} = (1/q)(\partial Q_{ss}/\partial \phi_s). \tag{5.61}$$

If the surface state density is constant, i.e., $\partial Q_{ss}/\partial \phi_s = 0$, then $C_{ss} = 0$ and the only voltage-dependent term of the total capacitance is the space charge capacitance:

$$C = \frac{1}{1/C_o + 1/C_{sc}}, \tag{5.62}$$

$$\frac{C}{C_o} = \frac{1}{1 + C_o/C_{sc}}. \tag{5.63}$$

The space charge capacitance C_{sc} depends upon surface potential (ϕ_s), semiconductor impurity concentration (C_B), and temperature (T). The surface state capacitance C_{ss} depends upon the particular spatial and energy distribution of surface states and their occupancy at a given external electric field.

5. FREQUENCY DEPENDENCE OF SURFACE CAPACITANCE

The space charge capacitance is independent of frequency (f) for accumulation and depletion if $f < 1/\tau_\varepsilon$, where τ_ε is the dielectric relaxation time of the semiconductor. In the case of inversion, C_{sc} depends upon frequency (low-frequency and high-frequency case), depending upon if the minority carriers can follow the variation of the electric field or not. Low-frequency and high-frequency behavior are distinguished at the transition frequency f_t.

(a) Low frequency:
If the signal frequency is low enough, the minority carriers will follow the variation of the voltage and contribute to the capacitance. Accordingly, with increasing voltage, C_s will rise again and approach C_o:

$$C_s = \varepsilon_s \varepsilon_o (p_S - n_S + C_B) q/Q_s, \tag{5.64}$$

where n_S, p_S = carrier concentrations at the surface

$$n_S = n_i \exp(u_S - u_F)$$
$$p_S = n_i \exp(u_F - u_S)$$
$$u_S = q^2 C_B x_d^2 / \varepsilon_s \varepsilon_o kT$$
$$u_F = (E_i - E_F)/kT = \ln(C_B/n_i).$$

(b) High frequency:
At sufficiently high frequency only the majority carriers are able to follow the variation of the measurement signal and the equilibrium theory cannot be applied. In this case

$$C_s = qC_B(dx_d/d\phi_s)$$
$$= \varepsilon_s \varepsilon_0/x_d. \quad (5.65)$$

Since x_d approaches $x_{d\,\text{max}}$ when the semiconductor surface becomes inverted, C_s will approach a constant value in the high-frequency model.

The transition frequency (f_t) from low-frequency to high-frequency behavior of the capacitance ratio C/C_o is a function of temperature, oxide thickness, semiconductor impurity concentration, and degree of light injection. The transition frequency has an activation energy of approximately half the semiconductor energy gap, i.e., for silicon

$$E_a \approx E_G/2 = 0.55 \text{ eV},$$

indicating the influence of mid-gap recombination centers on minority carrier generation.

For example, for n-type or p-type of silicon impurity concentration $C_B = 10^{16}$ cm^{-3} and oxide thickness $x_o = 0.2$ μm the transition frequency

$$f_t \approx 50 \text{ Hz}.$$

6. TEMPERATURE DEPENDENCE OF MIS CAPACITANCE

The temperature dependence of the C-V characteristics (see Figure 5.14) is due to

(a) variation of the intrinsic carrier concentration with temperature;
(b) variation of the carrier generation rate with temperature; this facilitates the transition from low- to high-frequency type behavior.

An increase in temperature results in an increase of the capacitance ratio C/C_o at a given gate voltage. For example, for an Si-SiO$_2$ system in which the silicon has an impurity concentration $C_B = 10^{15}$ cm^{-3} and $x_o = 0.1$ μm the variation of C/C_o from -50 to $150°$C is approximately 0.1.

7. REFERENCES

(1) G. Megla, *Dezimeterwellentechnik*, Fachbuchverlag, Leipzig, 1952.
(2) B. E. Deal et al., *J. El. Chem. Soc.*, **112**, 308 (1965).
(3) A. S. Grove et al., *Solid-State Electronics*, **8**, 145 (1965).
(4) F. P. Heiman, *IEEE Transact. Electr. Dev.*, **ED-13**, 855 (1966).
(5) A. Goetzberger, *BSTJ*, **45**, 1097 (1966).
(6) K. H. Zaininger and F. P. Heiman, *Solid State Technology*, **13**, 49 (May 1970).

5–4 Metal-Semiconductor Contacts

1. FREE ELECTRONS

Thermionic emission in metal–vacuum and metal–semiconductor systems involves the transition of free carriers (i.e., carriers which are not bound to individual atoms) whose availability and energy determine the electric characteristics of the vicinity of the interface and the transport of charge (i.e., the current density) across it.

Each electron carries the electric charge

$$q = -1.602 \cdot 10^{-19} \text{ Coulomb}$$

and has a rest mass in free space

$$m_o = 9.11 \cdot 10^{-28} \text{ g}.$$

The mass of an electron is different in a crystal lattice where the effective mass has to be used. The electron mass in a lattice is

$$m_n = h^2/8a^2 \, \Delta E \qquad (5.66)$$

where a is the lattice constant and ΔE is the width of the conduction band. Each gram of matter contains approximately $3 \cdot 10^{23}$ electrons which together have a charge of about 50,000 Coulomb. The energy of an electron (E_n) is related to its de Broglie wavelength (λ_n) by

$$E_n = h^2/2m_o \lambda_n, \qquad (5.67)$$

i.e., the wavelength of an electron is

$$\lambda_n \approx 12/E_n, \qquad (5.68)$$

where λ_n is expressed in Å and E_n in eV.

Although electrons are found in all matter, it is not easy to separate them from the positive charges, i.e., to isolate free electrons. If all electrons contained in one gram of matter were separated from the positive charges and placed one meter away, the two charge clouds would attract each other with a force of about 10^{21} Newtons.

An electron can be separated from matter only if there is sufficient energy available. In this case, a positive ion remains. If an electron escapes from an isolated atom or molecule (i.e., in the gaseous state), the required energy is called ionization energy; if it escapes from a liquid or a solid, the required energy is called work function. Free electrons are usually obtained from solids.

2. THE METAL–VACUUM SYSTEM

There are three ways to obtain free carriers from a solid, e.g., a metal.

(a) Temperature increase:
The surface of a heated solid will emit electrons; the energy required, i.e., the work function, will be taken from the energy of the molecular motion. The work function of the solid is defined as the energy required to remove an electron from the Fermi level to the vacuum level. The number of free electrons emitted per unit area of a uniform surface per second, i.e., the thermionic emission current density of electrons flowing from metal into vacuum, is, according to Richardson's law,

$$I_{MV} = A_{no}(1 - r_n)T^2 \exp(-q\phi_M/kT)$$
$$\approx A_{no} T^2 \exp(-q\phi_M/kT). \tag{5.69}$$

ϕ_M is the work function of the metal. It depends significantly on surface contamination and slightly on temperature and electric field at the surface. Work functions of clean metals range typically from 2 to 6 eV. In order to achieve a high density of free electrons at a given temperature, a small work function is desirable. (For this reason emitting cathodes are frequently coated with alkali oxides which have a small work function.)

The term r_n is the reflection coefficient for electrons crossing the potential barrier at the surface when the electric field just outside the surface is zero; r_n is practically independent of temperature and electric field at the surface. For metals and semiconductors, where $r_n \approx 0.05$, the reflection coefficient is usually negligible.

The term A_{no} is the Richardson constant and is

$$A_{no} = 4\pi m_o k^2 q/h^3 = 120 \text{ A/cm}^2 \, {}^\circ\text{K}^2, \tag{5.70}$$

where m_o is the free electron mass. The Richardson constant in a metal–semiconductor system (A_n) differs from that in a metal–vacuum system (A_{no}) mainly because in the former the effective electron mass enters the equations, whereas in the latter the free electron mass is used (see below).

(b) Illumination:
If a metal–vacuum interface is illuminated, free electrons will be emitted from the metal. The energy required to free an electron will be taken from the light. Each of the photons carries an energy

$$E_{\text{photon}} = hv, \tag{5.71}$$

where v is the frequency of light.

The energy of one photon must be sufficient to free one electron, i.e., the photon energy must be greater than the work function of the metal. The photon energy and the momentum will be divided between the lattice and the emitted electron. The energy balance of the photo effect is therefore

$$h\nu \geq h\nu_{\min} + m_o v^2/2.$$

The photo effect requires a minimum frequency $\nu_{\min}$; the kinetic energy of the free electrons increases with the frequency of light. For large metal work functions ultraviolet light is required to free electrons.

(c) Bombardment with fast particles:

Free electrons can also be obtained from a metal by bombardment of the surface with fast atoms, ions, or electrons, i.e., particles which have a high kinetic energy $mv^2/2$. Electrons generated in this way are called secondary electrons.

3. THE METAL–SEMICONDUCTOR SYSTEM

If a metal and a semiconductor are brought into intimate contact, the Fermi levels of the two components (which before contact and immediately after contact has been established are different by an amount equivalent to the built-in voltage) must become equal. Before equality is established electrons will flow from the region with the lower work function to the region with the higher one until both Fermi levels are equal. As a result, there will be no flow of carriers after equilibrium has been established. The equilbrium will be disturbed again if an electric field (bias) is applied; in this case a current will flow across the interface under steady-state conditions. Theoretically, the difference between the Fermi levels in conjunction with the energy gap determines the barrier height of a metal–semiconductor contact; in practice, discrepancies are observed because of dipole layers of trapped surface charges.

The changes in band structure after metal and semiconductor have been brought into contact are:

(a) The initial density and distribution of surface states of the semiconductor surface (intrinsic and defect states) will be modified due to the proximity of the metal atoms.
(b) The surface dipole layer, caused by a distortion of the electron cloud at the surface (which, together with the chemical electronegativity, determines the metal work function and semiconductor electron affinity), will be modified.
(c) In the ensuing barrier, surface charges induced in the metal will terminate on both (i) charges within the semiconductor within a space charge layer width and (ii) charges on surface states.

5-4 METAL-SEMICONDUCTOR CONTACTS

The band bending in the semiconductor near the metal–semiconductor interface (E_B) is related to the barrier height of the metal–semiconductor system as

$$E_B = q\phi_{MS} - (E_c - E_F), \qquad (5.72)$$

where $E_c - E_F$ is the energy difference between Fermi level and the bottom of the conduction band.

The saturation current which flows across the metal–semiconductor interface after establishing contact (assuming no external electric field) consists of two components:

(a) a minority carrier current generated in the semiconductor within the space charge layer or within a diffusion length of it;
(b) a current in the metal consisting of carriers with sufficient energy to surmount the barrier.

The current density across the metal–semiconductor interface is generally

$$I_{MS} = A_n T^2 \exp(-q\phi_{MS}/kT)\{\exp[q(V_D + V_a)/kT] - 1\} \qquad (5.73)$$

(where ϕ_{MS} is the barrier height and V_a is the applied voltage) and the following differences are observed compared to the metal–vacuum case.

(a) The Richardson constant A_n may differ considerably from the free space value $A_{no} = 120 \text{A/cm}^2 \, °\text{K}^2$.
(b) The Richardson constant A_n is not constant but is a function of the electric field (see Figure 5.19).
(c) The barrier height of the metal–semiconductor system (ϕ_{MS}) may vary more rapidly with electric field than the image force model predicts.

The Richardson constant of a metal–semiconductor contact differs from that in a metal–vacuum system for three reasons.

(a) Carriers which are thermally emitted over a potential barrier from metal into semiconductor have an effective mass which differs from the free carrier mass. In semiconductors with an isotropic effective mass (e.g., GaAs) the Richardson constant is reduced by a factor m_n/m_o; in semiconductors with several valleys (e.g., Si) the longitudinal and transverse electron masses have to be taken into account.
(b) Carriers which are emitted into the high-field region near the metal–semiconductor interface are scattered by optical phonons.
(c) The dielectric constants of most semiconductors are approximately an order of magnitude higher than that of vacuum, resulting in a decreased barrier width and an increased tunneling probability.

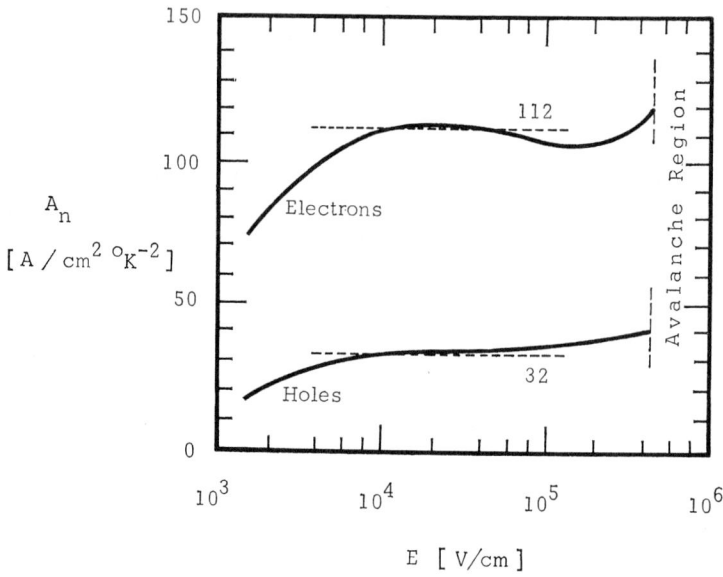

FIGURE 5.19

Effective Richardson constant (A_n) of majority carriers as function of electric field in n-type or p-type silicon of impurity concentration $C_B = 10^{16} \text{cm}^{-3}$. $300°\text{K}$.

For an anisotropic semiconductor the Richardson constant

$$A_n = A_{no}(l_I^2 m_j^* m_k^*)^{1/2}$$
$$= (4\pi q k^2/h^3)(l_I^2 m_j^* m_k^*)^{1/2}, \qquad (5.74)$$

where l_I = direction cosine of the current density relative to one of the principal axes of the effective mass tensor, represented by m_j^*

m_j^*, m_k^* = jth and kth component of the effective mass tensor.

Optical phonon scattering will affect A_n through the probability of electron or hole transmission over the potential barrier, so that an effective Richardson constant can be defined for anisotropic semiconductors which takes optical phonon scattering into account.

$$A_{n\text{ eff}} = A_n[f_P f_Q/(1 + f_P f_Q v_R/v_D)], \qquad (5.75)$$

where f_P, f_Q = probability of electron or hole transmission through a potential barrier in the presence of either scattering by optical phonons, or quantum mechanical reflection or tunneling, respectively

v_R = effective recombination velocity between potential barrier maximum and metal–semiconductor interface

v_D = effective diffusion velocity in the depletion layer at the metal–semiconductor interface.

Experimentally observed values of the Richardson constant as function of the orientation of silicon are given in Table 5.3.

TABLE 5.3

Richardson Constant in Silicon

Orientation	A_n[A/cm^2°K^2]	
	n-Type	p-Type
⟨100⟩	57	182
⟨111⟩	55	182

4. METAL AND SEMICONDUCTOR WORK FUNCTIONS

(a) Work function of a metal:

Conduction electrons can escape from a metal into free space only if they can overcome the thermionic work function of the metal (ϕ_M) which is defined as the energy required to remove an electron from the Fermi level of the metal to the vacuum energy level. The metal work function is sensitive to surface contamination. Work functions of selected clean metals are given in Table 5.4; the spread of values for a given metal is a result of different measurement methods.

(b) Work function of a semiconductor:

The escape of electrons from a semiconductor into free space requires a thermionic work function (ϕ_S) which is defined as the energy required to remove an electron from the Fermi level of the semiconductor to the vacuum energy level.

$$\phi_S = \chi_s + F_G/?q + \phi_F \tag{5.76}$$

where χ_s = electron affinity of the semiconductor, i.e., the energy required to remove an electron from the bottom of the conduction band to the vacuum energy level

ϕ_F = Fermi potential

$$\phi_F = -(E_F - E_i)/q;$$

for an n-type semiconductor $\phi_F < 0$,
for a p-type semiconductor $\phi_F > 0$.

TABLE 5.4
Work Functions of Selected Metals (300°K)

Metal	$q\phi_M$ [eV]
Ag	4.3–5.1
Al	4.1–4.3
Au	4.7–5.0
Ba	2.4–2.6
Ce	1.9–2.0
Cr	4.5–4.6
Cu	4.3–4.7
Mg	3.2–3.7
Mo	4.3–4.5
Ni	4.5–4.7
Pt	5.2–5.3
Si	5.1–5.2
W	4.5–4.6

The dependence of ϕ_S upon the impurity concentration of the semiconductor is a result of the concentration dependence of ϕ_F. Semiconductor work functions are given in Table 5.5.

(c) Metal–semiconductor work function:

The barrier height of a metal–semiconductor system (ϕ_{MS}) can be predicted by two models, depending upon the magnitude of the surface state density.

TABLE 5.5
Semiconductor Work Function as a Function of Impurity Concentration (300°K)

Semi-conductor	$q\phi_s$[eV]					
	n-type			p-type		
	10^{14} cm^{-3}	10^{15} cm^{-3}	10^{16} cm^{-3}	10^{14} cm^{-3}	10^{15} cm^{-3}	10^{16} cm^{-3}
Si	4.32	4.26	4.20	4.82	4.88	4.94
Ge	4.43	4.38	4.33	4.51	4.56	4.61
GaAs	4.44	4.37	4.31	5.14	5.21	5.27

(i) *Surface state model*:
In this model it is assumed that the surface state density of the semiconductor surface is large (i.e., $Q_{ss}/q > 10^{14}$ cm^{-2}). In this case the barrier height is independent of the properties of the metal and of the impurity concentration of the semiconductor. It is

for an *n*-type semiconductor $\quad \phi_{MS} \approx (2/3)E_G$,
for a *p*-type semiconductor $\quad \phi_{MS} \approx (1/3)E_G$.

Based on this model, the barrier heights given in Table 5.6 are expected. The temperature dependence of ϕ_{MS} is similar to that of the energy gap E_G.

TABLE 5.6

Metal-Semiconductor Work Functions for Selected Semiconductors Based on the Surface State Model (300°K)

Semi-conductor	$q\phi_{MS}$[eV]	
	n-type	*p*-type
Si	0.74	0.37
Ge	0.44	0.22
GaAs	0.96	0.48

(ii) *Work function difference model*:
In this model it is assumed that the surface state density is negligible and that the barrier height is determined by the difference in the work functions of metal (ϕ_M) and semiconductor (ϕ_S).

$$\phi_{MS} = \phi_M - \phi_S. \tag{5.77}$$

In this case the barrier height depends upon the nature of the metal and the impurity concentration of the semiconductor. Barrier heights based on this model are given in Table 5.7.

This model holds only qualitatively. Deviations of observed values of ϕ_{MS} from this ideal model are due to various effects, as surface states, image forces, contamination, etc. The work function model holds better, on a relative basis, for covalent semiconductors (e.g., Si) than for other semiconductors (e.g., GaAs). The metal–semiconductor work function for Si–Al and Si–Au contacts based on the work function model is shown in Figure 5.20.

TABLE 5.7

Metal-Semiconductor Work Functions for Selected Semiconductors Based on the Work Function Model; Semiconductor Impurity Concentration $C_B = 10^{15}$ cm^{-3} (300°K)

Semi-conductor	$q\phi_{MS}$[eV]							
	n-type				p-type			
	Al	Au	Pt	W	Al	Au	Pt	W
Si	−0.06	+0.54	+1.04	+0.34	−0.68	−0.08	+0.42	−0.28
Ge	−0.18	+0.42	+0.92	+0.24	−0.36	+0.24	+0.74	+0.04
GaAs	−0.17	+0.43	+0.93	+0.23	−1.01	−0.41	+0.09	−0.61

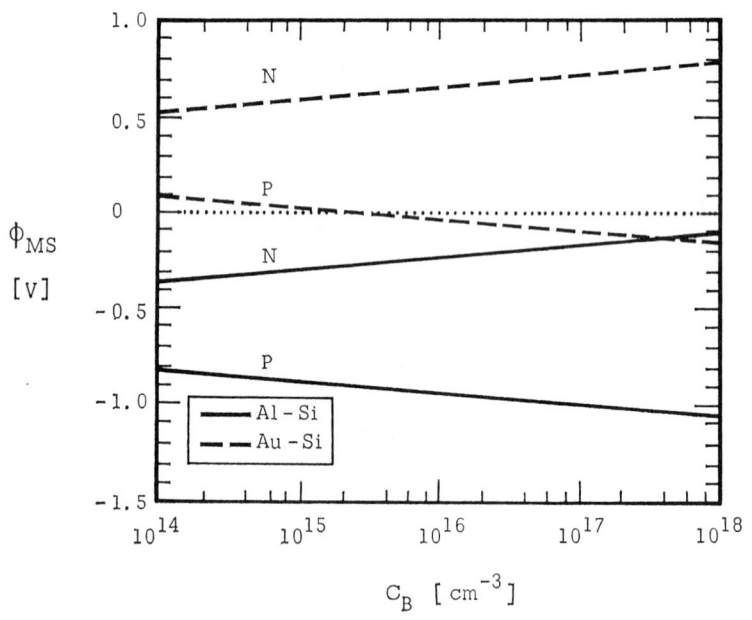

FIGURE 5.20

Metal-semiconductor contact potential (ϕ_{MS}) as function of semiconductor impurity concentration (C_B) for aluminum and gold contacts to n-type and p-type silicon. 300°K.

5. PROPERTIES OF THE METAL–SEMICONDUCTOR BARRIER

At the metal–semiconductor interface a potential barrier of height ϕ_{MS} and width x_b is established as a result of charges at the surface. These charges are located energetically close to the center of the energy gap. The presence of an electrostatic barrier results in the rectifying properties of a metal–semiconductor contact. A rectifying metal–semiconductor contact (junction) is generally known as a Schottky barrier diode. The condition for obtaining a rectifying contact is

for an *n*-type semiconductor $\quad \phi_M - \phi_S > 0,$
for a *p*-type semiconductor $\quad \phi_M - \phi_S < 0.$

The space charge within the barrier region is due to two contributions:

(a) The impurity atoms within the barrier are more ionized than those in the bulk and are not compensated by carriers as are those in the bulk. If $V_D + V_a \gg kT/q$, there are hardly any free electrons except at $x = x_b$. These ionized impurity atoms cause a space charge layer similar to that near a *p-n* junction.
(b) In the proximity of the surface an additional number of free carriers are found. The surface charge due to these carriers results from surface states which arise because of the termination of the crystal lattice.

The width of the barrier or depletion layer is

$$x_b = [\varepsilon_s(V_D + V_a)/2\pi q C_B]^{1/2}, \qquad (5.78)$$

where V_D is the built-in voltage of the contact; V_D is given in Figure 6.1 (where values for the one-sided step junction have to be used). The applied voltage is V_a, which in equilibrium is $V_a = 0$. The variation of the barrier (depletion layer) width with voltage is shown in Figure 5.21 (solid lines) as a function of impurity concentration and dielectric constant. In this treatment it is assumed that image effects and electronic space charges are neglected and that the charge continuity condition (dotted lines of Figure 5.21) holds, i.e.,

$$x_b(V_D + V_a) \geq 2.5 \cdot 10^5 \pi q/\varepsilon_s.$$

A relationship between barrier height and metal work function is as follows, if the surface state density is much greater than the charge in the space charge layer (surface state model):

$$\phi_{MS(n)} = \gamma_b(\phi_M - \chi_s) + (1 - \gamma_b)(E_G/q - \phi_{so}) \quad (n\text{-type}), \qquad (5.79a)$$
$$\phi_{MS(p)} = \gamma_b(\chi_s - \phi_M) + (1 - \gamma_b)\phi_{so} + \gamma_b E_G/q \quad (p\text{-type}), \qquad (5.79b)$$

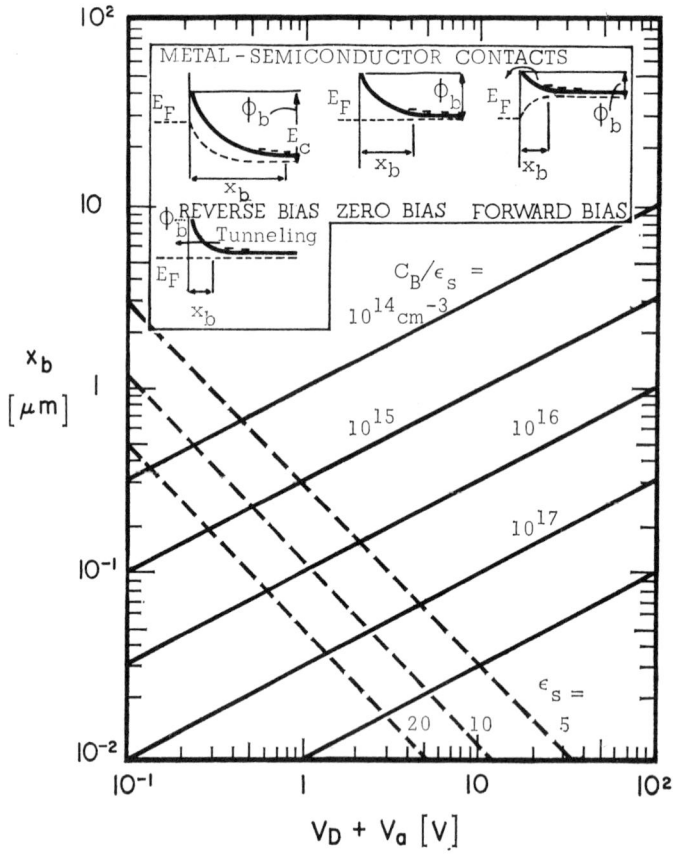

FIGURE 5.21

Width (x_b) of a Schottky barrier as function of voltage ($V_D + V_a$), semiconductor impurity concentration (C_B), and semiconductor dielectric constant (ϵ_s). 300°K. Electronic space charges are neglected and it is assumed that $V_D + V_a \gg kT/q$. The limitations due to charge continuity are indicated by the dashed lines. The inset shows the formation of a barrier for an n-type semiconductor (Schottky barrier; upper part) and an ohmic contact (lower part).

where $\gamma_b = x_i \varepsilon_o / (\varepsilon_i \varepsilon_o + x_b^* Q_{ss}/q)$
ε_i = dielectric constant of surface charge layer
x_b^* = width of surface charge layer
ϕ_{so} = neutral energy level.

The neutral energy level usually is within the energy gap such that if the surface states are occupied up to ϕ_{so} and empty above ϕ_{so}, the surface is electrically neutral. States below ϕ_{so} are donor-like (positive when empty) and states above ϕ_{so} are acceptor-like (negative when occupied). Values of ϕ_{so} are given in Table 5.8. For Si and GaAs the neutrality level lies within the energy gap; for Ge, GaSb, and InSb it lies below the valence band edge; and for InAs it lies above the conduction band edge.

TABLE 5.8
Energy of the Charge Neutrality Level (ϕ_{so}) Above the Semiconductor Valence Band Edge (E_v) for Selected Semiconductors (300°K)

Semiconductor	$q\phi_{so} - E_v$ [eV]
Si	0.30
Ge	<0
GaAs	0.76
GaSb	<0
InAs	0.41
InSb	<0

If the surface state density is small, for an n-type semiconductor the above equation reduces to

$$\phi_{MS} = \phi_M - \chi_s; \quad (5.80)$$

and if the surface state density is very high, it reduces to

$$\phi_{MS} = E_G/q - \phi_{so}, \quad (5.81)$$

i.e., the barrier becomes stabilized so that the Fermi level coincides with ϕ_{so} independent of ϕ_M.

For a given metal the sum of the barrier heights on n-type and p-type material of the same semiconductor is (in the work function model) equal to the band gap, i.e.,

$$\phi_{MS(n)} + \phi_{MS(p)} = E_G/q, \quad (5.82)$$

provided γ_b and ϕ_{so} are the same in both cases. Table 5.9 gives experimental

TABLE 5.9
Barrier Heights on Etched n-Type and p-Type Silicon Schottky Barriers

Metal	$q\phi_{MS(n)}$ [eV]	$q\phi_{MS(p)}$ [eV]	$q(\phi_{MS(n)} + \phi_{MS(p)})$ [eV]
Al	0.50	0.58	1.08
Au	0.81	0.35	1.16
Cu	0.69	0.46	1.15
Ni	0.67	0.51	1.18

data to substantiate this. (For comparison, the energy gap of silicon is $E_G = 1.12$ eV.). Other experimental data on Schottky barrier diodes are given in Tables 5.10 and 5.11.

TABLE 5.10
Optical Properties of n-Type Silicon Schottky Barriers (300°K)

Metal	Barrier height, $q\phi_{MS}$ [eV]	Reflectivity of Si-metal interface [%]	Photo yield at 1 eV [Electrons/incident photon]	Quantum efficiency at 1 eV [Electrons/absorbed photon]
Ag	0.63	97	0.04	1.3
Al	0.62		0.02	
Au	0.82	93	0.50	7.2
Cr	0.57		≈ 0	
Cu	0.59	84	1.20	7.5
Ni	0.58	43	0.70	1.2
Pd	0.75		0.40	
Pt	0.83	44	0.20	0.4

TABLE 5.11
Barrier Heights of Metal-Semiconductor Contacts (300°K)

Semiconductor	$q\phi_{MS}$[eV][a]			
	Al	Au	Pt	W
Si	0.62/	0.82/0.25	0.83/0.25	0.65/
Ge	0.48/	0.45/		0.48/
AlAs		1.20/	1.00/	
AlP				
AlSb		/0.54		
GaAs	0.80/0.56	0.93/0.45	0.90/0.48	0.75/
GaP	1.10/	1.30/1.30	1.50/	
GaSb		0.61/ohmic		
InAs		ohmic/0.47		
InP		0.50/0.76		
InSb		0.18/ohmic		

[a] The first value given for a particular metal corresponds to an n-type semiconductor, the second value to a p-type semiconductor.

5-4 METAL-SEMICONDUCTOR CONTACTS

Deviations from the ideal behavior of a Schottky diode are usually observed in experiments. This is due to surface contamination, surface charges, and others. If surface charges are taken into account, a double layer consisting of the barrier of width x_b and of a surface charge layer of width x_b^* can be assumed. The surface charge layer is only one to ten atom spacings wide on an atomically clean interface. The higher the surface state density, the less dependent ϕ_{MS} will be on work function differences. An empirical

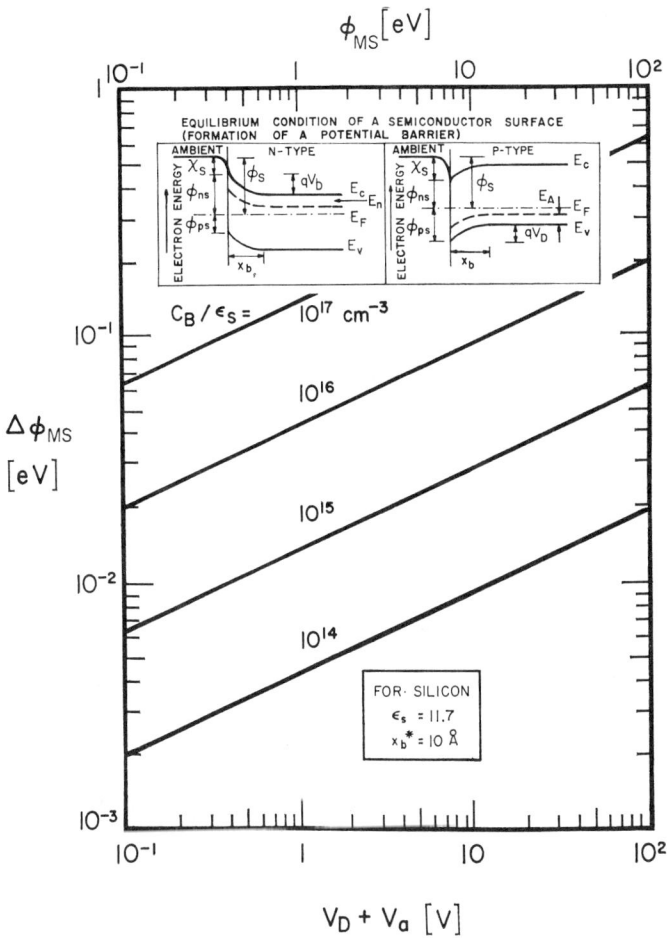

FIGURE 5.22

Barrier height reduction ($\Delta\phi_{MS}$) due to tunneling vs. barrier height (ϕ_{MS}), voltage ($V_D + V_a$), impurity concentration (C_B), and semiconductor dielectric constant (ϵ_s).

relationship between barrier height and semiconductor energy gap (E_G) and atomic number of the metal (Z) is for an n-type semiconductor

$$q\phi_{MS} = q\phi_{MS}^* + 3.0 \cdot 10^{-3}Z \quad \text{[eV]}, \tag{5.83}$$

where $\phi_{MS}^* \approx E_G/2q$.

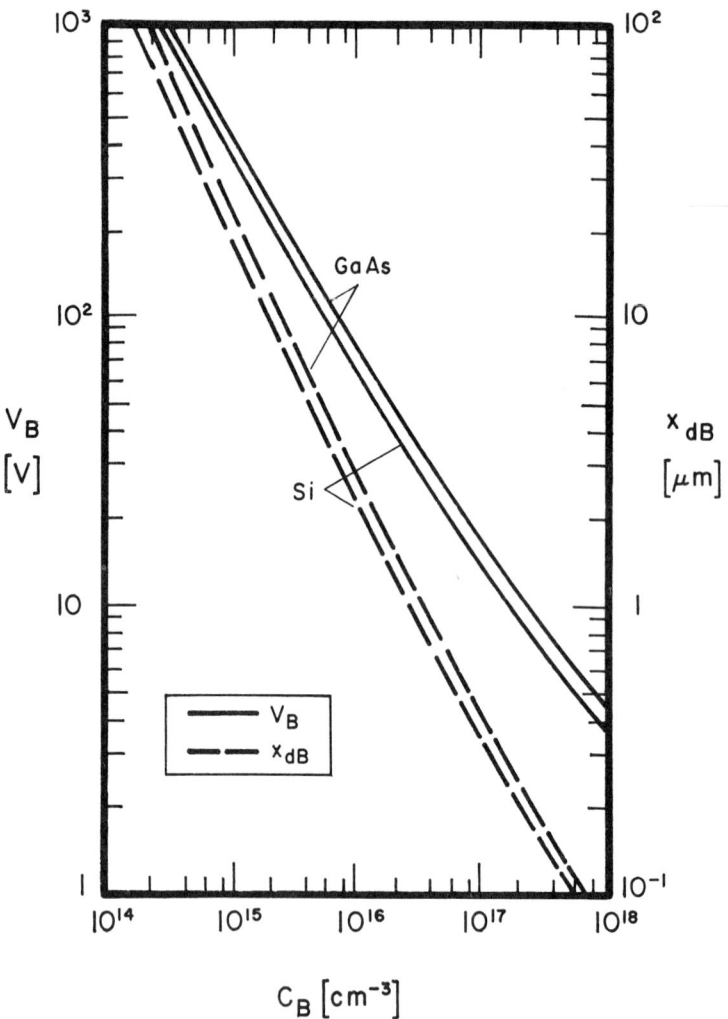

FIGURE 5.23

Breakdown voltage (V_B) and width of depletion layer within semiconductor at breakdown (x_{dB}) vs. semiconductor impurity concentration (C_B). Silicon and GaAs, semiconductor-metal junction, 300°K.

For an Al–Si contact ($q\phi_{MS}^* = 0.56$ eV, $Z = 13$) this relationship gives

$$q\phi_{MS} = 0.60 \text{ eV}.$$

In the case of tunneling through the barrier, the barrier height is reduced (see Figure 5.22). The amount of the reduction is

$$\Delta\phi_{MS} \approx 2q(V_D + V_a)(x_b^*/x_b)$$
$$= x_b^*[8\pi(V_D + V_a)q^3 C_B/\varepsilon_s]^{1/2}. \tag{5.84}$$

This assumes the existence of a critical thickness $x_b^* > 0$ above which the barrier can be considered to be opaque to charge carriers.

The capacitance per unit area of a metal–semiconductor barrier is

$$C_b = [(q/8\pi)C_B \varepsilon_s/(V_D + V_a)]^{1/2}. \tag{5.85}$$

The density of ionized impurity atoms is C_B. If the semiconductor bulk impurity concentration is uniform, a plot of $(1/C_b^2)$ vs. V_a is linear. The intercept of the straight line at $1/C_b^2 = 0$ can be used to determine the barrier height; the slope of the curve is a measure of the ionized impurity concentration. The breakdown voltage of a metal–semiconductor contact is shown in Figure 5.23.

6. CURRENT DENSITY THROUGH METAL–SEMICONDUCTOR CONTACT

The current transport across the barrier is mainly due to majority carriers, whereas in a *p-n* junction the current transport is due to minority carriers. At high temperature, however, the increase in the intrinsic carrier density causes the diffusion component due to the minority carriers to be quite significant.

A metal–semiconductor contact with a barrier height of approximately 1 eV has rectifying characteristics (owing to thermionic emission) if the semiconductor is lightly doped, and ohmic characteristics (owing to tunneling) if it is highly doped. There are three main conduction mechanisms in a metal–semiconductor system. In the case of an *n*-type semiconductor these are:

(a) transport of electrons from conduction band over the top of the barrier into the metal;
(b) injection of holes into the neutral region of the semiconductor;
(c) recombination or generation of electrons and holes in the semiconductor depletion region.

A fourth (minor) mechanism is due to tunneling of electrons from the conduction band into the metal.

Assuming the simple model of thermionic emission (i.e., the electrons go over the barrier), the forward and reverse current–voltage relationship of a metal–semiconductor contact is

$$I_{MS} = I_o[\exp(qV_a/kT) - 1]$$
$$\approx I_o \exp(qV_a/kT) \quad \text{if } V_a > 3kT/q. \quad (5.86)$$

I_o is the saturation current density which corresponds to the current density I_{MS} at $V_a = 0$. Generally,

$$I_o = A_n T^2 \exp(q\phi_{MS}/kT). \quad (5.87)$$

For an n-type semiconductor

$$I_o = -qn(kT/2\pi m_n)^{1/2} \exp(-qV_D/kT)[\exp(-q\phi_{MS}/kT) - 1] \quad (5.88a)$$

and for a p-type semiconductor

$$I_o = qp(kT/2\pi m_p)^{1/2} \exp(-qV_D/kT)[\exp(-q\phi_{MS}/kT) - 1]. \quad (5.88b)$$

If a forward voltage is applied such that $V_a = V_f > 3\ kT/q$, the above equation for the current across the metal–semiconductor interface predicts that $\ln I_f$ vs. V_f is linear. Extrapolating the straight line to $V_f = 0$, the saturation current I_o can be determined, from which ϕ_{MS} can be evaluated. If a reverse voltage is applied such that $V_a = V_r > 3kT/q$, the above equation predicts that a saturation characteristic, $I_r = I_o$, should be observed. In practice, several deviations occur (see Saxena, Ref. 16, for details). Large variations in the saturation current are observed owing to image force lowering of the barrier height.

If, under forward bias conditions, I_o has been determined by using the intercept of the straight line at $V_f = 0$, a plot of I_o/T^2 (activation energy plot) results in another straight line whose slope gives the value of the barrier height ϕ_{MS}. However, the I_o/T^2 curve is a straight line only for an ideal Schottky diode, i.e., if ϕ_{MS} is constant with respect to temperature or has a linear temperature dependence. In this case

$$\phi_{MS} = 0.1985/\Delta(1/T) \quad [\text{eV}]. \quad (5.89)$$

In a real Schottky diode (i.e., one in which the I_o/T^2 curve is not linear), introduction of an excess temperature T_* (which is additive to temperature T) will result in a straight line; in this case

$$I_{MS} = I_o\{\exp[qV_a/k(T + T_*)] - 1\}. \quad (5.90)$$

Experimental data indicate that for Au–GaAs Schottky diodes $T_* = 50°$K, for Cr–Si diodes $T_* = 34°$K, for Ni–Si diodes $T_* = 24°$K. Frequently, T_* is not a constant but varies with temperature as

$$T_* = qV_*/k - T, \quad (5.91)$$

where V_* is a corrective voltage characteristic of the system. When thermionic-field emission and field emission are significant, further modifications of the above equation are required.

In an ideal Schottky diode $V_* = kT/q$ and $T_* = 0$. In a diode which contains a surface charge layer and which obeys the relationship

$$I_{MS} = I_o[\exp(qV_a/n_b kT) - 1] \tag{5.92}$$

the corrective voltage is $V_* = n_b kT/q$, where n_b is independent of temperature. In a diode which is governed by the relationship

$$I_{MS} = I_o\{\exp[qV_a/k(T + T_*)] - 1\} \tag{5.93}$$

the corrective voltage is $V_* = k(T + T_*)/q$, where T_* is constant and independent of temperature.

The saturation current of a Schottky barrier diode can best be described by

$$I_o = A_n T^2 \exp[-\phi_{MSo}/k(T + T_*)] \tag{5.94}$$

where ϕ_{MSo} is the effective barrier height which is (contrary to ϕ_{MS}) independent of temperature. The effective barrier height, together with the thermal coefficient α_{MS}, describes the temperature variation of the barrier height.

$$\begin{aligned}\phi_{MS} &\approx \phi_{MSo} - \alpha_{MS} T \\ &= \phi_{MSo}[1 - 1/(1 + T/T_*)], \end{aligned} \tag{5.95}$$

where $\alpha_{MS} = \phi_{MSo}/T(1 + T/T_*)$.

Values of ϕ_{MSo} and α_{MS} are given for selected metal–semiconductor contacts in Table 5.12.

TABLE 5.12
Effective Barrier Height and Temperature Coefficient of Barrier Height of Selected Metal–Silicon Systems (300°K)

Metal	$q\phi_{MSo}$ [eV]	α_{MS} [10^{-4} eV/°K]
Au	0.90	3.0
Cr	0.68	1.8
Ni	0.68	1.4

If the metal–semiconductor contact is associated with a surface charge layer, the expression for the current density which assumed only thermionic emission has to be modified by the numerical factor n_b. In this case

$$I_{MS} = I_o[\exp(qV_a/n_b kT) - 1], \tag{5.96}$$

where usually $1.0 \leq n_b \leq 1.5$. The factor n_b can be approximated by

$$1/n_b \approx 1 - [(q/2)C_B \varepsilon_s/(V_D + V_a)]^{1/2}/(Q_{ss}/q + \varepsilon_i/x_b^*), \tag{5.97}$$

where x_b^* is the width of the layer and ε_i is its dielectric constant. In terms of the excess temperature T_*

$$n_b = 1 + T_*/T. \tag{5.98}$$

If $x_b^* \to 0$, the above relationship reduces to $n_b \to 1$.
Assuming $n_b = 1$, the total current density, according to thermionic emission theory, for $V_a > 3kT/q$,

$$I_{MS} = A_n T^2 \exp(-q\phi_{MS}/kT)[\exp(qV_a/kT) - 1]. \tag{5.99}$$

Deviations from this ideal relationship are due to

(a) generation–recombination current;
(b) image force lowering of the barrier (resulting in $n_b > 1$);
(c) tunneling;
(d) insulating interfacial layer (resulting in $n_b > 1$).

7. METAL–SEMICONDUCTOR INTERFACE FERMI LEVEL

Under steady-state, nonequilibrium conditions, the Fermi level varies with distance from the metal–semiconductor interface. The energy difference between semiconductor conduction band energy (E_c) and the Fermi level at the metal–semiconductor interface (E_{Fo}) is approximately proportional to the energy gap (E_G) of the semiconductor. The straight line in Figure 5.24 represents the relationship

$$E_c - E_{Fo} = (2/3)E_G; \tag{5.100}$$

it fits most n-type and p-type semiconductors, assuming Au as the metal. For metals other than Au the interface Fermi level lies slightly closer to E_c. The above relationship indicates that most semiconductor surfaces have a high peak density of surface states near one-third of the energy gap from the valence band edge.

The Fermi potential of the semiconductor is defined as

$$\phi_F = -(E_F - E_i)/q = -(E_{Fo} - E_{F\infty})/q \tag{5.101}$$

and the electrostatic potential as

$$\psi = -(E_{co} - E_{c\infty})/q. \tag{5.102}$$

Their difference

$$\phi_F - \psi = (E_{co} - E_{c\infty} - E_{Fo} + E_{F\infty})/q$$
$$= (kT/q) \ln(n/C_B) \tag{5.103}$$

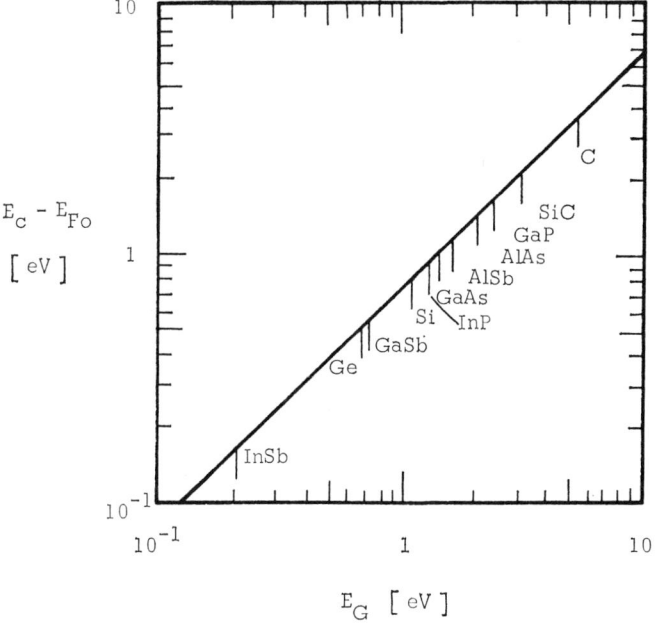

FIGURE 5.24
Location of Fermi level at the semiconductor surface (E_{Fo}) with respect to the conduction band edge (E_c) as function of the semiconductor energy gap (E_G). 300°K.

reduces to zero if all semiconductor impurity atoms are ionized, i.e., if $n = C_B$. In this case

$$\phi_F = \psi.$$

In the above equations subscript o refers to the metal–semiconductor interface ($x = 0$) and subscript ∞ refers to the semiconductor bulk ($x \gg 0$).

8. ELECTRON AFFINITY AND ELECTRONEGATIVITY

Electron affinity is defined as the energy required to remove an electron from the bottom of the conduction band of the semiconductor to the vacuum level. Materials which have a positive electron affinity readily form negative ions and are therefore considered to be electronegative. Materials which have a negative electron affinity (e.g., alkali elements) readily form positive ions and are called electropositive. Electronegativity is defined as the power of an atom in a molecule to attract electrons to itself. Electron affinities of selected semiconductors are given in Table 5.13 and electronegativities of selected elements are given in Table 1.5.

TABLE 5.13
Electron Affinities of Selected Semiconductors

Semi-conductor	Orientation	$q\chi_s$ [eV]
Si	⟨111⟩	4.05
Ge	⟨111⟩	4.13
GaAs	⟨110⟩	4.07
GaP	⟨110⟩	3.21
GaSb	⟨110⟩	4.06
InAs	⟨110⟩	4.90
InSb	⟨110⟩	4.59
CdS	⟨110⟩	3.90

Electron affinity (χ_s) and electronegativity (X_P) are related by the empirical relationship

$$q\chi_s + E_{ion} \approx 3 \cdot 10^3 \, X_p, \qquad (5.104)$$

where E_{ion} is the ionization energy of an atom.

9. PHOTOEMISSION

Photoemission (i.e., the emission of electrons owing to the incidence of photons) from metal into semiconductor or oxide is observed when photons are incident on the metal of such systems.

When photons are incident on the metal of a metal–semiconductor diode, electrons are excited over the barrier when the energy of the photon exceeds the barrier height, i.e., when

$$E_{photon} = hv > q\phi_{MS}.$$

By measuring the photoresponse R_{ph} of a metal–semiconductor contact, the barrier height can be measured in a direct way. A plot of $R_{ph}^{1/2}$ vs. E_{photon} is linear, and the intercept of the straight line on the energy axis gives ϕ_{MS}. Measuring the photoresponse at various reverse biases gives the barrier height and its lowering due to image forces.

Likewise the incidence of photons on the metal of a metal-oxide contact results in photoemission (see Figure 5.25). The associated energy is

$$\phi_{th} = \phi_M - \chi_o, \qquad (5.105)$$

where χ_o is the electron affinity of the oxide; for SiO_2 the electron affinity is $q\chi_o = 1.0$ eV. Values of photoelectric threshold energies for emission of electrons from selected metals into SiO_2 are given in Table 5.14.

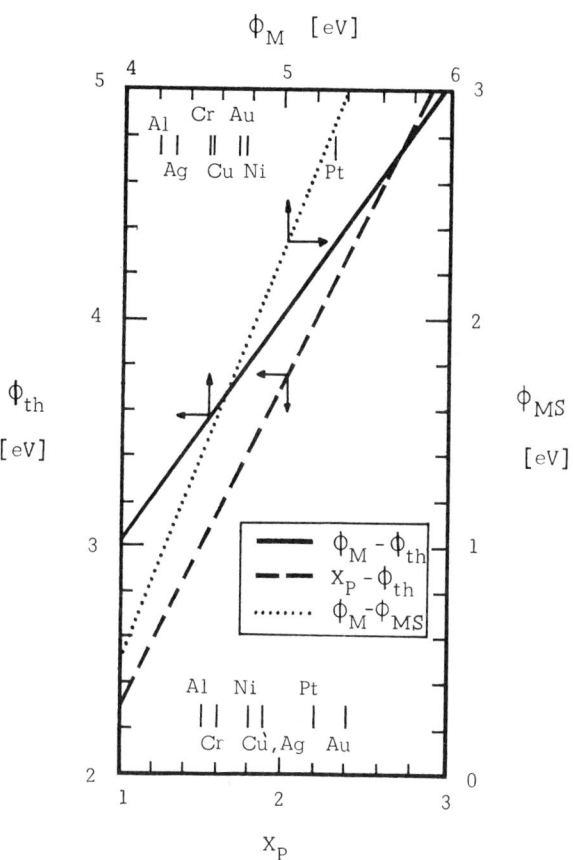

FIGURE 5.25

Relationship between metal work function (ϕ_M), photoelectric threshold energy (ϕ_{th}), barrier height (ϕ_{MS}), and electronegativity (X_P). The illustration is derived from experimental and theoretical data and is based on photoemission from evaporated thin film metal layers into SiO_2 at $300°K$; the substrate is n-type silicon of $\langle 111 \rangle$ orientation. The ϕ_M-ϕ_{th} curve represents a least average deviation fit with a slope 1.0. Values of ϕ_M of selected metals are shown in the upper portion of the illustration, values of X_P of these metals in the lower portion.

TABLE 5.14
Photoelectric Threshold Energies for Photoemission from Metal into SiO_2 (300°K)

Metal	$q\phi_{th}$ [eV]
Ag	3.7
Al	3.2
Au	3.9
Cr	3.6
Mo	3.4
Pt	4.3
W	3.6

10. REFERENCES

(1) H. K. Henisch, *Rectifying Semi-Conductor Contacts*, Clarendon Press, Oxford, 1957.
(2) L. Pauling, *The Nature of the Chemical Bond*, Cornell University Press, 1960.
(3) B. E. Deal et al., *J. Phys. Chem. Solids*, **27**, 1873 (1966).
(4) C. R. Crowell and S. M. Sze, *Solid-State Electronics*, **9**, 1035 (1966).
(5) C. A. Mead, *Solid-State Electronics*, **9**, 1023 (1966).
(6) D. V. Geppert et al. *J. Appl. Phys.*, **37**, 2438 (1966).
(7) A. M. Goodman and J. J. O'Neill, Jr., *J. Appl. Phys.*, **37**, 3580 (1966).
(8) S. M. Sze and G. Gibbons, *Appl. Phys. Lett.*, **8**, 111 (1966).
(9) M. M. Atalla, *Mikroelektronik*, **2**, Munich, Oct. 1966.
(10) E. Groschwitz, *Z. Angew. Physik*, **22**, 381 (1967).
(11) S. R. Kovel and G. Gibbons, *Proc. IEEE*, **55**, 2066 (1967).
(12) T. Sugano et al., *Jap. J. Appl. Phys.*, **7**, 1028 (1968).
(13) N. F. Mott, *Rev. Mod. Phys.*, **40**, 677 (1968).
(14) S. M. Sze et al., *Solid-State Electronics*, **12**, 107 (1969).
(15) H. Jäger and W. Kosak, *Solid-State Electronics*, **12**, 511 (1969).
(16) A. N. Saxena, *Surface Science*, **13**, 15 (1969).
(17) R. L. Bell, *Solid-State Electronics*, **13**, 397 (1970).
(18) A. M. Cowley, *Solid-State Electronics*, **13**, 403 (1970).
(19) C. Y. Chang and S. M. Sze, *Solid-State Electronics*, **13**, 727 (1970).
(20) E. H. Rhoderick, *Journal of Physics D/Applied Physics*, **3**, 1153 (1970).
(21) J. M. Andrews and M. P. Lepselter, *Solid-State Electronics*, **13**, 1011 (1970).

5–5 Examples

PROBLEMS AND SOLUTIONS

1. **Problem:** Determine the surface potential and the inversion layer thickness of a *p*-type silicon sample at 25°C if the bulk impurity concentration $C_B = 10^{16}$ cm^{-3}.

   Solution: $\phi_s = 0.26$ V, $u_S = 10$;
   $x_s = 0.19$ μm.

2. **Problem:** Determine the surface state density and temperature coefficient of the turn-on gate voltage of an Si-SiO$_2$ system consisting of *n*-type silicon of bulk concentration $C_B = 5 \cdot 10^{15}$ cm^{-3}, SiO$_2$ of thickness $x_o = 1000$ Å, and an aluminum contact. The measured turn-on voltage is $V_T = -0.8$ V.

   Solution: (a) Surface state density:

   $$\phi_{MS} = -0.34 \text{ V}, \ C_o = 3.5 \cdot 10^4 \text{ pF/cm}^2;$$
   $$V_T - \phi_{MS} + Q_{ss}/C_o = -1.5 \text{ V};$$

   i.e.,

   $$Q_{ss}/q = -2.5 \cdot 10^{11} \text{ cm}^{-2}.$$

   (b) Temperature coefficient of V_T:

   $$dV_T/dT = 2.8 \cdot 10^{-3} \text{ V/°K}.$$

3. **Problem:** Determine the variation of the capacitance minimum in the high-frequency model for a temperature variation from 25 to 250°C, assuming a silicon bulk concentration $C_B = 10^{16}$ cm^{-3} and an oxide thickness $x_o - 2000$ Å.

   Solution: At 25°C: $C_m = 0.67$,
   at 250°C: $C_m = 0.77$;

   i.e., there will be an increase of the minimum capacitance by 15%. The transition frequency from low-frequency to high-frequency model for the same temperature range increases from $f_t = 38$ Hz at 25°C to $f_t = 400$ kHz at 250°C, i.e., by a factor of approximately 10^4.

PROBLEMS FOR WHICH A SOLUTION IS NOT GIVEN

1. Determine the slope of the $C-V$ curve of an Si–SiO$_2$ system for which $C/C_o = 0.5$, $C_B = 10^{14}$ cm^{-3}, $x_o = 1000$ Å.

2. Determine the flat-band capacitance of an Si–SiO$_2$ system for which $C_B = 10^{15}$ cm^{-3} and $x_o = 0.1$ μm.

3. Assuming an Al contact to silicon, determine the contact potential ϕ_{MS} for n-type and p-type silicon of impurity concentrations $C_B = 10^{14}$ and 10^{16} cm^{-3}.

4. Assuming an MOS structure consisting of an Al–SiO$_2$–Si system, determine the photoelectric threshold energy, metal work function, and electronegativity.

5. Determine the electron mobility in a surface inversion layer of silicon whose bulk impurity concentration is $C_B = 10^{15}$ cm^{-3} for a 1000-Å thick oxide layer.

6
p-n JUNCTIONS

6–1 Diffusion (Built-in) Voltage
6–2 Depletion Layer Characteristics
6–3 Junction Characteristics at Breakdown
6–4 Examples

The previously discussed band structure of semiconductors consisting of conduction and valence bands, impurity energy levels, and Fermi level applies to an *n*-type or a *p*-type semiconductor (i.e., one in which the predominant impurities are either donors or acceptors). The distribution of donor and acceptor atoms in a semiconductor which, together with temperature, determines the carrier densities does not have to be uniform but may vary arbitrarily throughout the crystal. If the variation in impurity concentration is very gradual then the carrier densities are nearly equal to the equilibrium carrier densities which correspond to the impurity density at each point. If, however, the impurity density changes rapidly then nonequilibrium conditions are encountered and important physical phenomena occur.

A *p-n* junction is formed at the interface on an *n*-type and a *p*-type semiconductor brought into intimate contact; it is defined as a sharp boundary within a semiconductor crystal with predominantly donor impurities on one side and predominantly acceptor impurities on the other side. If there is no external voltage applied between the two sides then the Fermi level exists at a single energy value throughout the crystal. At the *p-n* junction conduction and valence bands are warped in such a way that the two majority carrier distributions are confined to their own areas, the warping being just sufficient to establish that no net current flows across the junction. In equilibrium the current flow across the junction is composed of two equal components of opposite sign (no net current in transition region); one component is due to carrier diffusion and the other is due to carrier drift as a result of the built-in electric field.

If an external voltage is applied to a *p-n* junction, this equilibrium is disturbed. If a forward bias is applied, the Fermi levels on both sides of the junction are different by an amount qV. The barrier to the flow of majority carriers is thus lowered so that more carriers are above the top of the barrier and a current can flow which increases exponentially with voltage. Holes are flowing through the *p*-type region and either recombine with electrons which have crossed the junction or move over into the *n*-type region and recombine there. The behavior of electrons is analogous.

If a reverse voltage is applied, the barrier is raised so that fewer majority carriers are above the top of the barrier and the flow of carriers across the junction is restricted and reaches, in the ideal case, a saturating value. The only current is carried by minority carriers which are easily swept across the region of the accelerating field. The magnitude of this current is ideally independent of the applied voltage and only determined by the abundance of minority carriers in the two regions. It results from the thermal generation of carriers and is, therefore, a function of temperature. At a high reverse voltage, avalanche breakdown sets in.

In the transition region between *n*-type and *p*-type regions a charge dipole region or depletion layer is created by the carriers diffusing out of the regions and leaving the ionized impurity atoms on either side unneutralized. The sum of built-in and applied voltage—the total voltage across the junction—charges the layer by repelling more majority carriers away from the junction and by exposing more impurity ions on both sides. Thus the depletion layer widens with voltage and behaves like a voltage-dependent capacitance. Essentially the entire drop of the applied voltage occurs across the depletion layer.

Outside of the depletion layer there exists a field-free carrier diffusion region (of length L) in which carriers can diffuse in either direction depending upon the polarity of the applied voltage. This diffusion length L is defined as the average distance the minority carriers travel before recombining. The associated recombination time (lifetime; τ), i.e., the average minority carrier diffusion time, is related to the width of the diffusion region and the carrier diffusion coefficient (for example, for the n-type region) by

$$L_p = (D_p \tau_p)^{1/2} \tag{6.1}$$

resulting in a minority carrier diffusion current

$$j_p = (q p_n D_p / L_p)[\exp(qV/kT) - 1]. \tag{6.2}$$

If the electric field reaches a critical value, ionization will set in and the junction will break down, i.e., any further attempt to increase the voltage across the junction will result in an avalanche-like increase in current density across the junction and will thus prevent a significant voltage increase over the breakdown voltage. In curved regions of a diffused junction a lower breakdown voltage is observed compared to the plane region of the junction due to a larger electric field. The influence of junction curvature is one of the main causes of deviation from expected avalanche breakdown characteristics. Other important causes are depletion layer width limitations and microplasma effects.

The simple theory of p-n junctions is based on the following assumptions which do not all necessarily apply in practical junctions.

(a) Contact potential and applied voltage are supported by a dipole layer with well-defined boundaries. Outside these boundaries the semiconductor is neutral.
(b) The Boltzmann relation is valid throughout the dipole layer.
(c) Electron and hole current are constant through the dipole layer.
(d) Minority carrier densities are small compared to majority carrier densities (low injection level).

Two types of p-n junctions are of superior importance since they are most frequently encountered in microelectronics; the abrupt or step junction (e.g., the emitter-base junction of a bipolar transistor) and the linearly graded junction (e.g., the base-collector junction of a transistor). *p-n* junctions form the basis of the physics of semiconductor devices. Simple *p-n* junctions are the constituents of most bipolar and MOS transistors; they can be made in a variety of doping profiles and geometries in order to display a wide spectrum of characteristics.

A heterojunction is a junction between two semiconductors of different energy gap. Although in an ideal heterojunction perfect matching of lattice constants and thermal expansion coefficients is required, this is usually not the case in practical heterojunctions and defects (e.g., interfacial dislocations) are present which act as trapping centers. In addition to a difference in energy gaps, the two sides of a heterojunction are usually characterized by differences in dielectric constant, work function, and electron affinity; work function and electron affinity are defined as the energy required to remove an electron from the Fermi level or from the bottom of the conduction band, respectively, to a position just outside the semiconductor (vacuum).

The *p-n* junction, being a two-terminal device, can perform various terminal functions which depend upon doping profiles, geometry, and bias conditions. The more important of these functions are:

(a) Rectifier:

This is a simple *p-n* junction which rectifies alternating current, i.e., it has a low resistance to current flow in one direction and a high resistance in the other direction. The forward and reverse current of a rectifying *p-n* junction is

$$I = I_s[\exp(qV/n_* kT) - 1] \tag{6.3}$$

where I_s is the saturation current and n_* is a constant; for diffusion current $n_* = 1$ and for recombination current $n_* = 2$.

$$I_s = qp_n D_p/(D_p \tau_p)^{1/2} \tag{6.4}$$

(b) Tunnel diode:

This is a *p-n* junction in which, contrary to ordinary junctions, both sides are degenerately doped; the Fermi level is located within the allowed bands, the impurity states have broadened into bands, and the intrinsic band gap is reduced (band-edge tailing). The current–voltage relationship shows a distinct current peak and valley leading to a negative resistance region. An externally applied voltage permits electrons to tunnel from valence to conduction band and vice versa if the following tunneling requirements are met:

(i) Occupied energy states exist on the side from which electrons tunnel.
(ii) Unoccupied energy states exist at the same energy levels as in (i) on the side to which electrons tunnel.

(iii) The tunneling potential barrier height is lowered and the barrier width is small enough to achieve a finite tunneling probability.
(iv) The momentum is conserved.
(c) Impatt diode:
A *p-n* junction can be operated in the *imp*act ionization *a*valanche *t*ransition *t*ime mode when it is biased into reverse avalanche breakdown and mounted in a cavity; this leads to a negative resistance characteristic. Application of a differential voltage results in a differential current which is out of phase with the voltage due to
 (i) the delay time with which the carrier population builds up toward a new level;
 (ii) the transit time during which the carriers are collected by the electrodes.
Impatt diodes usually consist of two effective regions—a drift region through which the carriers pass without avalanching and an avalanching region.
(d) Voltage regulator:
This is a *p-n* junction reverse-biased to avalanche breakdown.
(e) *p-i-n* Diode:
This is a *p-n* diode with a near-intrinsic (*i*) region between the *p*- and *n*-regions. Because of the low impurity concentration in the *i*-region most of the potential drop occurs across this region and consequently the breakdown voltage is very high.
(f) Fast recovery diode:
There are two types: the diffused *p-n* junction diode and the metal–semiconductor diode (Schottky barrier diode). These devices are characterized by ultrahigh switching speed.
(g) Charge-storage diode:
This device stores charge while conducting in the forward direction and conducts in the reverse direction for a short time; it then abruptly (within picoseconds) cuts off the current when the stored charges have been dispelled.

GENERAL REFERENCES

(1) W. W. Gärtner, *Transistors—Principles, Design, and Applications*, D. Van Nostrand, New York, 1960.
(2) J. L. Moll, *Physics of Semiconductors*, McGraw-Hill Book Co., New York, 1964.
(3) L. P. Hunter, *Semiconductor Phenomena and Devices*, Addison-Wesley, Reading, Massachusetts, 1966.
(4) S. M. Sze, *Physics of Semiconductor Devices*, John Wiley and Sons, New York, 1969.

6–1 Diffusion (Built-in) Voltage

1. ORIGIN OF BUILT-IN VOLTAGE

The built-in voltage (often also called the built-in potential or the diffusion voltage) of a *p-n* junction in equilibrium is the sum of the absolute values of the Fermi potentials on both sides of the junction.

$$V_D = \phi_T = \phi_{Fp} + |\phi_{Fn}|, \tag{6.5}$$

where
$$\phi_{Fp} = -(E_{Fp} - E_i)/q > 0 \tag{6.6a}$$

and
$$\phi_{Fn} = -(E_{Fn} - E_i)/q < 0. \tag{6.6b}$$

It is, in thermal equilibrium, a reverse potential across the junction and is a barrier to current flow and is closely related to the junction depletion layer.

If an *n*-type and a *p*-type semiconductor are physically separated, the Fermi levels will be different; upon intimate contact of the two semiconductors, however, the Fermi levels must be identical, since in thermal equilibrium carriers with the same energy are completely free to move from one semiconductor to the other without expending energy. This can be true only if the two Fermi levels are identical. Consequently, the band edges and the intrinsic Fermi levels of *n*-type and *p*-type regions will have different energy levels; they will be bent in the vicinity of the metallurgical junction, i.e., within the depletion layer. The difference in intrinsic Fermi levels is the built-in voltage if no external voltage is applied.

Upon contact between the *n*-type and *p*-type regions, holes and electrons from the two regions recombine in the vicinity of the metallurgical junction resulting in a depletion layer (which is depleted of all charge carriers). The departure from charge neutrality in this depletion layer results in an electric field. Once this depletion layer has been formed, carriers are prevented from further recombination.

Under reverse bias the built-in voltage, i.e., the barrier height of the junction, is increased and the depletion layer widens to accommodate this additional voltage. Under forward bias the built-in voltage is decreased, resulting in an increased flow of carriers across the junction. Under all conditions the barrier height remains finite; if it became zero, an infinite current flow would result.

6-1 DIFFUSION (BUILT-IN) VOLTAGE

Definitions:

(a) Step junction:

$$V_D = (kT/q) \ln (N_A N_D/n_i^2)$$
$$\approx (2\,kT/q) \ln (C_B/n_i)$$
$$= (1/2)x_{do} E_{max} \qquad (6.7)$$

where N_A and N_D correspond to impurity concentrations at the edges of the depletion region, E_{max} is the maximum electric field at the junction, and x_{do} is the width of the depletion layer at zero bias. For a two-sided step junction the built-in voltage (V_D) is the sum of the absolute values of the Fermi potentials on both sides of the junction; i.e., the total built-in voltage is the sum of the voltages on n- and p-side:

$$V_D = |V_{Dn}| + V_{Dp} \qquad (6.8)$$

where V_{Dn} and V_{Dp} correspond to the voltages taken from Figures 6.1

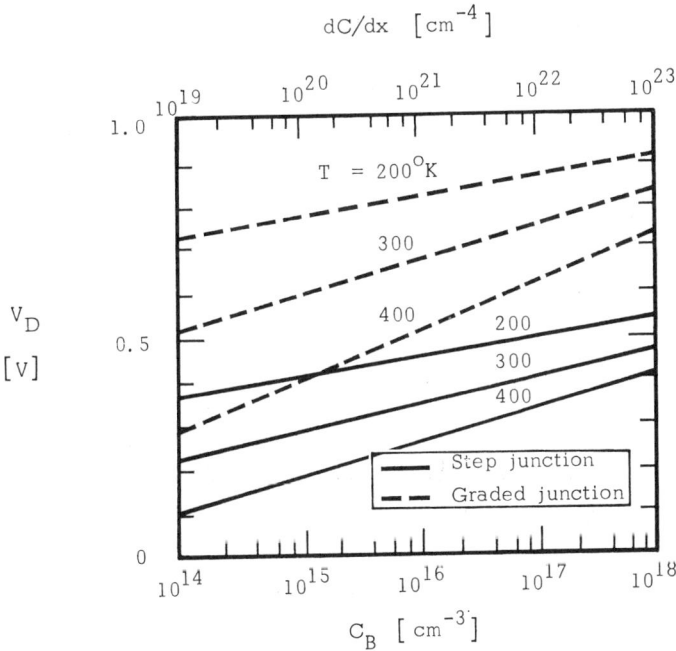

FIGURE 6.1

Diffusion or built-in voltage (V_D) as function of semiconductor impurity concentration (C_B) for a step junction or of the impurity concentration gradient at the junction (dC/dx) for a linearly graded junction and of temperature (T). Silicon. Note that for a step junction the total built-in voltage is the sum of the built-in voltages on both sides of the junction.

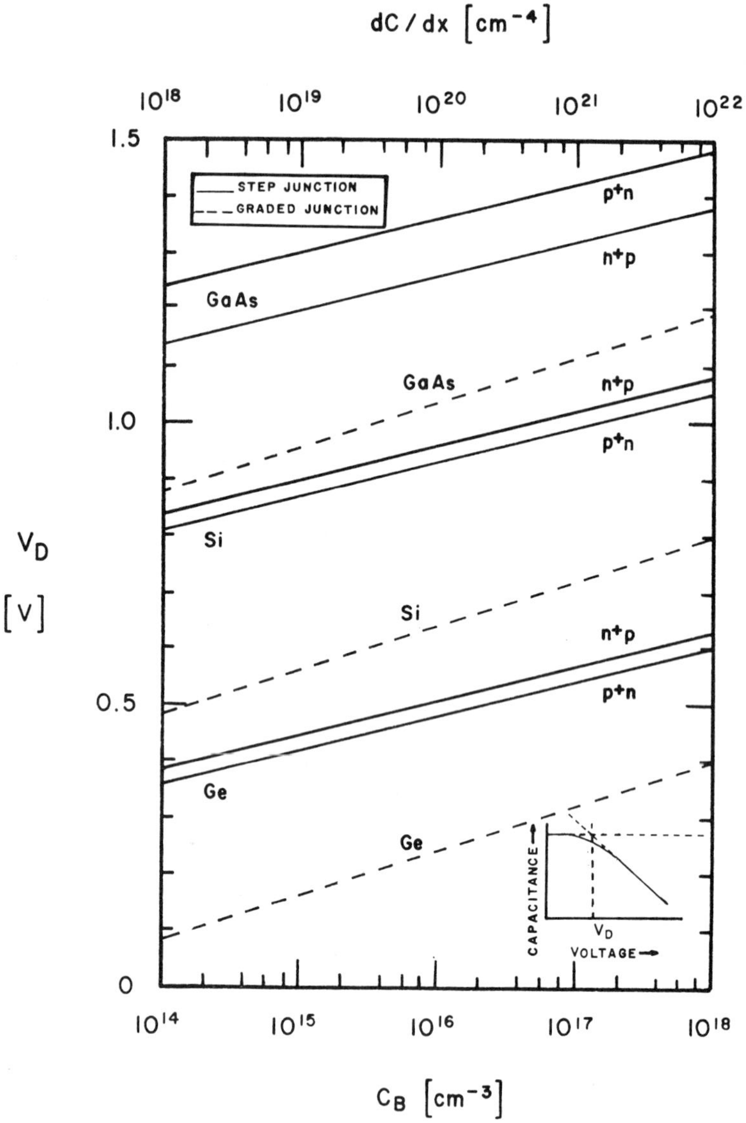

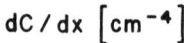

FIGURE 6.2

Diffusion voltage (V_D) vs. background impurity concentration (C_B) for step junction or impurity gradient (dC/dx) for linearly graded junction for Si, Ge, and GaAs. 300°K.

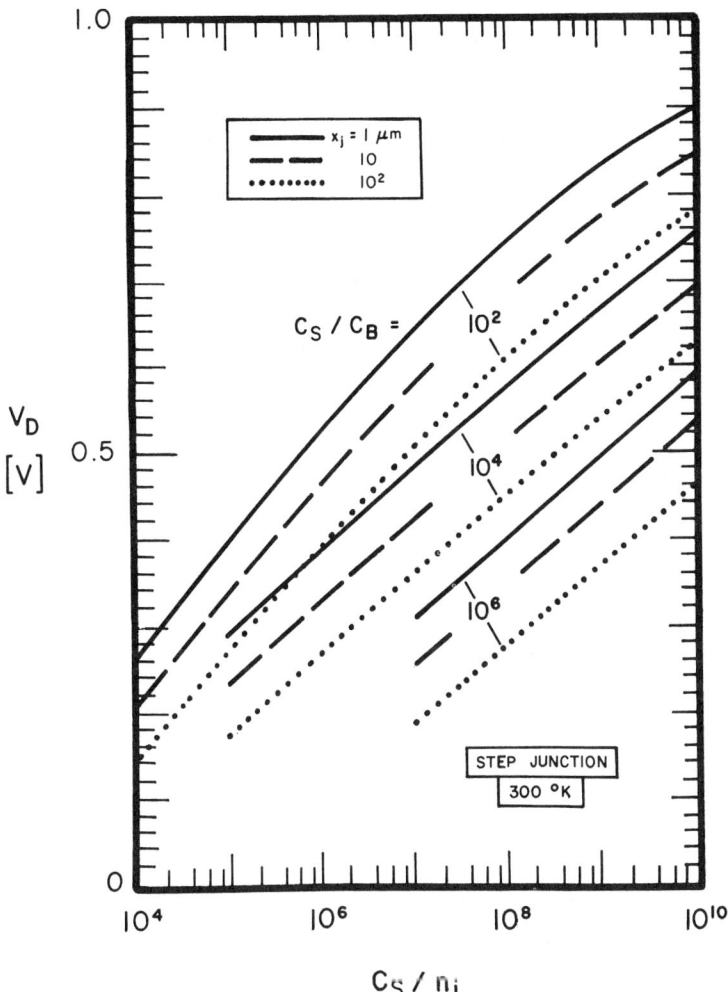

FIGURE 6.3

Diffusion voltage (V_D) of a step junction vs. ratio of surface impurity concentration (C_S) to intrinsic carrier concentration (n_i) vs. junction depth (x_j) and ratio C_S/C_B. Silicon, 300°K.

or 6.2 for the appropriate impurity concentrations on n- and p-side of the junction. The polarities of the portions of the built-in voltage are as follows.

$$V_{Dn} < 0, \; V_{Dp} > 0.$$

(b) Linearly graded junction:

$$\begin{aligned} V_D &= (2kT/q) \ln (ax_{do}/2n_i) \\ &\approx (4/3)(kT/q)[\ln (dC/dx) - 1] \\ &= (2/3)x_{do}\, E_{max} \end{aligned} \tag{6.9}$$

where $a = dC/dx$ (expressed in cm^{-4}).

2. NORMALIZED BUILT-IN VOLTAGE

The build-in voltage can be normalized with respect to thermal voltage (kT/q) as follows:

(a) Step junction:

$$\begin{aligned} V_D^* &= V_D(q/kT) \\ &= \ln (N_A N_D/n_i^2) \\ &\approx 2 \ln (C_B/n_i) \end{aligned} \tag{6.10}$$

(b) Linearly graded junction:

$$\begin{aligned} V_D^* &= V_D(q/kT) \\ &= 2 \ln x_{do}(dC/dx)/2n_i \\ &\approx (4/3)[\ln (dC/dx) - 1] \end{aligned} \tag{6.11}$$

3. REFERENCES

(1) C. T. Sah, *Proc. IRE*, **49**, 603 (1961).
(2) P. R. Wilson, *Solid-State Electronics*, **12**, 675 (1969).

6–2 Depletion Layer Characteristics

1. *p-n* JUNCTIONS

In semiconductor technology special *p-n* junctions with a particular impurity distribution near the metallurgical junction play an important role. These are abrupt or step junction, linearly graded junction, and *p-i-n* junction. Impurity distribution, space charge distribution, potential, and the capacitance-voltage relationship of these junctions are shown schematically in Fig. 6.4.

Generally, for arbitrary junction the following expressions apply for maximum electric field, built-in voltage, and depletion layer capacitance.

$$E_{max} = q/(\varepsilon_s \varepsilon_o) \int_{x_j}^{x_j+x_p} C(x)\, dx$$

$$= -q/(\varepsilon_s \varepsilon_o) \int_{x_j-x_n}^{x_j} C(x)\, dx \qquad (6.12)$$

$$V_D = -q/(\varepsilon_s \varepsilon_o) \int_{x_j-x_n}^{x_j+x_p} ds \int_{x_j+x_p}^{s} C(x)\, dx$$

$$= \frac{kT}{q}\left(\operatorname{ar\,sinh}\frac{C(x_j - x_n)}{2n_i} - \operatorname{ar\,sinh}\frac{C(x_j + x_p)}{2n_i}\right]$$

$$= \frac{kT}{q}\ln\left(\frac{C(x_j - x_n)\cdot C(x_j + x_p)}{n_i^2}\right) \quad \begin{cases} \text{for } n \gg n_i \text{ at} \\ x = x_j - x_n \\ \text{and } p \gg n_i \text{ at} \\ x = x_j + x_p. \end{cases} \qquad (6.13)$$

$$C = \varepsilon_s \varepsilon_o/(x_n + x_p)$$

$$= \varepsilon_s \varepsilon_o/x_d \qquad (6.14)$$

It is assumed that $x = 0$ at the semiconductor surface and $x = x_j$ at the *p-n* junction.

The following extreme cases for the impurity gradient are of interest.

(a) Steep gradient:
    This is the case if

$$dC/dx \gg q^2/(\varepsilon_s \varepsilon_o kT)[4n_i^2 + C(x)^2]^{3/2}.$$

The junction characteristics can effectively be described by a step junction.

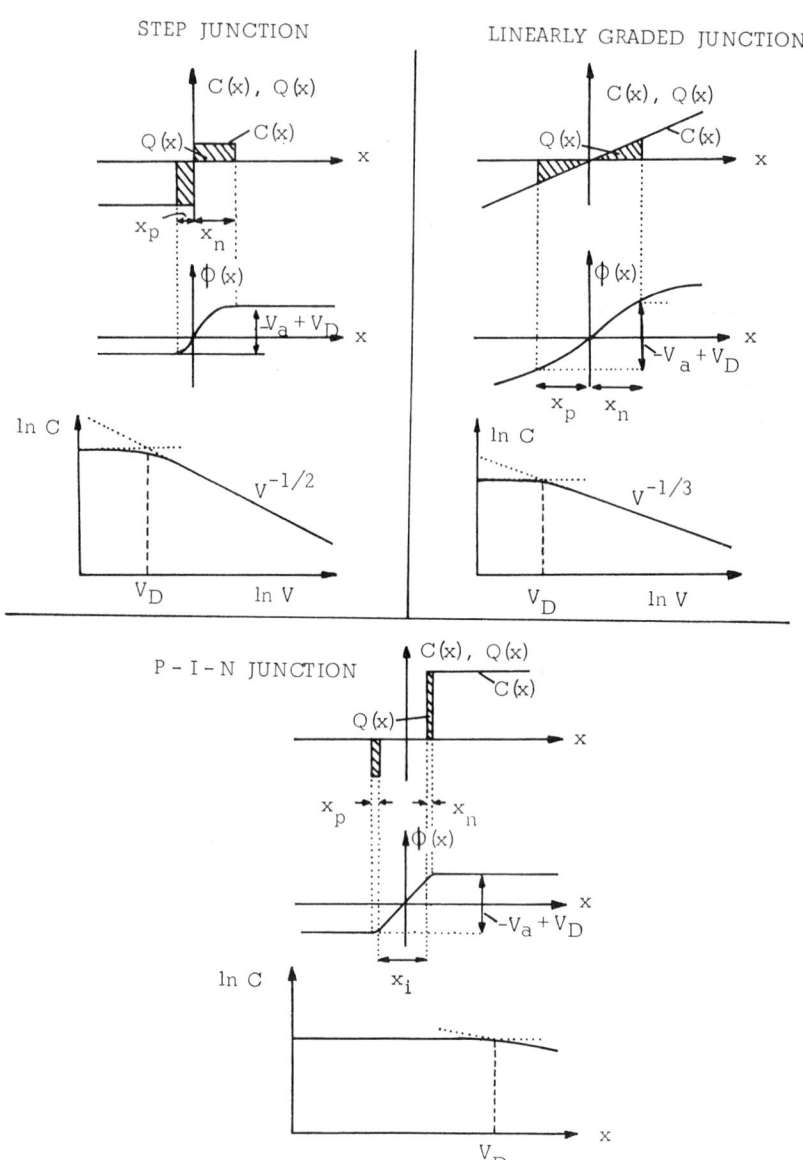

FIGURE 6.4

Impurity distribution, $C(x)$, space charge distribution, $Q(x)$, potential, $\phi(x)$, and capacitance-voltage relationship for step, linearly graded, and p-i-n junction. The illustration defines the built-in voltage (V_D) which for step and graded junctions is typically between 10^{-1} and 1 V and for a p-i-n junction between 10 and 10^2 V,

(b) Flat gradient:
   This is the case if

$$dC/dx \ll q^2/(\varepsilon_s \varepsilon_o kT)[4n_i^2 + C(x)^2]^{3/2}.$$

The junction characteristics can effectively be described by a linearly graded junction.

The special types of junctions have the following characteristics.

(a) Step junction:
   Under the assumption

$$C(x) = \begin{cases} n \text{ for } x - x_j > 0 \\ -p \text{ for } x - x_j < 0 \end{cases}$$

the general expressions (6.12) and (6.13) can be modified as follows:

$$\begin{aligned} E_{max} &= q/(\varepsilon_s \varepsilon_o) n x_n \\ &= q/(\varepsilon_s \varepsilon_o) p x_p \\ &= 2[q/(2\varepsilon_s \varepsilon_o) n p/(n+p)(V_a + V_D)]^{1/2} \end{aligned} \quad (6.15)$$

$$\begin{aligned} V_a + V_D &= q/(\varepsilon_s \varepsilon_o)[nx_n x_d - (nx_n^2 + px_p^2)/2] \\ &= q/(2\varepsilon_s \varepsilon_o)(n/p)(n+p)x_n^2 \\ &= q/(2\varepsilon_s \varepsilon_o)(p/n)(n+p)x_p^2 \end{aligned} \quad (6.16)$$

$$V_D = (kT/q) \ln (np/n_i^2) \quad (6.17)$$

$$x_n = [(2\varepsilon_s \varepsilon_o/q)(p/n)/(n+p)(V_a + V_D)]^{1/2} \quad (6.18)$$

$$x_p = [(2\varepsilon_s \varepsilon_o/q)(n/p)/(n+p)(V_a + V_D)]^{1/2}. \quad (6.19)$$

If, in a special case of a step junction, the junction is quite unsymmetrical (one-sided step junction), i.e., if for example $n \ll p$, the following simplifications can be made:

$$E_{max} = 2[q/(2\varepsilon_s \varepsilon_o)n(V_a + V_D)]^{1/2} \quad (6.20)$$

$$\begin{aligned} x_n &= [2(\varepsilon_s \varepsilon_o/q)(V_a + V_D)/n]^{1/2} \\ &= (p/n)x_p \gg x_p \end{aligned} \quad (6.21)$$

$$x_p = [2(\varepsilon_s \varepsilon_o/q)(n/p^2)(V_a + V_D)]^{1/2}. \quad (6.22)$$

If $n \gg p$, then the appropriate equations can be obtained by exchanging n and p.

The total depletion layer width ($x_d = x_n + x_p$) and the maximum electric field depend essentially only upon the impurity concentration on the lighter doped side of the junction, i.e., the characteristics of an unsymmetrical step junction are determined by the characteristics of the lighter doped side. Numerically, an unsymmetrical step junction behaves like a symmetrical step junction at twice the applied voltage whose impurity concentration ($n \approx p$) is equal to that of the lighter doped side of the unsymmetrical junction.

(b) Linearly graded junction:
Under the assumptions

$$C(x) > 2n_i x/L_D$$
$$x_a \equiv x n_i / C(x)$$

the general expressions (6.12) and (6.13) can be modified as follows:

$$E_{max} = q/(\varepsilon_s \varepsilon_o) n_i x_d^2 / 2 x_a$$
$$= (1/2)[q/(4\varepsilon_s \varepsilon_o)(n_i/x_a)]^{1/3}[3(V_a + V_D)]^{2/3} \quad (6.23)$$

$$V_a + V_D = (2/3)[q/(\varepsilon_s \varepsilon_o)](n_i/x_a) x_d^3 \quad (6.24)$$

$$x_d = (1/2)[12\varepsilon_s \varepsilon_o / q (x_a/n_i)(V_a + V_D)]^{1/3}. \quad (6.25)$$

Depletion layer width and maximum electric field increase and capacitance decreases with increasing voltage. With increasing impurity concentration gradient (dC/dx) the depletion layer width decreases, whereas capacitance and maximum electric field increase. The built-in voltage depends slightly upon the depletion layer width and therefore upon the applied voltage; usually $V_D < E_G/q$.

(c) *p-i-n* Junction:
A *p-i-n* junction is a step junction to which an intrinsic region of width x_i is added between the *n*- and the *p*-region; this intrinsic region is entirely free of impurities, hence it behaves like a dielectric and the electric field within this region is constant. The maximum electric field is the same as that of a *p-n* step junction,

$$E_{max} = q/(\varepsilon_s \varepsilon_o) n x_n$$
$$= q/(\varepsilon_s \varepsilon_o) p x_p. \quad (6.26)$$

The voltage has an additional term

$$E_{max} x_i = q/(\varepsilon_s \varepsilon_o) n x_n x_i$$
$$= q/(\varepsilon_s \varepsilon_o) p x_p x_i \quad (6.27)$$

which is additive to the corresponding expression for the step junction (6.16). If $x_i \gg x_n$ and $x_i \gg x_p$, then the additive term dominates so that

$$E_{max} = (V_a + V_D)/x_i \qquad (6.28)$$

$$x_n = (\varepsilon_s \varepsilon_o/q)(V_a + V_D)/(nx_i) \qquad (6.29)$$

$$x_p = (\varepsilon_s \varepsilon_o/q)(V_a + V_D)/(px_i). \qquad (6.30)$$

Compared to a normal step junction, a p-i-n junction has a significantly reduced capacitance and maximum electric field; furthermore, the capacitance is much less voltage-sensitive since the total depletion layer width ($x_d = x_n + x_p + x_i$) is only slightly larger than x_i (which is constant).

In the above equations the following terms have been used:

E_{max} = maximum electric field within the depletion layer; usually it coincides with the metallurgical junction; at the edges of the depletion layer the electric field has essentially decreased to zero

V_D = diffusion (built-in) voltage; it is the total electrostatic voltage drop across the depletion layer; if the electrostatic potential within the depletion layer were constant, electrons would diffuse into the p-region and holes into the n-region; such a carrier diffusion would create an electrostatic layer which would result in a compensation of the diffusion current by a field current of the same magnitude but of opposite sign

V_a = externally applied voltage

x_d = total width of the depletion layer

x_n, x_p = fractional widths of the depletion layer on n- or p-side of the junction, respectively

x_{d1}, x_{d2} = fractional widths of the depletion layer on heavier or lighter doped side of the junction, respectively

L_D = Debye length

2. TOTAL JUNCTION CAPACITANCE

The total capacitance (C_t) of a p-n junction is composed of the depletion layer capacitance (C_T) and the diffusion capacitance (C_D).

$$C_t = C_T + C_D. \qquad (6.31)$$

Both of these contributions are voltage-dependent. The total electrostatic potential variation across the junction is given by

$$V = V_a + V_D \text{ for reverse bias}$$

and

$$V = V_a - V_D \text{ for forward bias,}$$

where V_a and V_D are the externally applied voltage and the built-in voltage, respectively.

(a) Depletion-layer capacitance:
Since the depletion layer is essentially free of carriers, it acts as a dielectric of thickness x_d and is associated with a corresponding capacitance. The depletion layer (transition) capacitance accounts for most of the junction capacitance if the junction is reverse-biased; this capacitance is usually given as junction capacitance. For plane junctions the depletion layer capacitance is as follows:

(i) *Step junction:*

$$C_T = \varepsilon_s \varepsilon_o / x_d = (q\varepsilon_s \varepsilon_o C_B / 2V)^{1/2} \tag{6.32}$$

and

$$d(1/C_T^2)/dV = 2/q\varepsilon_s \varepsilon_o C_B. \tag{6.33}$$

(ii) *Linearly graded junction:*

$$C_T = \varepsilon_s \varepsilon_o / x_d = [q(\varepsilon_s \varepsilon_o)^2 (dC/dx)/12V]^{1/3}. \tag{6.34}$$

For spherical and cylindrical junctions the depletion layer capacitance is given in Figure 6.12.

The depletion layer capacitance consists of two contributions ($C_T = C_{Ti} + C_{Tc}$):

(i) The transition space charge capacitance (C_{Ti}) due to the change of the total number of ionized impurity space charges with voltage in half of the depletion layer.

$$C_{Ti} = (\varepsilon_s \varepsilon_o / x_d)(kT/q)(-dx_d^3/dV) \tag{6.35}$$

(ii) The transition carrier capacitance (C_{Tc}) due to the change of the total number of free carriers with voltage in the depletion layer.

$$C_{Tc} \approx (kT/q)(qn_i/6)x_d \exp(qV/2kT) \tag{6.36}$$

(b) Diffusion capacitance:
At forward bias the diffusion capacitance adds a significant contribution to the total junction capacitance. It arises from rearrangement of excess minority carriers outside the depletion region when the voltage across the junction is altered. Its influence is particularly important at low frequency.

For a linearly graded junction

$$C_D = q(q/kT)(L_p p_{no}/2 + L_n n_{po}/2) \exp(qV/kT)$$
$$\approx (x_d^2/3D_n)(q/kT)I_f, \qquad (6.37)$$

where L_n, L_p are the diffusion lengths (assuming L_n, $L_p \gg x_d$), n_{po} and p_{no} are the equilibrium minority carrier densities, I_f is the forward current across the junction. The diffusion capacitance increases with the current level.

In a linearly graded junction the diffusion capacitance dominates over the depletion layer capacitance in semiconductors with small energy gap or high intrinsic carrier concentration, large minority carrier lifetime, or small impurity gradient.

3. WIDTH AND CAPACITANCE OF DEPLETION LAYER

The total width of a p-n junction depletion layer and its associated capacitance per unit area are as follows (see Figures 6.5 and 6.6).

(a) Step junction:

$$x_d = [2\varepsilon_s \varepsilon_o (V_a + V_D)/(qC_B)]^{1/2} \qquad (6.38)$$

$$C = \{q\varepsilon_s \varepsilon_o C_B/[2(V_a + V_D)]\}^{1/2}. \qquad (6.39)$$

Depletion layer width increases and depletion layer capacitance decreases with the square root of voltage and approximately with the square root of resistivity on the lighter doped side of the junction.

For a two-sided junction C_B has to be replaced by

$$C_B = N_A N_D/(N_A + N_D). \qquad (6.40)$$

(b) Linearly graded junction:

$$x_d = \{12\varepsilon_s \varepsilon_o (V_a + V_D)/[q(dC/dx)]\}^{1/3} \qquad (6.41)$$

$$C = \{(\varepsilon_s \varepsilon_o)^2 q(dC/dx)/[12(V_a + V_D)]\}^{1/3}. \qquad (6.42)$$

Depletion layer width increases and depletion layer capacitance decreases with the cube root of voltage.

If the voltage across the junction is increased, the cube-root dependence of a graded junction tends to change to the square-root dependence of a step junction. The transition from linearly graded to step junction behavior of a diffused junction is facilitated by increasing V and C_S and by decreasing C_B and x_j.

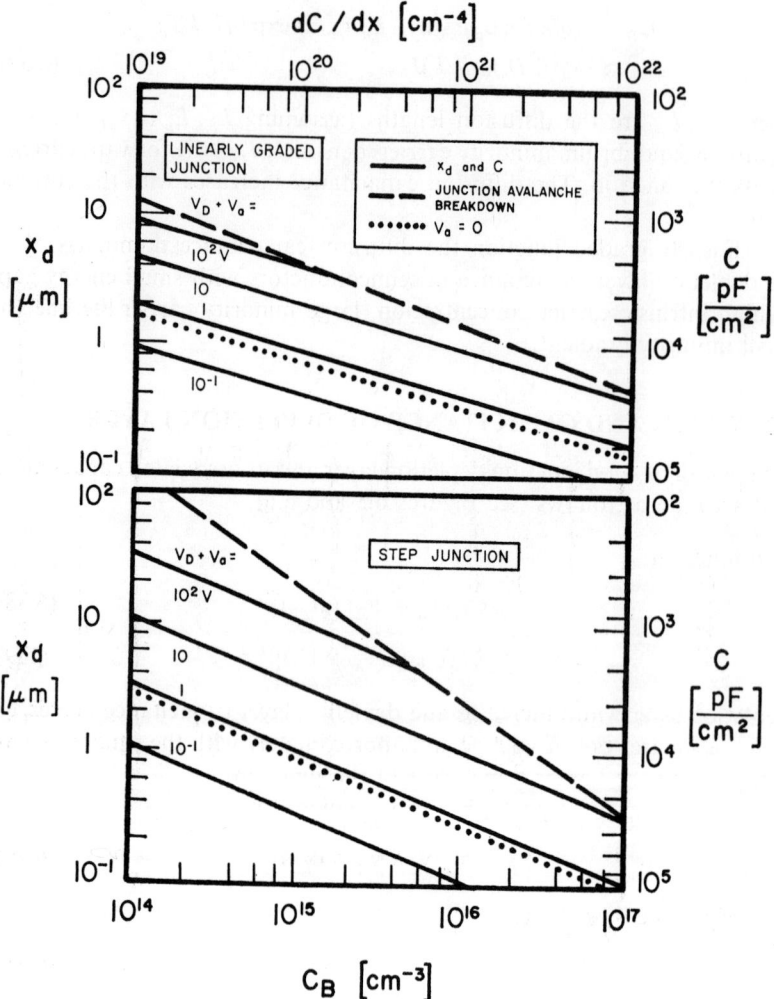

FIGURE 6.5

Depletion layer width (x_d) and capacitance per unit area (C) vs. impurity concentration (C_B) on lighter doped side of step junction or impurity gradient (dC/dx) of linearly graded junction, and total voltage across junction ($V_D + V_a$). Avalanche breakdown is indicated by the dashed line. Silicon, 300°K.

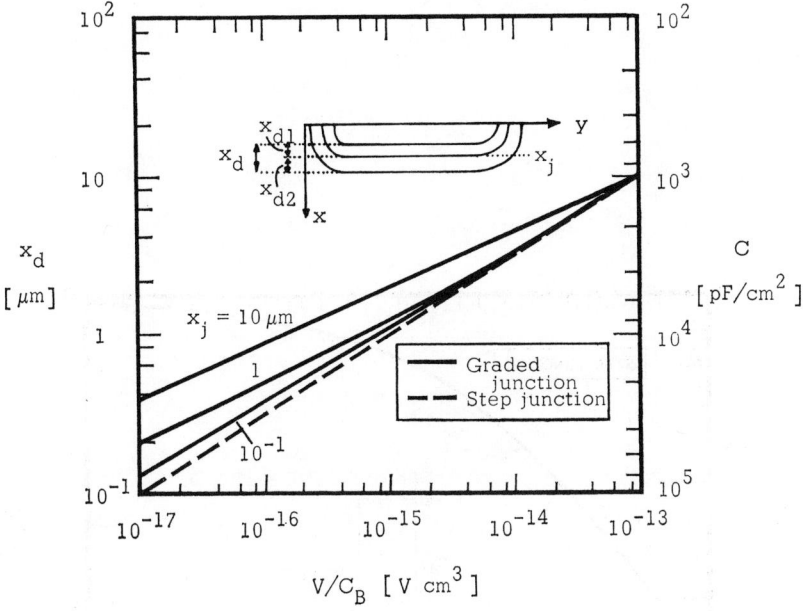

FIGURE 6.6
Depletion layer width (x_d) and depletion layer capacitance (C) per unit area as function of total voltage across junction (V), background impurity concentration (C_B), and junction depth (x_j) for $C_S/C_B = 10^3$. For a step junction x_d and C are independent of x_j. Silicon, 300°K. The inset defines the fractional depletion layer widths.

4. FRACTIONAL WIDTH OF DEPLETION LAYER

The depletion layer of a two-sided junction extends into both the lighter and the heavier doped side of the junction. The larger portion generally extends into the lighter doped side (see Figures 6.7 and 6.8).

(a) Step junction:
Although the width of the entire depletion layer is voltage-dependent, the ratio x_{d1}/x_{d2} remains constant with voltage. The ratios x_{d1}/x_d and x_{d2}/x_d are also independent of voltage. The quantities x_d, x_{d1}, and x_{d2} are independent of junction depth.

(b) Linearly graded junction:
The width of the entire depletion layer as well as the portions x_{d1} and x_{d2} and the ratios x_{d1}/x_d and x_{d2}/x_d are voltage-dependent. All of these quantities also depend upon junction depth, background impurity concentration, and impurity gradient at the junction.

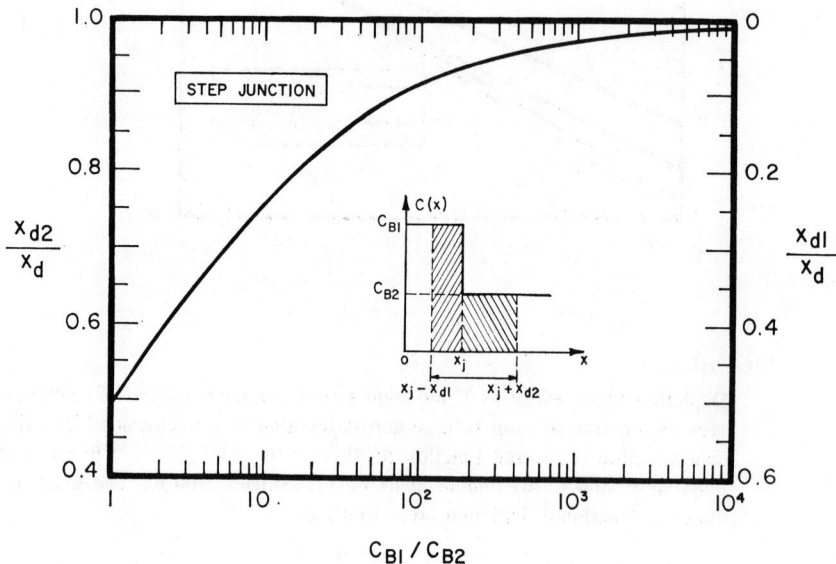

FIGURE 6.7
Fractional depletion layer widths (x_{d1}/x_d and x_{d2}/x_d) on heavier and on lighter doped sides of a step junction vs. ratio of impurity concentrations on heavier and lighter doped sides of the junction. Silicon, 300°K.

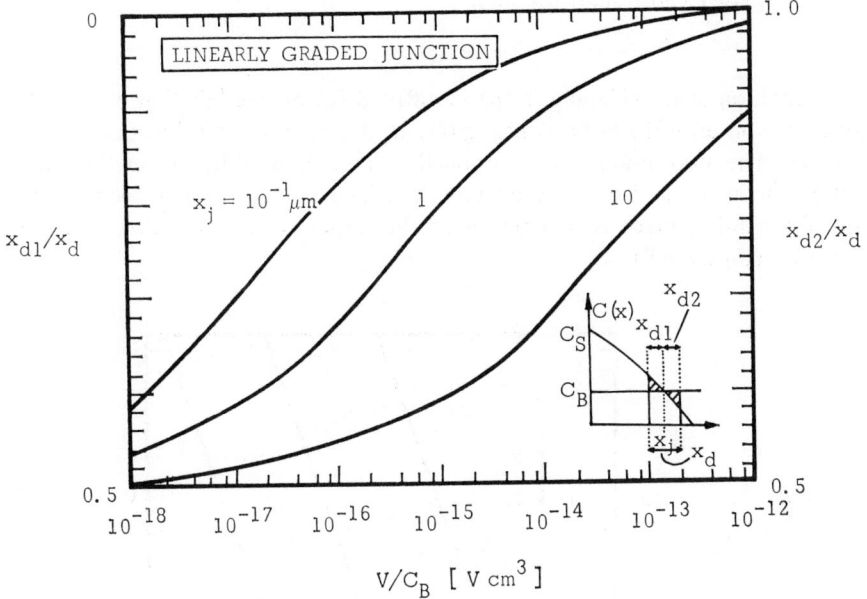

FIGURE 6.8

Fractional depletion layer widths (x_{d1}/x_d and x_{d2}/x_d) on heavier and on lighter doped sides of a linearly graded junction as function of voltage across junction (V), background impurity concentration (C_B), and junction depth (x_J). Silicon, $C_S/C_B = 10^3$, 300°K.

5. CROSS-OVER FROM LINEARLY GRADED TO STEP JUNCTION BEHAVIOR

At small reverse voltage, the depletion layer of a diffused junction is narrow, i.e.,

$$x_d < 2\sqrt{Dt},$$

and the linearly graded junction model applies. As voltage and consequently depletion layer width increase, i.e.,

$$x_d > 2\sqrt{Dt},$$

the depletion layer widening in the undiffused region exceeds that in the diffused region until the latter is negligible. In this case the step junction model applies. The step model can be applied to a shallow diffused junction, the linear model to a deep diffused junction. The voltage (V_c) at which the transition takes place is a function of the impurity concentration near the junction (Figure 6.9).

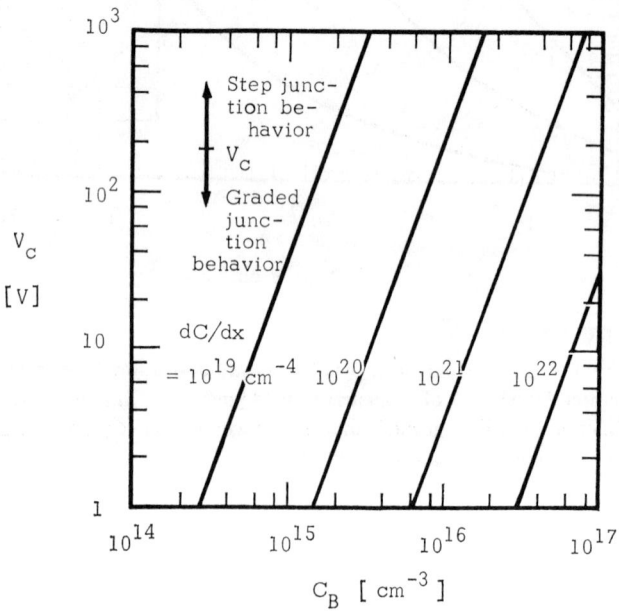

FIGURE 6.9

Voltage (V_c) at which the characteristics of a diffused p-n junction change from linearly graded behavior to step junction behavior (cross-over voltage) as function of semiconductor background impurity concentration (C_B) and impurity gradient (dC/dx) at the junction. Silicon, 300°K.

6. DEPLETION LAYER WIDTH AND CAPACITANCE AS A FUNCTION OF JUNCTION GEOMETRY

The depletion layer of a *p-n* junction is the region on both sides of the junction in which there exists an electric field. This field is at its maximum (E_{max}) at the metallurgical junction (x_j) and is zero at the edges of the depletion layer. The depletion layer extends from $x_j - x_{d1}$ to $x_j + x_{d2}$, where $x = 0$ at the semiconductor surface and x_{d1} and x_{d2} are the depletion layer widths on the highly and lowly doped sides of the junction.

The electric field across the junction can be obtained by integration of Poisson's equation which is

$$(1/x^g)\, d(x^g\, dV/dx)/dx = -Q(x)/\varepsilon_s \varepsilon_o. \tag{6.43}$$

A second integration yields a relationship between depletion layer width and junction voltage. The geometry factor g is

$g = 0$ for a plane junction,

$g = 1$ for a cylindrical junction,

$g = 2$ for a spherical junction.

The charge distribution for an erfc impurity distribution is

$$Q(x) = q[C_S \text{ erfc } (x/2\sqrt{Dt}) - C_B] \tag{6.44}$$

and the junction depth is defined by

$$C_B/C_S = \text{erfc } (x_j/2\sqrt{Dt}). \tag{6.45}$$

At low voltage (linearly graded junction model) the depletion layer width follows the 0.33 power law and is independent of junction geometry. At high voltage (step junction model) the dependence of the depletion layer width upon voltage is junction-dependent; i.e., for a plane junction it is a 0.5 power law, for a cylindrical junction a 0.46 law, and for a spherical junction a 0.31 law. The cross-over from low voltage to high voltage behavior of silicon junctions occurs approximately at

$$5 \cdot 10^{-10} < V/x_j^2 C_B < 5 \cdot 10^{-8} \text{ V cm}.$$

The maximum electric field is independent of junction geometry at small values of $V/x_j^2 C_B$; at higher values the influence of geometry becomes significant.

Figures 6.10, 6.11, and 6.12 apply to silicon *p-n* junctions at 300°K and were calculated. They are valid for both erfc and Gaussian impurity distributions; the differences are insignificant. The main difference between erfc and Gaussian curves is that the voltage at which the depletion layer reaches the

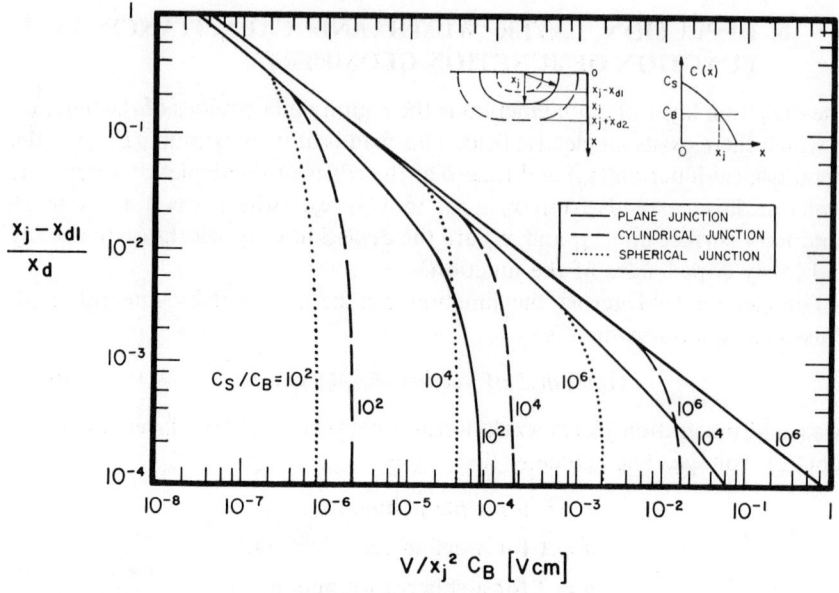

FIGURE 6.10

Relation between junction (x_j), depletion layer width (x_d), fractional depletion layer width (x_{d1}), voltage (V) across junction, and background impurity concentration (C_B), for various ratios C_S/C_B and different junction geometries. Silicon, 300°K.

semiconductor surface is for erfc distributions about half that for Gaussian distributions.

Although the illustrations apply to silicon, the use of the proper dielectric constants allows their application to other semiconductors. In Figures 6.10 through 6.12 x_j and x_d are given in cm, impurity concentrations in cm^{-3}, and voltage in V.

Figure 6.10 shows the fractional depletion layer width $(x_j - x_{d1})/x_d$ for various ratios of surface impurity concentration to background impurity concentration (C_S/C_B). The difference between different junction geometries is most significant at high values of $V/x_j^2 C_B$ and at low values of C_S/C_B. The curves fall to very small values of $(x_j - x_{d1})x_d$, i.e., $x_j = x_{d1}$, when the depletion layer reaches the semiconductor surface. The corresponding values of $V/x_j^2 C_B$ indicate the conditions for this behavior.

Figure 6.11 shows the ratio of depletion layer width to junction depth (x_d/x_j) and the maximum electric field at the junction. The curves are valid for the range $C_S/C_B = 10^2$ to 10^6 within the range of $V/x_j^2 C_B$ given. The influence of junction geometry is significant mainly at high values of $V/x_j^2 C_B$.

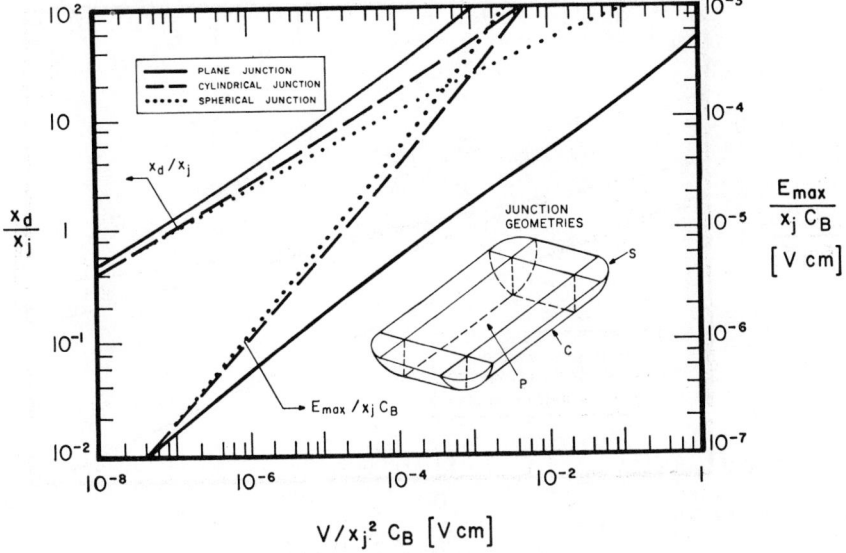

FIGURE 6.11
Ratio of depletion layer width (x_d) to junction depth (x_j) and maximum electric field (E_{max}) vs. voltage (V) and background impurity concentration (C_B) for various junction geometries. In the inset the junction geometries planar (p), cylindrical (c), and spherical (s) are defined. Silicon, 300°K.

Figure 6.12 shows the depletion layer capacitance per unit area (C) for various ratios C_S/C_B. For a plane junction, variation of C_S/C_B has no effect on C except at low values of $V/x_j^2 C_B$. For cylindrical and spherical junction, the influence of C_S/C_B on capacitance is significant at high values of $V/x_j^2 C_B$.

The capacitances (per unit area) of the various geometries are as follows:

(a) plane junction:

$$C = \frac{\varepsilon_s \varepsilon_o}{x_d} ; \tag{6.46}$$

(b) cylindrical junction:

$$C = \frac{\varepsilon_s \varepsilon_o}{x_d} \frac{\pi}{\ln\left[(x_j + x_{d2})/(x_j - x_{d1})\right]} ; \tag{6.47}$$

(c) spherical junction:

$$C = \frac{\varepsilon_s \varepsilon_o}{x_d} 2\pi \frac{(x_j - x_{d1})(x_j + x_{d2})}{x_j x_d} . \tag{6.48}$$

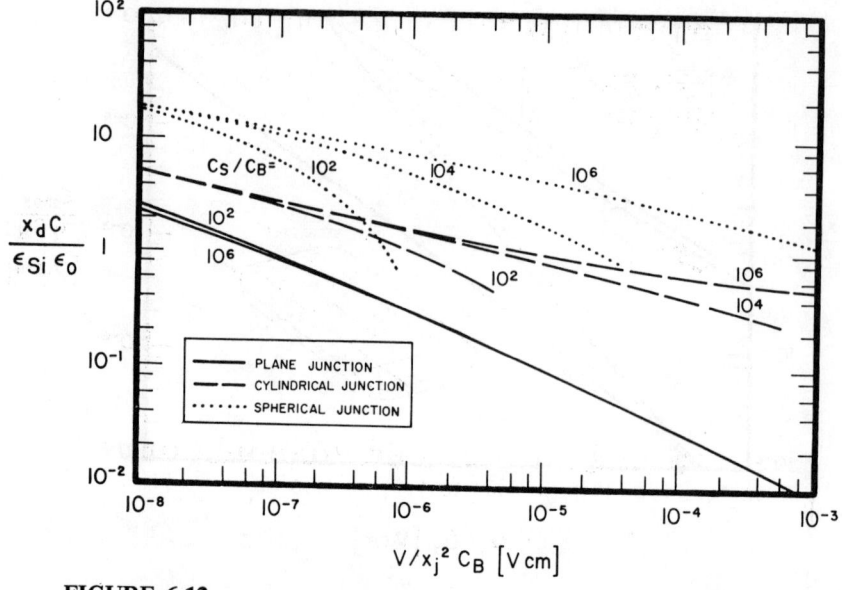

FIGURE 6.12

Relation between depletion layer width (x_d), depletion layer capacitance (C) per unit area, junction depth (x_j), voltage (V) across junction, and background impurity concentration (C_B) for various ratios C_S/C_B. Silicon, 300°K.

7. ELECTRIC FIELD NEAR p-n JUNCTION

The electric field (E) is the force acting on a unit charge and is given in an n-type semiconductor by

$$E = (kT/q)(1/n_n)(dn_n/dx) \tag{6.49a}$$

and in a p-type semiconductor

$$E = (kT/q)(1/p_p)(dp_p/dx). \tag{6.49b}$$

The variation of electric field with distance is

$$\frac{\partial E}{\partial x} = \frac{\partial^2 V}{\partial x^2} = \frac{\rho(x)}{\varepsilon_s \varepsilon_o}. \tag{6.50}$$

There is no electric field in a region of constant majority carrier density or constant impurity concentration; i.e., $E = 0$ if $dn/dx = 0$. The electric field aids the motion of minority carriers which are swept across the depletion layer very swiftly, with the peak velocity occurring at the metallurgical junction where the electric field is highest.

The current density due to electrons across the junction

$$j_n = qD_n(dn_p/dx) + q\mu_n n_p E. \tag{6.51}$$

The electric field reaches a maximum (E_{max}) at the metallurgical p-n junction and falls off toward the edges of the depletion layer. If the maximum electric field reaches a critical value (E_{crit}) the junction will break down by an avalanche mechanism.

The electric field near a p-n junction as function of distance from the junction ($x = x_j$) is given by the following expressions.

(a) Step junction:

$$E(x)/E_{max} = 1 - (x - x_j)/x_{d1} \quad \text{for } x < x_j \quad (6.52a)$$

$$E(x)/E_{max} = 1 - (x - x_j)/x_{d2} \quad \text{for } x > x_j \quad (6.52b)$$

$$E_{max} = 2V/x_d$$
$$= (2qC_B V/\varepsilon_s \varepsilon_o)^{1/2} \quad (6.53)$$

(b) Linearly graded junction:

$$E(x)/E_{max} = 1 - [(x - x_j)/x_{d1}]^2 \quad \text{for } x < x_j \quad (6.54a)$$

$$E(x)/E_{max} = 1 - [(x - x_j)/x_{d2}]^2 \quad \text{for } x > x_j \quad (6.54b)$$

$$E_{max} = (3/2)V/x_d$$
$$= (9/32)^{1/3} V^{2/3} [q(dC/dx)/\varepsilon_s \varepsilon_o]^{1/3}. \quad (6.55)$$

Extreme cases:

(a) at $x = x_j$, i.e., at the junction:

$$E = E_{max}.$$

(b) at $x = x_j - x_{d1}$ or at $x = x_j + x_{d2}$, i.e., at the edges of the depletion layer:

$$E = 0.$$

Figures 6.13 and 6.14 show the maximum electric field as a function of voltage, background impurity concentration, and junction depth. In Figure 6.14 the maximum electric field is given for an erfc-type impurity distribution and a ratio of surface concentration to background concentration $C_S/C_B = 10^5$. If $C_S/C_B \neq 10^5$, then the following errors are observed

C_S/C_B	Deviation from curves given [%]	
	erfc	Gaussian
10^2 to 10^4	10	10
10^5	0	5
10^4 to 10^6	5	5
10^6 to 10^8	10	10

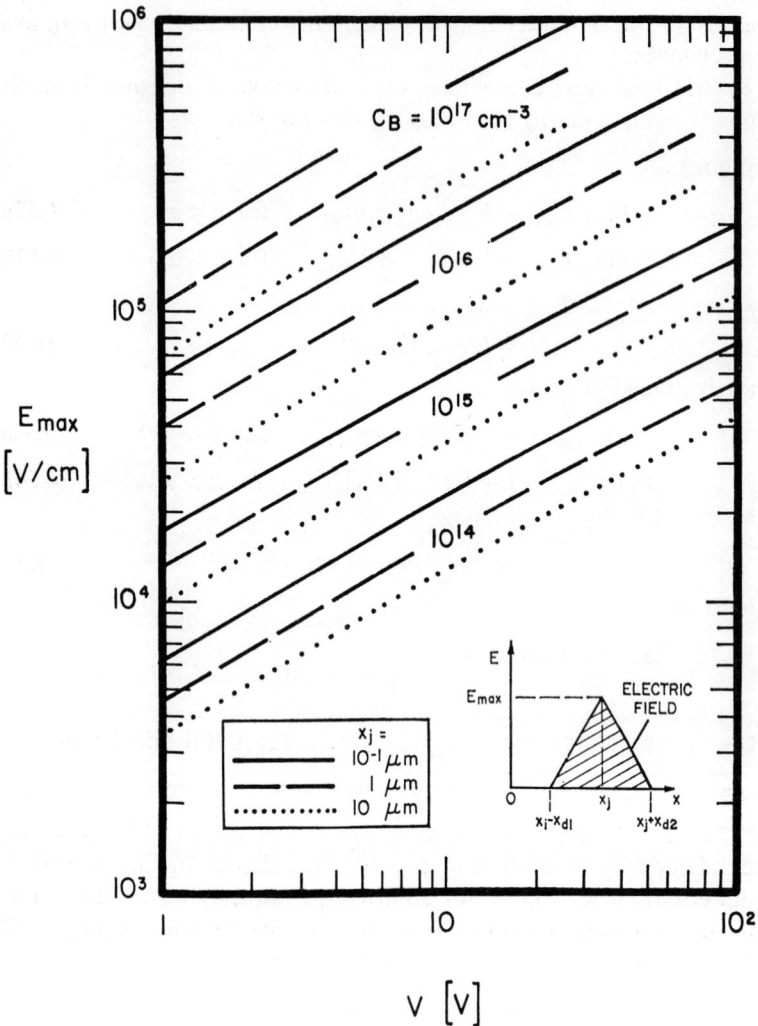

FIGURE 6.13

Maximum electric field (E_{max}) at metallurgical *p-n* junction vs. voltage (V) across junction, background impurity concentration (C_B), and junction depth (x_j). Silicon, 300°K.

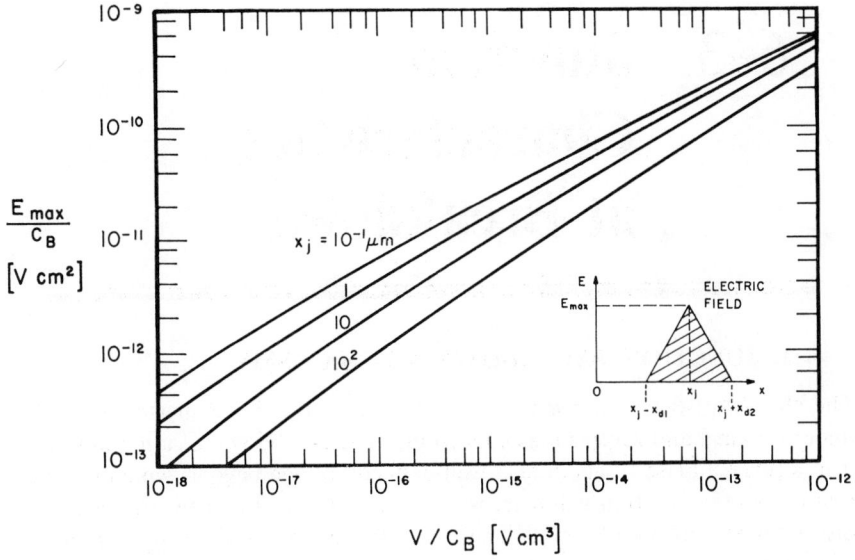

FIGURE 6.14
Ratio of maximum electric field (E_{max}) at a *p-n* junction to background impurity concentration (C_B) vs. voltage (V) across junction, background concentration, and junction depth (x_j). Silicon, 300°K.; $C_S/C_B = 10^5$.

8. REFERENCES

(1) H. Lawrence and R. M. Warner, *BSTJ*, **39**, 389 (1960).
(2) C. T. Sah, *Proc. IRE*, **49**, 603 (1961).
(3) A. B. Phillips, *Transistor Engineering*, McGraw-Hill Book Co., New York, 1962.
(4) H. Krömer, *Der Transistor*, Springer-Verlag, Berlin, 1963.
(5) L. K. Monteith, *RTI Report ASD-TDR-63-316*, Vol. II, Oct. 1963.
(6) S. M. Sze, *Physics of Semiconductor Devices*, John Wiley and Sons, New York, 1969.
(7) P. R. Wilson, *Solid-State Electronics*, **12**, 1 (1969).
(8) P. R. Wilson, *Solid-State Electronics*, **12**, 277 (1969).

6–3 Junction Characteristics at Breakdown

1. JUNCTION BREAKDOWN MECHANISMS

The characteristics of a *p-n* junction, due to its rectifying properties, differ under forward and reverse bias conditions. The breakdown of a *p-n* junction, if it is reverse biased, sets in at a well-defined voltage beyond which the current across the junction will increase quite rapidly. If it is forward biased, no distinct breakdown is observed. There are two reverse breakdown mechanisms of importance.

(a) Avalanche breakdown:
  In this case (which is the usual one and which precedes Zener breakdown), electrons and holes are thermally generated with the aid of intermediate levels and, due to impact ionization, additional electron-hole pairs are generated.
(b) Tunneling (Zener breakdown):
  In this case, due to the breaking of some covalent bonds between neighboring atoms within the depletion region, electrons and holes are generated and valence electrons will move from valence band to conduction band, i.e., tunneling through the energy gap will occur.

The breakdown voltage (V_B) is a function of impurity concentration, impurity gradient, junction depth, and the nature of the semiconductor; the energy gap (E_G) of the semiconductor depends also upon the nature of the semiconductor. The ratio V_B/E_G is indicative of the type of breakdown. Three regions of breakdown voltage are distinguished depending upon the magnitude of V_B/E_G.

(a) $V_B/E_G < 4/q = 4\text{V}/\text{eV}$:
  In this case there is band-to-band tunneling and the temperature coefficient of V_B is negative. For silicon this condition corresponds to breakdown voltages below about 5 V.
(b) $4/q < V_B/E_G < 6/q$:
  In this case the breakdown is due to a combination of band-to-band

6-3 JUNCTION CHARACTERISTICS AT BREAKDOWN

tunneling and avalanche multiplication. The temperature coefficient of V_B is at its minimum. For silicon this condition corresponds to breakdown voltages between 5 and 7 V.

(c) $V_B/E_G > 6q = 6$ V/eV:
In this case breakdown is mainly due to carrier avalanching and the temperature coefficient of V_B is positive. For silicon this condition corresponds to breakdown voltages greater than about 7 V.

The conditions under which avalanche or tunneling breakdown sets in are given in Table 6.1 for Si, Ge, and GaAs.

TABLE 6.1

Impurity Concentration (C_B) of Step Junction or Concentration Gradient (dC/dx) of Linearly Graded Junction Above Which Tunneling Breakdown or Below Which Avalanche Breakdown Set In (300°K)

Semi-conductor	E_G/q [V]	Tunneling breakdown			Avalanche breakdown		
		Step junction C_B [cm^{-3}]	Graded junction dC/dx [cm^{-4}]	V_B [V]	Step junction C_B [cm^{-3}]	Graded junction dC/dx [cm^{-4}]	V_B [V]
Si	1.12	$> 6 \cdot 10^{17}$	$> 5 \cdot 10^{23}$	< 4.5	$< 3 \cdot 10^{17}$	$< 1 \cdot 10^{23}$	> 6.7
Ge	0.67	$> 4 \cdot 10^{17}$	$> 2 \cdot 10^{23}$	< 2.7	$< 1 \cdot 10^{17}$	$< 4 \cdot 10^{22}$	> 4.0
GaAs	1.43	$> 7 \cdot 10^{17}$	$> 6 \cdot 10^{23}$	< 5.7	$< 3 \cdot 10^{17}$	$< 1 \cdot 10^{23}$	> 8.6

2. CARRIER MULTIPLICATION AND AVALANCHE BREAKDOWN

Carrier multiplication due to impact ionization takes place within the depletion layer of a *p-n* junction under the influence of an electric field. It is a function of the ionization rate, $\alpha_i(E)$.

When the electric field in a semiconductor is increased above a critical value (E_{crit}), the carriers gain sufficient energy to generate electron-hole pairs by impact ionization. The generation rate of electron-hole pairs due to impact ionization is

$$G_{np} = \alpha_{in} n \mu_n + \alpha_{ip} p \mu_p,$$

where α_{in} is the electron ionization rate, i.e., the number of electron-hole pairs generated by an electron per unit distance traveled, α_{ip} is the hole ionization rate due to hole impact, n and p are carrier densities, and μ_n and μ_p

are carrier mobilities. Both α_{in} and α_{ip} are strongly dependent upon electric field as follows.

$$\alpha_i = A_i^* \exp\left[-(a_i^*/E)^{m^*}\right] \tag{6.56}$$

where A_i^*, a_i^*, and m^* are parameters given in Table 6.2.

TABLE 6.2
Ionization Parameters for Semiconductors

		Si	Ge	GaAs
A_i^* [cm^{-1}]	Electrons	$3.80 \cdot 10^6$	$1.55 \cdot 10^7$	$1.34 \cdot 10^6$
	Holes	$2.25 \cdot 10^7$	$1.00 \cdot 10^6$	$1.34 \cdot 10^6$
a_i^* [V/cm]	Electrons	$1.75 \cdot 10^6$	$1.56 \cdot 10^6$	$2.03 \cdot 10^6$
	Holes	$3.26 \cdot 10^6$	$1.28 \cdot 10^6$	$2.03 \cdot 10^6$
m^*		1	1	2

The carrier multiplication factor M gives the ratio of the number of secondary carriers (n_2) due to ionization to the number of primary carriers (n_1).

$$M = n_2/n_1$$

$$= \frac{1}{1 - \int_{x_j - x_{d1}}^{x_j + x_{d2}} \alpha_i(E)\, dx} \tag{6.57}$$

where E = electric field across the depletion layer
E_{crit} = critical electric field at which avalanche breakdown occurs.
Ionization rates for electrons and holes differ slightly, whereas close to breakdown both are almost equal.

Avalanche breakdown of a junction occurs when the value of the integral in above equation approaches unity, i.e., when $M \to \infty$. In this case

$$V = V_B, \quad E = E_{\max} = E_{\text{crit}}.$$

Avalanche breakdown voltages in Si, Ge, and GaAs junctions are given in Figure 6.15.

Multiplication factor and avalanche breakdown voltage are related by

$$M = [1 - (V/V_B)^s]^{-1} \tag{6.58}$$

where s is an empirically determined parameter which depends upon the

6-3 JUNCTION CHARACTERISTICS AT BREAKDOWN

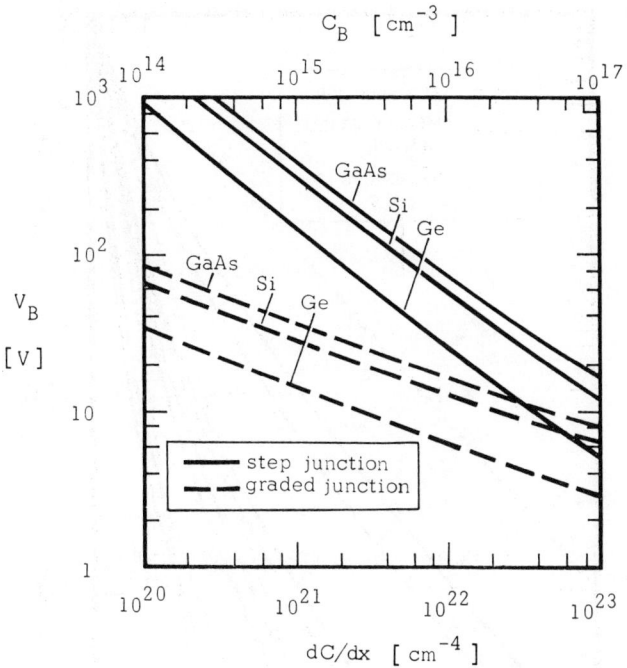

FIGURE 6.15

Avalanche breakdown voltage (V_B) as function of impurity concentration (C_B) on the lighter doped side of a step junction (upper scale) or impurity gradient at the junction (dC/dx) of a linearly graded junction (lower scale) for Si, Ge, and GaAs. 300°K.

impurity concentration on the lighter doped side of a step junction or upon the impurity gradient of a linearly graded junction and the ratio V/V_B. This is illustrated in Figures 6.16 and 6.17.

Since M is related to the current across a p-n junction, a plot of M vs. V/V_B describes the current-voltage relationship of the junction under reverse bias conditions. A higher impurity concentration (lower resistivity) or a steeper impurity gradient results in a sharper breakdown characteristic (i.e., a sharper "knee" of the I-V curve).

In presenting the parameters M and s the following assumptions are made:

(a) The junction is reverse-biased. If it were forward-biased, the depletion layer had a negligible width.
(b) The reverse current across the junction consists mainly of minority carriers that diffuse across the depletion layer.

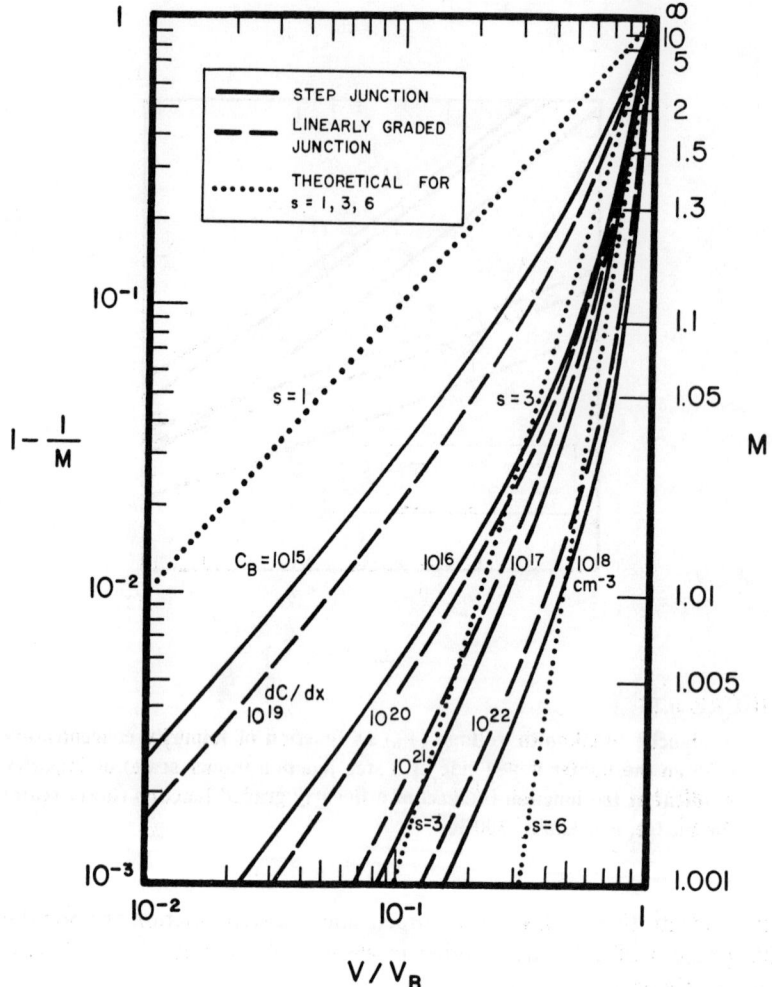

FIGURE 6.16

Minority carrier multiplication factor (M) vs. voltage (V) across junction, avalanche breakdown voltage (V_B), and background impurity concentration (C_B) for step junction or impurity gradient (dC/dx) for linearly graded junction. The dotted curves give the variation of M with V/V_B for various values of the multiplication parameter s. Silicon, 300°K. The solid and dashed curves are based on measurements. Superimposed are theoretical curves for $s = 1$, 2, and 6. The slope of the experimental curves shows that at $V \ll V_B$ parameter $s \approx 1$ and at $V \approx V_B$ parameter $s \gg 1$.

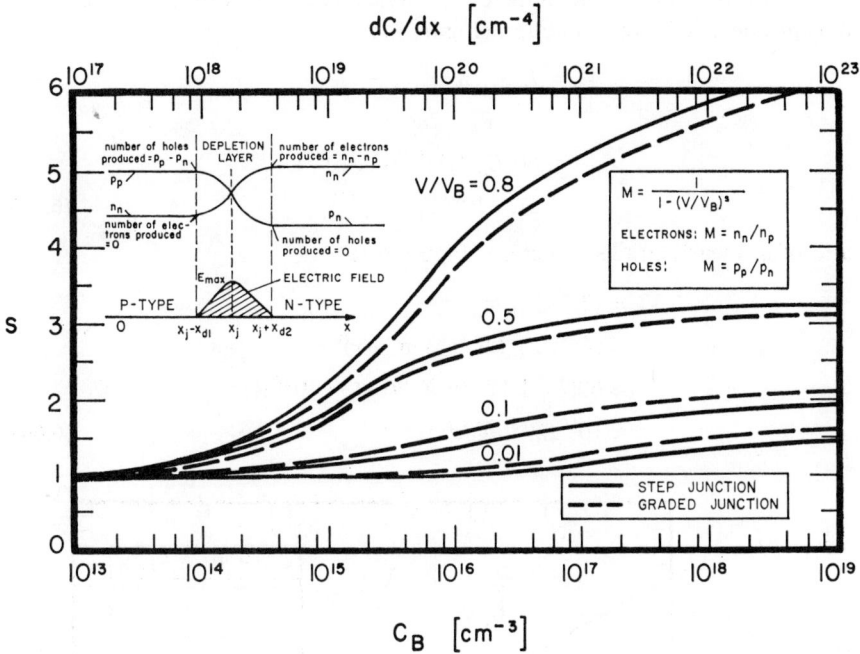

FIGURE 6.17

Multiplication parameter (s) vs. background impurity concentration (C_B) for step junction or impurity gradient (dC/dx) for linearly graded junction and the ratio V/V_B. Silicon, 300°K. The inset shows the distribution of majority and minority carriers in the vicinity of a p-n junction. The parameter s approaches unity for small background concentration or small impurity gradient. An upper limit of s under most conditions of interest is $s = 6$ at $V = 0.8 V_B$. No distinction is made between n-type and p-type semiconductors, i.e., the ionization rates of electrons and holes are assumed to be equal.

3. AVALANCHE BREAKDOWN VOLTAGE

Avalanche breakdown of a p-n junction occurs when the carrier multiplication factor (M) exceeds unity, i.e., when the number of secondary carriers exceeds that of the primary carriers due to ionization in an electric field. The field ($E = dV/dx$) at which avalanche breakdown sets in is the critical electric field ($E = E_{\text{crit}}$).

For an arbitrary semiconductor the avalanche breakdown voltage of an abrupt one-sided (step) junction (Figure 6.18)

$$\begin{aligned} V_B &= E_{\text{crit}} x_{dB}/2 \\ &= \varepsilon_s \varepsilon_o E_{\text{crit}}^2/2qC_B \\ &\approx 60(E_G/1.1)^{3/2}(C_B/10^{16})^{-3/4} \\ &\approx 5.2 \cdot 10^{13} E_G^{3/2} C_B^{-3/4} \end{aligned} \quad (6.59)$$

and of a linearly graded junction (Figure 6.19)

$$\begin{aligned} V_B &= (2/3)E_{\text{crit}} x_{dB} \\ &= (4/3)E_{\text{crit}}^{3/2}(2\varepsilon_s \varepsilon_o/q \, dC/dx)^{1/2} \\ &\approx 60(E_G/1.1)^{6/5}[(dC/dx)/(3 \cdot 10^{20})]^{-2/5} \\ &\approx 10^{10} E_G^{6/5}(dC/dx)^{-2/5} . \end{aligned} \quad (6.60)$$

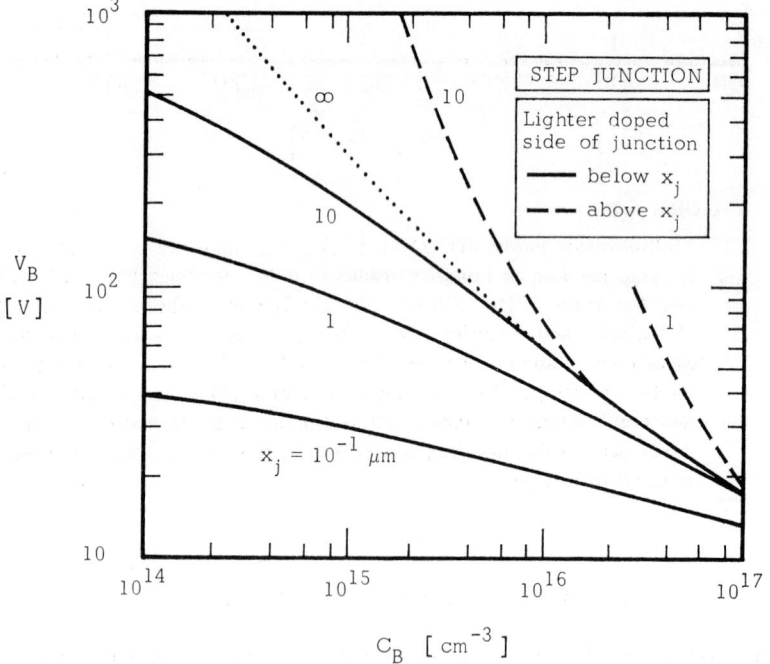

FIGURE 6.18

Breakdown voltage (V_B) of a step junction as function of impurity concentration on the lighter doped side of the junction (C_B) and junction depth (x_J). Silicon, 300°K. The dotted curve gives the avalanche breakdown voltage for infinitely deep junction.

6-3 JUNCTION CHARACTERISTICS AT BREAKDOWN

These equations are valid if
$$V_B > 6E_G/q.$$
In the above equations reach-through (effect of neighboring regions of different resistivity on the depletion layer), punch-through (depletion layer reaching the semiconductor surface), the effect of junction curvature, and surface effects—all reducing the avalanche breakdown voltage—have been neglected. It is also assumed that breakdown will occur uniformly over the entire junction area.

Ionization rate, carrier multiplication factor, and breakdown voltage of a p-n junction (step and linearly graded junction) are related by

$$1 - \frac{1}{M_n} \approx \frac{\ln(\alpha_{ip}/\alpha_{in})}{\alpha_{ip}/\alpha_{in} - 1}\left(\frac{V}{V_B}\right)^{S_n} \quad \text{(electrons)} \quad (6.61a)$$

$$1 - \frac{1}{M_p} \approx \frac{\ln(\alpha_{ip}/\alpha_{in})}{\alpha_{ip}/\alpha_{in} - 1}\left(\frac{V}{V_B}\right)^{S_p} \quad \text{(holes)} \quad (6.61b)$$

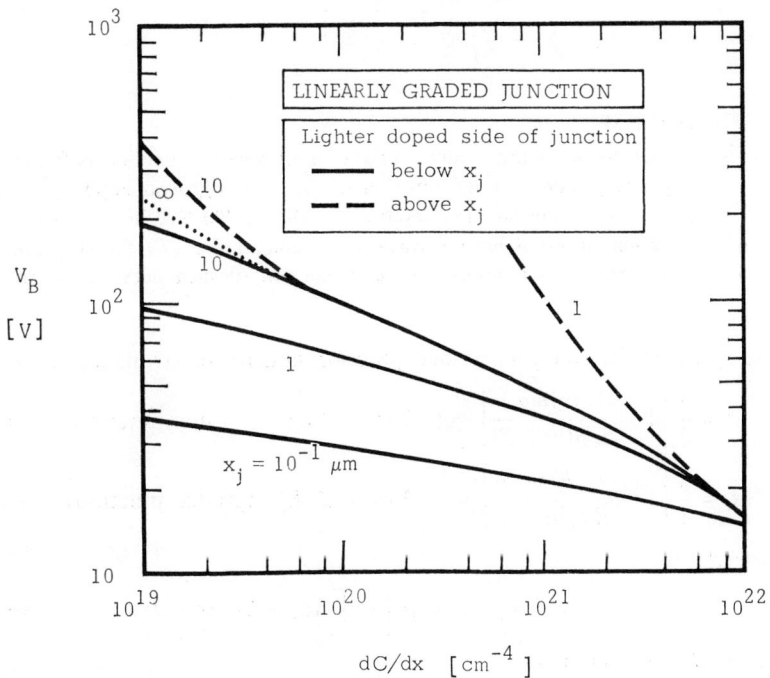

FIGURE 6.19

Breakdown voltage (V_B) of a linearly graded junction as function of impurity concentration gradient (dC/dx) at the junction and junction depth (x_j). Silicon, 300°K. The dotted curve gives the avalanche breakdown voltage for infinitely deep junction.

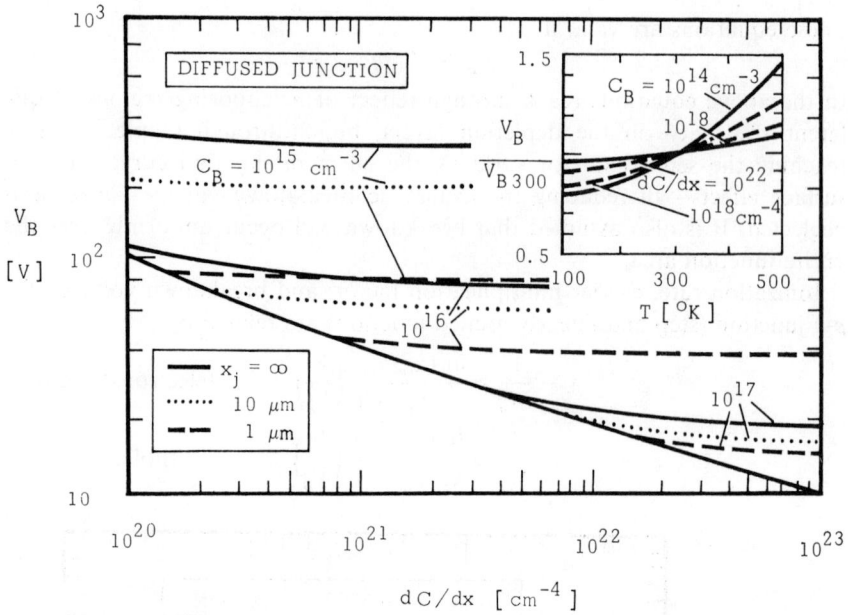

FIGURE 6.20

Avalanche breakdown voltage (V_B) of a diffused p-n junction as function of impurity gradient at the junction (dC/dx), background impurity concentration (C_B), and junction depth (x_j), Silicon, 300°K. The inset shows the variation of breakdown voltage with temperature (T) for step junction (solid curves) and linearly graded junction (dashed curves), referred to 300°K.

where s_i differs for step and graded junction and for electrons and holes:

$$s_i = \frac{1}{2}\left[1 - \frac{1}{E}\frac{d(\ln \alpha_i)}{d(\ln 1/E)}\right] = (1/2)(1 - a_i^* E) \quad \text{(step junction)} \quad (6.62a)$$

$$s_i = \frac{2}{3}\left[1 - \frac{1}{E}\frac{d(\ln \alpha_i)}{d(\ln 1/E)}\right] = (2/3)(1 - a_i^*/E) \quad \text{(graded junction)} \quad (6.62b)$$

Furthermore,

$$1 - 1/M = \int_0^{x_d} \alpha_i \exp\left[-\int_0^x (\alpha_i - \beta_i)\,dx'\right] dx \quad (6.63)$$

where $1/M = 1 - (V/V_B)^s$
  $x_d =$ junction depletion layer width
  $\alpha_i, \beta_i =$ ionization rate for initial, secondary particles, respectively;
    $\alpha_{in} =$ ionization rate of electrons, $\alpha_{ip} =$ ionization rate of holes
  $V, V_B =$ applied, breakdown voltage, respectively
    $s =$ impurity concentration-dependent parameter, which is related to the parameter s_i.

4. REDUCTION OF BREAKDOWN VOLTAGE

Deviations from theoretical, i.e., ideal breakdown characteristics occur because of junction curvature and surface breakdown phenomena; these effects are independent of each other, but both affect the field distribution in the depletion region. Junction curvature tends to increase the electric field in the region of curvature and thus leads to breakdown in this discrete region prior to bulk breakdown. Surface breakdown is due to the electric field near the semiconductor surface being higher than in the bulk as a result of surface states, therefore, the depletion layer at the surface will be narrower and breakdown at the surface will occur at a lower voltage than in the bulk.

(a) Reach-through:

If the depletion layer if a *p-n* junction extends to the vicinity of the substrate-film interface or any other region of higher impurity concentration (e.g., n^+-contact of a collector), a reduction of avalanche breakdown voltage is observed ("reach-through"). This reduction depends upon epitaxial film thickness (x_f) or its equivalent (i.e., the distance to the highly doped region), junction depth (x_j), and fractional depletion layer width (x_{d2}). The reduced value is $V_B'(<V_B)$.

True avalanche breakdown will take place if the distance between the edge of the depletion layer and the substrate-film interface ($x_{\text{eff}} = x_f - x_j - x_{d2}$) is greater than

$$x_{\min} = 6 \cdot 10^7 f_2 \, \varepsilon_s \varepsilon_o \, C_f^{f_1-1}/q. \tag{6.64}$$

Empirical values of $x_{\min}$ are given in Figure 6.21. In the above equation the impurity concentration of the epitaxial film (C_f) is expressed in cm^{-3} and $x_{\min}$ in μm; f_1 and f_2 are empirically determined quantities which are given in the inset of Figure 6.21. The actual breakdown voltage (V_B') is as follows.

(i) $x_{\text{eff}} > x_{\min}$:

$$V_B' = V_B.$$

(ii) $x_{\text{eff}} < x_{\min}$:

$$V_B' = x_{\text{eff}}\{[(2/3)(q/\varepsilon_s \varepsilon_o) f_2 \, C_f^{f_1+1}]^{1/2} - (q/6\varepsilon_s \varepsilon_o) C_f\} \tag{6.65}$$

(iii) $x_{\text{eff}} < x_{d2}$:

$$V_B' = V_B[(x_f - x_j)/x_{d2}][2 - (x_f - x_j)/x_{d2}]. \tag{6.66}$$

(b) Curvature of step junction:

A general expression, valid for various semiconductors, relating the breakdown voltage as affected by junction curvature of a step junction at 300°K to the energy gap E_G (in eV) of the semiconductor, background

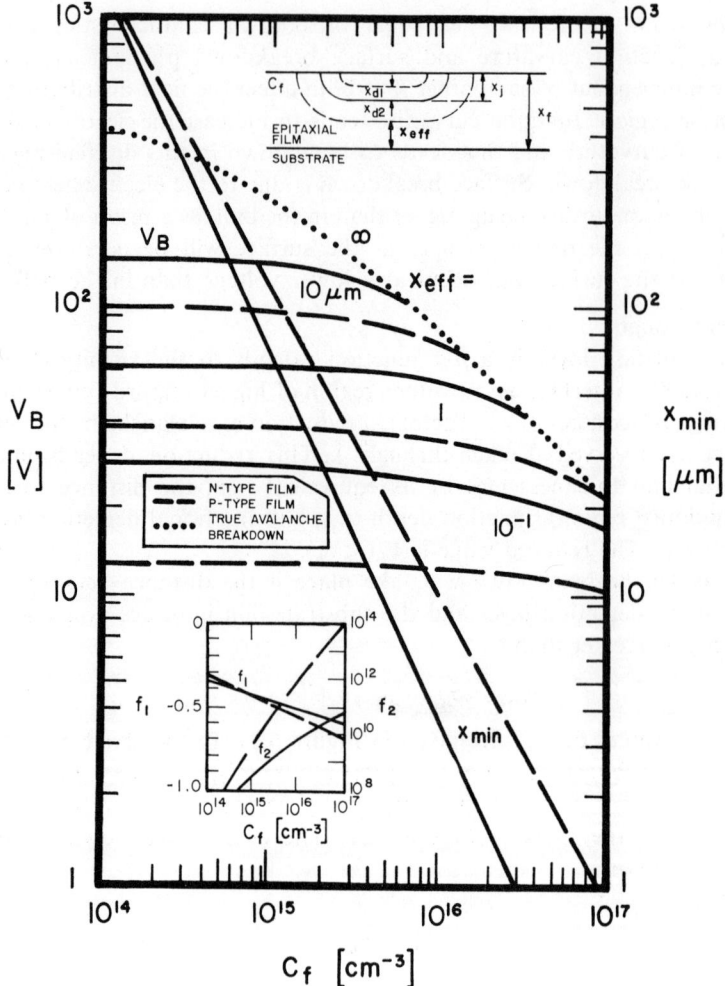

FIGURE 6.21

Reduction of avalanche breakdown voltage (V_B) due to "reach-through" vs. impurity concentration in epitaxial film (C_f) and film dimensions (x_{eff}); minimum epitaxial film thickness ($x_{\min}$) required to obtain true avalanche breakdown. Silicon, 300°K. The inset shows the variation of the parameters f_1 and f_2 with film impurity concentration.

impurity concentration on the lighter doped side of the junction (C_B), radius of curvature (for simplification assumed to be equal to the junction depth; given in μm), and depletion layer width x_d (in μm), is

$$V_B \approx 5.2 \cdot 10^{13} E_G^{3/2} C_B^{-3/4} A_B \qquad (6.67)$$

where

$$A_B = [(g + 1 + x_j/x_d)(x_j/x_d)^g]^{1/(g+1)} - x_j/x_d.$$

The geometry factor g is

for plane junction: $g = 0$,
for cylindrical junction: $g = 1$,
for spherical junction: $g = 2$.

The correction factor A_B gives the ratio of breakdown voltages of cylindrical or spherical junctions to that of a plane junction (for plane junction $A_B = 1$):

for cylindrical junction: $V_{Bc}/V_{Bp} = A_B$,
for spherical junction: $V_{Bs}/V_{Bp} = A_B$.

(c) Curvature of linearly graded junction:
A general expression, valid for various semiconductors, comparing the true breakdown voltage of a plane graded junction (V_{Bp}) to that of a cylindrical junction (V_{Bc}) is (see Figure 6.22)

$$\frac{V_{Bc}}{V_{Bp}} = 4 \frac{A_d^2 \ln A_d/(A_d^2 - 1) - 1/2}{A_d^2 - 1} \qquad (6.68)$$

where $A_d = 1 + x_{dcB}/x_j$.
The width of the depletion layer of a curved junction on the lighter doped side at breakdown (x_{dcB}) is related to the junction depth (x_j) by

$$x_{dcB}/x_j = [(2\varepsilon_s \varepsilon_v E_{crit}/qC_B x_j) + 1]^{1/2} - 1 \qquad (6.69)$$

where E_{crit} is the electric field at the junction at avalanche breakdown.
(d) Influence of stress:
The weight of a probe introduces localized stress resulting in an enhanced carrier density near the junction; this in turn results in a reduction of breakdown voltage. The reduction is more significant in p-type silicon than in n-type silicon. A probe weight of $w_p < 10^4$ g/cm^2 has an insignificant effect on breakdown voltage (of silicon junctions).

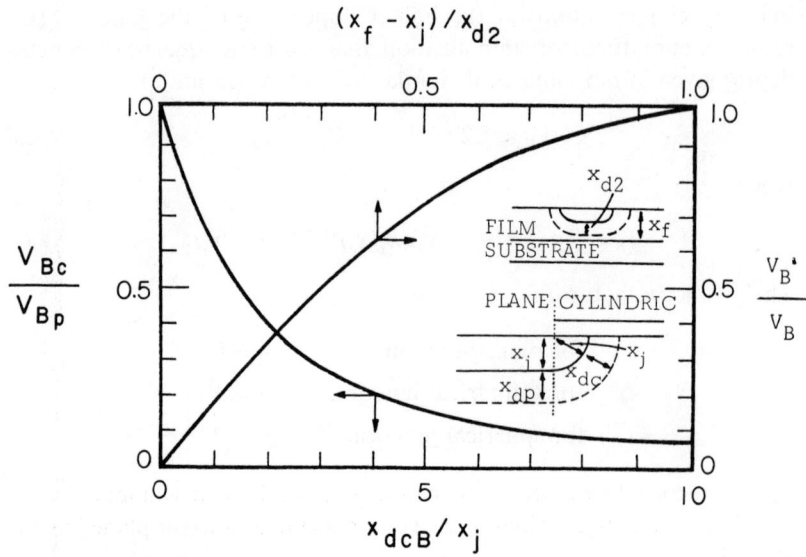

FIGURE 6.22

Reduction of avalanche breakdown voltage (V_B) of a linearly graded p-n junction due to "reach-through" as function of epitaxial film thickness (x_f), junction depth (x_j), and depletion layer width at junction breakdown (x_{dB}), curve V_B'/V_B vs. $(x_f-x_j)/x_{dB}$; and ratio of breakdown voltages in the cylindrical and plane portions (V_{Bc}, V_{Bp}) of a linearly graded p-n junction as function of depletion layer width at breakdown in the cylindrical portion (x_{dcB}) and junction depth (x_j), curve V_{Bc}/V_{Bp} vs. x_{dcB}/x_j. Silicon, 300°K.

Assumptions made in obtaining curves in Figure 6.23:
 (i) Probe weight is uniformly distributed over the entire probe area.
 (ii) The junction is deep ($x_j \gg 0$).
 (iii) Surface breakdown is neglected.
(e) Other effects on breakdown:
 In addition to breakdown reduction due to reach-through, junction curvature, and stress, other effects such as microplasma and surface field effects may reduce the breakdown voltage.
 The microplasma effect is a localized breakdown in small high-field regions and occurs at lattice defects and around metallic precipitations and results in breakdown voltage reduction if the defect density exceeds 100/cm².
 The surface field effect may result in a breakdown increase due to a reduction in edge fields facilitated by a separate metal electrode; this results in an effective increase in the radius of curvature and hence in an increase in V_B.

6-3 JUNCTION CHARACTERISTICS AT BREAKDOWN

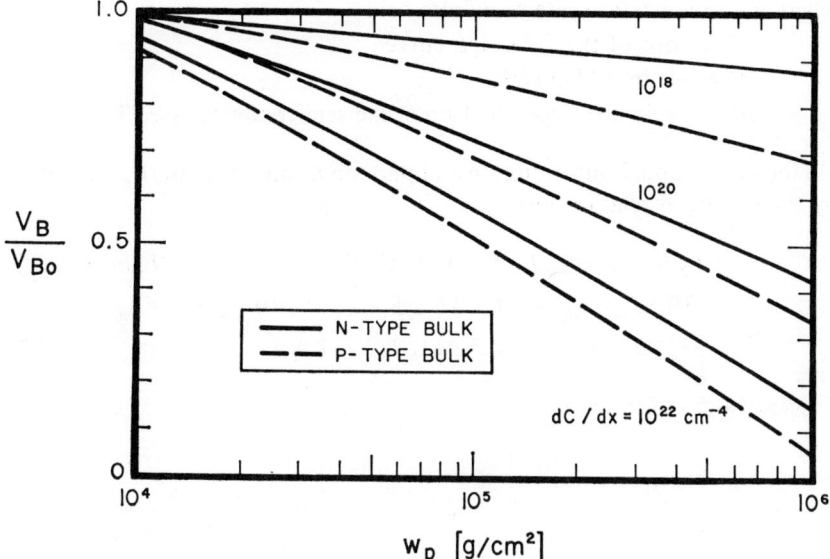

FIGURE 6.23
Reduction of avalanche breakdown voltage (V_B) due to weight of contact probe per unit area (w_p) and impurity gradient (dC/dx). Linearly graded p-n junction; silicon, 300°K.

5. TEMPERATURE DEPENDENCE OF BREAKDOWN VOLTAGE

At higher temperature V_B increases since the carriers passing through the depletion layer under high field lose part of their energy to optical phonons after traveling each electron-phonon mean free path which decreases with increasing temperature, as given below. At high temperature the carriers lose more energy to the lattice along a given distance at constant field and they must pass through a greater potential difference before acquiring sufficient energy to generate an electron-hole pair. This effectively increases the junction breakdown voltage.

A discussion of the dependence of avalanche breakdown voltage upon temperature considers the relationship between breakdown voltage and phonon mean free path; these two parameters are related through critical electric field and carrier ionization rate. A direct relationship between breakdown voltage and temperature is rather complex and will not be given here.

An empirical relationship between ionization rate at junction breakdown (α_i) and corresponding critical electric field (E_{crit}) is

$$\alpha_i = \alpha_{i\infty}(E/E_{crit})^{(a_i^*/E_{crit})^{m^*}} \exp\left[-(a_i^*/E_{crit})^{m^*}\right] \qquad (6.70)$$

where $\alpha_{i\infty}$ = ionization rate at infinite field
a_i^* = slope of the α_i vs. $1/E$ curve
$a_i^* = -d \ln \alpha_i / d \ln (1/E)$
m^* = parameter dependent upon the semiconductor (see Table 6.2).

On the other hand, ionization rate and phonon mean free path (l_{ph}) are related by the empirical relationship

$$\alpha_i = (1/l_{ph})\{[10.0(\Delta E_{ph}/E_G)^2 - 1.7(\Delta E_{ph}/E_G) + 8.7 \cdot 10^{-4}](E_G/qEl_{ph})^2 \\ + [31(\Delta E_{ph}/E_G)^2 - 11.9(\Delta E_{ph}/E_G) + 2.6 \cdot 10^{-2}](E_G/qEl_{ph}) \\ - 336(\Delta E_{ph}/E_G)^2 + 50.3(\Delta E_{ph}/E_G) - 1.9\}. \quad (6.71)$$

Typical values of interest are

$$10^{-2} \leq \Delta E_{ph}/E_G \leq 10^{-1}$$
$$1 \leq E_G/(qEl_{ph}) \leq 10.$$

The phonon mean free path (see also 1–7) is

$$l_{ph} = 3\kappa/(c_{ac} c_v)$$
$$= l_{pho} \tanh (E_{ph}/2kT)$$
$$\approx l_{pho}(E_{ph}/2kT)[1 - (E_{ph}/2kT)^2/3] \text{ if } E_{ph} < 3kT \quad (6.72)$$

and is the average distance a phonon travels from the point of generation to the point of annihilation or transformation. In the above equations

κ = thermal conductivity
c_{ac} = phonon velocity
c_v = specific heat
l_{pho} = high-energy, low-temperature asymptotic value of the phonon mean free path
E_{ph} = optical phonon energy.

The phonon mean free path is approximately inversely proportional to temperature. The average energy loss per phonon scattering event is

$$\Delta E_{ph} = E_{ph} \tanh (E_{ph}/2kT)$$
$$\approx (E_{ph}^2/2kT)[1 - (E_{ph}/2kT)^2/3] \text{ if } E_{ph} < 3kT. \quad (6.73)$$

Phonon energies and phonon mean free paths in selected semiconductors are given in Table 6.3.

6-3 JUNCTION CHARACTERISTICS AT BREAKDOWN

TABLE 6.3

Phonon Energies and Phonon Mean Free Paths (300°K)

Semiconductor	E_{ph} [eV]	l_{ph} [Å] electrons	l_{ph} [Å] holes	l_{pho} [Å] electrons	l_{pho} [Å] holes
Si	0.063	62	45	76	55
Ge	0.037	65	65	105	105
GaAs	0.035	35	35	58	58
GaP	0.050	32	32	42	42

The inset of Figure 6.20 shows the variation of avalanche breakdown voltage with junction temperature, referred to 300°K, for silicon step and linearly graded junctions. The avalanche breakdown voltage increases by about 0.1%/°C from -50 to $+150$°C.

6. DEPLETION LAYER WIDTH AT BREAKDOWN

For an arbitrary semiconductor the width of the depletion layer at avalanche breakdown (see also Figure 6.24a) of a step junction

$$x_{dB} = (2\varepsilon_s \varepsilon_o V_B/qC_B)^{1/2}$$
$$\approx (1.2 \cdot 10^{14} \varepsilon_s \varepsilon_o/q)^{1/2}(E_G/1.1)^{3/4}C_B^{-7/8} \qquad (6.74)$$

and of a linearly graded junction

$$x_{dB} = [12\varepsilon_s \varepsilon_o V_B/q(dC/dx)]^{1/3}$$
$$\approx (8.7 \cdot 10^{10} \varepsilon_s \varepsilon_o/q)^{1/3}(E_G/1.1)^{2/5}(dC/dx)^{-7/15}. \qquad (6.75)$$

If the depletion layer extends to both sides of the junction (as is usually the case), then modifications of Figures 6.18, 6.19 and 6.20 have to be made as follows. If $x_{d1} = x_{d2}$, i.e. if the depletion layer extends in equal amounts to both sides of the metallurgical junction, breakdown will be described by the curve $x_j = \infty$ and is independent of junction depth. If $x_{d1} \neq x_{d2}$, avalanche breakdown will occur at a voltage above or below the curve $x_j = \infty$ depending upon the ratio x_{d1}/x_{d2}, i.e., V_B will be between the upper and the lower curve corresponding to the same junction depth at the ratio x_{d1}/x_{d2}.

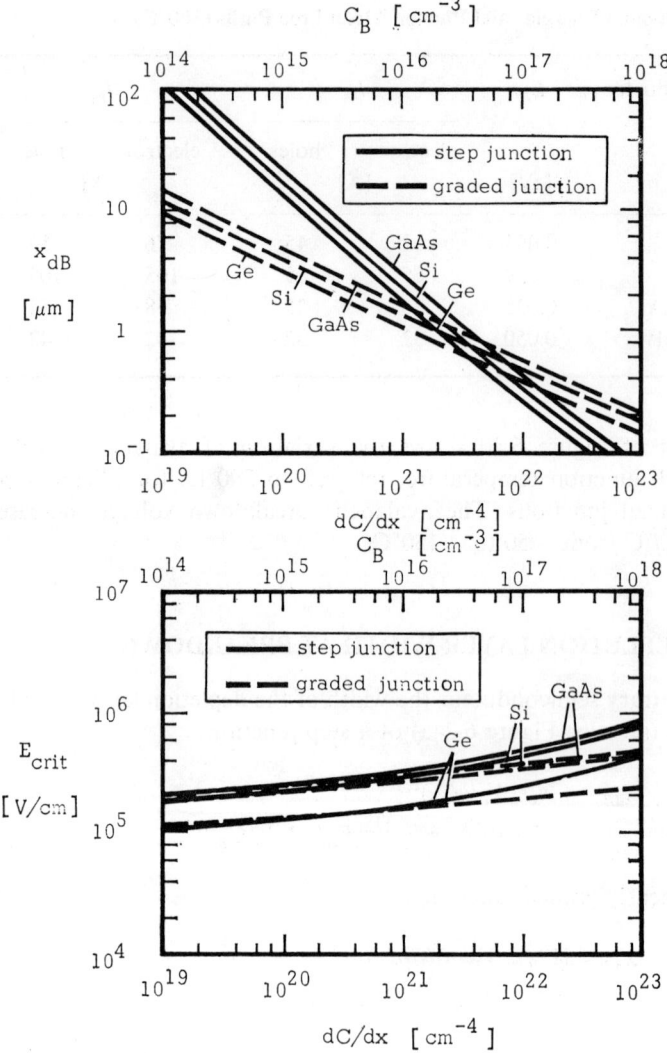

FIGURE 6.24

Depletion layer width at breakdown (x_{dB}) and corresponding electric field (E_{crit}) as function of background impurity concentration (C_B) for step junction or impurity concentration gradient (dC/dx) for linearly graded junction for Si, Ge, and GaAs. 300°K.

7. CRITICAL ELECTRIC FIELD

Junction breakdown due to carrier multiplication (avalanching) occurs when the maximum electric field in the depletion region reaches the critical field, i.e., when $E_{max} = E_{crit}$. Generally, the critical electric field at which avalanche multiplication sets in is

$$E_{crit} = (32/3\pi)(c_{ac}/\mu_o), \qquad (6.76)$$

where μ_o is the carrier mobility at small electric field. Due to the difference in mobilities, the critical field for electrons and holes differs.

For an arbitrary semiconductor the electric field at avalanche breakdown (see also Figure 6.24b) of a step junction

$$\begin{aligned}E_{crit} &= 2V_B/x_{dB} \\ &= (2qV_B C_B/\varepsilon_s \varepsilon_o)^{1/2} \\ &\approx 1.1 \cdot 10^7 (q/\varepsilon_s \varepsilon_o)^{1/2}(E_G/1.1)^{3/4} C_B^{1/8}\end{aligned} \qquad (6.77)$$

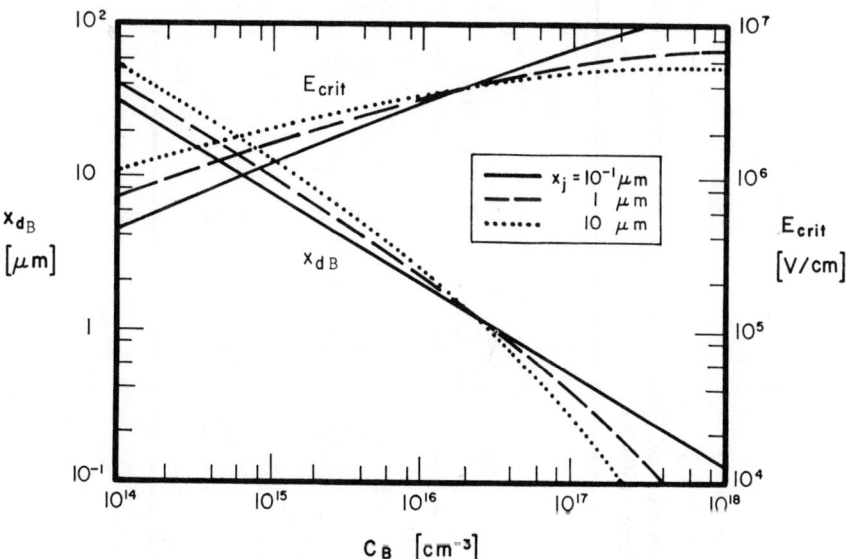

FIGURE 6.25

Depletion layer width at breakdown (x_{dB}) and corresponding electric field (E_{crit}) vs. background impurity concentration (C_B) and junction depth (x_j). Silicon 300°K.

TABLE 6.4
Critical Electric Field, Depletion Layer Width at Breakdown, and Avalanche Breakdown Voltage of Selected Semiconductors (300°K)

Junction type	Si			Ge			GaAs			GaP		
	E_{crit} [V/cm]	x_{dB} [μm]	V_B [V]	E_{crit} [V/cm]	x_{dB} [μm]	V_B [V]	E_{crit} [V/cm]	x_{dB} [μm]	V_B [V]	E_{crit} [V/cm]	x_{dB} [μm]	V_B [V]
One-sided step junction												
$C_B = 10^{14}$ cm^{-3}	$2.5 \cdot 10^5$	150	2000	$1.5 \cdot 10^5$	120	900	$3.0 \cdot 10^5$	180	2400	$4.5 \cdot 10^5$	250	6500
$= 10^{16}$ cm^{-3}	$4.0 \cdot 10^5$	2.5	58	$2.4 \cdot 10^5$	1.9	25	$4.4 \cdot 10^5$	2.9	72	$7.0 \cdot 10^5$	3.9	135
$= 10^{18}$ cm^{-3}	$1.2 \cdot 10^6$	0.07		$6.0 \cdot 10^5$	0.05		$1.2 \cdot 10^6$	0.08		$1.9 \cdot 10^6$	0.1	
Linearly graded junction												
$dC/dx = 10^{20}$ cm^{-4}	$3.2 \cdot 10^5$	4.2	92	$2.0 \cdot 10^5$	3.8	48	$3.6 \cdot 10^5$	4.5	110	$5.8 \cdot 10^5$	5.2	220
$= 10^{22}$ cm^{-4}	$4.8 \cdot 10^5$	0.5	17	$2.7 \cdot 10^5$	0.4	7	$5.2 \cdot 10^5$	0.5	19	$8.0 \cdot 10^5$	0.6	33
$= 10^{24}$ cm^{-4}	$8.0 \cdot 10^5$	0.06		$4.0 \cdot 10^5$	0.05		$9.0 \cdot 10^5$	0.07		$1.6 \cdot 10^6$	0.08	

6-3 JUNCTION CHARACTERISTICS AT BREAKDOWN

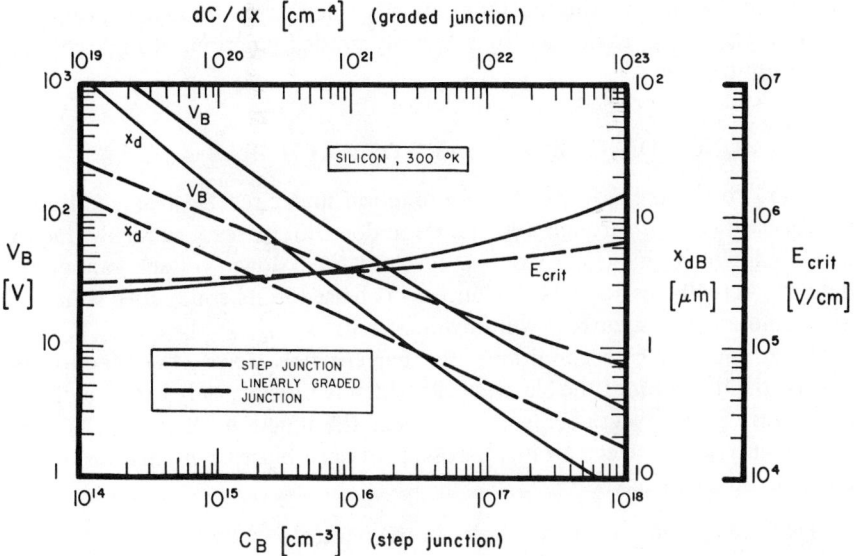

FIGURE 6.26

Avalanche breakdown voltage (V_B), depletion layer width at breakdown (x_{dB}), and critical (maximum) electric field (E_{crit}) at breakdown vs. background impurity concentration (C_B) or impurity concentration gradient (dC/dx). Silicon, 300°K.

and of a linearly graded junction

$$\begin{aligned}E_{\text{crit}} &= (3/2)V_B/x_{dB} \\ &= (9/32)^{1/3} V_B^{2/3}[q(dC/dx)/\varepsilon_s\varepsilon_o]^{1/3} \\ &\approx 1.5 \cdot 10^6 (q/\varepsilon_s\varepsilon_o)^{1/3}(E_G/1.1)^{4/5}(dC/dx)^{1/15}.\end{aligned} \qquad (6.78)$$

The critical electric field varies very slowly with either C_B or dC/dx owing to the strong dependence of ionization rate upon electric field.

Table 6.4 lists breakdown parameters of selected semiconductors at 300°K.

8. DIFFUSED JUNCTION

A diffused junction can be approximated by an exponential function. It describes quite accurately erfc and Gaussian impurity distributions. In most cases the impurity concentration profile of a diffused junction can be approximated by an exponential distribution on one side and a constant impurity level on the other side of the junction. There is a constant electric field on the exponentially graded side (which retards the flow of minority carriers in this region) and no field on the constant impurity side. At low voltage the diffused junction can be approximated by a linearly graded junction, at high voltage by a step junction.

9. JUNCTION CURRENT AT FORWARD BIAS

Contrary to the breakdown of a *p-n* junction under reverse bias condition, the forward current through a *p-n* junction does not increase suddenly due to a breakdown mechanism so that a forward breakdown voltage cannot be defined. Actually, as soon as the junction is biased in its conductive state, an appreciable forward current will flow across it.

Under forward bias conditions, the current flow across the junction is against the direction of the electric field which is possible only because of the large carrier density gradients existing near the junction. As in the case of reverse current, the electron flux across a forward biased *p-n* junction is due to three contributions:

(a) electrons recombining with holes in the neutral *n*-region;
(b) electrons recombining with holes in the neutral *p*-region;
(c) electrons recombining with holes in the space charge region.

Contributions (a) and (b) are the diffusion current and contribution (c) is the recombination current, i.e.,

$$I_f = I_{\text{diff}} + I_{\text{rec}}. \qquad (6.79)$$

Recombination in the neutral regions takes place within a diffusion length (L_n or L_p) of the depletion region. The ratio of diffusion current to recombination current is for electrons

$$I_{\text{diff}}/I_{\text{rec}} = 2(n_i/N_A)(L_n/x_d) \exp(q|V_a|/2kT) \qquad (6.80)$$

where L_n is the diffusion length of electrons in the *p*-region and V_a is the applied forward voltage ($qV_a = \phi_{Fn}$).

The total current across a forward biased *p-n* junction is

$$I_f = I_o \exp(qV_a/n_b^* kT) - 1$$
$$= (4\pi m_n k^2 q/h^3) T^2 \exp(-qV_D/kT)[\exp(qV_a/n_b^* kT) - 1] \qquad (6.81)$$

where I_o is the current if the externally applied voltage $V_a = 0$; the numerical factor $n_b^* = 1$ for pure carrier diffusion current and $n_b^* = 2$ for pure carrier recombination current. At low voltage the recombination component dominates whereas at high voltage the diffusion component dominates; the transition from recombination to diffusion dominance depends upon temperature and semiconductor band gap. For silicon at room temperature the transition takes place at about 0.4 to 0.5 V.

6-3 JUNCTION CHARACTERISTICS AT BREAKDOWN

The forward voltage of a *p-n* junction is related to the barrier height, i.e., the built-in voltage of the junction, V_D, which is the difference of the intrinsic Fermi levels on *n* and *p* sides of the junction and which is finite under all conditions. The barrier height is given for a step junction by

$$V_D \approx (2kT/q) \ln (C_B/n_i) \qquad (6.82)$$

and for a linearly graded junction by

$$V_D \approx (4/3)(kT/q)[\ln (dC/dx) - 1]. \qquad (6.83)$$

10. REFERENCES

(1) K. G. McKay and K. B. McAfee, *Phys. Rev.*, **91**, 1079 (1953).
(2) K. G. McKay, *Phys. Rev.*, **94**, 877 (1954).
(3) S. L. Miller, *Phys. Rev.*, **99**, 1234 (1955).
(4) A. B. Phillips, *Transistor Engineering*, McGraw-Hill Book Co., New York, 1962.
(5) C. C. Allen et al., *J. El. Chem. Soc.*, **113**, 508 (1966).
(6) S. M. Sze and G. Gibbons, *Solid-State Electronics*, **9**, 831 (1966).
(7) O. Leistiko, Jr., and A. S. Grove, *Solid-State Electronics*, **9**, 847 (1966).
(8) S. M. Sze and G. Gibbons, *Appl. Phys. Lett.*, **8**, 111 (1966).
(9) C. R. Crowell and S. M. Sze, *Appl. Phys. Lett.*, **9**, 242 (1966).
(10) D. P. Kennedy and R. R. O'Brien, *IBM J. Res. Develop.*, **10**, 213 (1966).
(11) S. M. Sze, *Physics of Semiconductor Devices*, John Wiley and Sons, New York, 1969.

6-4 Examples

PROBLEMS AND SOLUTIONS

1. Problem: Determine the built-in voltage at 25°C for a graded junction whose characteristics are:

$$C_B = 10^{15} \text{ cm}^{-3}, \ C_S = 10^{18} \text{ cm}^{-3}, \ x_j = 2.0 \ \mu\text{m}.$$

Solution: The impurity gradient is

$$dC/dx = 6.4 \cdot 10^{19} \text{ cm}^{-4},$$

hence the built-in voltage

$$V_D = 0.59 \text{ V}.$$

2. Problem: Compare total and fractional depletion layer width, maximum electric field, and depletion layer capacitance for plane, cylindrical, and spherical diffused junctions in silicon for which are given:

$$C_B = 10^{15} \text{ cm}^{-3}, \quad C_S = 10^{19} \text{ cm}^{-3}, \quad x_j = 0.5 \ \mu\text{m},$$
$$V = 25 \text{ V}.$$

Solution: For all junctions

$$C_S/C_B = 10^4, \quad V/x_j^2 C_B = 10^{-5}.$$

(a) Plane junction:

$$(x_j - x_{d1})/x_d = 6.5 \cdot 10^{-2}, \quad x_d/x_j = 10;$$

i.e.,

$$x_d = 5.0 \ \mu\text{m}, \quad x_{d1} = 0.2 \ \mu\text{m}, \quad x_{d2} = 4.8 \ \mu\text{m}.$$
$$E_{\max}/x_j C_B = 1.8 \cdot 10^{-6} \text{ V cm};$$

i.e.,

$$E_{\max} = 9 \cdot 10^4 \text{ V/cm}.$$
$$x_d C/\varepsilon_s \varepsilon_o = 7.5 \cdot 10^{-2};$$

i.e.,

$$C = 1.6 \cdot 10^2 \text{ pF/cm}^2.$$

(b) Cylindrical junction:
$$(x_j - x_{d1})/x_d = 6.5 \cdot 10^{-2}, \qquad x_d/x_j = 10.$$
i.e.,
$$x_d = 3.0 \ \mu m, \qquad x_{d1} = 0.3 \ \mu m, \qquad x_{d2} = 2.7 \ \mu m.$$
$$E_{max}/x_j C_B = 6 \cdot 10^{-6} \ V \ cm \ ;$$
i.e.,
$$E_{max} = 3.0 \cdot 10^5 \ V/cm.$$
$$x_d C/\varepsilon_s \varepsilon_o = 0.75 \ ;$$
i.e.
$$C = 2.7 \cdot 10^3 \ pF/cm^2.$$

(c) Spherical junction:
$$(x_j - x_{d1})/x_d = 6.5 \cdot 10^{-2}, \qquad x_d/x_j = 4.8 \ ;$$
i.e.,
$$x_d = 2.4 \ \mu m, \qquad x_{d1} = 0.3 \ \mu m, \qquad x_{d2} = 2.1 \ \mu m.$$
$$E_{max}/x_j C_B = 7.2 \cdot 10^{-6} \ V \ cm \ ;$$
i.e.,
$$E_{max} = 3.6 \cdot 10^5 \ V/cm.$$
$$x_d C/\varepsilon_s \varepsilon_o = 1.7 \ ;$$
i.e.,
$$C = 7.3 \cdot 10^3 \ pF/cm^2.$$

PROBLEMS FOR WHICH A SOLUTION IS NOT GIVEN

1. Determine depletion layer width and capacitance of a linearly graded silicon p-n junction for which $C_B = 10^{14}$ cm^{-3}, $C_S = 10^{17}$ cm^{-3}, $x_j = 1.0 \ \mu m$, and the total voltage across the junction $V = 10$ V.

2. Compare the widths of the fractional space charge layers which extend into the lighter doped side of 1 and 10 μm deep linearly graded silicon junctions if $C_B = 5 \cdot 10^{15}$ cm^{-3}, $C_S = 10^{18}$ cm^{-3}, $V = 10$ V.

3. Determine the depletion layer capacitance of a linearly graded silicon p-n junction for which $dC/dx = 10^{22}$ cm^{-4} if the applied voltage is $V_a = 0$ V, 10 V, and V_B.

4. Determine the variation of avalanche breakdown voltage with temperature for a linearly graded silicon p-n junction whose characteristics are:
$$C_B = 5 \cdot 10^{15} \text{ cm}^{-3}, \qquad C_S = 10^{19} \text{ cm}^{-3}, \qquad x_j = 2 \text{ μm}.$$
Assume a temperature variation from $-55°C$ to $+125°C$.

5. Determine the breakdown voltage of a linearly graded silicon p-n junction for which
$$C_B = 10^{15} \text{ cm}^{-3}, \qquad C_S = 10^{18} \text{ cm}^{-3}, \qquad x_j = 1 \text{ μm}.$$
Compare the values for plane and cylindrical cases.

6. Determine the carrier multiplication factor M for a linearly graded silicon junction for which
$$dC/dx = 10^{21} \text{ cm}^{-4}, \qquad x_j = 0.5 \text{ μm}, \qquad V_a = 10 \text{ V}.$$

7. Determine the magnitude of the electric field at a distance of 0.3 μm away from a linearly graded silicon junction for which
$$x_j = 1.0 \text{ μm}, \qquad x_{d1} = 0.7 \text{ μm}, \qquad x_{d2} = 1.5 \text{ μm}, \qquad V = 10 \text{ V}.$$

7
MEASUREMENT TECHNIQUES

7–1 Determination of Oxide Thickness and Junction Depth

7–2 In-Line Four-Point Probe Measurements

7–3 Other Measurement Techniques

7–1 Determination of Oxide Thickness and Junction Depth

1. MEASUREMENTS BASED ON LIGHT INTERFERENCE

The nondestructive measurement of the thickness of thin dielectric films (e.g., silicon dioxide) and the destructive measurement of junction depth are based on the use of the interference of visible light. In oxide thickness measurements, a tapered section of the oxide allows the determination of interference colors and interference order. In junction depth measurements, angle lapping in the proximity of the junction and subsequent staining is used to obtain interference fringes using monochromatic light.

The curves of Figures 7.1 and 7.2 allow the determination of the thickness of an SiO_2 layer or of junction depth in silicon, assuming

$$n_{Si} > n_g = n_{SiO_2} > n_{air}.$$

The refractive indices are assumed to be as follows.

$$\begin{aligned} \text{Silicon:} \quad & n_{Si} = 3.4 \\ \text{SiO}_2 : \quad & n_{SiO_2} = 1.5 \\ \text{Glass:} \quad & n_g = 1.5 \\ \text{Air} : \quad & n_{air} = 1.0. \end{aligned}$$

The determination of the thickness of thin films other than SiO_2 or of the junction depth in semiconductors other than silicon is possible using these illustrations if the proper refractive indices are used.

Interference of light reflected at two surfaces of different refractive indices results in the generation of optical fringes whose number (z_f) is indicative of depth, e.g., of the thickness of a semitransparent layer or of the depth of a stained and lapped semiconductor region.

2. OXIDE THICKNESS (x_o)

(a) Monochromatic light (Figure 7.1):
   Interference of light reflected at the oxide surface and at the oxide-semiconductor interface is considered.

7-1 OXIDE THICKNESS AND JUNCTION DEPTH

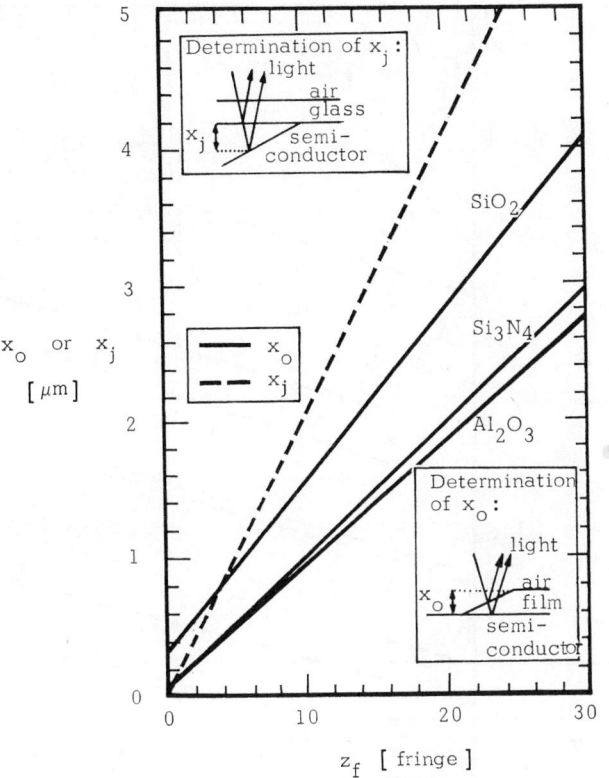

FIGURE 7.1

Thickness of a dielectric film (x_o) and junction depth (x_j) in a semiconductor as function of the number of fringes (z_f) due to light interference Monochromatic light, $\lambda = 0.546$ μm. It is assumed that $n_{SiO_2} = 2.0$, $n_{Si_3N_4} = 2.8$, and $n_{Al_2O_3} = 3.0$.

The thickness of the oxide is related to the number of fringes at light of wavelength λ_{Hg} by

$$x_o = (\lambda_{Hg}/n_{SiO_2})(2z_f + 1)/4 \tag{7.1}$$

where $z_f = 0, 1, 2, \ldots$.
(b) Polychromatic light (Figure 7.2):
Light interference is determined in a tapered oxide section in order to establish the interference order. The observed color corresponds to wavelength λ. If a nontapered oxide is used, a rough estimate of oxide thickness can be obtained from the oxide formation conditions. The

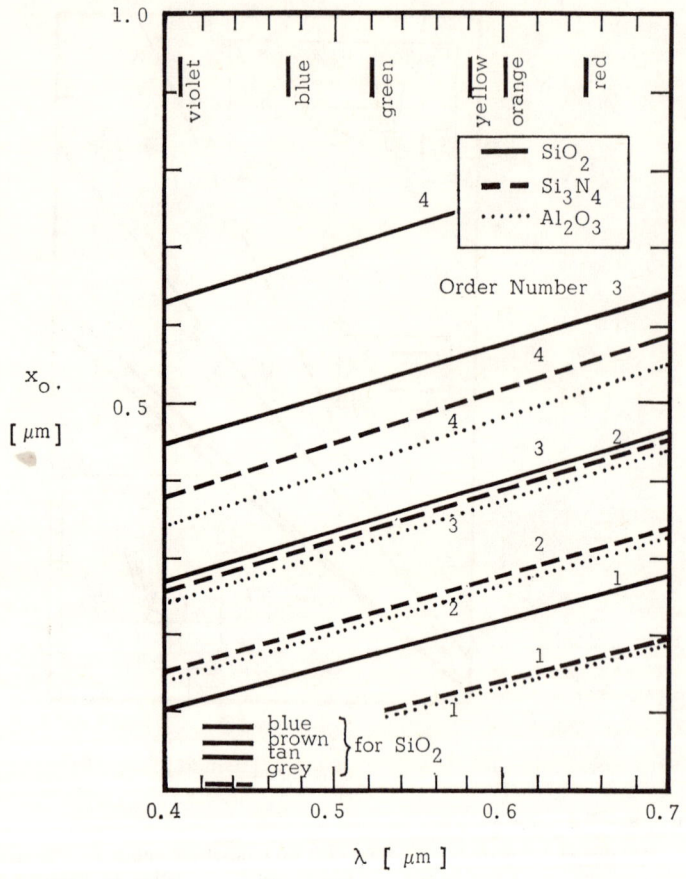

FIGURE 7.2

Thickness of dielectric layer (x_o) as function of wavelength or color of interference light (λ). Polychromatic light.

colors given are observed with the incident light perpendicular to the oxide surface. When observed obliquely, the oxide color changes, depending upon thickness, index of refraction, and angle of incidence. Intermediate colors, i.e., those observed between the ones given in the illustration, allow a more accurate determination of x_o.

This method assumes that light is reflected at the interface under the oxide in a perpendicular direction. If the measurement is made in polychromatic light, interference colors are observed; if made in monochromatic light, interference maxima are observed corresponding to a wavelength which is related to the oxide thickness.

3. JUNCTION DEPTH (x_j)

Interference of light reflected at the lower glass–air interface and at the semiconductor surface is considered.

The junction depth is related to the number of fringes at light of wavelength λ_{Hg} by

$$x_j = (\lambda_{Hg}/n_{air})(z_f/2) = \lambda_{Hg} z_f/2 \tag{7.2}$$

where $z_f = 0, 1, 2, \ldots$.

4. JUNCTION DELINEATION

A simple staining solution, to be applied to a silicon sample after lapping, is:

1 part HNO_3 + 999 parts HF (by volume).

This mixture stains p-type silicon regions in high-intensity light, whereas n-type regions remain unaffected. Very sharp delineations of p-n, n-n^+, and p-p^+ regions can be obtained by using the following staining composition:

2g HIO_4, 1ml HF (48%), 20ml H_2O.

The n-layer shows bright and the p-layer dark in the case of p-n junctions. In the case of n-n^+ and p-p^+ junctions, the n^+- and p-layers show bright. The delineation proceeds by the following reaction.

$$7Si + 2HIO_4 + 42HF \rightarrow 7H_2(SiF_6) + 8H_2O + 7H_2 + I_2.$$

A slight improvement in staining quality is obtained when using the following composition (which also reveals dislocations):

5g HIO_4, 10ml HF (48%), 30ml H_2O, 15ml CH_3COCH_3.

The overall reaction is

$$21Si + 7HIO_4 + 126HF + 7CH_3COCH_3$$
$$\longrightarrow 21H_2(SiF_6) + 7CH_2COCH_3 + 28H_2O + 21H_2.$$

5. EXAMPLES

(a) *Problem*: Determine the thickness of an SiO_2 layer and the junction depth in silicon if the interference patterns have yielded 3 fringes.
Answer: $x_o = 0.55$ μm; $x_j = 0.82$ μm
(b) *Problem*: Interference color green has been observed for the second order of an interference pattern in SiO_2. Determine the corresponding oxide thickness.
Answer: $x_o = 0.35$ μm

6. REFERENCES

(1) G. S. Fuller and J. A. Ditzenberger, *J. Appl. Phys.*, **27**, 544 (1956).
(2) E. Sirtl and A. Adler, *Z. Metallk.*, **52**, 532 (1961).
(3) H. Robbins, *J. El. Chem. Soc.*, **109**, 63 (1962).
(4) W. A. Pliskin and E. E. Conrad, *IBM J. Res. Develop.*, **8**, 43 (1964).
(5) B. M. Berry, *RTI Report ASD-TDR-63-316*, Vol. XII, September 1966.
(6) I. F. Nicolau, *Solid-State Electronics*, **12**, 446 (1969).

7–2 In-Line Four-Point Probe Measurements

1. RESISTIVITY OF SEMICONDUCTOR SAMPLE

A four-point probe where all points are in line can be used to determine the resistivity of a semiconductor wafer. The procedure consists of a voltage and current measurement and the application of a sample geometry factor and of appropriate correction factors where necessary.

Under ideal conditions, measurement of the voltage drop V between two adjacent points (inner points) if a current I flows between two outer points yields the ratio V/I. The term V/I is a surface resistance of the semiconductor sample and is indicative of the sample resistivity (ρ); it aids in the determination of impurity profiles of diffused layers.

The resistivity of the semiconductor is generally related to V/I by

$$\rho = A^*(V/I), \tag{7.3}$$

where the geometry factor A^* depends upon sample thickness (x_w), sample dimensions, and the spacing between points (S_p).

Two ranges of x_w are distinguished (Figure 7.3a):

(a) $x_w < 300\ \mu m$:

$$A^* = (\pi/\ln 2)x_w = 4.5324\ x_w. \tag{7.4}$$

In most practical cases this is the range of greatest interest. The geometry factor is independent of probe spacing.

(b) $x_w > 300\ \mu m$:
The geometry factor is substantially reduced and is a function of the probe spacing.

For a diffused layer of opposite conductivity type than the bulk and of thickness x_j, the sample thickness x_w has to be replaced by x_j.

$$\rho = A^*(V/I) = 4.5324\ x_j(V/I). \tag{7.5}$$

In case of an epitaxial film of thickness x_f on a substrate of opposite conductivity type the resistivity is

$$\rho = A^*(V/I) = 4.5324\ x_f\ (V/I). \tag{7.6}$$

The influence of the sample geometry is shown in Figure 7.3b.

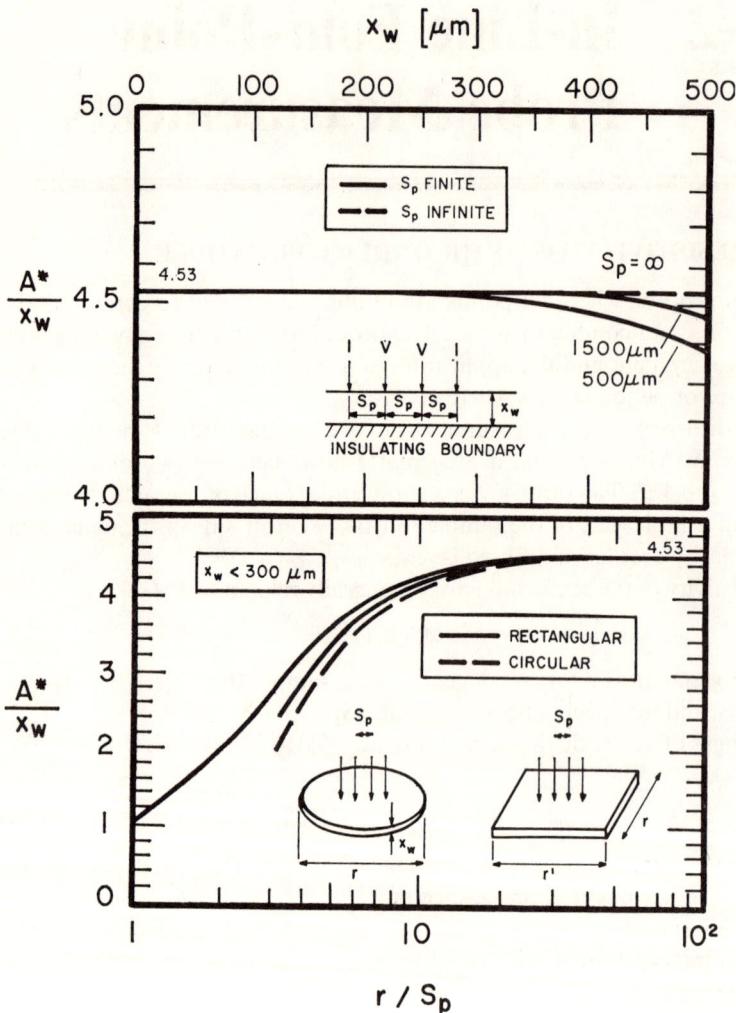

FIGURE 7.3

Geometry factor (A^*) vs. semiconductor sample thickness (x_w) and probe point spacing (S_p).

Geometry factor (A^*) vs. sample diameter (r, r') and point spacing (S_p). This figure shows the influence of sample geometry on V/I measurements.

2. REQUIREMENTS FOR CORRECT V/I MEASUREMENT

A measurement of V/I using an in-line four-point probe leads to an accurate value of resistivity only if the following requirements are met.

(a) The semiconductor resistivity in the proximity of the probe points is uniform.
(b) The minority carriers injected by the points recombine in the immediate proximity of the points.
(c) The point area is small compared to the spacing between the points.
(d) The thickness of the semiconductor sample is large compared to the point spacing ($x_w \gg S_p$).
(e) The four points are not applied close to the semiconductor edge.
(f) The spacing between the points is equal.
(g) The sample diameter is large compared to the point spacing ($r \gg S_p$).

If all of these conditions are met, the semiconductor resistivity can be calculated from

$$\rho = A^*(V/I)$$

$$= \frac{2\pi}{1/S_{p1} + 1/S_{p3} - 1/(S_{p1} + S_{p2}) - 1/(S_{p2} + S_{p3})} \cdot \frac{V}{I} \quad (7.7)$$

where S_{pi} is the spacing between two adjacent points. If

$$S_{p1} = S_{p2} = S_{p3} = S_p$$

(which is usually the case), then

$$\rho = \frac{V}{I} 2\pi \frac{S_p}{S_p/x_w + 1} \quad \text{for } S_p/x_w \ll 0.2, \quad (7.8a)$$

$$\rho = (V/I)(\pi/\ln 2)x_w \quad \text{for } S_p/x_w \gg 0.2. \quad (7.8b)$$

The following special cases exist:

(a) $x_w \to 0$, S_p finite:

$$\rho \to 0$$

(b) $x_w \to \infty$, S_p finite:

$$\rho \to (V/I)2\pi S_p$$

(c) x_w finite, $S_p \to 0$:

$$\rho \to 0$$

(d) x_w finite, $S_p \to \infty$:

$$\rho \to (V/I)2\pi x_w.$$

3. DEVIATIONS FROM IDEAL CONDITIONS

If the above conditions are not met, correction factors can be used to obtain the true sample resistivity. If several restrictions occur simultaneously, the measured resistivity has to be multiplied by all applicable correction factors. The true resistivity is

$$\rho = A^*(V/I)\eta_i \tag{7.9}$$

where η_i corresponds to one or the product of several correction factors given below.

(a) Figure 7.4a:
   The semiconductor is bordered by a conductive or a nonconductive plane on the opposite semiconductor surface. Correction factor η_i has to be used if the sample dimensions other than x_w are large compared to S_p.

$$\rho = A^*(V/I)\eta_1. \tag{7.10}$$

(b) Figure 7.4b:
   The semiconductor is limited on one side by a nonconductive plane close to one of the points. Correction factor η_2 has to be used if the nonconductive plane is perpendicular to an imaginary line connecting the four points.

$$\rho = A^*(V/I)\eta_2. \tag{7.11}$$

The correction factor approaches unity if

$$x_w/S_p \geq 2 \quad \text{or} \quad d_2/S_p \geq 2,$$

respectively. If

$$x_w/S_p \ll 0.2 \quad \text{or} \quad d_2/S_p \ll 0.2,$$

the correction factors η_1 and η_2 become inaccurate.

(c) Figure 7.5:
   If the center of the two points is displaced from the center of the sample by an amount d_3, the correction factor η_3 has to be used.

$$\rho = A^*(V/I)\eta_3. \tag{7.12}$$

Two types of displacement are considered:
(i) *In-line displacement*:
   The four points and the sample center lie on a straight line
(ii) *Perpendicular displacement*:
   The center of the four points and the center of the sample lie on a line perpendicular to a line connecting the four points

7-2 FOUR-POINT PROBE MEASUREMENTS 507

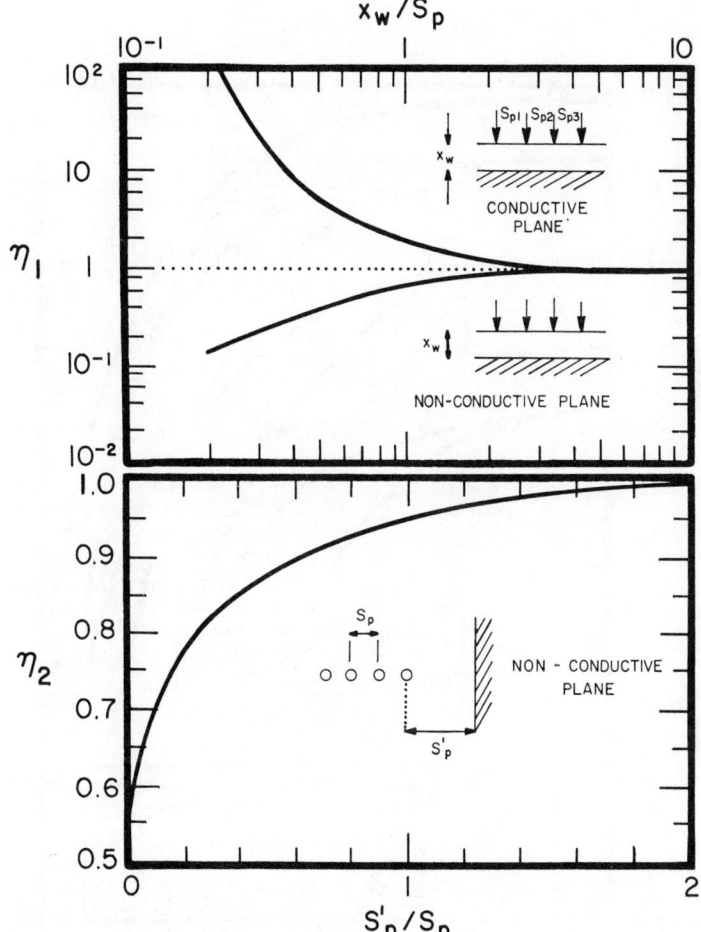

FIGURE 7.4

Correction factors (η_1 and η_2) applicable to four-point probe measurements if semiconductor sample is bordered by conductive or nonconductive planes close to the probe vs. sample thickness (x_w), probe spacing (S_p), and distance of probe from plane (d_2).

(d) Figure 7.6:

If a semiconductor sample of resistivity ρ_m and dimension $r \gg S_p$ encloses a circular region of resistivity $\rho_m + \Delta\rho$ and diameter r'' (which is comparable to probe size, i.e., $r'' \approx 3S_p$), then the true sample resistivity (ρ)

$$\rho = A^*(V/I)\eta_4. \tag{7.13}$$

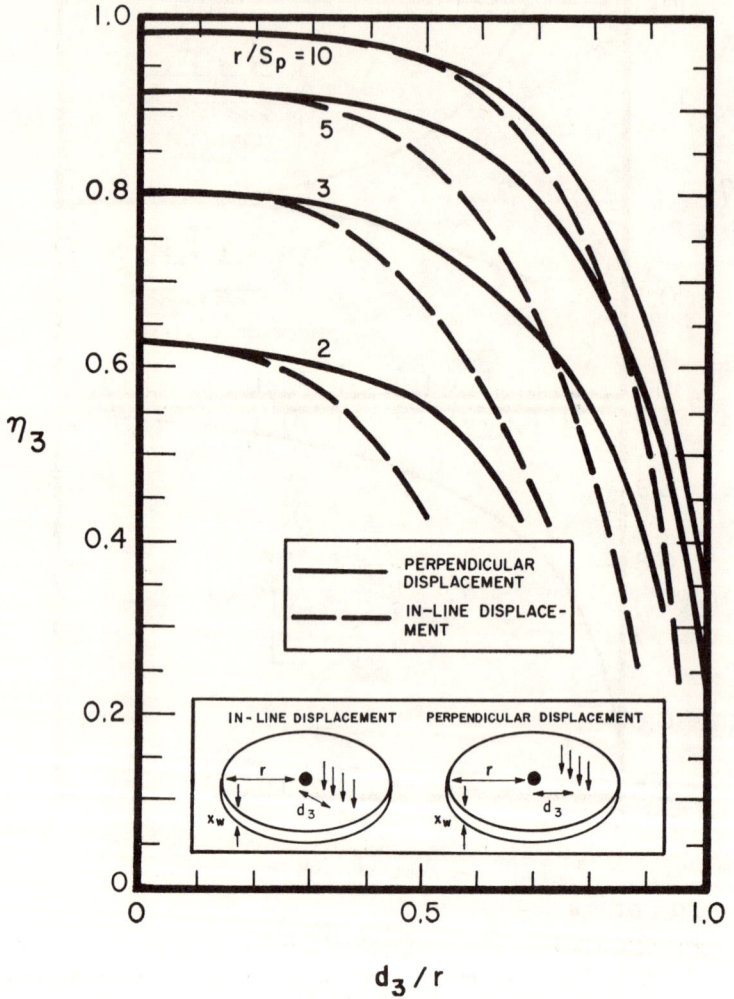

FIGURE 7.5

Correction factor (η_3) applicable to in-line four-point probe measurements if probe center is displaced from sample center vs. amount of displacement (d_3), sample radius (r), and probe spacing (S_p).

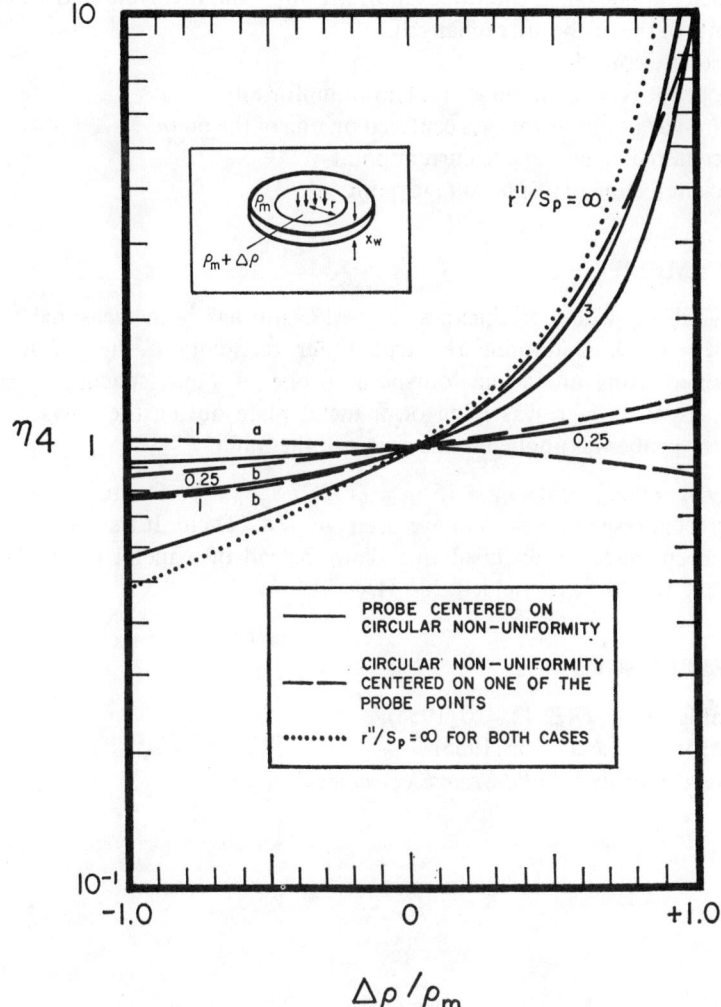

FIGURE 7.6

Correction factor (η_4) applicable to in-line four-point probe measurements if a non-uniformity in resistivity is located near the probe vs. sample resistivity (ρ_m), resistivity variation in non-uniform region ($\Delta\rho$), diameter of non-uniform region (r''), and probe spacing (S_p).

If $\Delta\rho = 0$, then $\eta_4 = 1$ and $\rho = \rho_m$.

Reciprocity applies; i.e., the values of η_4 are still valid if current and voltage points of the probe are exchanged.

Two cases are considered:
   (i) The probe is centered on a circular nonuniformity.
   (ii) A circular nonuniformity is centered on one of the points:
      a. centered on an outside current point
      b. centered on an inside voltage point.

4. EXAMPLE

Problem: A silicon wafer of thickness $x_w = 175$ μm has been measured to have a $V/I = 15$ Ω. Determine the true wafer resistivity if the V/I has been obtained using an in-line four-point probe of point spacing $S_p = 500$ μm where the wafer was lying on a metal plate during the measurement and the probe was applied in the center of the wafer.

Answer: $x_w/S_p = 0.35$, hence $\eta_1 = 15$, $\rho = (V/I)2\pi S_p \eta_1 = 71$ Ω cm.

The uncorrected resistivity would have been: $\rho_m = 4.7$ Ω cm. If the measurement had been made on an insulating plate instead of a metal plate, the measurement would have yielded 284 Ω cm.

5. REFERENCES

(1) L. B. Valdes, *Proc. IRE*, **42**, 420 (1954).
(2) A. Uhlir, Jr., *BSTJ*, **34**, 105 (1955).
(3) L. J. Swartzendruber, *Solid-State Electronics*, **7**, 413 (1964).

7–3 Other Measurement Techniques

1. GENERAL COMMENTS

The following brief description of some measurement methods used in semiconductor technology within the scope of this book is highly selective and gives only the most salient features of a particular technique. It is intended to be mainly a guide for further investigations. Although the descriptions given are not detailed, the most common measurements of physical and electrical properties of semiconductors have been included.

2. CRYSTALLOGRAPHIC ORIENTATION

(a) X-ray diffraction method:
   This method is useful for elemental and compound semiconductors.
(b) Optical method:
   This method is applicable to elemental semiconductors only.

3. CONDUCTIVITY TYPE

A simple method uses the fact that the sign of the thermal electromagnetic force, generated between an extrinsic semiconductor and two metal plates of different temperatures, is characteristic of the conductivity type. An alternate method uses the polarity of the Hall voltage, induced as a result of the deflection of carriers in a magnetic field, for the determination of the semiconductor conductivity type.

4. ENERGY GAP

The measurement of the energy gap of a semiconductor is based on the fact that absorption of monochromatic light increases rapidly when the wavelength is such that the corresponding energy hv is large enough to excite electrons from valence band to conduction band. This will take place at a critical frequency which corresponds to the absorption edge, i.e., to the wavelength at which the absorption coefficient displays a sudden sharp increase. The energy gap is then $E_G = hv$. Difficulties arise because the absorption edge is frequently not sharp enough and because other absorption mechanisms are present.

Alternate methods are the determination of the energy gap by measuring the Hall voltage or the semiconductor resistivity as a function of temperature.

5. THERMAL CONDUCTIVITY

(a) Below 300°K:

If the temperature gradient ∇T is measured between two ends of a rectangular bar of cross section A whose thermal conductivity is to be determined and where lateral heat flow due to electric power (VI) from one end to the other end which is connected to a heat sink is taking place, then the thermal conductivity can be calculated from

$$\kappa = (VI/A)/\nabla T. \tag{7.14}$$

This method cannot be used for small samples.

(b) Above 300°K:

(i) *Comparison method*:

The sample is sandwiched between two materials of known thermal conductivity. By measuring the temperature gradients along both the known and the unknown samples, a determination of the heat flow through the unknown sample can be obtained.

(ii) *Ångstrom method*:

A sinusoidal heat input is imposed at one end of the sample and the thermal conductivity calculated from the equation

$$\nabla \cdot [(\kappa/c_v d_s) \operatorname{grad} T] = \partial T/\partial t. \tag{7.15}$$

6. OPTICAL PROPERTIES

Reflection and absorption coefficients (which are wavelength-dependent) of a semiconductor can be measured by conventional means. The sum of reflectance, absorbance, and transmittance is unity. Derived from these parameters are refractive index and extinction coefficient (which is related to the absorption coefficient).

7. THIN FILM THICKNESS

The thickness of very thin films (in the order of a few Å) is usually measured by ellipsometric methods. The light used is polarized at 45° and the phase and amplitude ratio of the two reflected components are measured. Ellipsometry allows the simultaneous determination of the refractive index of the film.

7-3 OTHER MEASUREMENT TECHNIQUES

8. CARRIER DENSITY

Measurement of the Hall voltage can be used to determine carrier densities without knowledge of the carrier mobility. A less common method is the measurement of carrier densities by means of the electric conductivity or resistivity; this, however, requires the carrier mobility to be known. If mixed conduction prevails, i.e., if both electrons and holes contribute significantly to the electric conduction, the densities of two the carrier types have to be determined by two independent methods.

9. CARRIER MOBILITY

(a) Majority carrier mobility:
   It can be determined by measuring the semiconductor resistivity if the carrier density is known from other measurements, e.g., the Hall voltage.
(b) Minority carrier mobility:
   It can be determined by applying a sweeping pulsed field to the semiconductor; if the time it takes an injected pulse to travel from the injecting to the collecting contact is measured, the carrier velocity can be calculated which, divided by the field, gives the mobility.

10. CARRIER LIFETIME

(a) Morton-Haynes method:
   Electron-hole pairs are generated when an imaginary line on the semiconductor surface is illuminated by light pulses; if the carrier density is measured as a function of distance from the illuminated line, then carrier decay due to recombination can be determined from the carrier diffusion length, $L = \sqrt{D\tau}$. Measurements are affected by surface recombination and surface inversion layers.
(b) Stevenson-Keyes method:
   In this method a semiconductor sample is illuminated uniformly by light pulses; this results in the generation of excess carriers and consequently in a reduction of semiconductor resistivity. The photoconductive decay is a measure of carrier lifetime. This method allows the independent determination of bulk and surface recombination. The disadvantage of this method is the requirement that the light fall time must be short compared to the photoconductive decay time.
(c) Photomagnetoelectric (PME) method:
   If the surface of a semiconductor is illuminated and if the semiconductor is simultaneously brought into a magnetic field parallel to the surface, then the excess carriers (electrons and holes) generated by the light pulses

are deflected in opposite directions, resulting in voltage pulses whose magnitude is related to carrier lifetime and surface recombination. This method allows the determination of very short lifetimes.

11. RECOMBINATION RADIATION

The choice of a particular detection method depends upon the wavelength and the radiation intensity (Figure 7.7). The upper limit of the sensitivity of a radiation detector (minimum detectable power) is

$$S_R = (c_p m_{det} kT^2)^{1/2}/\tau_{det} \alpha_{det} \tag{7.16}$$

where c_p = specific heat of detector
m_{det} = mass of detector
τ_{det} = thermal time constant
α_{det} = absorption coefficient of surface of detector.

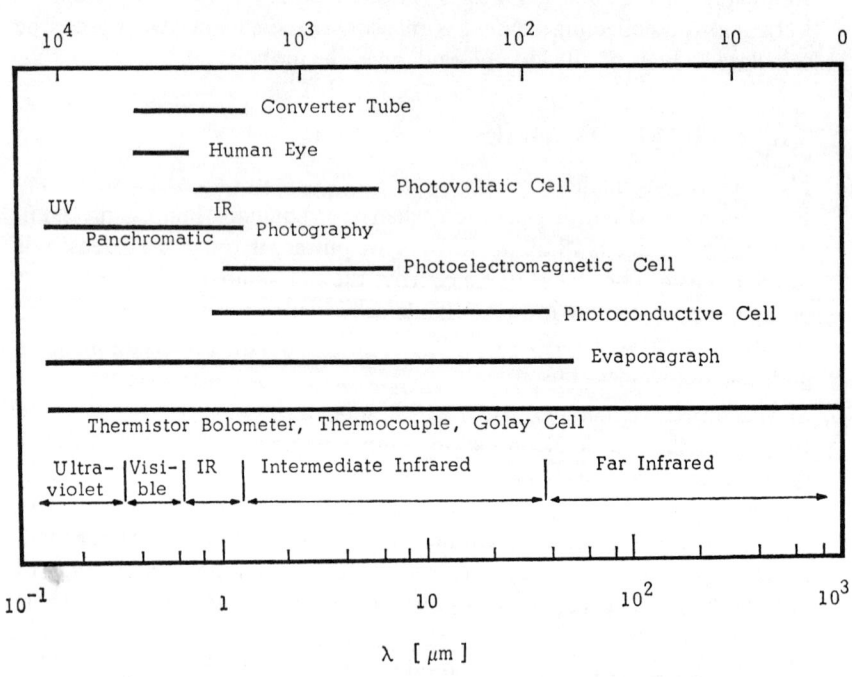

FIGURE 7.7

Common radiation detection methods. The temperature T_{bb} corresponds to the maximum black-body radiation.

7-3 OTHER MEASUREMENT TECHNIQUES

If the radiation lies in the infrared, there are four basic detection methods available.

(a) Image converter tube:
The infrared radiation is converted into an image in the visible range and can be easily viewed or photographed. This method uses a semitransparent photocathode on which the object to be viewed is imaged by means of an optical objective. The image on the photocathode is focused on the fluorescent screen at the other end of the tube by electron-optical methods and can be viewed with an optical magnifier.
  (i) *Advantages*:
  Relatively fast; useful for qualitative measurements; permanent and fast record with Polaroid film (ASA 10,000); display can be seen directly without photographic techniques; no focusing difficulties.
  (ii) *Disadvantages*:
  Low resolution; variation of resolution, magnification, distortion, and intensity of the tube with radial distance; fog on screen (emission of visible light from screen with no infrared emission from transistor); requires high supply voltage (20 kV); tube is sensitive to visible light—sensitivity decreases drastically after exposure to room light or decreases slightly after exposure to visible light used for focusing.

(b) Infrared spectroscopic plate:
Special photographic plates extend relatively far into the infrared. Class I-Z plates, which have to be hypersensitized immediately prior to exposure for increase in sensitivity, have a maximum sensitivity at 1.08 μm and can be used to at least 1.3 μm. No plates or film are available for longer wavelengths.
  (i) *Advantages:*
  High resolution; no distortion; no background fog if properly hypersensitized; useful for quantitative measurement; no change in distortion, magnification, resolution, and intensity over the whole area of the plate.
  (ii) *Disadvantages*:
  Necessity of hypersensitizing before and of development after exposure; long exposure times of several hours (ASA 12); focusing problems if a lens system is used for magnification.

(c) Thermal detectors:
Impinging infrared radiation heats the sensitive element and a temperature-dependent property of the element is monitored. A variety of thermal detectors is available, beginning in the simplest case with thermometer and thermocouple. A series arrangement of thermocouples, a "thermopile," is used in spectroscopy. In a Golay cell the radiation energy is

absorbed by the walls of a gas-filled chamber increasing temperature and gas pressure and thus producing a means of detection. Thermistors (semiconductor bolometers) are the most widely used thermal detectors; here the change of resistance with a small temperature change is being used for detection. In an Evaporagraph infrared radiation affects thermally the amount of condensation of oil on a thin membrane and yields a visible image of the radiation in terms of variations in oil thickness.

  (i) *Advantages*:
    Thermal detectors respond to long-wavelength radiation at ambient operating temperatures; sensitivity is independent of wavelength; they are small and consume usually little power.
  (ii) *Disadvantages*:
    Not very sensitive; slow since it takes time for the sensing element to heat; focusing difficulty.

(d) Photodetectors:
  Infrared photons react with bound electrons in the sensitive element, producing free charge carriers. Some electrical property which depends upon the number of free carriers in the sensitive element is monitored to determine changes in flux density of the incident infrared radiation. There are four different photoelectric phenomena which can be used: photoemission, photoconduction, the photovoltaic effect, and the photoelectromagnetic effect. Of the four, photoconducting and photovoltaic detectors are most frequently used. The sensitive element of both is a semiconductor. In photovoltaic detectors, photons create free carriers at a *p-n* junction producing a potential difference. Indium antimonide is the most commonly used material because of its speed and high sensitivity. Its sensitivity is close to the theoretical limit which is set by the random arrival of background photons. It is, however, sensitive only to 5.5 μm.

  (i) *Advantages*:
    Relatively high wavelength region makes photodetectors useful for thermal mapping of transistors and integrated circuits by use of a scanning system; high sensitivity; fast reaction; high signal-to-noise ratio.
  (ii) *Disadvantages*:
    Output signal is small and has to be amplified; need for cooling to liquid helium or liquid nitrogen temperatures; focusing difficulty.

12. RESISTIVITY

The four-point probe method described earlier is the most reliable and the most widely used method to determine the resistivity of a semiconductor.

A simple resistance measurement on a semiconductor of known dimensions, on the other hand, is not accurate enough because of the requirement for nonrectifying contacts.

13. SURFACE STATE DENSITY

In theoretical expressions involving the voltage dependence of the capacitance of MIS structures the voltage is usually

$$V = V_T - \phi_{MS} + Q_{ss}/C_o, \tag{7.17}$$

whereas in experiments only the turn-on voltage V_T is measured. The difference between theoretical and experimental capacitance-voltage curves of an MIS system provides therefore a means to determine the surface state density Q_{ss}. This voltage difference is

$$\Delta V = -\phi_{MS} + Q_{ss}/C_o \tag{7.18}$$

from which the surface state density

$$Q_{ss} = (\Delta V + \phi_{MS})C_o. \tag{7.19}$$

The voltage difference is usually taken at $(C_o + C_m)/2$.

14. REFERENCES

(1) R. C. Jones, *J. Opt. Soc. Am.*, **37**, 879 (1947).
(2) W. W. Gärtner, *Transistors—Principles, Design, and Applications*, D. Van Nostrand, New York, 1960.
(3) *Book of ASTM Standards*, ASTM Committee F-1 on Materials for Electron Devices and Microelectronics, Philadelphia, 1965.
(4) B. M. Berry, *RTI Report ASD-TDR-63-316*, Vol. XII, Sept. 1966.

APPENDIX

A-1 Properties of Metals and Other Data
A-2 List of Symbols

A-1 Properties of Metals and Other Data

1. SKIN EFFECT

At high frequency the performance of semiconductor structures may be limited by the frequency response of connecting metal films. In metal conductors the current distribution is uniform at low frequency; at high frequency, however, the current distribution within the conductor is nonuniform and, owing to inductive effects, the current is concentrated in the proximity of the surface. Similarly, the magnetic flux generated by the current has a maximum at the surface. The current density decreases exponentially with distance from the surface. Reduction of current and magnetic flux increases with frequency, conductivity, permeability, and thickness of the conductor. The nonuniform current distribution results in a significant resistance increase of the conductor.

The penetration depth (skin depth) is defined as the distance from the surface where the current density has fallen to $1/e$ ($=36\%$) of its value at the surface (Figure A.1). In practice, one considers the conductor as consisting of two layers: one is the conductive surface layer of thickness x_S which has an even current distribution, and the inner layer has a variable current and field distribution.

Assumptions:

(a) Film thickness is much greater than skin depth.
(b) Conductor has a rectangular crosscut.
(c) Permeability of all metals is $\mu_o = 1$.

Frequently deviations from above assumptions occur giving rise to variations in skin depth.

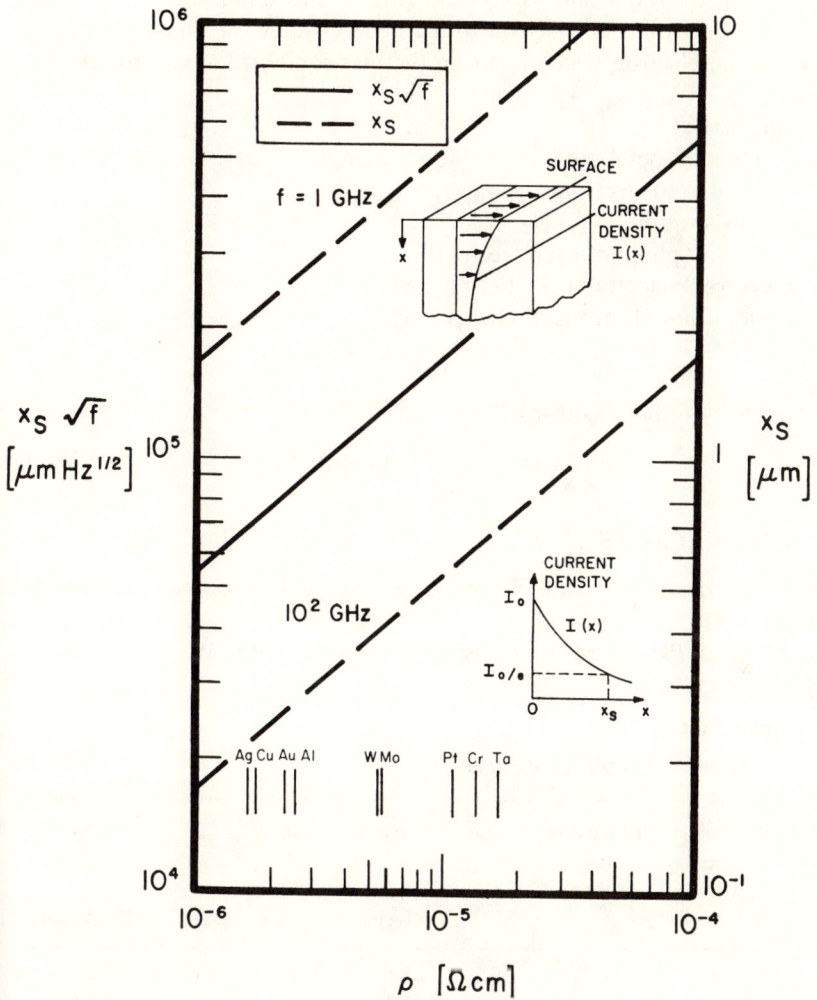

FIGURE A.1

Skin depth (x_S) vs. metal film resistivity (ρ) and frequency (f) for various metals.

2. ELECTRONIC PROPERTIES

Table A.1 lists the important parameters of metals used in hot electron structures. A hot electron is an electron with an energy a few kT above the Fermi energy, i.e., an electron which is not in thermal equilibrium with the crystal lattice.

Symbols used:
- r_m = radius of metal atom
- r_s = normalized electron spacing ($r_s \equiv r_m/a_o$)
- a_o = Bohr's radius
- l_{ph} = electron-phonon mean free path
- v_F = electron velocity at the Fermi level.

The lifetime of hot electrons of energy E is

$$\tau_n^* \approx r_s^{3/2}(E - E_F)^{-2} E_F^{3/2} E^{1/2} \tag{A.1}$$

and the density of hot electrons

$$n = (3/4\pi)(r_s a_o)^{-3} = (3/4\pi)r_m^{-3}. \tag{A.2}$$

3. REFERENCES

(1) *Reference Data for Radio Engineers*, ed. by H. P. Westman, ITT, New York, 1957.
(2) S. M. Sze, *Physics of Semiconductor Devices*, John Wiley and Sons, New York, 1969.

TABLE A.1

Electronic Properties of Metals

Metal	Density of free electrons, n [10^{22} cm^{-3}]	Electrons per atom	E_F [eV]	r_m [Å]	r_m/a_o	l_{ph} [Å]	v_F [10^8 cm/sec]
Ag	5.8	1	5.5	1.6	3.0	570	1.39
Al	18.4	3	11.7	1.1	2.0	4260	2.04
Au	5.9	1	5.5	1.6	3.0	406	1.40
Co	6.2	0.7	5.7	1.5	2.9	130	1.42
Cu	8.7	1	7.1	1.4	2.6	420	1.58
Fe	1.7	0.2	2.8	2.2	4.2	220	0.91
Ni	5.4	0.6	5.2	1.7	3.2	133	1.37
Pd	3.8	0.6	4.1	1.9	3.5	110	1.20
Pt	4.0	0.6	4.3	1.8	3.4	110	1.23

TABLE A.2
Periodic Table of the Elements (Metals are shown in boldface type)

Period	Group I a	I b	II a	II b	III a	III b	IV a	IV b	V a	V b	VI a	VI b	VII a	VII b	VIII a	VIII a	VIII a	b
1	1 H 1.01																	2 He 4.0
2	3 Li 6.9		4 Be 9.0		5 B 10.8		6 C 12.0		7 N 14.0		8 O 16.0		9 F 19.0					10 Ne 20.2
3	11 Na 23.0		12 Mg 24.3		13 Al 27.0		14 Si 28.1		15 P 31.0		16 S 32.1		17 Cl 35.5					18 Ar 39.9
4	19 K 39.1	29 Cu 63.5	20 Ca 40.1	30 Zn 65.4	21 Sc 45.0	31 Ga 69.7	22 Ti 47.9	32 Ge 72.6	23 V 50.9	33 As 74.9	24 Cr 52.0	34 Se 79.0	25 Mn 54.9	35 Br 79.9	26 Fe 55.8	27 Co 58.9	28 Ni 58.7	36 Kr 83.8
5	37 Rb 85.5	47 Ag 107.9	38 Sr 87.6	48 Cd 112.4	39 Y 88.9	49 In 114.8	40 Zr 91.2	50 Sn 118.7	41 Nb 92.9	51 Sb 121.8	42 Mo 95.9	52 Te 127.6	43 Tc 98.9	53 I 126.9	44 Ru 101.1	45 Rh 102.9	46 Pd 106.4	54 X 131.0
6	55 Cs 132.9	79 Au 197.0	56 Ba 137.3	30 Hg 200.6	57–71*	81 Tl 204.4	72 Hf 178.5	82 Pb 207.2	73 Ta 180.9	83 Bi 209.0	74 W 183.9	84 Po 210.0	75 Re 186.2	85 At 210	76 Os 190.2	77 Ir 192.2	78 Pt 195.1	86 Rn 222
7	87 Fr 223		88 Ra 226.0		89–103†													
Highest oxide	E_2O		EO		E_2O_3		EO_2		E_2O_5		EO_3		E_2O_6		EO_4			
Highest hydride	EH		EH_2		EH_3		EH_4		EH_3		EH_2		EH					

* Lanthanum Series: 57 **La**, 53 **Ce**, 59 **Pr**, 60 **Nd**, 61 **Pm**, 62 **Sm**, 63 **Eu**, 64 **Gd**, 65 **Tb**, 66 **Dy**, 67 **Ho**, 68 **Er**, 69 **Tm**, 70 **Yb**, 71 **Lu**.
† Actinium Series: 89 **Ac**, 90 **Th**, 91 **Pa**, 92 **U**, 93 **Np**, 94 **Pu**, 95 **Am**, 96 **Cm**, 97 **Bk**, 98 **Cf**, 99 **Es**, 100 **Fm**, 101 **Md**, 102 **No**, 103 **Lw**.

TABLE A.3
Important Physical Constants

Loschmidt number	$L = 6.025 \cdot 10^{23}$/mole
Gas constant (Avogadro number)	$R = 8.314 \cdot 10^7$ erg/°K mole
Boltzmann constant, $k = R/L$	$k = 1.380 \cdot 10^{-16}$ erg/°K
	$= 1.380 \cdot 10^{-23}$ W sec/°K
	$= 8.616 \cdot 10^{-5}$ eV/°K
Electronic charge, $q = F/L$	$q = 1.602 \cdot 10^{-19}$ A sec
	$= 1.000$ eV/V
Planck constant	$h = 6.625 \cdot 10^{-27}$ erg sec
	$= 4.134 \cdot 10^{-15}$ eV sec
	$\hbar (= h/2\pi) = 1.054 \cdot 10^{-27}$ erg sec
Faraday constant	$F = 9.652 \cdot 10^4$ A sec/g
Stefan-Boltzmann constant	$\sigma = 5.669 \cdot 10^{-5}$ erg/cm² sec°K⁴
Permittivity of free space	$\epsilon_o = 8.854 \cdot 10^{-14}$ F/cm
Electron rest mass	$m_o = 9.108 \cdot 10^{-28}$ g
Specific electron charge	$q/m_o = 1.759 \cdot 10^8$ A sec/g
Electron rest energy, $E_{no} = m_o c^2$	$E_{no} = 8.176 \cdot 10^{-7}$ erg
	$= 0.511 \cdot 10^6$ eV
Velocity of electromagnetic waves in vacuum	$c = 2.998 \cdot 10^{10}$ cm/sec
Bohr's radius	$a_o = 0.529 \cdot 10^{-8}$ cm
Compton wavelength	$\lambda_c = 2.427 \cdot 10^{-10}$ cm
Wavelength associated with 1 eV	$\lambda_{eV} = 1.237$ μm
Wave number associated with 1 eV	$1/\lambda_{eV} = 8066$ cm^{-1}
Frequency associated with 1 eV	$\nu_{eV} = 2.424 \cdot 10^{14}$ Hz
Temperature associated with 1 eV	$T_{eV} = 11{,}605$ °K
Energy associated with 1 eV	$E_{eV} = 1.602 \cdot 10^{-12}$ erg
Other useful numbers	
	$\epsilon_{Si} \epsilon_o = 1.04 \cdot 10^{-12}$ F/cm
	$\epsilon_{SiO_2} \epsilon_o = 0.35 \cdot 10^{-12}$ F/cm
	$e = 2.71828\ldots$
	$\pi = 3.14159\ldots$
	$kT(300°K) = 2.586 \cdot 10^{-2}$ eV
	$kT/q(300°K) = 2.586 \cdot 10^{-2}$ V

TABLE A.4
Useful Conversions of Units

1 dyne	$= 1$ cm g/sec^2
1 erg	$= 1$ dyne cm
	$= 1$ cm^2 g/sec^2
	$= 6.24 \cdot 10^{11}$ eV
1 dyne/cm^2	$= 1$ g/cm sec^2
1 g (gram)	$= 6.24 \cdot 10^{11}$ eV sec^2/cm^2
1 kcal/mole	$= 4.33 \cdot 10^{-2}$ eV
1 atm	$= 760$ mm Hg
	$= 760$ Torr
	$= 1.01 \cdot 10^6$ dyne/cm^2
1 bar	$= 10^6$ dyne/cm
	$= 0.987$ atm
1 cal	$= 4.185 \cdot 10^7$ erg
1 Joule	$= 10^7$ erg
	$= 1$ W sec
	$= 1$ VA sec
	$= 0.238$ cal
1 Coulomb	$= 1$ A sec
1 eV	$= 1.6 \cdot 10^{-12}$ erg
	$= 1.6 \cdot 10^{-12}$ dyne cm
	$= 23.06$ kcal/mole
1 Farad (F)	$= 1$ A sec/V
1 Gauss (G)	$= 10^{-8}$ V sec/cm^2
1 A	$= 1.036 \cdot 10^{-5}$ Faraday/sec
1 cm	$= 10^4$ μm
1 cm^2/sec	$= 3.6 \cdot 10^{11}$ μm^2/hr
1 μm/min	$= 1.67 \cdot 10^2$ Å/sec
1 mil	$= 25.4$ μm
log x	$= 0.4343$ ln x
	(log 10 = 1, ln e = 1; ln 10 = 2.30)

A–2 List of Symbols

1. FREQUENTLY AND GENERALLY USED SYMBOLS

(These may be supplemented by various subscripts and superscripts)

C_B	Uniform impurity concentration of semiconductor bulk
D	Diffusion coefficient
n	Electron, carrier concentration
T	Temperature
t	Time
V	Voltage
v	Velocity
x	Distance normal to semiconductor surface
y	Distance parallel to semiconductor surface
λ	Wavelength
μ	Mobility
ν	Frequency
ρ	Resistivity
τ	Carrier lifetime

2. NOTATIONS

A	Linear oxidation rate constant of semiconductor
A	Area
a	Crystal lattice constant
a	Impurity concentration gradient at a *p-n* junction
A^*	Geometry-dependent correction factor for resistivity measurements using an in-line four-point probe
A_o	Amplitude of electromagnetic wave without attenuation
a_o	Bohr's radius
a_D	Correction factor of impurity diffusion coefficient
a_h	Radius of irradiated area at semiconductor surface
a_i	Components of one-dimensional thermal expansion coefficient
a_r	Reflection coefficient

526

A-2 LIST OF SYMBOLS

a_T	Temperature-dependent parameter affecting the effective impurity gradient
a_t	Transmission coefficient
$A(x)$	Amplitude of electromagnetic wave at distance x from surface of attenuating medium
B	Parabolic oxidation rate constant of semiconductor
b	Ratio of the thickness of the original semiconductor that is being converted to oxide to the total thickness of the oxide being formed during thermal oxidation
C	Externally measured capacitance of MOS or diffused structure per unit area
c	Velocity of an electromagnetic wave (light)
C^*	Equilibrium concentration of oxidant in oxide
c_{ac}	Velocity of sound
C_{Au}	Gold concentration in semiconductor at saturation
C_B	Background impurity concentration, bulk impurity concentration
C_B^*	Impurity concentration at which the μ–C_B curve displays an inflection
C_b	Barrier capacitance
$C_{B\,max}$	Maximum solid solubility of impurity in semiconductor
C_{Bo}	Impurity concentration above which a semiconductor is degenerate
C_D	Minority carrier diffusion capacitance of a p-n junction
C_d	Average impurity concentration in the depletion layer of a semiconductor surface
$C_e(x, t)$	Concentration of impurities originally contained in the epitaxial film after the completion of epitaxial growth
C_f	(Uniform) impurity concentration in epitaxial film at a distance greater than $2\sqrt{Dt}$ away from the metallurgical interface of film and substrate
C_G	Concentration of dopant in the bulk of the gas in an epitaxial reactor
c_{ik}	Elastic constants
C_L	Concentration of semiconductor atom sites
C_m	Minimum capacitance of an MOS structure per unit area
C_o	Oxide capacitance per unit area

$\overline{C_O}$	Average impurity concentration in oxide
C'_O	Impurity concentration at the outside surface of the oxide (oxide-gas interface)
C''_O	Impurity concentration at the inside surface of the oxide (oxide-semiconductor interface)
C_P	Peak impurity concentration after ion implantation
c_p	Specific heat at constant pressure
C_S	Surface concentration (impurity concentration at semiconductor surface)
C_S^*	True impurity concentration at the surface of a semiconductor taking into account the reduction of concentration due to the electric field generated by the impurity ions
C_s	Semiconductor space charge capacitance per unit area
C_{sc}	Space charge capacitance per unit area
$C_{sc(inv)}$	Space charge capacitance minimum at onset of strong inversion
C_{sco}	Space charge capacitance per unit area for $\phi_{s(inv)} = 0$
C_{ss}	Surface space charge capacitance per unit area
C_{sub}	(Uniform) concentration of impurities in the substrate before epitaxial growth (or after epitaxial growth at a distance greater than $2\sqrt{Dt}$ away from the metallurgical interface of substrate and film)
C_T	Transition capacitance of a p-n junction
C_T	Total number of molecules per unit volume in a gas
C_t	Total capacitance of a p-n junction
C_{Tc}	Transition carrier capacitance due to the change in the total number of free carriers in the space charge region of a junction
C_{Ti}	Transition space charge capacitance due to the change in the total number of ionized impurity charges at a junction
c_v	Specific heat at constant volume
$C(x)$	Total density of impurity atoms at depth x
$C(x, t)$	Impurity concentration dependent upon depth normal to semiconductor surface and time
$C(y, t)$	Impurity concentration dependent upon distance from diffusion source parallel to semiconductor surface and time
$C_1(x, t)$	Concentration of impurities originally contained in the substrate after epitaxial growth
$C_1(x', t)$	Impurity concentration within the oxide

$C_1(x'_*, t)$	Impurity concentration within the oxide at the semiconductor-oxide surface
$C_2(x, t)$	Impurity concentration within the semiconductor
$C_2(x_*, t)$	Impurity concentration within the semiconductor at the semiconductor-oxide surface
D	Impurity diffusion coefficient in general
d	Projection of lattice spacing on lattice axes (apparent lattice spacing)
D_{amb}	Ambipolar diffusion coefficient
D_c	Diffusion coefficient of carriers
$D(C_B)$	Diffusion coefficient of impurity in semiconductor of impurity concentration C_B
D_d	Diffusion coefficient of impurity during drive-in diffusion
D_{dis}	Diffusion coefficient of impurity along a dislocation
D_{eff}	Effective diffusion coefficient
D_G	Diffusivity of carrier gas in epitaxial reactor
d_G	Density of gas
D_h	Thermal diffusivity (diffusion coefficient)
D_I	Diffusion coefficient of an interstitial impurity
D_i	Diffusion coefficient of impurity in an intrinsic semiconductor
d_I	Average distance between neighboring scattering centers
D_n	Electron diffusion coefficient
D_o	Diffusion coefficient of impurity within oxide
D_P	Pseudo-diffusion coefficient of an impurity in a liquid semiconductor
D_p	Hole diffusion coefficient
D_p	Diffusion coefficient of impurity during predeposition
D_S	Diffusion coefficient of a substitutional impurity
D_s	Diffusion coefficient of impurity in a semiconductor
d_s	Density of semiconductor
D_{sm}	Diffusion coefficient of a substitutional impurity at the melting point of the semiconductor
D_∞	Apparent impurity diffusion coefficient at $T = \infty$
E	Electric field
e	Basis of natural logarithms
E_A	Energy of an acceptor
E_a	Activation energy

APPENDIX

E_a^*	Activation energy required for diffusion jumps of impurity atom
E_{aI}^*	Activation energy for an interstitially located impurity atom
E_{aS}^*	Activation energy for a substitutionally located impurity atom
E_c	Lower edge of conduction band
E_{crit}	Electric field above which carrier drift velocity is no longer proportional to field
E_D	Energy of a donor
E_{dis}	Activation energy for the motion of a dislocation
E_F	Extrinsic Fermi level
E_{Fo}	Fermi level at a semiconductor surface
$E_{F\infty}$	Fermi level deep inside a semiconductor
E_G	Energy band gap
$E_{G,\text{opt}}$	Optical energy band gap
E_{Go}	Energy gap in the absence of stress
E_H	Activation energy of a hydrogen atom
E_I	Energy of an impurity ion during ion implantation
E_i	Intrinsic Fermi level
E_I^*	Energy of ion at a point where nuclear scattering is equal to electronic scattering
E_{ion}	Ionization threshold energy
E_{Io}	Energy of impurity ion at a semiconductor surface
$E_{\max}$	Maximum electric field at a p-n junction
E_n	Electron energy
E_p	Height of potential barrier
E_{ph}	Phonon energy
E_R	Repulsive energy
E_s	Energy of forming a Schottky defect in a crystal
E_v	Upper edge of valence band
F	Dimensionless electric field
f	Frequency
f_D	Term describing the effect of the built-in electric field on impurity diffusion
f_d	Fraction of dislocation density to density of atom sites
F_h	Heat flux at semiconductor surface into semiconductor upon absorption of thermal irradiation
f_I	Fraction of interstitial sites
f_L	Fractional density of available lattice positions

A-2 LIST OF SYMBOLS

f_l	Atom fraction of an impurity element in the liquid phase
f_s	Atom fraction of an impurity element in the solid phase
f_t	Frequency at which transition from low-frequency to high-frequency behavior of MOS structure occurs
$F(x)$	Flux density of diffusing impurity atoms
$F_{1/2}$	Modified Fermi potential (Fermi-Dirac integral)
g	Term dependent upon junction geometry
G_n	Electron generation rate
G_p	Hole generation rate
G_{np}	Generation rate of electron-hole pairs due to impact ionization
H	Hardness
H	Magnetic field
h	Planck's constant
H_E	Heat of evaporation of solvent
H_F	Heat of fusion
H_G	Constant in Henry's Law
h_G	Gas-phase mass-transfer coefficient in terms of concentration in the gas
H_S	Heat of sublimation
h_S	Gas-phase mass-transfer coefficient in terms of concentration in the solid
$H_{Si(l)}$	Partial molar heat of mixing silicon in liquid solution
H_V	Heat of vaporization
I	Ion beam current during ion implantation
i	Imaginary constant ($i = \sqrt{-1}$)
I_o	Saturation current density
j_n	Electron diffusion current in a p-type semiconductor
j_p	Hole diffusion current in an n-type semiconductor
j_{Sn}	Total drift and diffusion current density of electrons into surface of p-type semiconductor
j_{Sp}	Total drift and diffusion current density of holes into surface of n-type semiconductor
j_U	Generation-recombination current density
K	Effective diffusion coefficient
k	Boltzmann's constant
k_I	Empirical factor important for electronic stopping

k_i	Miller indices of a crystal
k_o	Equilibrium distribution coefficient at melting point
k_S	Chemical surface reaction rate constant
k_w	Wave vector
L	Diffusion length
l_{ac}	Mean free path of carriers due to acoustical scattering
l_B	Mean free path of carriers in the semiconductor bulk
l_c	Mean free path of carriers between collisions
L_D	Debye length
l_{dis}	Length of a dislocation line between pinning points
L_f	Latent heat of fusion
l_I	Mean free path of carriers owing to impurity (ion) scattering
l_L	Mean free path of carriers owing to lattice vibrations
L_n	Electron diffusion length in a p-type semiconductor
l_N	Mean free path of carrier due to scattering on neutral atoms
l_{ph}	Mean free path of carriers due to optical phonon scattering
l_s	Crystallographic slip distance of a dislocation
l_o	Length
M	Carrier multiplication factor
M	Atomic mass
m	Segregation coefficient
M_H	Hall effect material figure of merit
m_k	Direction cosine of direction l_o with respect to principal crystallographic directions
$m_{n,p}$	Mass of electron or hole. For an electron $m_{n,p} = m_n$, for a hole $m_{n,p} = m_p$
m_n	Mass of an electron in a lattice
m_o	Mass of an electron in free space
m_p	Mass of a hole in a lattice
m_{ph}	Mass of phonon
n	Density of electrons per unit volume
N^*	Complex refractive index
n^*	Refractive index (real part)
N_A	Concentration of acceptor ions
n_{air}	Refractive index of air
n_B	Integer associated with the lattice Brillouin zones
N_c	Effective density of states in conduction band

A-2 LIST OF SYMBOLS

N_D	Concentration of donor ions
N_d	Dislocation density
n_d	Electron concentration within space charge region
n_{do}	Electron concentration outside space charge region
n_g	Refractive index of glass
N_H	Impurity concentration-dependent term used in Hall effect equation
n_i	Carrier concentration in intrinsic semiconductor
N_l	Concentration of impurity atoms in the liquid phase
N_N	Density of neutral scattering centers
n_n	Equilibrium electron concentration in an n-type semiconductor
N_O	Number of oxidant molecules incorporated into a unit volume of oxide
n_p	Equilibrium electron concentration in a p-type semiconductor
N_{ph}	Number of optical phonon collisions per unit length
N_{pho}	Number of optical phonon collisions per unit length at low energy
N_s	Concentration of impurity atoms in the solid phase
n_S	Electron density in the immediate vicinity of the semiconductor surface
n_s	Electron concentration under stress
n_{Si}	Refractive index of silicon
n_{SiO_2}	Refractive index of silicon dioxide
N_t	Density of recombination centers
n_t	Electron concentration at time t
n_{to}	Electron concentration at time zero
N_v	Effective density of states in valence band
n_z	Number of electrons in a noncontinuous energy band
n_o	Electron concentration in the absence of stress
n_1	Density of electrons in conduction band when Fermi level coincides with recombination level
O_s	Oxidation coefficient
P	Poisson ratio
p	Pressure
p	Hole density under equilibrium conditions
p_c	Critical pressure
p_d	Hole concentration within space charge region

p_G	Partial pressure of the oxidant in the bulk of the gas
P_i	Ionization probability
P_{ion}	Probability of an electron to gain sufficient energy to ionize
p_n	Equilibrium hole concentration in an n-type semiconductor
p_p	Equilibrium hole concentration in a p-type semiconductor
P_r	Ionization probability
p_S	Hole concentration in the immediate vicinity of the semiconductor surface
p_S	Partial pressure of the oxidant at the outer oxide surface (oxide-ambient interface)
p_s	Hole concentration under stress
p_t	Hole concentration at time t
p_{to}	Hole concentration at time zero
p_v	Vapor pressure
p_o	Hole concentration in the absence of stress
p_1	Density of holes in valence band when Fermi level coincides with recombination level
Q	Number of impurity atoms per unit area (impurity dosage, "charge")
q	Electric charge of an electron
Q_B	Charge density in a semiconductor surface layer due to impurity atoms
Q_h	Amount of heat
Q_i	Dosage of implanted ions (number of implanted ions per unit of area)
Q_j	Number of impurity atoms per unit area between semiconductor surface and junction
Q_n	Charge density of electrons within inversion region of a p-type semiconductor
Q_p	Charge density of holes within inversion region of an n-type semiconductor
Q_s	Total semiconductor space charge density
Q_{sm}	Space charge density at minimum capacity of MIS structure
Q_{ss}	State density at semiconductor surface
Q_{th}	Seebeck coefficient
Q_{12}	Number of impurity ions per unit area within a subsurface impurity layer

A–2 LIST OF SYMBOLS

R	Gas constant
r	Semiconductor sample diameter
r'	Semiconductor sample length
r_a	Radius of semiconductor atom
r_c	Capture radius for which $E_a^* = kT$
r_e	Etch rate
r_F	Radius of Fermi surface
r_g	Growth rate of epitaxial film
R_H	Hall coefficient
r_i	Covalent radius of impurity atom
r_l	Ratio of probability of phonon emission to probability of ionizing collision
r_m	Radius of metal atom
r_s	Tetrahedral radius
R_{th}	Thermal resistance of a semiconductor
S	Entropy
S	Surface recombination velocity in the presence of a space charge region
s	Impurity concentration-dependent parameter related to carrier avalanche mechanism
s^*	Parameter depending upon the carrier scattering mechanism
S_e, S_n	Stopping power for electronic, nuclear scattering of impurity ion during ion implantation
s_i	Parameter related to the avalanche mechanism
S_o	Surface recombination velocity in the absence of a space charge region
S_p	Point spacing of four-point probe
s_r	Relaxation length (distance of electron travel for which energy variation is equal to kT)
$S_{\text{Si}(l)}$	Partial molar excess entropy of mixing of silicon in liquid solution
T	Temperature
t	Time
t^*	Corrective oxidation time constant
T_{Au}	Temperature at which gold is diffused to saturate semiconductor
T_B	Water bath temperature

536 APPENDIX

T_b	Boiling point
T_c	Critical temperature
T_D	Debye temperature
t_d	Drive-in diffusion time (denoted t if not confused with predeposition time)
T_e	Electron temperature
t_e	Etching time
T_H	Temperature of Hall effect device
T_m	Melting point
T_o	Temperature at which transition from nondegenerate to degenerate behavior of a semiconductor occurs
t_o	Oxidation time
t_p	Predeposition time (denoted t if not confused with diffusion time)
T_S	Temperature of Hall generator heat sink
T_x	Temperature of semiconductor at distance x from surface after thermal irradiation
T_1	Initial uniform temperature of semiconductor
T_2	Temperature of semiconductor surface after thermal irradiation
U	Steady-state carrier recombination rate
u	Dimensionless potential
u_b	Dimensionless potential referring to the bulk
u_d	Dimensionless potential referring to the depletion layer
u_F	Fermi potential
U_n	Electron generation rate
U_p	Hole generation rate
u_S	Surface potential
u_v	Volume
V	Voltage across junction
V^*	Normalized junction voltage
V_a	Externally applied voltage across a junction
V_a^*	Normalized applied voltage
V_a'	Applied polarizing voltage
V_B	Avalanche breakdown voltage
V_B'	Breakdown voltage under the influence of "reachthrough" ($V_B' \leq V_B$)
V_{Bc}	Avalanche breakdown voltage of a cylindrical junction

A-2 LIST OF SYMBOLS

V_{Bp}	Avalanche breakdown voltage of a plane junction
V_c	Cross-over voltage from linearly graded to step behavior of a diffused junction
V_D	"Built-in" voltage, contact voltage, contact potential, diffusion voltage
V_D^*	Normalized diffusion voltage
v_d	Carrier drift velocity
v_{dis}	Velocity of a dislocation line in a crystal
V_{Dp}	Diffusion voltage of a p-type semiconductor region
V_f	Forward bias of a junction
v_f	Electron velocity at Fermi level
V_G	Gate voltage of MIS structure
v_G	Velocity of gas stream in epitaxial reactor
V/I	Term (measured in units of voltage and current and described in units of resistance) characteristic of resistivity of a semiconductor
V_m	Voltage at minimum capacitance of MIS structure
$V_{\max}$	Voltage at maximum capacitance of MIS structure
v_n	Velocity of electron
V_o	Voltage drop in oxide
v_o	Velocity at which the semiconductor-oxide interface moves into the semiconductor during thermal oxidation
V_{01}	Voltage difference between $C/C_o = 1$ and $C/C_o = 0$
v_{ph}	Velocity of phonon
V_r	Reverse bias of a junction
v_R	Rotation velocity
v_S	Carrier velocity at semiconductor surface
V_T	Turn-on (threshold) gate voltage of MIS transistor
V_z	Voltage difference between the two current control terminals of a Hall generator
V_μ	Effective bulk charge voltage
w_{at}	Atomic weight
w_p	Weight of probe
x	Distance within semiconductor normal to semiconductor surface
x'	Distance within oxide normal to semiconductor-oxide interface
x_*	Distance within the semiconductor at the semiconductor-oxide interface

x'_*	Distance within oxide at the semiconductor-oxide interface
x_b	Schottky barrier thickness
x_c	Portion of the semiconductor which during oxidation is converted to oxide
x_{cS}	Average distance of carriers from semiconductor surface
x_d	Depletion layer width, space charge layer width
x_{dB}	Depletion layer width at avalanche breakdown
x_{dc}	Depletion layer width of a cylindrical p-n junction
$x_{d\,\text{max}}$	Maximum width of depletion layer
x_{dp}	Depletion layer width of a plane p-n junction
x_{d1}	Portion of the junction depletion layer which extends into the heavier doped side of the p-n junction
x_{d2}	Portion of the junction depletion layer which extends into the lighter doped side of the p-n junction
x_e	Semiconductor depth etched away during anisotropic etching
x_{eff}	Effective thickness of epitaxial layer taking junction depth and depletion layer into account
x_f	Epitaxial film thickness
x_h	Length of heat flow in a semiconductor
x_i	Width of an intrinsic region in a p-i-n junction
x_i^*	Apparent oxide thickness before the onset of oxidation
x_j	Junction depth normal to semiconductor surface
x_{jB}	Collector-base junction depth
x_{jE}	Emitter-base junction depth
$x_{j\,\text{min}}$	Minimum junction depth necessary to achieve true avalanche breakdown prior to surface penetration of the depletion layer
x_l	Penetration depth of light
x_m	Minimum oxide thickness required to mask against an impurity
x_{min}	Minimum thickness of epitaxial film required to obtain true avalanche breakdown
x_N	Normalized junction depth
x_n, x_p	Portion of the junction depletion layer which extends into the n-type, p-type semiconductor, respectively
x_o	Oxide thickness
x_ε	Thickness of dielectric
X_P	Electronegativity
x_P	Distance from semiconductor surface at which impurity concentration is highest (C_P), projected range of implanted ions

A-2 LIST OF SYMBOLS

x_S	Thickness of surface charge layer of semiconductor-metal contact
x_s	Thickness of inversion layer
x_w	Semiconductor thickness
Y	Mole fraction
y	Distance parallel to semiconductor surface
Y_B	Bulk modulus
y_e	Lateral spread of anisotropic semiconductor etching
y_j	Junction depth parallel to semiconductor surface
Y_l	Young's modulus
y_m	Width of mask opening
Z	Atomic number
z	Distance along the semiconductor surface normal to the y-axis
z_d	Density of lattice defects
z_F	Density of Frenkel defects
z_f	Fringes of light interference
z_i	Density of possible interstitial crystal sites
z_S	Density of Schottky defects
α	Absorption coefficient
α'	Coefficient of linear thermal expansion
α_{fc}	Free carrier absorption coefficient
α_i	Ionization rate of carriers
α_{in}	Ionization rate of electrons
α_{ip}	Ionization rate of holes
α_n	Probability of electron capture by a neutral gold atom
α_{no}	Probability of electron capture by a recombination center
α_p	Probability of hole capture by a negative gold atom
α_{po}	Probability of hole capture by a recombination center
α_t	Angle of total reflection of light with respect to normal to material surface
$\alpha_x, \alpha_y, \alpha_z$	Components of three-dimensional thermal expansion coefficient
β	Term dependent upon atomic density and radius
β_i	Ionization rate of secondary particles
β_n	Probability of electron capture by a positive gold atom
β_p	Probability of hole capture by a neutral gold atom
γ	Grüneisen constant
γ'	Three-dimensional thermal expansion coefficient

γ_E	Energy transfer during collision of implanted ion with target nucleus or electron
γ_i	Carrier injection ratio (ratio of number of injected carriers to total number of equilibrium minority carriers)
δ	Imaginary part of refractive index (extinction coefficient)
Δc	Density of injected carriers
ΔE	Width of energy band
$\Delta\phi_n$	Image force lowering
ε	Dielectric constant
ε_D	Deformation potential
$\varepsilon_n, \varepsilon_p$	Dielectric constant of n-type, p-type semiconductor, respectively
ε_s	Dielectric constant of a semiconductor
ε_{Si}	Dielectric constant of silicon
ε_{SiO_2}	Dielectric constant of SiO_2
ε_o	Permittivity of free space
η_G	Viscosity of gas
η_i	Correction factors for four-point probe measurement
η^*	Viscosity
η_c	Energy difference between an electronic level and the conduction band
η_n	Energy difference between Fermi level and conduction band
η_p	Energy difference between Fermi level and valence band
η_v	Energy difference between an electronic level and the valence band
Θ	Parameter describing the impurity concentration dependence of the diffusion coefficient
κ	Thermal conductivity
κ_m	Magnetic susceptibility
κ_n	Thermal conductivity of electrons in n-type semiconductor
κ_p	Thermal conductivity of holes in p-type semiconductor
κ_v^*	Compressibility
κ_{vS}^*	Adiabatic compressibility
κ_{vT}^*	Isothermal compressibility
λ	Wavelength
λ_E	Wavelength of dominant light emission
μ	Carrier mobility
μ_B	Carrier mobility in the semiconductor bulk

A–2 LIST OF SYMBOLS

μ_c	Charge carrier scattering mobility
μ_D	Dislocation scattering carrier mobility
μ_{eff}	Effective carrier mobility in inversion or surface layer of semiconductor combining surface and bulk mobilities
μ_H	Hall mobility
μ_I	Impurity scattering mobility
μ_L	Lattice scattering mobility
μ_n	Electron mobility
μ_o	Carrier mobility at small electric field
μ_{opt}	Optical mode mobility
μ_p	Hole mobility
μ_S	Carrier mobility at semiconductor surface neglecting bulk mobility
v^*	Frequency of jumping of an impurity atom from one site to an adjacent one
v_o	Frequency of lattice vibrations
v_{ph}	Phonon frequency
ξ	Degeneracy correction factor
π	Peltier coefficient
π_{ij}	Tensor component of resistivity change
π_l	Longitudinal piezoresistance coefficient
ρ	Resistivity
$\bar{\rho}$	Average resistivity of a diffused layer
ρ_a	Semiconductor resistivity after diffusion of gold to saturation
ρ_b	Semiconductor resistivity in the absence of gold (or before diffusion of gold)
ρ_e	Resistivity of extrinsic semiconductor
ρ_i	Resistivity of intrinsic semiconductor
ρ_m	Apparent (measured) resistivity
ρ_o	Resistivity in the absence of stress
ρ_s	Sheet resistance of a diffused layer
ρ_σ	Resistivity of intrinsic semiconductor under stress
σ	Conductivity
σ_c	Carrier capture cross-section
σ_l^*	Longitudinal stress in direction l_o
σ_n	Electron capture cross-section of a recombination center

σ_p	Hole capture cross-section of a recombination center
σ_R	Standard deviation in projected range of implanted ion
τ	Minority carrier lifetime (recombination time)
τ_B	Carrier lifetime within semiconductor bulk
τ_c	Average time between carrier collisions
τ_n	Electron lifetime in a p-type semiconductor
τ_n^*	Lifetime of hot electrons
τ_{no}	Lifetime of excess electrons in heavily doped p-type semiconductor
τ_p	Hole lifetime in an n-type semiconductor
τ_{po}	Lifetime of excess holes in heavily doped n-type semiconductor
τ_S	Carrier lifetime at semiconductor surface
τ_t	Total minority carrier lifetime
τ_ε	Dielectric relaxation time of semiconductor
ϕ_F	Fermi potential of semiconductor
ϕ_i	Fermi potential of intrinsic semiconductor
ϕ_M	Vacuum work function of metal
ϕ_{MS}	Metal-semiconductor work function difference, barrier height
ϕ_S	Vacuum work function of semiconductor
ϕ_s	Surface potential
$\phi_{s(\text{inv})}$	Surface potential at the onset of strong inversion
ϕ_{th}	Photoelectric threshold energy
χ_o	Electron affinity of semiconductor oxide
χ_s	Electron affinity of semiconductor
ω	Frequency ($\omega = 2\pi f$ or $2\pi v$)

Author Index

Adams, C. M., 103
Adcock, W., 47
Adler, A., 502
Adler, R. B., 119
Allen. C. C., 493
Andreath, Jr., P., 129
Andrews, J. M., 440
Ashley, K. L., 332
Atalla, M. M., 383, 405, 440

Bakeman, Jr., P. E., 194
Barber, H. D., 264
Bardeen, J., 228, 296, 306
Bean, K. E., 47, 103, 362
Beer, A. C., 40, 119
Bell, R. L., 440
Bemski, G., 332
Bergstresser, T. K., 57
Berry, B. M., 502, 517
Berry, R., 194
Blakemore, J. S., 69, 82, 103, 119, 405
Boltaks, B. I., 143
Borrego, J. M., 194
Bracken, R. C., 306
Brice, J. C., 162
Brient, Jr., S. J., 69
Brooks, H., 119
Bruckner, R., 376
Bullis, W. M., 208, 264, 317
Bullough, R., 200
Burger, R. M., 332, 338, 341
Busen, K. M., 193

Cagnina, S. F., 208
Carslaw, H. S., 103, 193
Caughey, D. M., 296
Chan, T. C., 162

Chang, C. Y., 440
Cho, C. C., 296
Choo, S. C., 332
Clauser, H., 40
Clawson, A. R., 296
Cohen, M. H., 24
Cohen, M. L., 57
Collet, M. C., 332
Collins, D. R., 208, 370
Conrad, E. E., 502
Conwell, E. M., 47, 228, 253, 264, 296
Cotton, F. A., 14
Cowley, A. M., 440
Crank, J., 143, 193
Crowell, C. R., 440, 493
Currin, C. G., 47
Cuttriss, D. B., 316

Das, M. B., 406
Dash, S., 129
Davies, D. E., 222
Davies, J. A., 222
Davis, E. A., 24, 296
Deal, B. E., 362, 376, 387, 406, 417, 440
DeMan, H., 273
Derick, L., 370
DeVries, D. B., 162
Ditzenberger, J. A., 162, 502
Donovan, B., 162, 306
Donovan, R. P., 200, 317, 338, 341, 362
D'Stefan, D. J., 135
Duh, C. Y., 273

Earleywine, E., 40, 47
Ehrenreich, H., 14

AUTHOR INDEX

Eriksson, L., 143, 222
Evans, R. A., 162

Fairbanks, R. H., 103
Fairfield, J. M., 332
Finkelnburg, W., 14
Fistul', V. I., 58, 69, 228, 306
Fitzgerald, D. J., 405, 406
Fontana, E. H., 376
Frankl, D. R., 82, 253
Fritzsche, H., 24
Frosch, C. J., 370
Fuller, C. S., 162, 502

Gärtner, W. W. 264, 296, 447, 517
Geppert, D. V., 440
Gereth, R., 162
Ghandhi, S. K., 162, 200, 338, 362
Gibbon, C. F., 194
Gibbons, D. F., 103
Gibbons, G., 440, 493
Gibbons, J. F., 222, 273
Gleim, P. S., 47, 362
Goetzberger, A., 387, 417
Gokhale, B. V., 332
Goldberg, C., 405
Goldstein, Y., 383
Goodman, A. M., 440
Gray, P. V., 387
Grebene, A. B., 332
Greenough, K. F., 376
Groschwitz, E., 440
Grove, A. S., 40, 82, 193, 293, 317, 362, 370, 383, 387, 405, 406, 417, 493
Grover, N. B., 383
Gwyn, C. W., 338, 341

Hall, R., 405
Hall, R. N., 69, 264
Hamilton, D. M., 40
Hannay, N. B., 24, 376
Haynes, J. R., 57
Heiman, F. P., 406, 417
Henisch, H. K., 82, 296, 306, 440
Heywang, W., 40

Hilton, A. R., 119
Hippel, A. R. von, 14
Ho, B. L., 296
Hogarth, C. A., 40, 47
Holmes, P. J., 135
Honig, R. E., 103
Horiuchi, S., 370
Hower, P. L., 273
Hu, S. M., 194
Hunter, L. P., 162, 383, 447

Irvin, J. C., 162, 193, 228, 296, 316, 317

Jaeger, J. C., 103, 193
Jäger, H., 440
Jain, V. K., 239
Johnson, W. S., 222
Jones, C. E., 119
Jones, R. C., 517
Jonscher, A. K., 24
Joshi, M. L., 129
Jost, W., 143, 193

Kass, W., 193
Kendall, D. L., 162
Kennedy, D. P., 193, 194, 493
Kern, E. L., 40, 47
Kieffer, R., 242
Kino, G. S., 273
Kittel, C., 24
Kleen, W., 40
Klein, D. L., 135
Kosak, W., 440
Kovel, S. R., 440
Kowalchik, M., 200
Kramer, D. A., 103
Krembs, D. M., 306
Krömer, H., 471
Ku, H. Y., 362
Kudman, L., 103

LaChapelle, T. J., 208
Landsberg, P. T., 58
Lane, C. H., 362
Lange, N. A., 103

AUTHOR INDEX

Larach, S., 119
Lawrence, H., 471
Lawrence, J. E., 129
Lee, C. A., 272
Legat, W. H., 129
Lehovec, K., 193
Leistiko, O., 162, 406, 493
Lepselter, M. P., 440
Ligenza, J. R., 361
Lindhard, J., 222
Logan, R. A., 272
Long, D., 58
Looney, J. C., 376
Ludwig, G. W., 316

McAfee, K. B., 272, 493
McKay, K. G., 272, 493
McSkimin, H. J., 129, 376
MacFarlane, G. G., 57
Madelung, O., 24, 40, 58, 103, 129
Madigan, J. R., 332
Mai, C. C., 162, 376
Maissel, L., 103
Maita, J. P., 57, 264, 306
Manchester, K. E., 222
Many, A., 383
Mataré, H. F., 306
Maycock, P. D., 103
Mayer, H., 296
Mayer, J. W., 222
Mead, C. A., 440
Megla, G., 417
Mendel, E., 135
Millea, M. F., 162
Miller, S. L., 272, 493
Milnes, A. G., 332
Moll, J. L., 208, 253, 272, 273, 447
Mönch, W., 194
Monteith, L. K., 193, 471
Morin, F. J., 57, 264, 306
Mott, N. F., 440
Murley, P. C., 193

Nayer, P. S., 103
Neuberger, M., 40
Newman, R. C., 200

Nicholas, K. H., 162
Nicolau, I. F., 502
Nicolet, M. A., 273
Nicollian, E. H., 387
Norris, C. B., 273

O'Brien, R. R., 193, 493
O'Keefe, M., 193
O'Neill, J. J., 440
Overstraeten, R. Van, 273
Ovshinsky, S. R., 24

Padnos, B. N., 135, 242
Pao, H. C., 406
Patrick, W. J., 306
Paul, W., 119
Pauling, L., 440
Pearson, G. L., 228, 296, 306
Phillips, A. B., 69, 332, 471, 493
Phillips, J. C., 58
Pliskin, W. A., 362, 502
Plummer, W. A., 376
Polata, B., 194
Prior, A. C., 272

Racette, J. H., 69, 264
Read, W. T., 332
Reddi, V. G. K., 273
Reid, R. C., 239
Revesz, A. G., 362, 406
Rhoderick, E. H., 440
Robbins, H., 502
Rodriguez, V., 273
Rosenberg, A. J., 69
Rosi, F. D., 103
Ross, B., 332
Rosvold, W. C., 135
Ruch, J. G., 273
Runyan, W. R., 47, 103, 129, 239
Russell, L. K., 129
Ryder, E. J., 272

Sah, C. T., 332, 370, 405, 406, 452, 471
Sanderson, A. C., 332
Saxena, A. N., 440

Scharfetter, D. L., 273
Schmidt, S., 194
Schmidt-Tiedemann, K. J., 119
Seidel, T. E., 273
Sevin, L. J., 383
Sharma, S. K., 239
Shaw, R. F., 24, 296
Shepherd, W. H., 239
Sherwood, T. K., 239
Shirn, G. A., 193
Shockley, W., 69, 129, 208, 296, 332
Sirtl, E., 502
Sklar, M., 362
Slack, G. A., 103
Slobodskoy, A., 193
Smith, A. M., 162, 200
Smith, C. S., 129
Smith, R. C. T., 193
Snow, E. H., 376
Spenke, E., 47
Stegmeier, E. F., 103
Struthers, J. D., 208
Stull, D. R., 103
Sugano, T., 440
Swartzendruber, L. J., 510

Tannenbaum, E., 162
Teichthesen, L. A., 47
Thai, N. D., 129, 194
Theuerer, H. C., 239
Thomas, R. E., 296
Thurmond, C. D., 200
Trumbore, F. A., 162, 200

Uhlir, Jr., A., 510

Vadasz, L., 406
Valdes, L. B., 510

Wagini, H., 103
Wallmark, J. T., 383
Wang, C. C., 376
Warner, R. M., 471
Watters, R. L., 316
Weiss, H., 306
Westman, H. P., 522
White, H. G., 272
White, J. P., 405
Wieder, H. H., 239, 296
Wilcox, W. R., 208
Wilkinson, G., 14
Willardson, R. K., 40, 119
Williams, R., 341
Wilson, C. L., 69
Wilson, P. R., 452, 471
Wolf, H. F., 58, 228, 370, 376
Wortman, J. J., 129

Yamaguchi, J., 370
Yang, K.-H., 135
Yarbrough, D. W., 162

Zaininger, K. H., 362, 376, 406, 417
Ziman, J. M., 58

Subject Index

Absorption coefficient, 104-114
Absorption edge, 54, 105-109, 114
Acceptors, 31, 76-81, 154, 223, 250,
  279, 282, 325, 344, 444
Accumulation, 384, 386, 390,
  396, 404
Acoustoelectric effect, 129
Activation energy, 14, 59, 76-81,
  146-148, 153, 156, 195,
  223-228, 318, 363, 417
Ambipolar diffusion coefficient, 290
Amorphous semiconductors, 2, 21-24,
  41, 209, 339
Amphoteric impurity, 201
Annealing, 209, 210, 342
Anodic oxidation, 342
Anti-Schottky defect, 7
Atomic density, 34, 123, 128,
  156, 199
Atomic radius, 25, 199
Atomic weight, 27, 34, 89, 90, 128, 199
Atomically clean surface, 388
Attractive energy, 17
Auger recombination, 318, 328
Avalanche breakdown, 36, 445,
  472-493
Average resistivity, 312-316

Barrier height of metal-semiconductor
  contact, 385, 421-436
Barrier height of p-n junction, 80, 444,
  448, 493
Bohr's radius, 14
Boltzmann distribution, 10, 59, 72
Bond, 4, 16-21, 25, 56, 251, 339, 353,
  400, 472
Box-type impurity distribution, 166

Breakdown voltage, 375, 472-491
Brillouin zone, 49
Built-in electric field, 140, 180-184
Built-in voltage, 427, 448-457, 493
Bulk modulus, 97, 127
Buried layer (subsurface layer), 174,
  209, 313-315

Capacitance of p-n junction, 453-468
Capacitance of semiconductor surface,
  407, 415, 416
Capture probability, 324, 325, 331
Carrier density, 61-64, 74-79, 206,
  207, 227, 254-263, 308, 310,
  319, 513
Carrier injection, 22, 289, 308,
  321, 323
Carrier mass, 30, 31, 33, 36, 75,
  226, 418
Carrier mean free path, 13, 22,
  274-277, 283, 285
Carrier multiplication, 473-480
Carrier relaxation time, 275, 285
Chalcogenide glasses, 23
Charge neutrality level, 429
Charge-storage diode, 447
Chemical liquid phase synthesis, 229
Chemical vapor phase synthesis, 229
Cohesive energy, 4, 25-29, 112, 251
Complementary error function,
  163-165, 172, 367, 369
Compressibility, 371, 373
Conductivity (electrical), 11, 13, 23,
  87, 91, 251, 258, 278, 307-317,
  337, 374, 382, 503-511, 516
Conductivity (thermal), 13, 35, 83-93,
  99, 375, 512

547

548 SUBJECT INDEX

Conductivity modulation, 307
Constant energy surfaces, 52
Constrictions in lattice, 145
Covalent cohesion (bond), 5, 18
Critical electric field, 266, 292-296, 445, 469, 474-491
Critical impurity concentration, 169
Cross-over voltage, 464
Crystal damage, 7, 8, 23, 24, 31, 112, 132, 195, 209-211, 320, 322
Crystal orientation, 42, 209, 353, 400, 511
Crystal structure, 2, 15, 21, 34, 381, 400
Cubic lattice, 2-4, 16, 145, 299
Curvature of p-n junction, 465-468, 481-484
Cylindrical junction, 465-468, 483

Damage ratio, 211
Debye length, 107, 114, 115, 338
Debye temperature, 88, 98, 116, 364
Deep impurities, 201-208, 225-227, 250, 319, 320, 325
Deformation potential, 121, 279
Degenerate semiconductor, 36, 51, 59-69, 76-79, 93, 105, 224, 227, 260, 284, 300-302, 309, 327
Density, 33, 35, 339
Density of states, 9, 35, 75, 251, 257, 261, 262
Depletion layer capacitance, 457-459
Depletion layer of a p-n junction, 444, 448-471, 487-491
Depletion of semiconductor surface, 386, 390-405, 409
Diamond lattice, 3, 15, 16, 25
Dielectric constant, 33, 35, 107, 110, 111, 115, 226, 374
Dielectric relaxation time, 115, 116, 271, 416
Diffuse carrier reflection, 285, 286
Diffused junction, 491
Diffusion capacitance, 457-459

Diffusion coefficient of carriers, 115, 289
Diffusion coefficient of impurities, 144, 148-155, 201, 350, 363-366
Diffusion current, 289
Diffusion length, 149, 445
Diffusion of impurities, 140-200, 209, 363-366
Dipole cohesion (bond), 4, 18
Direct transition, 48-51, 106, 108, 114, 319
Dislocation scattering, 280
Dislocations, 8, 125-127, 132, 152, 250, 280, 322, 446
Distribution coefficient, 32, 158-160, 199
Donors, 31, 76-81, 154, 223, 250, 279, 282, 325, 342, 444
Drift velocity, 265-268, 271, 285, 292

Edge effects, 184-187
Effective charge, 30
Einstein relationship, 289
Elasticity, 127
Electric field, 35, 140, 183, 389, 404, 405, 453-457, 467-471, 474-491
Electroluminescence, 37, 330, 514-516
Electron affinity, 36, 385, 423, 437
Electron energy, 10, 49-52, 418
Electron mean free path, 13, 22, 274-277, 285
Electron temperature, 265
Electron velocity, 265-268, 271, 285, 292, 522
Electron wavelength, 88, 418
Electronegativity, 17-20, 25, 437-439
Electronic stopping, 211-213
Energy bands, 9, 30, 38, 48-57, 70-82, 250, 256, 257, 340, 420
Energy gap, 19, 24-30, 33, 48, 52-56, 60, 107-109, 113, 226, 258, 311, 320, 340, 341, 375, 472, 511
Energy loss of particle, 211

SUBJECT INDEX 549

Epitaxial growth, 142, 229-239
Equilibrium, 17, 81, 251, 318, 321, 373, 420, 444
Escape of impurities, 239, 358
Etching, 130-135, 231, 371
Eutectic temperature, 240-242
Excess temperature of Schottky barrier, 434-436
Exciton, 114
Exponential impurity distribution, 164, 165
External rate limitation, 177-179

Fast recovery diode, 447
Fermi distribution, 10, 59, 65, 72, 81
Fermi integral, 64, 65, 81, 289
Fermi level, 61-67, 70-82, 210, 259-263, 310, 325, 382, 396, 399, 420, 436, 437, 448, 522
Ferromagnetic scattering, 286
Fick's laws, 96, 140, 144
Flash evaporation, 229
Flat-band condition, 384, 386, 410, 415
Forward current across p-n junction, 492
Four-point probe, 189-193, 503-510, 516
Fractional density of lattice positions, 148-152
Fractional width of depletion layer, 455, 457, 461-467, 487
Frank loop, 124
Free electrons, 418
Freeze-out, 310
Frenkel defect, 7

Galvanomagnetic effect, 37
Gamma irradiation, 320
Gaussian impurity distribution, 163-168, 217, 367, 369
Generation of carriers, 37, 117-119, 252, 318, 321-323, 329, 473
Graded junction, 446, 449-491
Growth rate of epitaxial film, 229-234
Gunn effect, 37

Hall coefficient, 297-301
Hall effect, 292, 297-306
Hall generator, 303-305
Hall mobility, 300-303
Hard-sphere model, 145
Heat of sublimation, 160
Henry's law, 177, 353, 361
Heterojunction, 446
Heteropolar bond, 6, 18
Homopolar bond, 6, 19
Hopping of carriers, 22
Hopping of impurities, 145
Hot electrons, 37, 522
Hydrogen model, 14, 26, 223

Impatt diode, 447
Impurity depletion, 360
Impurity dosage (charge), 173-176
Impurity flux, 140, 144, 181, 346
Impurity gradient, 141, 163-167, 173-176, 215, 217-222
Impurity pile-up, 360
Impurity redistribution, 141, 142, 232-234, 236-239, 338, 356-361, 398
Impurity scattering, 276-279, 307, 310
Indirect transition, 48-51, 106, 108, 114, 319
Injection, 22, 289, 308, 321, 323
Interchange diffusion, 147
Interface states, 329, 342, 382, 391
Interstitial diffusion, 144-146, 201, 250, 363
Intervalley scattering, 281
Intravalley scattering, 280
Intrinsic carrier concentration, 35, 53, 75, 181-183, 254-258, 260
Inversion of surface, 386, 390-405, 410, 415
Ionic cohesion (bond), 5, 17-20
Ionization energy, 14, 59, 76-81, 146-148, 153, 156, 195, 223-228, 318, 363, 419
Ionization potential, 25
Ionization probability, 269
Ionization rate of carriers, 267-271,

SUBJECT INDEX

Ionization *(continued)*
   445, 473-480, 485

Junction delineation, 452
Junction depth, 166-168, 187-189,
   450, 452

Laser effect, 37
Lateral diffusion, 142, 184-187, 371
Lattice constant, 3, 27, 33, 60, 99,
   120, 126, 145, 153, 375
Lattice defects, 7, 8, 23, 24, 31, 112,
   132, 195, 209-211, 320,
   322, 342
Lattice scattering, 276, 278, 307, 311
Lattice vibrations, 13, 98, 99, 153,
   278, 364
Layer resistivity, 189-193
Lifetime, 252, 318-332, 513
Linear impurity distribution, 164, 166
Liquid semiconductors, 158-162

Majority carriers, 259
Masking, 141, 184-187, 195, 336,
   344, 366-370
Mass-transfer coefficient, 177, 230,
   234, 349
Mass-transfer control, 230
Matthiessen's rule, 286
Maximum electric field, 449, 452-457,
   467-471
Melting point, 19, 25, 27, 33, 35, 83,
   89, 99, 156, 158, 199,
   240-242, 339, 375
Metallic cohesion (bond), 5, 18
Metal-semiconductor interface, 382,
   383, 418-440
Microplasma, 445, 484
Miller indices, 6, 147
Minority carriers, 259
MIS capacitance, 407-417
MIS system, 381-417
Misfit factor, 126, 196, 199
Mobility, 22-29, 31, 33, 35, 265-267,
   274-296, 513
Momentum, 14, 129, 319

Network formers, 337, 364
Network modifiers, 337, 363
Neutrality condition, 255, 263
Neutron irradiation, 320
Nonequilibrium, 81, 318
Nuclear stopping, 211-213

Optical mode scattering, 280, 422
Optical properties, 24, 104-119, 512
Oxidation coefficient, 354-356
Oxidation rate constants, 346-353
Oxide capacitance, 408, 414
Oxide charge, 392
Oxide formation, 337, 342-361
Oxide growth rate, 348
Oxide states, 391
Oxide thickness, 344, 346-356,
   498-500

Penetration depth, 105, 111-113
Phase diagrams, 240-242
Phonon, 14, 83-90, 97-99, 265-272,
   278, 320, 485
Phonon drag, 98
Phonon emission probability, 270
Phonon energy, 36, 88, 108
Phonon mean free path, 36, 266-272,
   485-487, 522
Photo yield, 430
Photoconductivity, 116-118, 308, 419
Photoelectric threshold energy,
   438-440
Photoemission, 438
Photon, 14, 318, 383, 419
Photovoltaic effect, 118
Piezoelectricity, 37, 38, 120-122
p-i-n junction, 447, 454-457
Plane junction, 465-468, 483
Plasma waves, 37
p-n junction, 237, 383, 444-493
Poisson's ratio, 123, 127
Polishing of surface, 135
Polycrystalline semiconductors, 21-24,
   118, 231, 288, 289
Polymers, 339
Practical surface, 388

SUBJECT INDEX 551

Proton irradiation, 320
Pseudodiffusion coefficient, 159-162

Quantum efficiency, 430

Range of particle, 212-216
Raoult's law, 158
Reach-through, 481-484
Recombination, 117, 207-210, 250, 308, 318-332, 382, 448
Recombination radiation, 37, 330, 514-516
Rectifier, 446
Redistribution of impurities, 142, 233, 236-239, 338, 356-361, 398
Reflection, 113, 430
Refractive index, 29, 104-111, 375
Repulsive energy, 17
Resistivity, 11, 13, 23, 87, 91, 251, 258, 278, 307-317, 337, 374, 503-511, 516
Reynolds number, 234
Richardson constant, 419-422, 434-436
Rutherford scattering, 212

Schottky barrier, 382, 383, 420-437
Schottky defect, 7, 147
Secondary electrons, 420
Seebeck coefficient, 93-96
Segregation coefficient, 359-361, 369
Semiconductor charge, 392-396
Shallow impurities, 223-227, 310
Sheet resistivity, 312 316
Skin effect, 520
Solubility, 24, 61, 112, 156, 170, 201-205, 346
Space charge capacitance, 407, 415
Specific heat, 35, 83, 85, 92, 99
Specular carrier reflection, 285, 286
Spherical junction, 465-468, 483
Spin orbit splitting, 30, 31, 50
Sputtering yield, 211
Standard deviation, 213, 216

Steady-state condition, 322
Step junction, 445, 449-491
Stress, 52, 120-129, 184, 195, 196, 371, 483
Substitutional diffusion, 146, 201, 250, 364
Suhl effect, 308
Surface charge, 375
Surface layer, 312-316, 381-417
Surface mobility, 285-289, 400-404
Surface potential, 396, 397
Surface-reaction control, 230
Surface recombination, 252, 329
Surface recombination velocity, 329
Surface relaxation time, 285
Surface resistance (V/I), 189-193, 503-510
Surface state capacitance, 407, 416
Surface states, 118, 329, 382, 389-392, 400, 425, 517

Temperature limitation, 57
Tetrahedral radius, 145
Thermal conductivity, 13, 35, 83-93, 99, 375, 512
Thermal energy, 91
Thermal expansion, 35, 87, 99-101, 375
Thermal oxidation, 337, 342-361
Thermal radiation, 92, 97
Thermal resistance, 92, 96
Thermal voltage, 91
Thermionic emission, 22, 418, 419, 423, 433-436
Thermoelectric power, 93-96
Thin films, 284-289, 302, 512
Thomas-Fermi scattering, 212
Three-temperature method, 229
Transition frequency of MIS structure, 387, 416
Transmission (optical), 113
Transparency, 29, 111-114
Tunneling through p-n junction, 36, 446, 472, 473
Tunneling through Schottky barrier, 382, 431, 433

Turn-on voltage, 394-399, 410
Two-step diffusion, 170

Vacuum evaporation of
 semiconductor, 229
Van der Waals cohesion (bond), 4, 18
Vapor phase oxidation, 342
Vapor pressure, 35, 101-103
Viscosity, 235, 373
Voids in lattice, 145
Voltage regulator, 447

Wave vector, 49-51
Welker's rule, 28
Wiedemann-Franz law, 93
Work function, 36, 385, 419-436
Wurtzite lattice, 15

Young's modulus, 123, 127

Zener breakdown, 472
Zincblende lattice, 3, 15, 16, 26